Characterisation of Soft Magnetic Materials Under Rotational Magnetisation

T0179181

Characterisation of Soft Magnetic Materials Under Rotational Magnetisation

Stanislaw Zurek

CRC Press
Taylor & Francis Group
Boca Raton London New York

CRC Press is an imprint of the
Taylor & Francis Group, an **informa** business

CRC Press
Taylor & Francis Group
6000 Broken Sound Parkway NW, Suite 300
Boca Raton, FL 33487-2742

First issued in paperback 2019

© 2018 by Taylor & Francis Group, LLC
CRC Press is an imprint of Taylor & Francis Group, an Informa business

No claim to original U.S. Government works

ISBN-13: 978-1-138-30436-9 (hbk)
ISBN-13: 978-0-367-89157-2 (pbk)

Library of Congress Cataloging-in-Publication Data

Names: Zurek, Stanislaw (Electrical engineer), author.
Title: Characterisation of soft magnetic materials under rotational magnetisation / Stanislaw Zurek.
Description: Boca Raton : Taylor & Francis, CRC Press, 2017. | Includes bibliographical references and index.
Identifiers: LCCN 2017033847| ISBN 9781138304369 (hardback : alk. paper) | ISBN 9780203730126 (ebook)
Subjects: LCSH: Electronics--Materials. | Magnetic materials. | Magnetization. | Plasticity.
Classification: LCC TK7871.15.M3 Z87 2017 | DDC 620.1/1297--dc23
LC record available at https://lccn.loc.gov/2017033847

Visit the Taylor & Francis Web site at
http://www.taylorandfrancis.com

and the CRC Press Web site at
http://www.crcpress.com

This book is dedicated to the most resourceful, the wisest, the most hard-working and the bravest person I ever knew – my mother Janina Żurek who used to say:

We learn our whole life, and we still die stupid.

Contents

Symbols, Acronyms and Abbreviations

$\odot$	a circle with a dot represents a vector, whose direction is perpendicular to the surface of the page and the sense is pointing towards the reader
$\otimes$	a circle with a cross represents a vector, whose direction is perpendicular to the surface of the page and the sense is pointing away from the reader
1&2DM	International Workshop on 1&2 Dimensional Magnetic Measurement and Testing
1D	one-dimensional, uni-directional
2D	two-dimensional (including rotational)
2DM	International Workshop on 1&2 Dimensional Magnetic Measurement and Testing
$3d$	electron shell important for ferromagnetism
3D	three-dimensional
a	interatomic distance (m)
a_{rec}	range of interval in a rectangular distribution (various units)
a_{tri}	range of interval in a triangular distribution (various units)
a.u.	arbitrary units
A	location 'A', or ammeter
A	area (m^2)
A_1, A_2	area (m^2)
A_C	area of a conductive pad (m^2)
A_{coil}	active area of a coil (m^2)
A_i	numerical value of 'partial area' of i-th trapezoid in mathematical integration of digital signal (V · s)
AI+	positive analogue input
AI−	negative analogue input
AI GND	analogue input ground
AI SENSE	analogue input sense, reference for non-referenced single-ended mode input
AISN	*see* AI SENSE
A_{sample}	cross-section area of a sample (m^2)
A_{yoke}	cross-section area of a yoke (m^2)
ADC	analogue-to-digital converter
AVG	average of rectified voltage signal (V)
B	location 'B'
B	flux density (T)
$\overline{B}$	amplitude of purely circular B (T)
$\vec{B}$	flux density vector

B_1	flux density in medium '1'
B_{1n}	normal component of B_1
B_{1t}	tangential component of B_1
B_2	flux density in medium '2'
B_{2n}	normal component of B_2
B_{2t}	tangential component of B_2
B_A	(amplitude of) alternating flux density (T)
B_C	(modulus of vector of) circular flux density (T)
B_{mean}	flux density averaged over an area (T)
BN	Barkhausen noise
B_p	peak flux density (T)
B_{peak}	peak flux density (T)
B_r	remanence (T)
B_R	rotational flux density (T)
B_{sample}	flux density in a sample (T)
B_{yoke}	flux density in a yoke (T)
B_x	x component of B (T)
B_y	y component of B (T)
B_z	z component of B (T)
c	specific heat (J/(kg · K))
C	capacitance (F)
C_0	constant value in a given equation (various units)
C_1	constant value in a given equation (various units)
C_2	constant value in a given equation (various units)
C_a	material constant in additional component of power loss (J · $\sqrt{s}$/kg)
C_e	material constant in eddy current component of power loss (J · s/kg)
C_h	material constant in hysteresis component of power loss (J/kg)
$C_{integral}$	constant value resulting from integration of any function (various units)
C_n	constant value in a given equation (various units)
CNC	computer numerical control
C_x	constant value in a given equation (various units)
CGO	conventional grain-oriented electrical steel
d	thickness (m)
d_f	degrees of freedom (unitless)
$_e d_f$	effective degrees of freedom (unitless)
d_r	digit of resolution (various units)
D	specific density (kg/m³)
DAC	digital-to-analogue converter
DAQ	data acquisition device
DGO	double grain-oriented electrical steel
DMM	digital multimeter
E	energy in (J), (J/kg) or (J/m³)
E_{circle}	energy in circular motion (J)
E_{line}	energy in linear motion (J)
E	electric field (V/m)

$\vec{E}$	electric field vector
EDM	electrical discharge machining
EMF	electromotive force (V)
ENOB	effective number of bits
E_x	x component of electric field (V/m)
E_y	y component of electric field (V/m)
E_Y	Young's modulus of elasticity (Pa)
E_z	z component of electric field (V/m)
f	frequency (Hz)
F	mechanical force (N)
FEM	finite-element method
FF	form factor (unitless)
FFF	flux fringing factor (unitless)
FSO	full-scale output
g	gravity (m/s^2)
g_a	acceleration (m/s^2)
G	gain (unitless)
G_{corr}	correction gain (unitless)
G_{DF}	digital feedback gain (unitless)
G_{sys}	system gain (unitless)
GF	gauge factor (unitless)
h	generic function (various units)
h_0, h_1, h_2, h_i, h_n	amplitudes of harmonics (various units)
H	magnetic field strength (A/m)
$\vec{H}$	magnetic field vector
H_1	magnetic field strength in medium '1'
H_{1n}	normal component of H_1
H_{1t}	tangential component of H_1
H_2	magnetic field strength in medium '2'
H_{2n}	normal component of H_2
H_{2t}	tangential component of H_2
H_a	applied magnetic field (A/m)
H_c	coercivity (A/m)
$_BH_c$	coercivity value derived from a B–H loop (A/m)
$_JH_c$	coercivity value derived from a J–H loop (A/m)
H_d	demagnetising magnetic field (A/m)
H_{mean}	magnetic field strength averaged over an area (A/m)
HGO	high-permeability grain-oriented electrical steel
H_p	peak magnetic field strength (A/m)
H_x	x component of H (A/m)
H_y	y component of H (A/m)
H_z	z component of H (A/m)
i	subsequent item in a set, series or array
I	electric current (A)
I_1	primary current (A)

I_2	secondary current (A)
I_e	eddy current (A)
I_p	peak current (A)
I_x	current for the x channel (A)
I_y	current for the y channel (A)
IM	intermodulation
IMD	intermodulation
j	subsequent item in a set, series or array
J	magnetic polarisation (T)
$J_{\parallel}$	parallel polarisation (T)
$J_{\perp}$	perpendicular polarisation (T)
J_p	peak polarisation (T)
J_{sat}	saturation polarisation (T)
k	coefficient of mechanical friction (unitless)
k_Q	coverage factor (unitless)
$k_{Q,t}$	Student's distribution correction factor (unitless)
K	Poynting–Umov vector (W)
K_0	magnetocrystalline anisotropy constant (J/m^3)
K_1	magnetocrystalline anisotropy constant (J/m^3)
K_2	magnetocrystalline anisotropy constant (J/m^3)
K_t	heat transfer coefficient (W/(kg · K))
l	length or distance (m)
l_1	length or distance (m)
l_2	length or distance (m)
l_{coil}	length or path of the coil (m)
l_{gap}	length of air gap (m)
l_{lag}	lagging distance in circular motion (m)
$l_{surface}$	length or path on the surface of the sample (m)
l_{yoke}	length of yoke (m)
L	inductance (H)
m	mass (kg)
M	magnetisation (A/m)
MAE	magneto-acoustic emissions
MMF	magnetomotive force (A)
MOKE	magneto-optic Kerr effect
M_S	magnetisation saturation (A/m)
n	number of items, e.g. turns or points (unitless)
N	number of items, e.g. turns or points (unitless)
N_1	primary turns (unitless)
N_2	secondary turns (unitless)
N_{BN}	number of elements (values) in the digitised Barkhausen noise
N_d	demagnetising coefficient (unitless)
NO	non-oriented electrical steel
nom	nominal value (various units)
NTC	negative temperature coefficient
opamp	operational amplifier

p	instantaneous power loss (W)
P	power in (W), (W/kg) or (W/m^3)
P_a	additional component of power loss, in (W), (W/kg) or (W/m^3)
P_{circle}	power in circular motion (W)
P_e	eddy current component of power loss, in (W), (W/kg) or (W/m^3)
P_h	hysteresis component of power loss, in (W), (W/kg) or (W/m^3)
P_h	hysteresis component of power loss, in (W), (W/kg) or (W/m^3)
PA	partial area (various units)
PID	proportional–integral–derivative (controller)
P_{rot}	rotational power loss, in (W), (W/kg) or (W/m^3)
P_x	component in calculation of rotational power loss, in (W), (W/kg) or (W/m^3)
P_y	component in calculation of rotational power loss, in (W), (W/kg) or (W/m^3)
PLC	power line cycles (unitless)
PS	power spectrum of Barkhausen noise signal
PTC	positive temperature coefficient
Q	probability (unitless)
Q_t	probability in Student's t-distribution (unitless)
r	radius (m)
R	resistance (Ω)
R_0	additional resistance (Ω)
R_1, R_2	resistance values (Ω)
R_{in}	input resistance (Ω)
R_m	measured resistance (Ω)
R_{ref}	reference resistance (Ω)
R_s	resistance of a shunt resistor (Ω)
RS− and RS+	remote sensing connections for a half-bridge circuit
RCP	Rogowski–Chattock potentiometer
RMS	root mean square
s	system gain, equivalent to G_{sys} (unitless)
S	sensitivity or sensitivity coefficient (various units)
$S_{control}$	controlled signal (various units)
S_{corr}	correction signal (various units)
S_{diff}	difference signal (various units)
S_{gen}	generated signal (various units)
S_H	signal from sensor of H (various units)
S_{target}	target signal (various units)
SC− and SC+	shunt calibration connections for a half-bridge circuit
SI	International System of units (*Le Système International d'unités*)
t	time (s)
t_D	delay time (s)
t_L	triggering correction time (s)
t_{on}	time instant of a start of a process (s)
t_{off}	time instant of an end of a process (s)
t_S	sampling time interval (s)

T	length of cycle or period (s)
T_0	starting temperature (K)
THD	total harmonic distortion (%)
T_m	measured temperature (K)
T_{ref}	reference temperature (K)
TNP	total number of peaks of Barkhausen noise signal (unitless)
TSA	total sum of amplitudes of Barkhausen noise signal (V)
TX	transformer
u	uncertainty (various units)
u_c	combined uncertainty (various units)
$_e u_c$	expanded combined uncertainty (various units)
u_s	standard uncertainty (various units)
$u_{s,S}$	scaled standard uncertainty (various units)
v	variable (various units)
$\underline{v}$	average of a variable v (various units)
v_{-1}, v_0, v_1, etc.	subsequent values (various units)
v_i	ith value (various units)
v'_{-1}, v'_0, v'_1, etc.	subsequent approximated values (various units)
v_E	expected value (V)
v_L	triggering level (V)
v_n	nth value (various units)
v_{nom}	nominal value (various units)
v_R	real or actual value (V)
$v(t)$	approximating function (V)
V	voltage (V)
V	voltmeter
V_0	voltage (V)
V_1	voltage (V)
V_2	voltage (V)
V_3	voltage (V)
V_a	voltage at a point 'a' (V)
V_A	voltage in part 'A' (V)
V_{avg}	average voltage (V)
V_{AVG}	average voltage (V)
V_b	voltage at a point 'b' (V)
V_B	voltage in part 'B' (V)
$V_{B\text{-}coil}$	B-coil voltage (V)
V_{BN}	output voltage of a Barkhausen noise sensor (V)
V_H	output voltage of a Hall-effect sensor (V)
V_i	ith numerical voltage value (V)
V_{in}	input voltage (V)
$V_{indirect}$	voltage induced in an indirect coil (V)
V_m	measurand (various units)
V_M	magnetic potential (A/m)
$V_{M(A,B)}$	magnetic potential difference between points 'A' and 'B' (A/m)
V_{max}	maximum voltage (V)

$|V|_{mean}$	mean value of rectified voltage (V)
V_{min}	minimum voltage (V)
V_{NP}	needle probe sensor voltage (V)
V_{out}	output voltage (V)
$V_{pk\text{-}pk}$	voltage peak-to-peak (V)
V_r	voltage across a shunt resistor (V)
V_{ref}	reference voltage (V)
V_{res}	voltage resolution (V)
V_{RMS}	RMS value of voltage (V)
V_s	sample volume (m³)
V_{shunt}	voltage across shunt resistor (V)
V_{supply}	supply voltage (V)
V_{th}	thermal noise voltage (V)
V_x	voltage for the x channel (V)
V_y	voltage for the y channel (V)
W	wattmeter
W_R	rotational loss (W)
WCM	Wolfson Centre for Magnetics, Cardiff University, Cardiff, UK
x	direction in orthogonal system of coordinates
x'	apparent direction, deviated from x
y	direction in orthogonal system of coordinates
y'	apparent direction, deviated from y
z	direction in orthogonal system of coordinates
$\vec{x}$	unit vector in direction x
$\vec{y}$	unit vector in direction y
$\vec{z}$	unit vector in direction z

GREEK CHARACTERS

α	arbitrary direction of magnetostriction (rad)
α iron	ferromagnetic form of iron
α_k	angle of rotation of polarised light beam in the longitudinal Kerr effect (rad)
γ	electric conductivity (S/m)
γ iron	non-ferromagnetic form of iron
Γ	function described by Equation 5.9
Δ	change or difference of any parameter
Δl	change of length (m)
Δt	change of time (s)
δ	error of angular positioning (rad)
θ	angle of rotation (rad)
θ_x	angle of rotation of apparent axis x' (rad)
θ_y	angle of rotation of apparent axis y' (rad)
ϑ	angle of incident beam in Kerr microscopy (rad)
λ	magnetostriction (m/m or unitless)
$\lambda_{pk\text{-}pk}$	peak-to-peak magnetostriction (m/m or unitless)

μ	magnetic permeability, $\mu = \mu_r \cdot \mu_0$ (H/m)
μ_0	magnetic constant, magnetic permeability of free space $\mu_0 = 4 \cdot \pi \cdot 10^{-7}$ (H/m)
μ_1	magnetic permeability of medium '1'
μ_2	magnetic permeability of medium '2'
μ_r	relative magnetic permeability $\mu_r = \mu/\mu_0$ (unitless)
ρ	correlation coefficient (unitless)
ς	tangential stress (Pa)
σ	stress (Pa)
σ_s	standard deviation (various units)
$_v\sigma_s^2$	variance (various units)
$\sigma_{s,rec}$	standard deviation in rectangular distribution (various units)
$\sigma_{s,tri}$	standard deviation in triangular distribution (various units)
σ_{sdom}	standard deviation of the mean (various units)
τ	torque (N · m)
ϕ	phase angle or angular lag (rad)
ω	angular frequency (rad/s)
Φ	magnetic flux (Wb)
Φ_{coil}	magnetic flux in a coil (Wb)
Φ_{sample}	magnetic flux in a sample (Wb)
Φ_{yoke}	magnetic flux in a yoke (Wb)
χ	magnetic susceptibility, $\chi = \mu_r - 1$ (unitless)

Preface

Many moons ago, I came across a mocking text from an unknown source and author, translating what is apparently *really* meant by the various statements used in scientific papers. Recalling a few lines from my memory, it went more or less like this:

Statement	What it Means
in my experience	it happened once
time after time	it happened twice
in a series of experiments	it happened three times
thank you to Dr. X for the help with experiments and Prof. Y for valuable discussion	Dr. X carried out the experiments and Prof. Y explained to me what they meant
it is well known	we couldn't be bothered to look up the original research
typical data	the prettiest graphs we could make

The point of bringing up this anecdote was to remind that there is a great disparity between the *published science* and the way the input data were *produced* for the later analysis so that the results could be published.

The 'science' typically deals with the effects, properties, characteristics, material behaviour and so on, without going into too deep details of how the data were actually collected. This is normal because papers are limited in length, so without exception, their content must be accordingly condensed by the authors to the most important or relevant findings.

The measurement techniques are often 'well known' and perceived as 'just' a tool for doing the *real* science. However, without appropriate tools, it would be impossible to do *any* science. Therefore, from a practical viewpoint, the methodology used for a given research is always *fundamental* even though it seats somewhere in the background.

Sometimes there are papers describing the measurement system, apparatus or method. But again, such papers must be limited in length so that not all details can be presented. And especially, the focus would be almost entirely on the positive aspects, whereas difficulties would be at best just mentioned somewhere in a single sentence or so.

In the subject of magnetic properties of soft magnetic materials, it is quite interesting to see how rarely the measurement uncertainties are discussed. These are only discussed if the paper is *specifically* about measurement errors. This is certainly the case for more than 500 references used in this book. Most of these references do not even mention uncertainties or measurement errors.

In real laboratory life, the practical measurement problems are multiple, they are everywhere, they are difficult to solve, and not all the answers are known by The Tribe Elders (also known as the Professors). Each researcher in every experimental laboratory has, because it must have, a tremendous amount of know-how and practical insight for carrying out the given measurements. Most of this knowledge is gained the hard way, by learning 'on the job'. And a lot of this knowledge is not published *anywhere* because it is 'well known' or not deemed worthy of publishing.

This mountain must be climbed by every researcher at the beginning of the journey on magnetic measurements. Is there a better path to learn of such things? Would it be useful to have some sort of publication that combines all the relevant information so that it can be used both as a starting point *and* as a reference to come back to?

Typical scientific books must cover a vast span of the whole subject so that they cannot talk about the details. Scientific papers must be condensed so that they cannot talk about the details either.

Perhaps the best source of the information about *practical* difficulties are PhD theses. But even these would contain only limited breadth because of the specific topic of a given research, such as only related to *just* the rotational power loss, or *just* the Barkhausen noise, or *just* the magnetostriction and so on.

It should also be remembered that there is a certain amount of competition between research laboratories, so many 'trade secrets' are simply not published because the know-how is considered as being at least slightly confidential (and even strictly confidential in some cases). Some of these problems would be sometimes alluded to in private discussion, perhaps at the coffee breaks of an appropriate conference. But again, these would not be published.

***

The topic of rotational power loss measurement (and other associated characterisation) is sufficiently wide that an international conference was started in 1992 to have a proper forum for discussion: *International Workshop on 1&2 Dimensional Magnetic Measurement and Testing.*

There is a lot of cross-over between rotational measurements (1D, 2D and 3D) and other standardised methods typically used in the industry, such as the Epstein frame or single-sheet tester.

This book focuses completely on the various *practical* measurement methods for the characterisation of magnetic materials under rotational magnetisation, as well as other non-standard 1D, 2D and 3D excitation. The *measurement* of rotational power loss is the main topic, but the measurement of other properties such as magnetostriction and domain observation is also included for completeness.

Because of the practical approach taken in this book, there are more references to PhD theses and 1&2DM workshop presentations – where much more practical measurement details are shared than it is the case for typical scientific papers. For the same reason, some of the drawings and images used in this book are less 'pretty' because they were meant to show the 'raw' data, which are not always published.

There is a certain amount of overlap between several topics discussed in different chapters. This was done on purpose because it allows studying each chapter separately, depending on the needs of the reader.

The reader is briefly introduced to the topic of magnetism. The later chapters deal with the definition of the various measurement methods, with the sensing techniques and detailed description of magnetising apparatus as presented in the literature. The last part of the main text introduces the uncertainty analysis, which is usually quite a problematic topic. This chapter hopefully is written in a way that clarifies many of the confusing concepts of uncertainty, and thus it should help students or early-career researchers to better absorb the ideas, and also serve as a reference for more mature scientists.

One of the goals of the 1&2DM workshop is to find a path for the standardisation of the rotational power loss measurement. At the time of writing this book, this goal has not been achieved yet and many problems remain to be solved.

However, it is obvious that great progress has been made since the first 1&2DM workshop, and in my opinion, we get closer to defining such standard, as judged by the state-of-the-art-of-the knowledge in the subject. Therefore, this book also presents a synthesis of the optimum magnetising apparatus, which should bring us closer to the standardised method of rotational power loss measurement.

Nevertheless, there are still many technical problems to be solved and research to be carried out. This book is written to help with the future research, to make life easier for the legions of future students and researchers tackling the difficult subject of *measurement* of power loss under rotational magnetisation, as well as other 1D, 2D and 3D excitation regimes.

Stanisław Żurek
Canterbury, the United Kingdom

MATLAB® is a registered trademark of The MathWorks, Inc. For product information, please contact:

The MathWorks, Inc.
3 Apple Hill Drive
Natick, MA 01760-2098, USA Tel: 508 647 7000
Fax: 508-647-7001
E-mail: info@mathworks.com
Web: www.mathworks.com

Acknowledgements

I would like to thank my very knowledgeable brother Sławomir Żurek for igniting my interest in all things magnetic, and for providing the first opportunity for hands-on experience with electric motors and transformers.

I am indebted to my first mentors from Czestochowa University of Technology and KBR Magneto, Poland – Professor Marian Soiński and Dr. Roman Rygał – who made my professional career possible. Together with Dr. Wojciech Pluta, they have been encouraging me throughout the years, and provided magnetic samples as well as thought-provoking forum for discussion.

My time at the Wolfson Centre for Magnetics, at Cardiff University, the United Kingdom taught me a lot about magnetics, professionalism as well as friendship. I bow with respect before Professor Anthony Moses, Professor Phil Beckley and Professor David Jiles, with whom I had the privilege of working. The names of all the friends and colleagues from the Wolfson Centre are too numerous to be mentioned here, but the following countries were proudly involved: Brazil, China, England, Germany, Ghana, Greece, India, Iraq, Japan, Myanmar, Poland, Romania, Spain, Thailand, Ukraine, the United States, Wales and probably several more. I have learnt a great deal from each and every one of you, and I hope I reciprocated at least a little. You will find a lot of your work cited in this book, and I am thankful for meeting you and experiencing the interesting research projects in various areas of magnetics. Specifically, I would like to mention Richard Wakely for machine-winding of the H-coils, which were fundamental to my PhD studies.

I am more than obliged to Dr. Piotr Klimczyk of Brockhaus Measurements, Germany for providing the expensive prototypes of the 'ideal' PCB-based H-coils.

I thank Professor Sławomir Tumański, editor in chief of *Przegląd Elektrotechniczny* (*Electrical Review*; international journal published in Poland), for the permission for reusing many of the published copyrighted materials.

I am very grateful to my current company Megger Instruments Ltd, the United Kingdom, who kindly allowed me to use their equipment for some illustrations in this book and for performing a few fundamental magnetic studies. I thank my colleagues Andrew Gilham, who created fantastic 3D images of the proposed configuration of rotational magnetic yokes, Christopher Waller, who precision-drilled several samples, and Jeffrey Jones, who helped me with the topic of analogue feedback.

Several samples were kindly annealed by Magnetic Shields Ltd, the United Kingdom.

***

Most importantly, I am thankful to my wife Joanna for her perpetual support and for taking many wonderful photographs used here. The work on this book would have taken much longer if it had not been for her delicious cakes, which seem to be capable of fuelling my brain and speeding up my thoughts.

Dziękuję!

Author

Stanislaw Zurek completed his MSc at the Czestochowa University of Technology, Poland in 2000, and in 2005 graduated as PhD at Wolfson Centre for Magnetics, Cardiff University, UK. In 2008, he joined Megger Instruments Ltd, where he currently holds the position of Manager of Magnetic Development and in his free time continues to pursue the topic of rotational power loss measurement. He became an IEEE Senior Member in 2010. He has authored and co-authored more than 80 scientific papers, as well as an additional 20 technical papers. He also is a co-inventor of ten patent applications related to electromagnetic technology.

1 Introduction

1.1 BRIEF INTRODUCTION TO MAGNETISM

A *magnetic field* is produced whenever there are electric charges in motion, which represent the *electric current, I*, measured in amperes* (A). A magnetic field can also be produced by other magnetic sources (permanent magnets, magnetic dipoles, etc.), but these are primarily also always generated by the macroscopic motion of electric charges or fundamental subatomic properties such as *spin magnetic moment* and *orbital magnetic moment*.

For a visual representation, the magnetic field is often depicted by *magnetic field lines* (Figure 1.1). The lines encircle the conductor with current, and the *magnetic field strength, H*, is always generated in a direction (and plane) perpendicular to the current causing it.

The strength of the magnetic field at a given point in space depends on the *magnetic path length, l*, measured in metres (m), through which the field travels.

Looking at the example shown in Figure 1.1, we can see that the magnetic field lines follow a path, whose length for a given radius, *r*, is

$$l = 2 \cdot \pi \cdot r \quad \text{(m)} \tag{1.1}$$

Therefore, the path *l* increases proportionally with the distance *r*. At any point in space, a magnetic field can be described quantitatively by a value[†] of *magnetic field strength, H*. Because the field strength is produced by the current *I* (A) over a path *l* (m), the resulting unit is ampere per metre (A/m).

From the Maxwell equations (describing the laws of electromagnetism), the magnetic field strength can be derived for the configuration shown in Figure 1.1, and for an infinitely long conductor, the magnetic field strength at a given point is

$$H = \frac{I}{l} = \frac{I}{2 \cdot \pi \cdot r} \quad \text{(A/m)} \tag{1.2}$$

Hence, the value of *H* is directly proportional to the current generating it but inversely proportional to the distance, or more precisely to the magnetic path length. This means that the field is strongest near the source (in this case, current-carrying conductor) and reduces with the distance from it.

---

* Throughout this book, SI units will be used.
[†] Depending on the type of measurement or analysis, the magnetic field strength can be expressed either as a single-valued scalar H or a vector $\vec{H}$ oriented in a two- or three-dimensional space. The scalar notation will be used wherever possible throughout this book for simplification and clarity. The text will also define that the described entity is a 'vector of H' (if the arrow notation is not explicitly used).

1

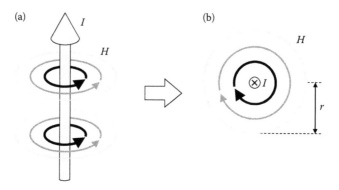

FIGURE 1.1   An electric current I in a wire generates a magnetic field measurable as magnetic field strength H: (a) general view, (b) cross-sectional view.

If instead of a single wire, there are N closely positioned parallel wires carrying the same current I (e.g. as in a multi-stranded wire), this is equivalent to a single wire with the equivalent total cross-sectional area and a combined current $N \cdot I$; therefore,

$$H = \frac{N \cdot I}{l} \quad \text{(A/m)} \tag{1.3}$$

which can be rearranged taking a form of simplified Ampere's law (one of the Maxwell equations) in a scalar notation for a uniform medium:

$$N \cdot I = H \cdot l \quad \text{(A)} \tag{1.4}$$

where $N \cdot I$ represents the so-called *magnetomotive force*, or *MMF*, with the unit of ampere (A), commonly referred to as *ampere–turn* product (N is unitless). The magnetomotive force represents a source generating magnetic field and the magnetic field strength H can be a measure of the so-called *applied magnetic field*.

When the current is 'wrapped' in a coil, the field inside such a loop is intensified due to the contributions from all the sides (Figure 1.2a).

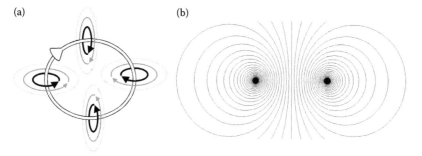

FIGURE 1.2   Magnetic field around a loop of current (a) and a computer simulation of the cross section showing the increased density of magnetic field lines inside the current loop (b).

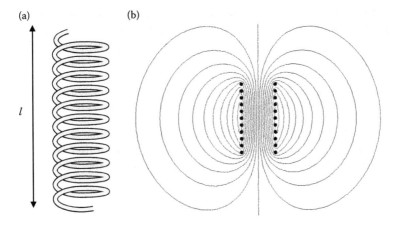

(a)   (b)

FIGURE 1.3   An example of a coil: solenoid (a) and a computer simulation of the cross section showing the magnetic field lines inside the solenoid (b).

It can be calculated that, at the centre of such single-turn flat loop with radius r, the magnetic field strength is equal to

$$H = \frac{I}{2 \cdot r} \quad (A/m) \tag{1.5}$$

The current loops can be placed one after another in a series, creating a long coil or a *solenoid* (Figure 1.3). The magnetic field lines are concentrated inside the solenoid where the resulting magnetic field is the most uniform.

It can be derived mathematically that, for an infinitely long solenoid wound with a negligibly thin wire, the magnetic field strength at the geometrical centre (inside the solenoid) is equal to (Chikazumi, 2002)

$$H = \frac{N \cdot I}{l} \quad (A/m) \tag{1.6}$$

where N is the number of turns, I is the current (A) and l is the length of the solenoid.

PRACTICAL COMMENT

In practice, an 'infinitely long solenoid' means that its length l is 'much greater' than its radius r. Also, 'thin wire' means that the diameter of the wire should be 'much smaller' than the diameter of the solenoid. For solenoids with a finite length, the value of H calculated from Equation 1.6 differs from the real value, for example, a ratio $l/r = 10$ results with an error of around 2% (Fiorillo, 2004). Therefore, a solenoid with $r = 100$ mm and $l = 1000$ mm will produce a field, which is 2% lower than the ideal value calculated from Equation 1.6.

Long solenoids are commonly used as a source of known magnetic field, for example, for calibration of sensors in research laboratories. For shorter or multi-layer coils, special correction coefficients are used (Langford-Smith, 1953).

For traceable metrological sources of magnetic field, other devices can be used because solenoids might not guarantee the required accuracy. Another typical source of uniform magnetic field is a Helmholtz coil (Fiorillo, 2004).

Any medium exposed to a magnetic field is magnetised – that is, it responds with *flux density*, B, whose unit is tesla (T). In the case of vacuum (free space), the relationship between B and H is intrinsically linear and can be expressed[*] as

$$B = \mu_0 \cdot H \quad \text{(T)} \tag{1.7}$$

where μ_0 is the *magnetic permeability* of free space and it is a physical constant with a numerical value defined exactly as $\mu_0 = 4 \cdot \pi \cdot 10^{-7}$ (H/m).

Substances consist of atoms and these are spaced out by some distances so that a certain amount of 'free space' is present in any matter. Hence, for a medium other than free space, the correlation between B and H should be written as a combination of the contributions from the free space (the $\mu_0 \cdot H$ component) and the response of the material or substance (the $\mu_0 \cdot M$ component) (Chikazumi, 2002):

$$B = \mu_0 \cdot H + \mu_0 \cdot M = \mu_0 \cdot (H + M) \quad \text{(T)} \tag{1.8}$$

The value of *magnetisation*, M, has the unit (A/m), which is the same as that for the magnetic field strength H. The value of M represents the combined magnetic contribution of all atoms inside the material. Equation 1.8 can also be written as

$$B = \mu_0 \cdot H + J \quad \text{(T)} \tag{1.9}$$

The *polarisation*, J, with unit of tesla (T) is therefore directly linked to magnetisation M such that

$$J = \mu_0 \cdot M \quad \text{(T)} \tag{1.10}$$

Equations 1.8 and 1.9 are commonly further modified to the following form:[†]

$$B = \mu \cdot H = \mu_r \cdot \mu_0 \cdot H \quad \text{(T)} \tag{1.11}$$

---

[*] Because the magnetic field strength can be a vector $\vec{H}$, the flux density $\vec{B}$ can also be a vector. For a three-dimensional representation of anisotropic problems, the magnetic permeability μ_r can take a form of a tensor. The calculations are similar, but they need to be carried out in an appropriate matrix notation.

[†] Equation 1.11 is used especially in engineering where the value of μ_r can be applied conveniently as a 'figure of merit' for various magnetic materials.

where $\mu = \mu_r \cdot \mu_0$ and μ_r is the unitless *relative magnetic permeability* encompassing the combined contributions of free space ($\mu_0 \cdot H$) and magnetic response of atoms – the polarisation J or magnetisation M. The relative permeability μ_r is calculated simply as a ratio of the permeability of the given medium to the permeability of free space. Therefore, by definition for free space $\mu_r = 1$, Equation 1.11 simplifies to Equation 1.7.

It is sometimes convenient to use the notion of *magnetic susceptibility, χ*. By assuming

$$M = \chi \cdot H \quad (\text{A/m}) \tag{1.12}$$

It can be shown that

$$\chi = \mu_r - 1 \quad (\text{unitless}) \tag{1.13}$$

so that, by definition for free space, the susceptibility $\chi = 0$.

PRACTICAL COMMENT

It should be noted here that 'free space' is synonymous with the absence of any medium. In the previous system of units (CGS), permeability was unitless, which meant that both B and H could be expressed in the same units and for free space $\mu_r = 1$ (unitless), so it could be simply written that 1 gauss = 1 oersted (in the similar sense as 1 inch = 25.4 mm) and therefore $B = H$. This led to a situation where both values were used interchangeably, especially in theoretical physics (where in many cases, the free space was the analysed medium in theoretical calculations).

In the currently used system of units (SI), such approach is *incorrect* and B and H have clearly defined separate units of (T) and (A/m), respectively, so that $B \neq H$ always (SI, 2006). Of course, for linear media, one can be derived easily from the other, but this is true only for *exactly* linear media.

Some of the misunderstandings are aggravated by different names commonly used in different science and technology branches. As a general concept, it can be assumed that the term *magnetic field* denotes space or volume with *applied magnetic field*. However, *magnetic field* does not refer to any particular quantity with SI units (H or B) as such. At least two variables are required, for example, B and H, to fully quantify the magnetic effects analogically as the electric effects require both current I and voltage V for full quantification (White, 2011). In this analogy, it is also true that $V \neq I$ because each of the variables represents a different physical quantity.

In the opinion of the author, it can be conceptually understood that the magnetic field can be quantified by *magnetic field strength H*, which can be thought of as being the magnetising 'source'. This causes the medium to respond with *flux density B*. Such response comprises two components: the contribution of free space (the 'empty' space between the atoms, the $\mu_0 \cdot H$ component)

TABLE 1.1

Various Names of 'Magnetic Field' Commonly and Sometimes Incorrectly Used in Science and Technology (Non-Exhaustive List)

SI Name	Magnetic Field Strength[a]	Flux Density	Magnetisation[b]	Polarisation
SI Symbol and Unit	H (A/m)	B (T)	M (A/m)	J (T)
Other names or meanings used in engineering	Magnetising field Magnetic field[c] Magnetic field intensity Applied magnetic field	Induction[d]	Process of magnetising	Intrinsic induction[d] (ASTM, 2009) Intrinsic flux density (Drake, 1982) True flux density (Drake, 1982)
Names or meanings used in other sciences (physics, applied mathematics, etc.)	Magnetising field Magnetic field Auxiliary magnetic field (Elmore et al., 1985)	Magnetic field (Gubbins et al., 2007) Fundamental magnetic field (Wachter et al., 2006) Auxiliary magnetic field (Crowell, 2005)		

[a] 'Magnetic field strength' is usually referred to by engineers as 'magnetic field' or 'magnetising field'.

[b] The word 'magnetisation' commonly means also the process of magnetising (e.g. applying the magnetic field to a material).

[c] As a general concept, 'magnetic field' refers to the field produced by a flowing current. It does not describe any particular quantity with defined SI units (neither H nor B).

[d] 'Induction' can be used when referring to a process of inducing the voltage in a magnetically linked circuit (Faraday's law of induction). For this reason, it is 'safer' to avoid the word 'induction' and instead refer to B always as 'flux density' to avoid the ambiguity.

and the contribution of the atoms or *polarisation J* (or depending on the notation *magnetisation M*).

Table 1.1 lists some names and meanings commonly encountered in various branches of science and technology. *Magnetic field strength* is often commonly shortened to 'magnetic field'. Moreover, the term 'magnetisation' also means the 'process of magnetising', and 'magnetic induction' could denote the 'process of inducing a voltage' in a magnetic circuit.

To complicate the matter even further, due to non-SI system of units, standards such as ASTM use the name of *intrinsic induction* when referring to *polarisation* (ASTM, 2009).

Therefore, confusion can arise easily and great care must be taken to fully understand the meaning from the context before analysing or importing any descriptions from one discipline of science to another. For these reasons, it is always 'safest' to use the terminology as defined by SI units. It is of prime importance to know whether a given publication uses SI or some other system of units.

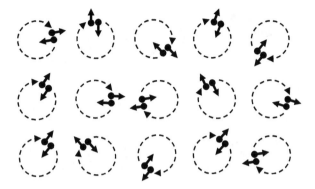

FIGURE 1.4 In *diamagnets*, in each atom, the spin moments (shown here conceptually as arrows) are opposing each other and electron orbits (dashed circles) are all oriented randomly (for simplicity shown here only in two dimensions).

In substances, each electron can contribute to the overall magnetic behaviour. Electrons follow the orbits – such moving charges effectively represent the electric current and hence generate some magnetic field (*orbital magnetic moment*). Even stronger effect is made by the *spin magnetic moment* (intrinsic property of electrons). Depending on the contribution of the orbital and spin magnetic moments in a given substance, we can distinguish three major types of magnetic behaviour.

Diamagnetic materials have even number of electrons in the atoms and the spin magnetic moments are paired in an anti-parallel manner, such that their contributions cancel each other out (Figure 1.4).

However, the movement of electrons in their orbits can be affected by applying an external magnetic field, which is equivalent to inducing a current generating an opposing magnetic field to counteract the changes (Chikazumi, 2002). As a consequence, the resulting B is somewhat weakened inside the material, so that there is a measurable effect of $\mu_r < 1$.

For most substances, the diamagnetic effect is very small,[*] with the maximum known occurring in bismuth for which $\mu_r = 0.99983$ (a commonly used copper exhibits $\mu_r = 0.99999$). For this reason, in everyday engineering applications, the diamagnets are simply assumed as 'non-magnetic' and magnetically equivalent to free space so that $\mu_r \approx \mu_0$. A diamagnetic material is always repelled by magnetic field, regardless of the polarity, albeit with a miniscule force.

Paramagnetic materials usually have odd number of electrons (Jiles, 1998). Without an external magnetic field, all moments are aligned randomly (Figure 1.5). But with the magnetic field present, each unpaired electron responds by aligning its spin with the applied field. As a result, the overall B is increased and the permeability is slightly greater than unity, although the effect is also very weak (1.00002 for aluminium and 1.0000004 for air). The permeability depends somewhat on temperature, but the effect

---

[*] With the notable exception of superconductors approaching the behaviour of ideal diamagnets for which $\mu_r = 0$ ($\chi = -1$). Diamagnets are repelled from the magnetic field and for the superconductors, the effect is so strong that they can levitate above the permanent magnets against the gravity.

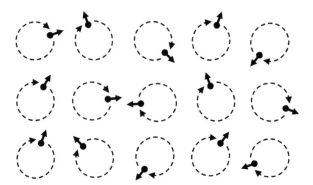

FIGURE 1.5 In *paramagnets*, each atom has one electron with contributing spin moment, but all are randomly oriented.

is so small that it can be neglected in most engineering applications around room temperature. Hence, the paramagnets are also treated as 'non-magnetic' and the approximation $\mu_r \approx 1$ is also commonly applied in engineering with a negligible practical error.

In paramagnets, the magnetic moments of electrons are normally oriented randomly in space and do not interact with each other, or their interaction is inhibited by some other factors such as thermally caused vibrations. Therefore, without an externally applied magnetic field, paramagnets do not exhibit any significant resultant magnetisation because all random contributions cancel each other out.

Both diamagnets and paramagnets normally exhibit response with practically a constant value of permeability. This property is often used in magnetic sensors, as will be described in the following chapters.

A third important group of materials exhibits *ferromagnetism*. The phenomenon differs from the two abovementioned effects because the unpaired electrons are packed so closely that they magnetically interact with each other and the interaction is very strong. In order to minimise the internal energy, the neighbouring magnetic moments align spontaneously in the same direction (Figure 1.6).

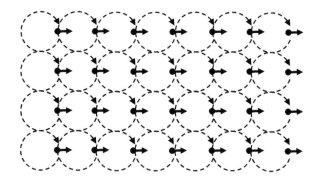

FIGURE 1.6 In *ferromagnets*, each atom has one electron with contributing spin moment – the atoms are packed so closely that the neighbours are coupled magnetically and they align over large volumes, so that there is a resultant spontaneous magnetisation.

Depending on the strength of the neighbouring interaction, we can distinguish the following sub-phenomena:

- *Ferromagnetism* – alignment of neighbouring magnetic moments occurs over large volumes, called *magnetic domains*. When an external magnetic field is applied, then whole magnetic domains are affected, due to strong coupling between the neighbours. As a result, the material responds with large changes of B. Thus, μ_r can reach very high values, even up to 1,000,000 for some materials.
- *Ferrimagnetism* – some neighbours are aligned parallel, and some anti-parallel (e.g. when a complex crystal comprises two mutually intersecting sub-lattices). The partial contribution from the anti-parallel neighbours reduces the magnetic response to some degree (Figure 1.7), but externally the material still exhibits ferromagnetic-type characteristics (e.g. large values of μ_r).
- *Antiferromagnetism* – neighbours contributing to exactly half of the inter-action are aligned parallel, and in the other half, they are anti-parallel, so that the resulting μ_r is low (comparable to paramagnetic materials (Chikazumi, 2002)).
- There are also other types: superparamagnetism (Colliss et al., 1968), heli-magnetism (Ballou et al., 1987), spin glass (Young, 1998), etc. However, their practical technological significance for energy conversion is marginal, and being beyond the scope of this book, they will not be discussed here.

All types of ferromagnetism have one common feature – strong magnetic interaction between the neighbouring atoms (Hilzinger et al., 2013). The properties are not necessarily related to the chemical elements involved[*] but rather result from the physical structure in which all the atoms are placed and how they interact with each other (Figure 1.6).

If the structure is disturbed enough as to upset the magnetic neighbour-to-neighbour interactions, then the effect can be reduced or disappear completely. All ferromag-

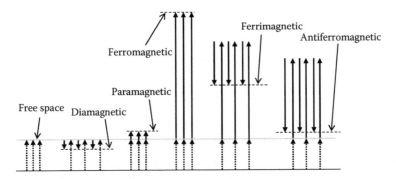

FIGURE 1.7   Qualitative comparison of magnetic permeability for various types of materials (not to scale).

---

[*] Ferromagnetism has also been demonstrated for completely non-metallic polymers (Nicholson, 2006).

netic substances can be made paramagnetic simply by heating up the material. Increased temperature causes vibration of atoms within the structure, and if enough thermal energy is applied, then eventually the magnetic energy is overpowered and the material becomes paramagnetic above a certain threshold called *Curie temperature* (or the *Néel temperature* for antiferromagnets). Therefore, the group name 'ferromagnetic-like' can be used to describe all such effects.*

Another common feature is that the strong ferromagnetic-type response is limited – the material saturates if the excitation is high enough. This is not the case for non-ferromagnetic materials. Currently, it is impossible to generate a magnetic field with an amplitude large enough to cause the saturation of paramagnets (although theoretically it should be possible).

The highest known value of ferromagnetic saturation is 2.35 T (for a type of Co–Fe alloy) (Fiorillo et al., 2016), but paramagnets do not saturate even when subjected to over 1000 T (Miura, 2008).

For simplicity, unless stated otherwise, in the remaining part of this book, the word 'non-magnetic' will refer to both diamagnetic and paramagnetic substances, and 'magnetic' to widely understood ferromagnets (ferromagnetic and ferrimagnetic, etc.)

1.2   MAGNETIC DOMAINS AND DOMAIN WALLS

The neighbouring coupling in ferromagnets results in large, magnetically ordered regions called *magnetic domains*. If the whole body of a ferromagnet was to be occupied by a single domain, then magnetic field lines would have to close through the surrounding air, which requires a lot of energy stored in the magnetic field (Figure 1.8a).

In order to minimise this energy, the domains reconfigure themselves in such a way as to contain more energy inside the material. The two-domain system from Figure 1.8b has less field lines (less energy) in the air, and the introduction of *closure*

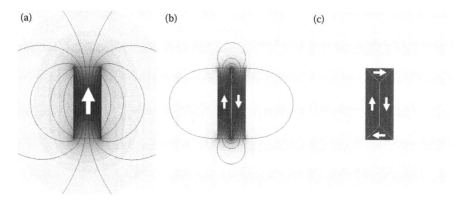

(a)   (b)   (c)

FIGURE 1.8   Appropriate configuration of magnetic domains allows the reduction of stored energy, white lines show the domain walls. (a) single domain, (b) two anti-parallel domains, (c) system with closure domains.

---

* A similar approach is used to describe strong non-linear interactions in a different class of materials called 'ferroelectrics' even though they are not magnetic and do not contain iron (from Latin *ferrum* – iron).

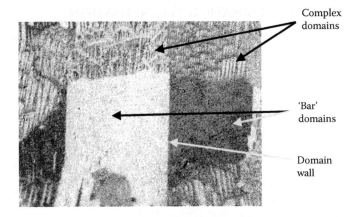

FIGURE 1.9 Typical domain structure in non-oriented electrical steel – visible wide 'bar' domains as well as very complex domains outside of the main crystal grain (the width of the image is around 0.2 mm). (Courtesy of Wolfson Centre for Magnetics, Cardiff University, Cardiff, the United Kingdom.)

domains in Figure 1.8c leads to containing almost all energy within the material (the lowest energy state as compared with one- or two-domain system). The boundaries between the domains (white lines in Figure 1.8b,c) are called *domain walls*.

The total energy of a given domain system depends on many factors: mechanical stress, non-magnetic inclusions, roughness of the surface, crystal anisotropy, size of polycrystalline grains, type of excitation and so on. Therefore, in reality, the domain structures can be very complex. A typical domain structure in non-oriented electrical steel is shown in Figure 1.9.

The domain walls also require some energy to exist. For this reason, the domains cannot be infinitely small because the domain walls would store more energy than would be saved from the creation of small domains. Hence, depending on many conditions, the domain structure will reorganise itself as to find the minimum of energy for a given state.

The transition from one domain to another is gradual, and it is called a *domain wall* (Figure 1.10, see also Figure 1.8b,c). It is estimated to occur over around 150 atoms (Jiles, 1998).

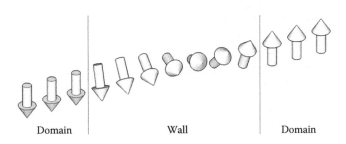

Domain   Wall   Domain

FIGURE 1.10 Schematic illustration of a gradual change of ordering within a domain wall (not to scale).

1.3 PROCESS OF ALTERNATING MAGNETISATION

As mentioned above, when static H is applied, then the medium responds with B. For non-magnetic materials, it can be assumed that the response is practically linear and the increase in B is directly proportional to H, with permeability $\mu = \mu_r \cdot \mu_0$ being the proportionality factor (slope of the curve). Moreover, in isotropic materials, the response is oriented in the same direction as the cause, so that if the H vector is rotated or reversed, the B vector follows accordingly. When the magnetising field is removed, then the material has nothing to respond to, B reduces to zero and the material returns to a completely 'demagnetised' state (Figure 1.11).

In ferromagnetic materials, the process is different. As discussed above, the whole volume of such materials is occupied by magnetic domains. With an applied H, the material must respond by changes in B, which is achieved by changes in the configuration of the domain structure (Figure 1.12).

When describing the magnetisation process of ferromagnets, a special reference point should be mentioned. This point is the *demagnetised state* in which all the magnetic domains are organised in such a way that most of the field lines close

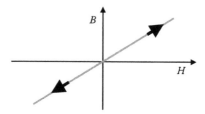

FIGURE 1.11   Linear response or static $B = f(H)$ characteristics loop for non-magnetic materials (e.g. air or copper).

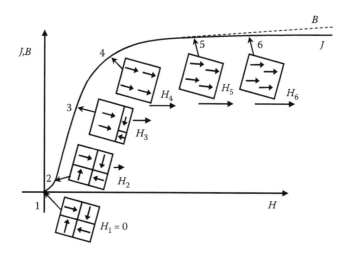

FIGURE 1.12   Magnetisation curve $J = f(H)$, or $B = f(H)$ with schematic representation of behaviour of magnetic domains. (Adapted from Brailsford, F., *Magnetic Materials*, Methuen & Co. Ltd., London, UK, 1954.)

within the material and the flux density averaged over volume of the material is zero. This state is shown as point 1 in Figure 1.12.

When H is applied, then initially, domains whose orientations coincide with the direction of H grow at the expense of other domains (point 2), eventually assimilating the neighbouring domains (point 3) so that ultimately the whole volume is occupied by a single domain (point 4).

Because there are no more differently oriented domains to be consumed, the increase in B slows down (permeability reduces because the slope of the curve is less steep). Further response occurs by the rotation of the single-domain orientation towards the direction of the applied field (point 5). Eventually, the ferromagnet saturates completely when a strong enough excitation is applied (point 6) because there are no more possible changes to the domain structure – the polarisation J (or magnetisation M) achieves the highest possible value called *magnetic saturation*.

However, the response of ferromagnets consists of two components: the polarisation J and the free space component $\mu_0 \cdot H$, which continues to grow linearly with the applied H because the non-magnetic contribution will *not* saturate. This is indicated by the difference between B and J at high values of H in Figure 1.12. For small and intermediate values of H (below the 'knee' of the B–H curve), the difference between B and J is so small that it can sometimes be neglected in engineering applications.

The process described above takes place for increasing H from zero to some positive value. Under alternating excitation, the process of magnetisation occurs as shown in Figure 1.13. The points from 1 to 6 in Figure 1.13 are the same as in Figure 1.12. However, when H is decreased from saturation, then the response of the material follows a different path. Once H is decreased to zero, the material retains some of the flux density (point 7) resulting in a value called *remanence* (and usually denoted as B_r). A certain negative excitation, called coercivity H_c, needs to be applied to bring the flux density to zero (point 8). It should be noted that this is not equivalent to the demagnetised state (for which both $H = 0$ and $B = 0$, point 1).

When the negative H is increased further, eventually the opposite saturation is reached (point 6'). If the excitation changes again from its maximum negative value, it will cause the response to go through the 'mirror reflection' of the points of

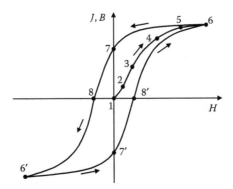

FIGURE 1.13 B–H loop.

remanence and coercivity (points 7′ and 8′) back to the positive saturation (point 6). A full cycle therefore creates a loop called *B–H loop*,* or *hysteresis loop*.†

Domain movements are impeded by non-magnetic inclusions, surface roughness, mechanical stresses, etc. This causes some energy to be dissipated, and if that happens, then the process is locally irreversible. This energy loss is directly proportional to the area of such loop. So the measurement of *H* and *B* is important because knowing the point-by-point values allows the calculation of the *B–H* loop area, and thus measurement of magnetic losses.

The magnetisation processes for non-magnetic and ferromagnetic materials described above are true for a slowly changing excitation (so-called *quasi-static*‡). This is not the case under 'dynamic' excitation, for example, magnetisation at 50 or 60 Hz.

According to Faraday's law of induction,§ varying *B* induces electromotive force *EMF(t)* and hence also voltage *V(t)* in a magnetically coupled electric circuit (Figure 1.14) and for a simple case can be calculated as

$$EMF(t) = -N \cdot A \cdot \frac{dB(t)}{dt} \quad (V) \tag{1.14}$$

where *t* is the time (s), *N* is the number of turns (unitless) and *A* is the cross-sectional area (m²). The notation '(*t*)' means that the variables are a function of time *t*.

If this circuit is loaded with some resistance, then an electric current will flow through it, causing energy dissipation or loss. The induced *EMF* can be measured across the coil as voltage *V* (Figure 1.14).

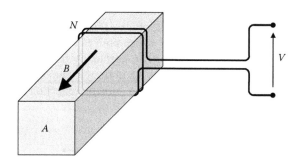

FIGURE 1.14   Illustration of Faraday's law of induction – induced voltage across the terminals of the coil is proportional to the number of turns *N*, area *A* and rate of change of *B*.

---

* Because the difference between *B* and *J* is often neglected under most practical conditions, the term '*J–H* loop' is rarely used, unless necessary (e.g. in hard magnetic materials).
† The word 'hysteresis' can be used to describe two different meanings – introducing even more confusion. This is explained in detail later.
‡ 'Quasi-static' means that the process is as such variable but the changes occur so slowly that other effects can be treated as negligibly small (i.e. no significant dynamic effects).
§ 'Induction' here means 'the process of inducing' – and this example demonstrates clearly why using precise names of SI system (e.g. 'flux density' rather than 'induction') is so important for avoiding the ambiguity.

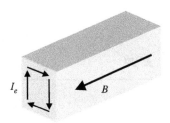

FIGURE 1.15   Eddy currents I_e flowing inside a conductor subjected to varying B.

Even in non-magnetic conductors, such as copper or aluminium, the body of the metal is also effectively one turn of electric circuit with some, usually very low, resistance. If the material is exposed to a varying B, then according to Equation 1.14, a voltage will be induced, and as a consequence a current will flow (Figure 1.15). These so-called *eddy currents* are strongest at the surface. The currents inside are weaker for several reasons: the outer 'turn' encloses the greatest cross-sectional area, the inside currents partially cancel each other out, and for higher frequencies, there is also a phenomenon called *skin effect*.

PRACTICAL COMMENT

In all materials, for a stationary value of H, the value of B does not change and the eddy currents are not induced because $dB/dt = 0$, so $EMF(t) = 0$, and therefore $I_e = 0$. Hence, if, for very low-magnetisation frequency, it can be assumed that $I_e \approx 0$, then this would correspond to a quasi-static excitation because the dynamic effects could be neglected.

If H changes, for instance, sinusoidally, then at any instance of time, B also changes. In non-conducting non-magnetic materials (e.g. air), the resistivity is so great that there are no electric currents induced (or they are negligibly small) and the dynamic $B–H$ characteristic is the same for static and dynamic excitation (see Figure 1.11).

In conducting non-magnetic materials (e.g. copper), the varying B induces eddy currents, which dissipate the energy inside the material. As a result, the $B–H$ characteristic becomes an elliptical loop (Figure 1.16), whose area is directly proportional to the energy lost for the eddy currents.

However, it should be borne in mind that the $B–H$ loop from Figure 1.16 is caused solely by the eddy currents due to dynamic magnetisation of conducting *non-magnetic* material, whereas the $B–H$ loop from Figure 1.13 is a result of quasi-static domain movements. This conceptual difference will be clarified later.

Most magnetic materials are also fairly good conductors. Hence, the eddy currents will also be present during alternating magnetisation. The effect is relatively strong because due to high permeability for the same exciting H, the response B is much greater in magnetic materials than in non-magnetic materials. However, as

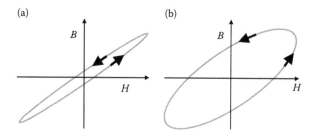

FIGURE 1.16   Dynamic $B–H$ loop for a conducting non-magnetic material with the same amplitude of B under sinusoidal excitation for (a) lower frequency and (b) higher frequency (compare with Figure 1.11).

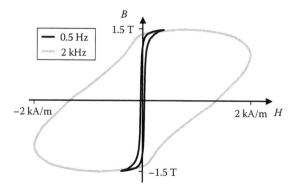

FIGURE 1.17   Typical quasi-static (0.5 Hz) and dynamic (2 kHz) $B–H$ loops measured for the same peak flux density (sinusoidally controlled 1.5 T) for non-oriented electrical steel. (Adapted from Zurek S. et al., *Przeglad Elektrotechniczny (Electrical Review)*, (NR 12/2005), 2005b, 5.)

discussed above, there is also another component contributing to energy loss – the magnetic hysteresis.

As a result, the dynamic $B–H$ loop in magnetic materials is a combination of the 'static' and 'saturable' $B–H$ loop (Figure 1.13) and the contribution from eddy currents (Figure 1.16). An example of a dynamic $B–H$ loop for non-oriented electrical steel is shown in Figure 1.17. As can be seen at the higher excitation frequency, the loop resembles a somewhat elliptical shape, which indicates that the energy is lost mostly for eddy currents.

1.4   SOFT, HARD AND SEMI-HARD MAGNETIC MATERIALS

Figure 1.18 depicts a $B–H$ loop with the coercivity shown as H_c and remanence as B_r. It is evident that H_c has a direct influence on the area of the loop, which in turn is proportional to the power (and energy) involved in the magnetisation process.

We can see from Figure 1.18 that after reducing H to zero, the material stays magnetised with the value B_r and some energy remains stored in the material.

The combination of H_c and B_r values informs about the capability of the material to store energy. In fact, the value of H_c is used as means of classification of all

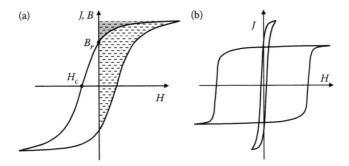

FIGURE 1.18   Energy in the *B–H* loop over half of a magnetising cycle – dotted area represents total energy stored in magnetic material, grey area represents energy returned to the exciting system (a) and illustration of hysteresis loops for soft and hard ferromagnets (b).

magnetic materials into three main groups, called, respectively, as magnetically *soft*, *hard* and *semi-hard*.*

The international standard IEC 60404–1 (IEC, 2000) classifies, somewhat arbitrarily, the magnetically soft materials as those with coercivity less than 1 kA/m. They are 'soft' because they can be magnetised to high values of *B* with low excitation. Their *B–H* loop is 'narrow' (Figure 1.18b), so they do not store or dissipate much energy and can be used for efficient transformation of energy or signals.

Materials with coercivity roughly between 1 kA/m and 100 kA/m are referred to as 'semi-hard' (Soinski, 2001). They are more difficult to magnetise, but also to demagnetise, so they are capable of storing some energy. The state of their magnetisation can be changed without excessive excitation. Because of these properties, they are used in magnetic memory and magnetic storage devices, for example, in computer hard drives. It is important that the stored state of magnetisation is stable (written data must be reliably retained), but also that it is equally easy to change the state (e.g. erase the data) and store a new value.

The magnetically hard materials (or permanent magnets) are those with coercivity above 100 kA/m. There is no upper limit – the best materials available commercially have coercivity even above 1 MA/m (Leonowicz et al., 2005). The permanent magnets are difficult to magnetise and when the external field is removed, they retain the supplied energy and themselves become a strong source of magnetic field. If a bar made out of appropriate magnetically hard material is magnetised, then it is capable of forcing the field lines outside of its body, so they close through the surrounding air (Figure 1.19).

This field surrounding a permanent magnet can magnetise other objects in the vicinity, but it also affects the magnetised body itself. The internal polarisation of the

---

* The terms commonly used are actually 'soft magnetic materials' (or 'soft magnets'), 'hard magnetic materials' ('permanent magnets', or simply 'magnets'), and 'semi-hard magnetic materials'. This can be misleading, because the 'softness' or 'hardness' refers only to magnetic, and not to *mechanical*, properties. All the three types of materials can be manufactured as mechanically soft (e.g. metallised rubber) or hard (solid metal). Thus, 'magnetically soft' or 'magnetically hard' are more precise terms (IEC, 2000). However, the 'imprecise' terms are so commonly used that there is even an international scientific conference named *Soft Magnetic Materials* (Wikipedia, 2011).

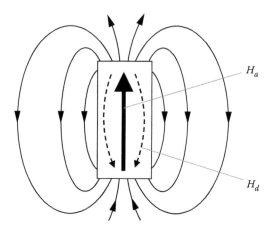

FIGURE 1.19 Components of magnetic field strength in and around a permanent magnet.

magnet causes the field lines to close from one end of the magnet to the opposite end. However, the permeability of the body of the magnet is actually slightly higher than that of the surrounding air. Some lines will close through air, but some will close back through the body of the magnet. As can be seen from Figure 1.19, these latter lines are oriented such as to oppose the direction of the original magnetisation.

Therefore, the 'applied' field H_a (generated by the energy stored in the magnet) is reduced by the *demagnetising* component H_d so that the resultant H is

$$H = H_a - H_d \quad \text{(A/m)} \tag{1.15}$$

which in effect also reduces B at the given point in space.

This demagnetising effect occurs for any body magnetised in a non-closed magnetic circuit (for instance, with an air gap) – including the magnetically soft materials (Soinski, 2001). The demagnetising effect is described usually by a unitless demagnetising coefficient N_d, which depends on the geometrical shape of the specimen and for a closed magnetic circuit (e.g. toroidal core), the reduction of the applied field is practically negligible.

On the other hand, for example, if a thin plate was to be magnetised through its thickness (i.e. perpendicularly to its surface), the demagnetising coefficient would be at its maximum and the resulting relative permeability of the material would be reduced almost to that of free space, even if it was made from a material with very high permeability. Therefore, in magnetic materials, a distinction has to be made between the actual properties of the *material* (shape independent) and the apparent properties of the *sample* (might be shape dependent).

In a general case, the factor N_d cannot be analytically calculated – this is possible only for ellipsoids or other simple shapes. More details about the effects of demagnetising field are given in Chapter 4.

Magnetically semi-hard and hard materials are beyond the scope of this book, so they will not be further discussed.

PRACTICAL COMMENT

As it was mentioned above, flux density B and polarisation J differ from each other. Let us roughly analyse a hypothetical magnetically soft material which at $H = 1000$ A/m would respond with $B = 1.9$ T. We can therefore calculate that the $\mu_0 \cdot H$ component is less than 1.3 mT so the difference between $B = 1.9000$ T and $J = 1.8987$ T is less than 0.1%, which is often negligible when compared to other sources of measurement errors.

However, for magnetically hard materials, there can be $H = 1$ MA/m so that $\mu_0 \cdot H$ can be on the order of 1.25 T, which is certainly *not* a negligible value. For the same reason, there are even two types of coercivity defined for permanent magnets $_BH_c$ and $_JH_c$ depending whether the value is derived from the B or J curve (Leonowicz et al., 2005).

There is at least one notable exception when the difference between B and J must be recognised and treated properly also in the case of magnetically soft materials, namely, if the B sensing coil is much larger than the actual cross-sectional area of the sample. Equation 1.9 cannot be directly applied because the J component contributes with the area of the sample, and the $\mu_0 \cdot H$ with the (much larger) area of the coil, so the calculation would result in an erroneously high value of B as compared to J. In order to reduce this error, a compensation of the 'air flux' is introduced, by applying a 'dummy coil' or a 'mutual inductor' whose role is to induce voltage of equal magnitude but opposite sign to the $\mu_0 \cdot H$ component. The 'dummy coil' and the main search coil are then connected in series opposition so that the $\mu_0 \cdot H$ component is compensated out, and thus not measured at all. Hence, the J contribution is measured 'directly'.

This technique is described in more detail in Chapter 4, and it is used in the well-known Epstein frame, in which all results are measured with known J and not B values (IEC, 2008). Therefore, care should be taken when comparing the values obtained with the Epstein frame to those measured, for instance, on a toroidal sample, for which normally the B value is detected (IEC, 2003), especially for cases with very high excitation.

1.5 FERROMAGNETIC MATERIALS USED IN ENGINEERING

Ferromagnets are very important in modern technology and engineering, mostly for conversion of electrical (transformers) and mechanical energy (motors and generators), but also for other applications such as sensors and data storage (Soinski, 2001). These materials are continuously being improved, bringing their properties closer to the limits imposed by the laws of physics and chemistry of elements.

As mentioned above, ferromagnetism is linked to strong interactions between neighbouring atoms in the structure. This interaction can be affected by other forms of energy, for example, thermal. For pure elements, the chemistry (atomic number) and physics (atomic forces) decide the specific density of a given substance. In all gases, the atoms move too quickly to interact magnetically with

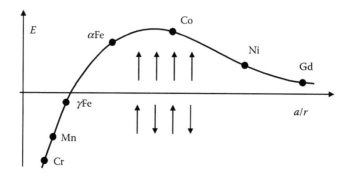

FIGURE 1.20   The Bethe–Slater curve showing some ferromagnetic (top) and antiferro-magnetic elements (bottom). (Adapted from Soinski M., *Magnetic Materials in Technology* (in Polish: *Materialy magnetyczne w technice*), Biblioteka Centralnego Osrodka Szkolenia i Wydawnictw SEP, 2001, ISBN 83-915103-5-2 and Buschow K.H.J. et al., *Physics of Magnetism and Magnetic Materials*, Plenum Publishers, 2003, ISBN 0306474212.)

their neighbours, so they are either paramagnetic or diamagnetic. On the other hand, if the nuclei are too big, then they cannot be packed densely enough to ensure strong interaction, even in solid materials. Also, the neighbouring interactions depend on which *electron shells* are occupied. For ferromagnetism, the shell 3*d* is the most important.

The atom–atom interactions can be analysed with the help of the so-called *exchange integral*, whose result can be graphically represented by the Bethe–Slater curve (Figure 1.20) (Buschow et al., 2003). The interaction is calculated for a given element taking into account the ratio of interatomic distance *a* to the radius *r* of the electron shell 3*d*.

At room temperature, only three elements exhibit classic ferromagnetism: iron, cobalt and nickel. Owing to the weaker interaction, for gadolinium, the Curie temperature is 20°C, and for dysprosium, it is as low as −185°C, so they become ferromagnetic only below these values.

On the other hand, if the ratio *a/r* is too small, then substances can be antiferromagnetic (see also Figure 1.7) with their apparent properties similar to that of non-magnetic elements. This can also be used as a positive feature, as in the case of γ *iron*, which is used as the basis for non-magnetic stainless steel.

However, ferromagnetism also occurs in more complex chemical compounds. In fact, the magnetism itself was first discovered in natural iron ore called *magnetite* (also known as *lodestone*), which is an iron oxide and actually exhibits ferrimagnetism (see also Figure 1.7).

The so-called *ferrites* are the most widely known man-made ferrimagnetic compounds used in engineering. These are also based on iron oxides, with the addition of manganese, zinc, nickel, cobalt, strontium and barium.

Magnetic cores of high-power transformers (operated usually at 50 and 60 Hz) are made of grain-oriented electrical steel, which is an alloy of over 96% iron, 3% silicon and small additions of other metals (e.g. aluminium). The magnetic properties are improved by appropriate crystallographic structure.

Medium-size power transformers can also be made from amorphous ribbons. Again, they are based on iron (85%–95%) with the addition of silicon (5%–10%) and boron (1%–5%) (Metglas, 2011a). Silicon increases electrical resistivity (reduces losses) and boron improves the process of obtaining the amorphous state in a rapidly solidified metal. Small, low-frequency transformers are made either from grain-oriented or non-oriented electrical steel (similar chemical composition but different crystallographic structure).

Magnetic cores (rotors and stators) of small- and medium-power electric motors and generators are made of non-oriented electrical steel. Large rotating machines use grain-oriented electrical steel as well.

High-frequency low-power transformers are made predominantly from MnZn soft ferrites, and for higher-power applications, amorphous and nanocrystalline ribbons continue to gain popularity, especially for high-power switch-mode power supplies (Soinski, 2014).

There is a whole range of nickel–iron alloys, with the nickel content varying mostly between 20% and 80% depending on applications. These alloys (with trade names such as Permalloy, Supermalloy, Mumetal, Radiometal, Hypernik) are used for medium-frequency applications where low losses and high-magnetic permeability are important – for instance, in instrumentation transformers, magnetic shielding and sensors. The highest known relative permeability (up to 1,000,000) for polycrystalline metal alloys is with around 76%–78% nickel, 14%–16% iron, 4%–5% copper, 2%–3% chromium or molybdenum (also with some additions of manganese and silicon) (Jiles, 1998; Vacumschmelze GmbH & Co. KG, 2011; Hilzinger et al., 2013).

Cobalt–iron alloys (48%–50% cobalt) exhibit the highest currently known saturation polarisation of 2.35 T, for any ferromagnets. The other class of Co–Fe alloys contains only around 20% of cobalt and is therefore less expensive, but its saturation polarisation is lower (2.22 T). The Curie temperature is around 920–950°C, so these materials are used for the most demanding conditions, where weight and performance are the most critical factors and a much higher price can be tolerated (e.g. aerospace applications) (Vacumschmelze GmbH & Co. KG, 2011; Hilzinger et al., 2013).

There is also a niche application for cobalt-based amorphous ribbon, which is used for magnetic amplifiers, where the so-called 'squareness' of the $B–H$ loop is required.

Combined usage of electrical steel (grain-oriented and non-oriented) constitutes over 80% of all magnetically soft materials (Tumanski, 2011) with the remainder comprising soft ferrites, nickel–iron alloys, iron powder, iron-based amorphous ribbons and others.

Magnetically hard materials (permanent magnets) are used in smaller volumes. There are a few most popular types. Neodymium–iron–boron magnets exhibit the highest known energy density and are commonly used in rotating machines such as wind turbine generators or efficient DC motors. The highest known coercivity (but lower remanence) is for samarium–cobalt magnets. Because both these types of magnets use rare-earth elements, they are relatively expensive and used only for high-performance devices.

When lower performance can be tolerated, either magnetically hard ferrites (ferromagnetic, with hexagonal crystallography) or so-called Alnico magnets (comprising 35%–60% of iron), which can be made in the final form by casting, are used.

Magnetically semi-hard materials are used for sensors and data storage applications. Magnetic tapes (audio and video cassettes, floppy disks) and the surface of hard disks are made with a magnetisable layer, whose magnetisation can be changed locally by the writing head. Although some of these processes do involve the rotation of magnetisation, but due to dimensions involved, this occurs at the micromagnetic level and thus is beyond the scope of this book. The semi-hard materials will not be discussed further.

On the other hand, the magnetically hard materials in theory are not re-magnetised during operation – ideally they would remain as single-domain magnets. However, in practice, in very demanding conditions (high-magnetic field strength, high temperature), some demagnetisation and re-magnetisation occur, at least partially. This creates losses in such magnets (either for eddy current or for static hysteresis) but calculation, modelling and measurement of such phenomena are still very difficult to carry out accurately (Yamazaki et al., 2005). Nevertheless, the question of rotation losses in permanent magnets is pertinent and will gain even more significance when more magnet-based electric motors and generators of electric cars will be developed.

However, measurements on hard magnetic materials are beyond the scope of this book and will not be discussed here.

1.6 ISOTROPY AND ANISOTROPY

Some equations used later in this book can be derived only for a simplified case of an ideal *isotropic* material. The term 'isotropic' means that the properties of such material are the same in all directions, for instance, most liquids are isotropic. But most solid materials exhibit some *anisotropy* (different properties in different directions). For example, from a mechanical viewpoint, wood is significantly *anisotropic* because it splits easily along the grain but not in other directions.

Magnetic materials usually exhibit some anisotropy of various interrelated magnetic properties such as permeability and power loss (Soinski et al., 1995), coercivity (Soinski, 1987), magnetostriction (Soinski, 1989), and also other parameters such as resistivity (Soinski, 1985). In some cases, a given type of anisotropy is especially made to be as high as possible.

In order to better illustrate the concept of anisotropy and its implications, let us consider one of the simplest crystallographic structures – a simple cubic *monocrystal** as shown in Figure 1.21.

In a simple cubic structure, the shortest distance is between any two molecules residing in neighbouring corners. The edge joining the two neighbours defines a crystallographic direction, and as can be seen from Figure 1.21, it can be described as *[100], [010]* or *[001]*. The planes perpendicular (normal) to these directions are (100), (010) and (001), respectively.[†]

---

* 'Monocrystal' means that the whole body is a single crystal made of the same repeating basing structure (in this case, a simple cube) – there are no defects, misoriented grains, etc. (Soinski, 2001).

† This definition of crystallographic directions and planes is usually referred to in the literature as *Miller indices* (Soinski, 2001). The type of brackets used for denoting directions and planes can vary depending on the author. Most frequently, square brackets [ ] are used for crystallographic directions, but the planes can be marked with round ( ), curly { } or triangular < >. This is usually clear from the context.

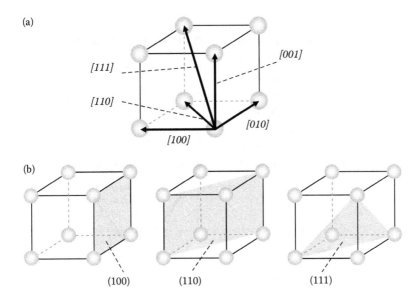

FIGURE 1.21   Crystallographic directions (a) and planes (b) for a simple cubic structure (the orientation of axes was chosen for clarity of illustration).

For directions, the digit '1' denotes the given mathematical direction in a Cartesian system of coordinates [x, y, z]. So, for instance, the longest diagonal of the cube is *[111]* because this requires 'shift' in all three coordinates so that $x = 1$, $y = 1$ and $z = 1$, hence *[xyz] = [111]*.

On the other hand, the planes are defined by the vector *normal* (i.e. perpendicular) to the surface of the plane. So, for instance, the plane (100) is the one to which the normal vector is aligned along the direction *[100]*, as shown in Figure 1.21.

For more complex crystals (for instance, with hexagonal base), more directions and planes can be defined, but the underlying concept is the same.

When the cube is dissected with a given plane, it can be seen that the distances between the molecules within that plane are different, as shown schematically in Figure 1.22a. Therefore, there are also differences of properties in various directions on such a plane.

Many forces in nature (e.g. electrostatics or gravity) are proportional to the square of the distance between the acting bodies. So, in order to achieve the same response in the less 'dense' direction, proportionally stronger excitation must be produced (depending on square of distance). If directional values of such excitation are plotted for each pair of neighbouring atoms, a simplistic hypothetical anisotropy is obtained. This is shown for the corresponding planar 'molecular density' in Figure 1.22b. There are characteristic 'butterfly' shapes for the planes (100) and (110) and it will be shown in the following chapters that such anisotropy shapes are quite prominent for materials based on cubic structures. However, most magnetic materials do not crystallise in simple cubic structures, so the shapes shown in Figure 1.22 are used only for a very simplistic illustration.

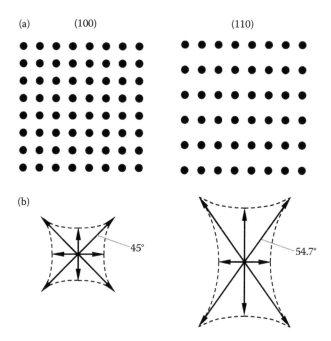

FIGURE 1.22   Examples of 'molecular density' for a simple cubic crystal dissected with a given plane (a) and corresponding hypothetical anisotropy for main directions (b).

There are many other factors contributing to the character and shape of anisotropy. Very important is the way in which a given material was produced because this influences to a large degree the orientation of individual crystallites. For instance, the structure of iron α (ferromagnetic) has one extra atom at the centre of the cube, nickel has an extra atom at the centre of each face, the crystals are hexagonal in cobalt, and so on (Soinski, 2001).

If sufficiently high temperature is applied, then the internal structure of metal can change – grains of crystals grow within the whole volume of the material. If this process is not controlled, then the grains grow either in random or in some other preferred direction, for example, induced by internal mechanical stresses.

For instance, grain-oriented electrical steel is manufactured in such a special way that the grains can be very large (20 mm or more, Figure 1.23) and they are all aligned along the same direction, coinciding with the rolling direction for the sheet production. Hence, the anisotropy of properties in grain-oriented electrical steels is significant (Soinski et al., 1995).

The elementary cubic structures (Figure 1.21) are aligned such that the plane (110) coincides with the plane of the sheet, and the direction *[100]* with the rolling direction (Figure 1.23). Usually, such configuration is referred to in a form *[100]*(110), or similar, or with the name Goss texture, after the surname of the inventor of such a production technology (Goss, 1934, 1935).

For Fe–Si monocrystal, the *J–H* curves measured along each of the directions look like those shown in Figure 1.24. We can see that the easiest direction to be

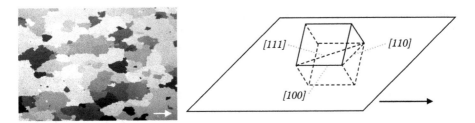

FIGURE 1.23 Surface of grain-oriented electrical steel with removed coating, showing individual grains, the arrow shows the rolling direction and schematic representation of alignment of the crystals in the grains.

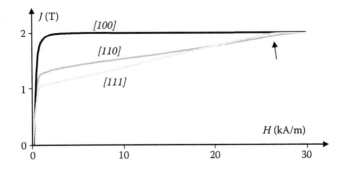

FIGURE 1.24 Magnetisation curves measured in various directions in single-crystal electrical steel. (Adapted from Fulmek P.L. et al., Energetic model of ferromagnetic hysteresis: Magnetization of grain oriented steel sheets in asymmetric directions, *Proceedings for 5th 1&2DM Workshop*, Grenoble, France, 1997, p. 17 and Bozorth R.M., *Ferromagnetism*, IEEE Press, 2003, ISBN 0-7803-1032-2.)

magnetised is *[100]* (rolling direction), then *[110]* (transverse to rolling direction) and finally *[111]*. It can be noted that this corresponds to the images shown in Figure 1.22b – these are: *[100]* = *easy* (0°), *[110]* = *hard* (90°) and *[111]* = *very hard* (≈55°) directions as shown.

The grain-oriented steel sheet is therefore cut along the 'easy' direction and placed in the limbs of medium and large transformers so that the magnetic field can travel efficiently along the easiest path. For 50 Hz (and 60 Hz) applications, the thickness of these laminations is usually between 0.18 and 0.5 mm (Cogent, 2002a).

It should be noted that there is a very interesting part in the graph shown in Figure 1.24, which is pointed by a dashed arrow, namely, towards very high fields, the 'hardest' direction *[111]* becomes easier to magnetise than the 'hard' direction *[110]*. Of course, this also occurs during rotational magnetisation at sufficiently high-magnetic field strength, and the corresponding results will be shown in the following chapters (Zurek, 2017b).

In non-oriented electrical steel, the grains are fairly small, typically around 0.05–0.15 mm. The manufacturing process is such as to obtain random orientation of grains in order to create fully isotropic material. In practice, this is not possible and some anisotropy always exists, mostly because of mechanical processing (rolling). This is important in small- and medium-size rotating machines in which entire laminations are cut out of the same sheet. Owing to the three-phase rotation, all parts of the magnetic core are exposed to the magnetic field at all angles, hence isotropy is very much desirable. For 50 and 60 Hz, the thickness of laminations is usually between 0.35 and 0.5 mm, and for higher frequencies, it must be lower, for example, 0.127 mm (Cogent, 2002b). Higher-effective frequencies occur, for instance, in electric motors used in electric cars where the speed of rotation is continuously variable by appropriate motor drives.

Even lower anisotropy is exhibited by iron-based amorphous ribbons. In ideal amorphous materials, there are no crystals and all atoms are distributed randomly within the volume of the material. These materials are made by very rapid (10^6 of K/s) cooling of molten alloy. With such fast cooling, the atoms do not have time to crystallise and they are 'frozen' at their random positions (similarly to as they were in a liquid). From a physical point of view, they have a structure similar to ordinary glass, and for this reason, they are sometimes referred to as *metallic glasses*. The most known company and product in the market is actually known as Metglas, which is a registered trademark (Metglas, 2011b).

The concept of amorphous structure and its excellent magnetic properties (especially low losses, high permeability and isotropy) was used as a basis for a new class of materials – nanocrystalline ribbons. An amorphous material with appropriate composition (containing also silicon, boron and copper) is annealed in a controlled way so that very small crystals are formed. Their size is between 10 and 20 nm (Soinski, 2001; Vacumschmelze GmbH & Co. KG, 2011; Hilzinger et al., 2013); hence the name 'nanocrystalline'. The remainder of the volume remains amorphous; hence there is also a quite good isotropy of such material. Because they are used mostly for high-frequency transformers and inductors, there is very little overlap with the topic of rotational magnetisation; hence they will not be discussed here.

High-frequency materials such as ferrites and iron powders are quite isotropic, and the small anisotropy results from the directional pressure during sintering, which allows the production of very complex shapes in one piece (Hoganas, 2009). These materials are produced as bulky items rather than laminations; therefore the theoretical analysis and measurements of rotational magnetisation need to be carried out in three dimensions, which greatly complicates the problem. Nevertheless, there are some publications on the subject showing both measured and simulated data.

In aerospace applications, the Co–Fe alloys are exposed to the rotational field. The presence of hexagonal crystals reduces anisotropy, which is advantageous for rotating machines. However, owing to high cost and narrower practical applications, there are fewer publications on rotational magnetisation in these alloys.

Data for the Ni–Fe alloys are even less available because these materials are used for transformers and sensors, not motors. This is mostly because the mechanical force is proportional to the square of B and since the saturation of Ni-based alloys is around half of Fe-based alloys, then the attainable mechanical force is significantly lower.

1.7   LOSSES UNDER ONE-DIMENSIONAL MAGNETISATION

The area of the B–H loop represents the energy dissipated during cyclic excitation of the material. In this section, the mathematical description of the physical processes will be derived in detail for a simple case of one-directional, alternating excitation. The losses dissipated during the process of magnetisation are commonly referred to as 'magnetic loss' or even 'iron loss' (as opposed to 'copper loss' in electric circuits).[*]

Let us consider the following configuration (Figure 1.25) for a thin isotropic plate, for which thickness d is negligibly smaller than its length and width l (i.e. approaching the condition of being infinitely thin, so that $l \gg d$). The top and bottom areas are $A' = A'' = l^2$.

The plate is exposed to a time-varying but spatially uniform magnetic field strength with only component H_x present (so that $H_y = H_z = 0$), which is a condition commonly occurring, for instance, in transformers. Because the material is isotropic, it responds uniformly only with component B_x (therefore, $B_y = B_z = 0$).

The combined instantaneous power flowing through the total area $A' + A''$ is therefore an integral over the whole surface A (and $d\vec{A}$ oriented along the z axis represents the vector normal to the surface A') – this approach is called as Poynting–Umov or Poynting vector (Anuszczyk et al., 2009):

$$p = \oint_A (\vec{E} \cdot \vec{H}) d\vec{A} \quad \text{(W)} \tag{1.16}$$

where $\vec{H}$ denotes the contribution of the magnetic field and $\vec{E}$ the contribution of the electric field.

It was assumed that the magnetising field has only one component $\vec{H}_x$ (the excitation is applied only in this direction) and therefore the material responds only with one component of flux density $\vec{B}_x$ (the material is isotropic so there is no contribution from crystallographic anisotropy). In order to follow with further calculations, we also need to know what components of the electric field are present (Figure 1.26).

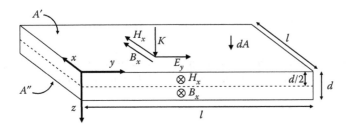

FIGURE 1.25   A schematic representation of an isotropic thin plate magnetised uniformly by magnetic field strength H_x (drawing not to scale because the assumption is that $l \gg d$).

---

[*] The terms 'iron loss' and 'copper loss' are used especially by designers of transformers and rotating machines.

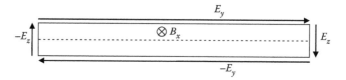

FIGURE 1.26   Representation of B and E component vectors.

Faraday's law of induction (compare also with the simplified version of Equation 1.14) links the vectors* of B and E:

$$\nabla \cdot \vec{E} = -\frac{\partial \vec{B}}{\partial t} \tag{1.17}$$

Using the full notation, Equation 1.17 is equivalent to

$$\begin{vmatrix} \vec{x} & \vec{y} & \vec{z} \\ \dfrac{\partial}{\partial x} & \dfrac{\partial}{\partial y} & \dfrac{\partial}{\partial z} \\ E_x & E_y & E_z \end{vmatrix} = -\frac{\partial B_x}{\partial t} \cdot \vec{x} - \frac{\partial B_y}{\partial t} \cdot \vec{y} - \frac{\partial B_z}{\partial t} \cdot \vec{z} \tag{1.18}$$

which can be expanded as

$$\frac{\partial E_z}{\partial y} \cdot \vec{x} + \frac{\partial E_y}{\partial x} \cdot \vec{z} + \frac{\partial E_x}{\partial z} \cdot \vec{y} - \frac{\partial E_x}{\partial y} \cdot \vec{z} - \frac{\partial E_y}{\partial z} \cdot \vec{x} - \frac{\partial E_z}{\partial x} \cdot \vec{y} = -\frac{\partial B_x}{\partial t} \cdot \vec{x} - \frac{\partial B_y}{\partial t} \cdot \vec{y} - \frac{\partial B_z}{\partial t} \cdot \vec{z} \tag{1.19}$$

But we know that only B_x is present, so $B_y = B_z = 0$; hence the equation simplifies to

$$\frac{\partial E_z}{\partial y} \cdot \vec{x} + \frac{\partial E_y}{\partial x} \cdot \vec{z} + \frac{\partial E_x}{\partial z} \cdot \vec{y} - \frac{\partial E_x}{\partial y} \cdot \vec{z} - \frac{\partial E_y}{\partial z} \cdot \vec{x} - \frac{\partial E_z}{\partial x} \cdot \vec{y} = -\frac{dB_x}{dt} \cdot \vec{x} \tag{1.20}$$

We can see that only the component $\vec{x}$ is present on the right-hand side of the equation, so the $\vec{y}$ and $\vec{z}$ counterparts on the left-hand side must also be equal to zero; hence

$$\frac{\partial E_z}{\partial y} \cdot \vec{x} - \frac{\partial E_y}{\partial z} \cdot \vec{x} = -\frac{\partial B_x}{\partial t} \cdot \vec{x} \tag{1.21}$$

---

* Appropriate vector and matrix notations must be used here because the analysed problem comprises components in all three dimensions.

According to initial assumptions, the magnetisation is uniform, but the thickness is much smaller than the length of the plate $d \ll l$, so the E_z component can be neglected (d is a dimension along the z axis) (Anuszczyk et al., 2009):

$$\frac{\partial E_z}{\partial y} \cdot \vec{x} \ll \frac{\partial E_y}{\partial z} \cdot \vec{x} \tag{1.22}$$

and therefore

$$\frac{\partial E_y}{\partial z} \cdot \vec{x} = -\frac{\partial B_x}{\partial t} \cdot \vec{x} \tag{1.23}$$

At this point, only one component is present; we can change from partial to normal derivatives:

$$\frac{dE_y}{dz} = -\frac{dB_x}{dt} \tag{1.24}$$

In order to calculate the value of E_y, we can integrate over half of the thickness (the system is symmetrical):

$$E_y = \int_0^{d/2} \left(-\frac{dB_x}{dt} \right) dz = -\frac{d}{2} \cdot \frac{dB_x}{dt} \quad \text{(V)} \tag{1.25}$$

It was assumed that the magnetisation is uniform with H_x in the x direction such that $H_y = H_z = 0$ and it was approximated that $E_x \approx E_z \approx 0$. Equation 1.16 allows the calculation of the instantaneous power flowing through the total area ($A = A' + A''$). Neglecting the side areas (because also $d \cdot l \ll l \cdot l$), the instantaneous power $p(t)$ is equivalent to (Pfutzner, 1994)

$$p(t) = 2 \cdot A \cdot (\vec{E} \cdot \vec{H}) = \begin{vmatrix} 0 & H_x & 0 \\ E_y & 0 & 0 \\ 0 & 0 & 2 \cdot A \end{vmatrix} = -2 \cdot A \cdot E_y \cdot H_x \quad \text{(W)} \tag{1.26}$$

Combining Equations 1.25 and 1.26, we obtain

$$p(t) = -A \cdot d \cdot \frac{dB_x}{dt} \cdot H_x \quad \text{(W)} \tag{1.27}$$

Average power over a given time T (e.g. one cycle of sinusoidal magnetisation) can be calculated as

$$P = \frac{1}{T} \int_0^T p(t) dt \quad \text{(W)} \tag{1.28}$$

Equation 1.27 can be substituted into Equation 1.28:

$$P = \frac{1}{T} \int_0^T \left(-A \cdot d \cdot \frac{dB_x}{dt} \cdot H_x \right) dt = -\frac{A \cdot d}{T} \int_0^T \left(\frac{dB_x}{dt} \cdot H_x \right) dt \qquad \text{(W)} \qquad (1.29)$$

It is possible to express average power per sample volume V_s because from Figure 1.25, it can be seen that $V_s = A \cdot d$ (m³):

$$P = -\frac{A \cdot d}{T \cdot V_s} \int_0^T \left(\frac{dB_x}{dt} \cdot H_x \right) dt = -\frac{1}{T} \int_0^T \left(\frac{dB_x}{dt} \cdot H_x \right) dt \qquad \text{(W/m}^3\text{)} \qquad (1.30)$$

and average power per unit mass where D (kg/m³) is the specific density of the material:

$$P = -\frac{1}{T \cdot D} \int_0^T \left(\frac{dB_x}{dt} \cdot H_x \right) dt \qquad \text{(W/kg)} \qquad (1.31)$$

As can be seen, Equations 1.30 and 1.31 are based on H and B values, which shows the practical significance of measuring the B–H characteristics. The results are directly proportional to the area of the B–H loop and because the H and B values can be measured conveniently by appropriate sensors, the calculation of power loss can be carried out easily from the measured values.

PRACTICAL COMMENT

There is a minus sign before the integral in Equations 1.30 and 1.31. This is simply a matter of initial assumption of the orientation of the Poynting–Umov vector.

The minus denotes that the energy is dissipated from the magnetised plate to the surrounding medium (Anuszczyk et al., 2009), which is the case in real life because the lost energy is radiated *out* as heat. In the literature, it can be usually found that the integral of the Poynting–Umov vector (in Equation 1.16) is negative, hence the final calculation returns a positive value of power loss (Pfutzner, 1994), that is

$$P = \frac{f}{D} \int_0^T \left(\frac{dB_x}{dt} \cdot H_x \right) dt \qquad \text{(W/kg)} \qquad (1.32)$$

where f is the magnetising frequency (Hz), equal to $1/T$.

Derivation of such equations must also take into account whether the system of coordinates is mathematically 'right-handed' or 'left-handed'. The exact details are beyond the scope of this book, but the reader should be aware of the implications.

The calculations of the Poynting–Umov vector shown above were carried out for a uniform excitation so that the surface values of H and B vectors were synonymous with their spatially averaged values. Similar theoretical calculations can be carried out also taking into account a non-uniform distribution of magnetic field through the thickness of the plate. In such case, when the surface values are used (which are generally different from average values), it can be shown that (Pfutzner, 1994; Anuszczyk et al., 2009)

$$\oint_A (\vec{E} \cdot \vec{H})\, d\vec{A} = \int_{V_s} (\gamma \cdot \vec{E}^2)\, dV + \int_{V_s} \left(\frac{d\vec{B}}{dt} \cdot \vec{H} \right) dV_s \qquad \text{(W)} \qquad (1.33)$$

where γ is the conductivity of the material in (S/m).

It can be seen that there are two components on the right-hand side of Equation 1.33. Conceptually, this is explained as that the first term $\gamma \cdot \vec{E}^2$ represents losses on eddy currents, whereas the second term $(d\vec{B}/dt) \cdot \vec{H}$ means the quasi-static hysteresis loss. However, the real distribution of fields through the plate thickness is generally not known (and as yet cannot be measured with sufficient accuracy), so the values cannot be calculated. Moreover, the two components are affected by each other – for instance, eddy currents can change the distribution of the magnetic field inside the material, so that all the values become a function of the distance from the surface of the plate, as well as time. In reality, the situation is even more complicated because the magnetisation process in ferromagnetic materials is associated with localised domain walls, which generate additional microscopic eddy currents, locally around the moving domain walls (Zirka et al., 2015).

As a result, the total loss cannot be separated easily into eddy current and hysteresis loss by *measuring* the appropriate components. In fact, a third component 'excess loss' (initially introduced as 'anomalous') was introduced in the analysis of magnetic losses, as will be described later. Nevertheless, Equation 1.33 shows that some validity of the analysis separating various components of magnetic losses.

In any case, the Poynting–Umov vector does include all the magnetic losses (hysteresis, eddy current and excess), when the average values of flux density and magnetic field strength are used, as, for instance, in Equation 1.32.

1.8   LOSS COMPONENTS: HYSTERESIS, EDDY CURRENTS AND EXCESS

Historically, total magnetic loss was conceptually separated into three components: hysteresis, eddy currents and excess. There are phenomenological models that approximate the measurable loss behaviour with a sufficiently good accuracy. Such approach is referred to as *statistical model* (Bertotti, 1988). Latest research shows that it might not be completely correct (Zirka et al., 2015), but it will be described here for completeness and to provide explanation of the terms used in many research papers.

The hysteresis loss is equivalent to the area of the *B–H* loop under quasi-static or very slow (e.g. 0.001 Hz) alternating magnetisation. Under such conditions, the induced global eddy currents (Figure 1.15) are negligibly small and it is assumed that the excess loss is also negligible.

The frequency of magnetisation at which the process can be described as *quasi-static* can be found in an empirical way, namely, a sample is excited at some frequency, for instance, 50 Hz. Then, the measurement frequency is lowered (e.g. to 25 Hz). If the area of the *B–H* loop is reduced, then it means that eddy currents were present and even lower frequency is required. The process is repeated at ever-decreasing frequency until the *B–H* loop area stops decreasing. For each material, there exists such a low frequency, below which the area of the loop does not decrease (Gozdur, 2004), as shown in Figure 1.27.

However, it is evident from Equation 1.32 that the average power loss is directly proportional to the frequency of excitation. This is a consequence of the fact that for higher frequency, the loop is 'drawn' more times per second; hence more 'area' will be enclosed within the same unit of time. For instance, at a frequency of 0.017 Hz (so that eddy currents can still be neglected – see Figure 1.27), the loop is traversed just once per minute. But when the frequency is doubled to 0.033 Hz (still without significant global eddy currents), the loop will be repeated twice per minute dissipating twice the amount of *energy* (the same per each cycle); hence the *power* lost per minute will also double. For this reason, this quasi-static hysteresis loss is assumed to be directly proportional to the frequency, simply due to the f term before the integral in Equation 1.32.

This direct proportionality can be summarised in a constant C_h:

$$P_h = C_h \cdot f \quad \text{(W/kg)} \tag{1.34}$$

where P_h is the quasi-static, hysteresis loss (W/kg) and C_h is the material constant[*] (J/kg).

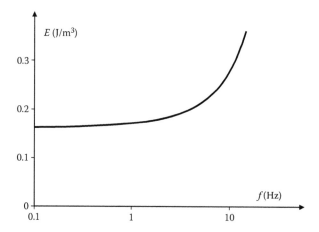

FIGURE 1.27   Area of the *B–H* loop (proportional to the dissipated energy) becomes constant below a certain frequency of excitation. (Adapted from Gozdur R., *Przeglad Elektrotechniczny (Electrical Review)*, R. 80 (Nr 2/2004), 2004, 147.)

---

[*] It was actually shown by Steinmetz in 1892 (Ewing, 1900) that this coefficient can be approximated as $C_h = C_x \cdot B_p^{1.6}$, where C_x is the material constant and B_p is the peak flux density. However, for some modern materials, B_p might be required to be raised to the power of 2.0 or some other value to obtain a better fit.

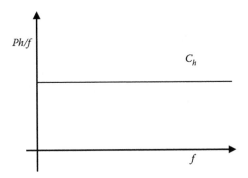

FIGURE 1.28   A typical plot of P_h/f versus f for a magnetically soft material.

The constant C_h can be found easily by calculating the area of the $B–H$ loop at a frequency that is low enough as to not to induce global eddy currents (Figure 1.27). Of course, all the points must be measured at the same amplitude of B as to ensure comparable conditions for energy dissipation.

In order to simplify calculations, the losses are often 'pre-processed' by dividing them by the frequency value, as in Equation 1.35. This is graphically presented in Figure 1.28.

The hysteresis loss is associated with domain walls. During magnetisation, they are created and annihilated as well as their movements are impeded by pinning sites such as impurities and crystal dislocations.

$$P/f = P_h/f = C_h \qquad \text{(J/kg)} \qquad (1.35)$$

PRACTICAL COMMENT

The energy dissipated under very slow, quasi-static magnetisation is the *hysteresis component* of loss (or *hysteretic component*). However, it is common that the whole $B–H$ loop, measured under 'fast' dynamic magnetisation, is also referred to as 'hysteresis loop'.

Therefore, once again, there can be confusion between the two meanings. Nevertheless, in most of the literature on the subject, it should be clear from the context what the author meant – the whole loop, or just its quasi-static hysteretic component.

As mentioned above, the eddy currents are always generated inside conducting material exposed to a varying magnetic field (Figure 1.15). For an idealised simplified case, such loss can be calculated from the first principles; hence this is often referred to as the *classical* eddy current loss.

If it is assumed that an isotropic material is in a form of a thin sheet (i.e. the thickness much smaller than the width and the length), the flux density is sinusoidal, and the magnetisation is uniform (no domain walls and no skin effect), then the losses resulting from eddy currents can be calculated as (Bozorth, 2003)

$$P_e = \frac{\pi^2 \cdot d^2 \cdot B_p^2 \cdot f^2}{6 \cdot \rho \cdot D} = C_e \cdot f^2 \quad \text{(W/kg)} \tag{1.36}$$

where P_e is the eddy current loss (W/kg), d is the sheet thickness (m), B_p is the peak of sinusoidal flux density (T), ρ is the material resistivity ($\Omega \cdot$ m), D is the material specific density (kg/m³) and C_e is a constant (J · s/kg).

Using the P/f notation, we can write that

$$P_e/f = C_e \cdot f \quad \text{(J/kg)} \tag{1.37}$$

Therefore, if we consider a hypothetical material in which magnetic losses comprise only hysteresis and eddy current components (see also Equation 1.33), we can write that

$$P = P_h + P_e \quad \text{(W/kg)} \tag{1.38}$$

or using the P/f notation

$$P/f = C_h + C_e \cdot f \quad \text{(J/kg)} \tag{1.39}$$

which can be graphically presented as in Figure 1.29.

However, when this approach was employed in the first half of the twentieth century, it was found that the total measured loss significantly exceeds the value estimated by Equation 1.38 (Bozorth, 2003). The differences between the calculated and measured values were so large that they could not be attributed to any measurement or calculation errors. This discrepancy became even more pronounced with the introduction of a new type of magnetically soft material – the grain-oriented electrical steel, and later on iron-based amorphous ribbons, where the 'error' can exceed 90% of the total measured value. At the time when this effect was first observed, there was no available theoretical explanation. The difference was named as *anomalous*

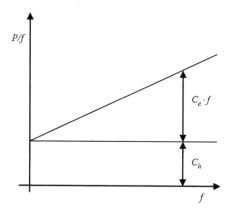

FIGURE 1.29   A graphical representation of the hysteresis and eddy current losses in a *P/f* notation.

loss because the measured values did not follow the then used theoretical model, and there was yet no theoretical basis to explain them.

It was then shown that the total loss can be split into three components (Figure 1.30) and a comprehensive description of such separation method (so-called statistical model) was given by Bertotti (Bertotti, 1988).

Quite good approximation can be obtained basing the calculations on the measured values with controlled sinusoidal B with fixed amplitude (Figure 1.31), with the following equations:

$$P = C_h \cdot f + C_e \cdot f^2 + C_a \cdot f^{3/2} \quad \text{(W/kg)} \quad (1.40)$$

where C_a is the material constant $(J \cdot \sqrt{s}/kg)$.

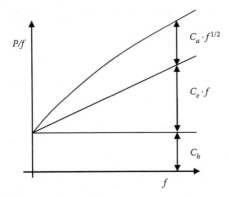

FIGURE 1.30   The concept of 'anomalous loss' or 'excess loss'.

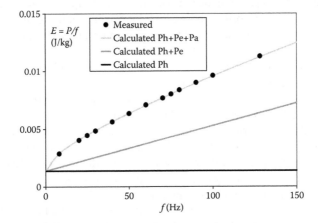

FIGURE 1.31   An example of power loss separation at 1.0 T for a typical grade of grain-oriented electrical steel; the corresponding values of the constants are $C_h = 1.37 \cdot 10^{-3}$ J/kg (extrapolated for $f = 0$ Hz), $C_e = 3.92 \cdot 10^{-5}$ J · s/kg (calculated from Equation 1.36), $C_a = 4.29 \cdot 10^{-5}$ J · $\sqrt{s}$/kg (estimated by non-linear regression).

And with the P/f notation:

$$P/f = C_h + C_e \cdot f + C_a \cdot f^{1/2} \qquad \text{(J/kg)} \qquad (1.41)$$

As mentioned above, the discrepancy between the measured and calculated losses could not be explained on the basis of uniform magnetisation. So even though the phenomenon was known, the two-component approach (hysteresis + eddy currents) was still widely used, even in large power transformers (Jezierski, 1975). Owing to its simplicity, it is still sometimes used in much more modern applications if the errors resulting from such approximations can be neglected (Suss et al., 2006).

At the time, the remaining difference was not understood and therefore referred to as *anomalous* loss. However, since then, the phenomenon was studied extensively and it is now generally accepted that non-uniform flux distribution and large-scale domain wall movements contribute to this loss (Fiorillo, 2004). Therefore, this component is not so 'anomalous' anymore and the term *excess loss* (or *additional loss*) is now used.

However, there are many assumptions made for such three-component loss separation to be carried out. For instance, the eddy current loss equation is valid only for ideal sinusoidal B, and for a completely uniform material (without domains).

Such assumption is not completely valid for materials with large grains, and with long domain walls or domain walls penetrating the full depth of the lamination thickness. Domain walls can bow during movement so that there can be an additional time dependence of the magnetisation at a given depth through the thickness of a lamination. Thus, the assumption of 'uniform magnetisation' is no longer valid because the resulting variation is non-sinusoidal at a given point in volume, or the magnitude differs through the depth of the lamination.

As a result, the three-component model is capable of giving some information about the total loss dissipated in the material at a given frequency and amplitude, but not about the shape of the corresponding B–H loop, which indicates its limited validity. Some theoretical models are capable of much better prediction also of the shape of the B–H loop, but in most cases, there is a large degree of semi-empirical derivation, with as yet unknown analytical relationship of the equation parameters to the actual material parameters (Zirka et al., 2015).

Also, other mechanisms should be taken into account. For example, from a microscopic viewpoint, the hysteresis loss can be treated as loss due to micro eddy currents generated locally by Barkhausen jumps of domain walls (Chikazumi, 2002). Therefore, from a physical viewpoint, all components of the loss can be generating heat by eddy currents, over smaller or larger distances in the sample. In other words, even hysteretic loss is eventually caused by eddy currents, but on a very small scale.

For materials with large domains, the skin effect might not take place in the same way because the domain walls penetrate the full depth of the thickness to begin with. Equations such as Equation 1.33 might be theoretically correct, but providing accurate information about the exact temporal and spatial distribution of all the required components will be extremely difficult. There is a continuing research in the understanding of such phenomena but the understanding is still limited (Zirka et al., 2015).

1.9 PRACTICAL IMPORTANCE OF POWER LOSS

As mentioned above, power loss dissipated in the material increases its temperature, which has crucial practical implications.

The foremost concern is that the device can overheat to the point of catastrophic failure. As a rule of a thumb, an increase of temperature by around 10°C decreases the operational life of the electrical insulation by half. Therefore, an electromagnetic machine must be designed accordingly so that the operating conditions for the insulation material are taken into account. In high-power machines, well-designed cooling is required, for example, by submerging the transformer in liquid, which can be additionally pumped around the circuit so that the heat dissipation is further improved, or generators in power plants are cooled by pumping dry hydrogen through the structure.

Dissipated power loss immediately impacts on the energy efficiency of the machine in question, be it motor, generator or transformer. Higher efficiency usually calls for somewhat *larger* machines in order to reduce the level of excitation, or power density in the magnetic cores and windings. And conversely, smaller sizes can be achieved with better cooling, but only at the expense of efficiency (not taking into account frequency of operation).

In any case, the design of magnetic cores requires knowledge of their properties so that the efficiency and thermal operating conditions can be calculated precisely before the machine is built. This is fundamental from a commercial viewpoint because large transformer cores can weigh in excess of 100 tons, so even a small error in calculation has large financial implications.

Therefore, one of the most important parameters of magnetically soft materials is the amount of power loss dissipation as a function of the excitation level. The designer can accommodate a more lossy material by reducing the magnetisation level, or conversely the operating point can be optimised towards a higher level if the material exhibits lower loss. Better-quality materials are usually more expensive, so the cost of the given structure goes hand in hand with the power loss characteristics.

There are three main types of international standards for measuring power loss under uni-directional magnetisation: toroidal sample (IEC, 2003; ASTM, 2011), Epstein frame (IEC, 2008; ASTM, 2014) and single-sheet tester (SST) (ASTM, 2009; IEC, 2010).

These three methods have been known for many decades, and they are used widely in the industry for the characterisation of magnetically soft materials. They are defined very well in the relevant international standards such as IEC or ASTM.

In terms of weight and volume, electrical steel sheet represents the vast majority of magnetic materials used in technology. These steels are classified in accordance to measurements performed with the Epstein frame or SST method (IEC, 2000).

Researchers, scientists and metrologists worldwide continue the work on improving such methods. For example, the Epstein frame method was re-defined many times so that smaller samples and better reproducibility were achieved.

On the other hand, the manufacturers continue to improve the quality of electrical steels (lower losses) (Moses, 2012) and the measurement methods are at the heart of the research into improvement of materials and manufacturing.

However, modern switch-mode power supplies can expose the magnetic materials to complex magnetising conditions that are not covered by the abovementioned standardised methods. The metrological community would welcome such methods but so far it was not possible to define them.

The same applies to *rotational power loss*. This book presents several measurement methods but none of them is standardised, nationally or internationally. At the time of writing this book, the reproducibility as well as ease of application was insufficient to begin the process of standardisation. And without the well-defined measurements, it is not possible to take into account the data because the data cannot be produced with sufficient accuracy.

1.10 ROTATIONAL MAGNETISATION

So far, we discussed only the process of alternating magnetisation – the excitation is applied along one constant direction and only the magnitude and sense* of the vector vary (see also Figure 1.25).

Such alternating excitation is widely used in technology and is the very basis of the operation of most transformers, magnetic actuators, sensors, etc.

However, in magnetic cores of rotating machines, the material can be subjected to excitation, whose direction varies, be it rotating in the plane of a lamination or in some other arbitrary two-dimensional (2D) or even three-dimensional (3D) manner. Since the rotating machines and their laminated cores are of great practical importance, let us focus on the case of 2D, rotational magnetisation.

Under alternating excitation, it is assumed that the material is always magnetised in the same direction, and only the amplitude of the alternating flux density B_A changes (including the changes from negative to positive values), as shown in Figure 1.32a.

Under 2D excitation, the B vector can be rotated within a given plane, for instance, within the plane of lamination. If the material is isotropic, it is relatively easy to set

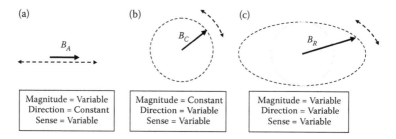

FIGURE 1.32   Three types of magnetisation: (a) alternating B_A with variable magnitude and constant direction, (b) circular B_C where only direction changes and the magnitude of circular flux density vector is constant and (c) rotational B_R where the magnitude and direction of rotational flux density vector changes.

---

* The 'direction' of the vector is defined by a straight line aligned with the axis of the vector. Therefore, the direction does not define which way the vector points. This is why all vectors are represented by arrows, whose tip defines the 'sense' of the vector.

FIGURE 1.33 Rotational magnetisation in three-phase machines: in turbogenerator stator core (a). (Adapted from Radley B. et al., *IEEE Transactions on Magnetics*, 17 (3), 1981, 1311), in electric motor core (b). (Adapted from Anuszczyk J., *Analysis of localisation of flux density and power loss in rotational magnetisation in magnetic circuits of electric machines* (in Polish: *Analiza rozkladu indukcji i strat mocy przy przemagnesowaniu obrotowym w obwodach magnetycznych maszyn elektrycznych*), Zeszyty Naukowe Politechniki Lodzkiej, Nr 629, Rozprawy Naukowe z. 158, Lodz, 1991), and at the T-join of three-limb power transformer (c). (Adapted from Kanada T. et al., *IEEE Transactions on Magnetics*, 32 (5), 1996, 4797.)

up a circular flux density B_C – the magnitude of the vector can remain constant, while its direction can be continuously varying due to the rotation of the vector – Figure 1.32b.

Anisotropy of the material is synonymous with different response of the material to excitation in different directions. The resultant *loci* of the B vector (the path drawn by the tip of the vector) can be elliptical or distorted in some other way. As can be deduced from Figure 1.32c, such rotational flux density B_R might comprise a combination of both: the circular B_C and the alternating B_A.

Of course, the rotational magnetisation does not have to result with elliptical flux density – all the values can vary arbitrarily: magnitude, direction and sense.

All these three types of magnetisation can be identified in magnetic cores of three-phase electric machines. The alternating flux density B_A occurs in the main limbs of transformers and in the teeth of generators and motors, whereas the (mostly) elliptical or even circular magnetisation B_C can be found at the back of the teeth (Figure 1.33a,b). A distorted rotational magnetisation (lozenge shapes) was found to occur in the T-joints of three-limb transformers (Figure 1.33c).

It can be seen from Figure 1.33 that the rotational losses can occupy relatively large portions of the cores. If elliptical magnetisation is included, then rotational loss (including all elliptical shapes) could be contributing up to 50% of total magnetic loss (Moses, 1992). In three-phase power transformers, the contribution of rotational loss varies depending on the configuration and construction of T-joints or corners (Jezierski, 1975).

For this reason, the topic of rotational magnetisation and rotational losses remains in the interest of scientists, researchers and machine designers.

1.11 LOSSES UNDER ROTATIONAL MAGNETISATION

During one cycle of alternating excitation, the magnetic material is subjected to a positive magnetic field strength, then to negative and back to positive and so on. Regardless of the amplitude of this excitation, the B–H loop is always created, and for increased excitation, the response will always increase. Even at full saturation, where J reaches maximum, B can always increase, and under dynamic magnetisation, it will contribute to higher eddy current loss. As a result, the total losses under alternating excitation always increase with excitation as shown in Figure 1.34.

The situation is different when the circular flux density occurs. Under moderate excitation, the material responds with some changes in the domain structure (see, for instance, point 3 in Figure 1.12). The domains whose directions coincide with the applied excitation grow. However, when the excitation changes its direction, the domains must respond accordingly and different domains must increase their size for the net flux density vector to be set up at a different angle. So, when the direction of excitation rotates, the domain structure must also change continuously, and in a much more complex way than under alternating conditions. Therefore, the rotational losses will be higher.

However, as the comparison from Figure 1.34 shows, above a certain level of excitation, the rotational loss decreases rather sharply. This occurs because at a sufficiently high excitation, the domains begin to merge and hence some domain walls are eliminated (see point 4 in Figure 1.12). Because hysteresis and excess loss components are linked to domain wall movements, these two components are appropriately diminished (there are no domain walls left to cause losses) and the total loss is reduced significantly.

Under dynamic excitation, the eddy currents are still present, but under quasi-static conditions, the total loss can reduce to very small values at saturation. In theory, at full saturation, the quasi-static losses should decrease to zero.

The derivation of the formula for the rotational losses in the so-called fieldmetric technique is carried out similarly to that shown in Section 1.7 for alternating magnetisation (Pluta, 2001). But of course this time both in-plane components of flux density and magnetic field strength must be used. For brevity, the calculations will not be repeated here, but we will just give the final Equation 1.42 (Tumanski, 2011) (compare with Equation 1.31 for alternating conditions):

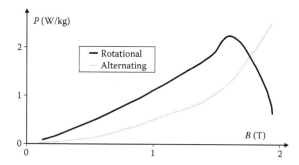

FIGURE 1.34 Comparison of power loss characteristics for conventional grain-oriented silicon-iron electrical steel sheet for alternating and circular magnetisation. (Adapted from Bozorth R.M., *Ferromagnetism*, IEEE Press, 2003, ISBN 0-7803-1032-2.)

$$P = \frac{1}{T \cdot D} \int_0^T \left(\frac{dB_x}{dt} \cdot H_x + \frac{dB_y}{dt} \cdot H_y \right) dt \qquad \text{(W/kg)} \qquad (1.42)$$

Once again, we can see that the equation refers to the directional values of flux density B_x, B_y and magnetic field strength H_x, H_y. If these values are measured with suitable sensors, then the power loss can be calculated.

PRACTICAL COMMENT

In the previous full derivation, it was assumed that $H_y = 0$ (see Figure 1.25). If that assumption was not made, then the y component would not be eliminated from the calculations and hence the result would lead to that of Equation 1.42. For 3D problem, all three components are present.

There is also another expression for rotational losses (Pfutzner, 1994):

$$P = \frac{2 \cdot \pi}{T^2 \cdot D} \int_0^T (H_y \cdot \overline{B_x} - H_x \cdot \overline{B_y}) dt \qquad \text{(W/kg)} \qquad (1.43)$$

where $\overline{B_x} = \overline{B} \cdot \cos(2 \cdot \pi \cdot f \cdot t)$, $\overline{B_y} = \overline{B} \cdot \sin(2 \cdot \pi \cdot f \cdot t)$ and $\overline{B}$ is the amplitude of purely circular flux density (T).

However, it was mathematically and experimentally shown that Equation 1.43 is valid only for purely circular rotational magnetisation (Pfutzner, 1994), whereas Equation 1.42 is valid for any monotonous rotational magnetisation, for example, elliptical (Sievert, 1992a).

Of course, if different physical processes are used as the basis for loss measurement, then different equations for rotational losses can be derived, for example, based on torque or temperature. These will be described in detail in the following chapters.

1.12   BRIEF HISTORY OF ROTATIONAL STUDIES

When Maxwell published his equations describing the laws of electromagnetism, there was still no explanation how ferromagnetism works. Various models were proposed, but it was suspected that the phenomenon was occurring at a molecular level (Ewing, 1900).

Two main concepts were proposed. In Poisson's view, the excitation was affecting all 'molecules' in such a way that they were becoming magnetic, which could easily explain the paramagnetic effect and partially ferromagnetic effect. However, Poisson's model could not easily explain the saturation or even the effect of hysteresis.

Another approach was taken by Weber. In his model, it was assumed that all molecules are already magnetic, but they all point in random directions, so no net flux density is detected without excitation. When the material is magnetised, the molecules orient themselves along the field, and when they are all oriented, the saturation occurs. Some experimental work by Beetz confirmed that electrodeposition of mag-

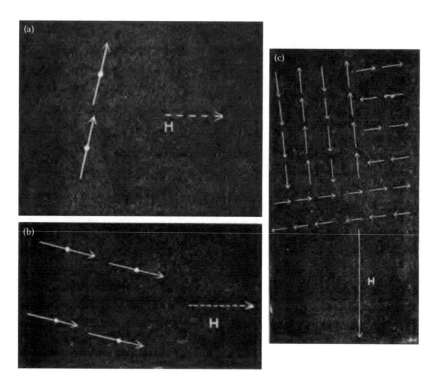

FIGURE 1.35   Images from Ewing's book depicting analysis of a pair (a), a group of four (b) and a group of multiple 'molecular magnets' (c). (Courtesy of Ewing J.A., *Magnetic Induction in Iron and Other Metals*, 3rd edition, Van Nostrand Company, 1900.)

netically oriented molecules results in material which remains close to saturation and it is difficult to demagnetise it (Ewing, 1900).

This model was further developed by Ewing, who carried out mathematical analysis on 'molecules' positioned closely enough so that they experienced mutual magnetic coupling (Figure 1.35).

Each molecule was assumed to be a small permanent magnet bar (Figure 1.36). By calculating forces between the molecules and the external magnetic field, it was possible to show that at low fields, the process is reversible, at moderate fields, the permeability increases rapidly (irreversible magnetisation), and at high fields, saturation occurs, but it is very difficult to fully saturate the material magnetically. Ewing carried out his analysis only for alternating magnetisation.

Baily was the first to actually measure the rotational loss in a sample and to show the then very strange effect of losses reducing with excitation (Figure 1.37) (Baily, 1896). The accuracy of his measurement method was initially criticised because his peers did not believe that such reduction was real. However, using Ewing's model, it was theoretically shown by Swinburne that such effect would indeed appear (Ewing, 1900).

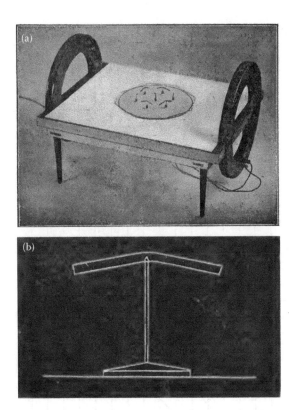

FIGURE 1.36 Experimental setup used by Ewing: (a) A group of seven 'molecular' freely rotating magnets and exciting coils at the ends of the table. (b) A side view on a single 'molecular' magnet. (Courtesy of Ewing J.A., *Magnetic Induction in Iron and Other Metals*, 3rd edition, Van Nostrand Company, 1900.)

All these results were obtained over a century ago, before the year 1900. Initially, the measurements were carried out by a physical rotation of the sample or the magnetising yoke. The method relies on measuring the mechanical torque exerted on the specimen.

The research on rotational losses was continued. Brailsford (1938) used a similar mechanical measurement method, which he later modified to allow dynamic conditions (1947), so that the eddy current loss could be included (Brailsford 1948).

Some developments in theoretical modelling were published in 1955 by Kornetzki and Lucas (Kornetzki et al., 1955). With the assumption that the material has 180° as well as 90° magnetic domains, they calculated that the rotational loss should be in moderate fields up to two times higher than alternating loss, which was confirmed to some degree with measured data. However, owing to *Barkhausen jumps* (sudden movements of domain walls) and other effects, the model was generally expected not to be accurate enough.

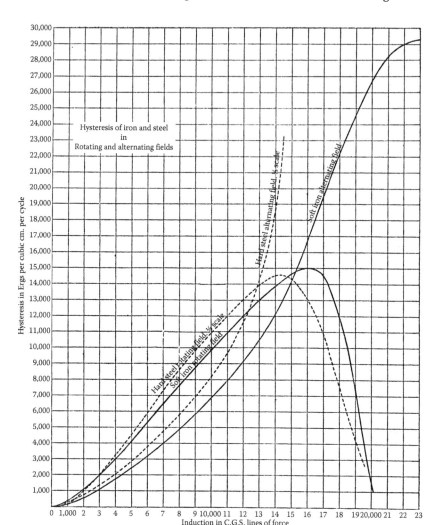

FIGURE 1.37   Alternating and rotational losses measured by Baily. (Reproduced from Baily F.G., *Philosophical Transactions of Royal Society of London, Series A*, 187, 1896, 715. With permission of the Royal Society.)

Boon and Thompson (Boon et al., 1965) published results on one of the first measurement apparatus employing the so-called *fieldmetric technique*. In 1973, Moses and Thomas showed a different approach to the same technique (Moses et al., 1973), and Radley and Moses carried out more studies on large cores of rotating machines (Radley et al., 1981).

Brix et al. published results measured with the fieldmetric method. The study was one of the first to employ digital devices (computer) in the rotational measurements (Brix et al., 1982).

Sasaki et al. developed a system for measuring the rotational losses on a single strip of Epstein frame sample (Sasaki et al., 1985).

Enokizono and his colleagues published results measured on a computerised system with a planar yoke (Enokizono et al., 1990a). In the same year, Sievert published a significant review paper on rotational research – he referred to 55 publications (Sievert, 1990). He highlighted the influence of the sensor misalignment in the field-metric method on the accuracy of measurement.

The international researchers realised the scientific weight of the topic and decided to establish a regular international meeting for all involved in the rotational magnetisation, strongly applicable to electrical steels. The first *International Workshop on Magnetic Properties of Electrical Sheet Steel under Two-Dimensional Excitation* was held in 1991 in Braunschweig, Germany. The conference was set up by several of the most active laboratories on the subject at the time (in alphabetical order):

- Gifu University, Gifu, Japan
- Institut fur Werkstoffe der Elektrotechnik, Aachen University, Aachen, Germany
- Istituto Elettrotecnico Nazionale Galileo Ferraris, Turin, Italy
- Laboratoire d'Electrotechnique de Grenoble, Grenoble, France
- Oita University, Oita, Japan
- Okayama University, Okayama, Japan
- Physicalisch-Technische Bundesanstalt, Braunschweig, Germany
- Wolfson Centre for Magnetics Technology, Cardiff University, Cardiff, the United Kingdom

It was decided that this international conference will be held biannually. So far, it was held in several countries: Austria, Belgium, China, France, Germany, Italy, Japan, Poland and the United Kingdom. At each of the workshops, tens of papers are published on the topic. The scope of the conference was widened to also include alternating and other non-rotational measurements, and now it is known as *International Workshop on One- and Two-Dimensional Magnetic Measurement and Testing*, referred shortly by those involved as *1&2DM* or even simply as *2DM*.*

Full peer-reviewed articles from several *1&2DM* Workshops were published in *Electrical Review*. This technical and scientific journal was established in Poland in 1919 under the name *Przeglad Elektrotechniczny* (literally: '*Electrotechnical Review*'), and nowadays also known as *Electrical Review* (SIGMA-NOT, 2011). Publications also appeared in *IEEE Transactions on Magnetics, International Journal of Applied Electromagnetics and Mechanics*, other journals, as well as just in locally published conference proceedings (especially for the first few conferences).

At the first Workshop in 1991, yet another measurement technique was applied by Fiorillo and Rietto (Fiorillo et al., 1992a). The losses were measured by detecting the changes of temperature of the sample in adiabatic conditions. Magnetic domain observation was also reported in this first workshop (Radley, 1992).

---

* The 1&2DM workshop has a changing website with a permanent address: http://www.2-dm.com. The author has been involved in maintaining this website address and also the content of the website for two workshops: Cardiff 2008 and Oita 2010.

The rotation of magnetisation can be applied either in clockwise or in anticlockwise rotation. Majority of the researchers noticed that the values measured with the fieldmetric method in both directions differ from each other. It became generally accepted that the result of power loss should be averaged from both directions of rotation (Sievert et al., 1995, 1996).

By the 1990s, the computer technology improved significantly so that most of the laboratories were able to employ computerised measurement systems. In 1995, an intercomparison was attempted between measurements of six European laboratories, in order to investigate the possibility of standardisation of the rotational power loss measurement (Sievert, 1995). The laboratories used various shapes and sizes of magnetising yokes and the results differed rather substantially. Because of the significant uncertainties involved, none of the methods was found to be intrinsically superior. In some cases, negative losses were reported (for either clockwise or anticlockwise, but the average value agreed reasonably well with other measurements). No further attempt on standardisation of rotational measurements was made till the time of writing of this book.

In the years 1991–1997, the first finite element computer modelling of the yokes and samples was shown (Von Musil, 1992; Nencib et al., 1995; Xu et al., 1997). One of the conclusions was that a square sample is better for field homogeneity than a round sample. This had a strong influence on further studies, which for many years were carried out focused mostly on square or hexagonal samples.

In the years 2000–2003 (and also later), there was a series of articles by Gorican and his colleagues from Maribor University in Slovenia, with data measured on a round sample, magnetised in a yoke resembling stator core of an induction motor (Gorican et al., 2000a, 2002, 2003; Jesenik et al., 2003a). The results were measured under controlled conditions with circular flux density up to impressive 2.0 T and clearly showed a drop in the power loss above 1.7 T. It was postulated in these articles that the detection of B and H components might not be reliable enough and that there could be a substantial asymmetry of power loss anisotropy when measured under alternating excitation at various inclination angles. Later articles by the same group showed that the magnetisation homogeneity in the round sample was better than previously thought.

Similar measurements were later performed by the author of this book (while carrying out research at the Wolfson Centre for Magnetics, Cardiff University, Cardiff, the United Kingdom) showing comparable results up to 2.0 T (Zurek et al., 2006a). The author also confirmed that the commonly used electrical steel has negligible asymmetry of power loss anisotropy, and that the asymmetry is caused mostly by the angular misalignment of the sensors used in the 2D measurement system (Zurek et al., 2006a,b).

Zhu and his colleagues from University of Technology, Sydney, Australia and Hebei University of Technology, Tianjin, China published numerous papers on 3D rotational measurements (Zhu et al., 2003, 2009; Zhong et al., 2005; Li et al., 2010, 2011), and many more.

There were also several recent papers on the topic of measurements of rotational magnetisation published by the author of this book. These include qualitative analysis of the P_x and P_y components of rotational losses (Zurek, 2014), discussion on measurement uncertainties of B-coils (Zurek, in press) and further investigation of H sensors and differences between rotational loss measured with the fieldmetric method in clockwise and anticlockwise directions (Zurek, 2017c,d,e).

There are also many publications describing attempts at modelling of the rotational losses. The idea is to be able to include this phenomenon into the design process of magnetic cores. There were several doctoral theses written on this problem. However, to date, these losses are not taken into account by designers of neither rotating machines (Benedetti, 2012) nor transformers (Loizos, 2012). The models are just too complex and not practically usable to be included in commonly used design procedures for magnetic cores of electric machines.

A great number of scientific papers were published so far on the subject. Not all of these will be mentioned here because all aspects linked to rotational studies are simply too numerous. Nevertheless, the main various measurement methods and magnetising systems are discussed in the following chapters.

2 Principles of Rotational Power Measurement

There are three main approaches to the measurement of rotational losses in magnetically soft materials: torquemetric (based on mechanical effects), thermometric (temperature effects) and fieldmetric (electromagnetic effects).

As such, none of them is metrologically superior since they all merely employ different physical quantities related to the same underlying phenomenon. They are not competing but rather complementing each other and allow scientific cross-validation of the measured values and achieved reproducibility of the results. Under carefully controlled conditions, the agreement between different methods is good, considering the variety of the measured physical quantities.

A fourth method that could be described as wattmetric (electromagnetic effects detected as electric effects) was attempted by some researchers. Nowadays, this method does not seem to be used due to problems with excitation level and magnetisation uniformity. This method is briefly explained for the sake of completeness because it evolved from the standardised measurements under alternating magnetisation.

Also, some of these methods can be simultaneously combined with measurements of other quantities. For instance, magnetostriction can contribute to acoustic noise generated by magnetic cores. Some of these were studied in more detail with connection to rotational magnetisation and will be discussed briefly in this chapter.

2.1 MECHANICAL

Many measurement apparatus employ mechanical rotation of either the sample under test or the armature that generates the magnetising field, or even both. The relative rotation of the magnetising field can also be achieved by a multi-phase coil so that no mechanical movement of the sample or the yoke is actually occurring.

However, the main principle of mechanical approach is that the rotational power loss is carried out by the measurement of a basic mechanical quantity, such as force, in a direct or indirect way. It is irrelevant how the relative rotation of sample or magnetic field is achieved.

This concept can be illustrated by using an analogy with mechanical friction (Figure 2.1). A mass m is dragged in a straight line and with a constant velocity v, on a surface with a friction coefficient k_1. A certain positive force F_1 is required to overcome the force of the friction, reacting with the force $-F_1$ on the spring. The friction dissipates some energy. The force is transferred through the spring, which is stretched over the length l_1.

If the same mass m, with the same spring, is dragged at the same velocity v but on a different surface with higher friction coefficient $k_2 > k_1$, then more energy is

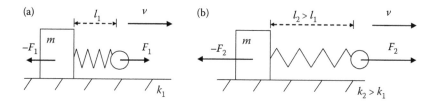

FIGURE 2.1 Mechanical friction as an example of energy loss, with lower (a) and higher (b) friction coefficient.

dissipated. Therefore, higher force F_2 is now required to drag the body. The spring will be stretched out to a greater length l_2 in order to balance the forces.

It can therefore be seen that the length of the spring introduces a 'lag' between the causing force and the position of the body. Nevertheless, in both cases, the whole system moves with the same velocity so that the only difference is the 'lag' caused by the higher energy loss. If the surface was completely frictionless (e.g. in vacuum and with weightlessness), then no energy would be lost, no force would be required and spring would not be stretched at all (in a steady-state condition); hence there would be no 'lag' between the two bodies because for a linear motion, no force would be required (after the body was accelerated to velocity v).

If force F acts over a linear distance l, then some energy E_{line} is transferred, according to the following equation:

$$E_{line} = F \cdot l \quad \text{(J)} \tag{2.1}$$

Similar behaviour would take place if the mass was dragged not in a straight line, but in a circle (if the centrifugal force is neglected), again on a flat surface. If the movement was caused by a rotating arm, then the dragging force would exert a certain torque on the arm (Figure 2.2).

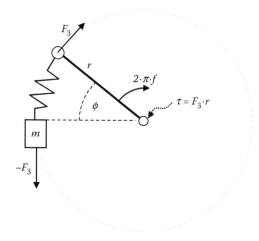

FIGURE 2.2 Angular lag ϕ is proportional to the torque τ exerted on the driving arm and the energy loss (top view).

In such a case, the 'lagging distance' can be expressed as the length of the arc of the circular path proportional to angle ϕ, such that

$$l_{lag} = \phi \cdot r \quad \text{(m)} \tag{2.2}$$

If the 'friction' is constant, then the lagging angle is also constant. And if one full revolution is made, then a constant force F would have worked over the whole perimeter of the circle. Hence, during one full rotation, the energy dissipated per one revolution can be expressed as

$$E_{circle} = F \cdot 2 \cdot \pi \cdot r \quad \text{(J)} \tag{2.3}$$

Knowing time T required for completing one revolution, we can also calculate the power:

$$P_{circle} = \frac{F \cdot 2 \cdot \pi \cdot r}{T} = F \cdot 2 \cdot \pi \cdot r \cdot f \quad \text{(W)} \tag{2.4}$$

where $f = 1/T$ is the frequency of rotation (Hz).

Torque τ can be expressed as a force with a value F acting on an arm with length r, so that $\tau = F \cdot r$, which has the unit of (N · m). Therefore, Equation 2.4 can be rewritten as

$$P_{circle} = \tau \cdot 2 \cdot \pi \cdot f = \tau \cdot \omega \quad \text{(W)} \tag{2.5}$$

where ω is the angular frequency (rad/s).

Hence, we can see that in such a circular system, the 'lag' of the spring is a direct measure of the force (torque), which in turn is a direct measure of the associated energy loss.

It can be shown mathematically that similar principles can be applied even if the 'lag' is not constant. The dissipated power will be then equal to the appropriate average (integral) over the full cycle.

A similar approach is taken in a mechanical loss-measuring apparatus. The excitation is applied as a rotating vector of magnetic field strength H. The sample responds with a vector of flux density B and the lag between the cause H and the effect B can then be found. This lag corresponds to the lost energy, whatever the cause of the loss (hysteresis, eddy currents, etc.).

In some experiments, the value of the torque can be derived indirectly, but the underlying principle is the same. An example of such a measurement will be explained in detail in the following sections.

PRACTICAL COMMENT

Equation 2.5 includes the value of rotation frequency because the calculation concerns power. This results from the definition of power – in a more powerful process, the same amount of energy (Equation 2.3) can be transferred during shorter time because $E = P \cdot t$.

So, in order to compare losses more easily at various frequencies, the measured values are usually divided by the frequency (as shown in Chapter 1 in Figure 1.31).

Therefore, the notation *P/f* (hence *P · t*) used in Chapter 1 simply means comparison of energy, but the calculation is based on the measured values of power. The international standards define other magnetic measurements of power, not energy (IEC, 2003), although one value can be derived from the other if the frequency is known.

2.1.1   TORQUE MAGNETOMETER – ISOTROPIC DISC SAMPLE

In this type of magnetometer, the sample should be circular in order to avoid additional effects caused by the shape itself, due to shape anisotropy. Otherwise, there are other forces acting on the sample and they can make the measurement much less accurate, if at all possible (Pluta, 2013).

Many variants of such apparatus were constructed. Some authors also use the name 'torsion magnetometer' (Zbroszczyk et al., 1981) or 'tensometric method' (Anuszczyk et al., 2009). But the description given below describes the underlying concept, which is common to all such devices.

The main components of the torque magnetometer are depicted in Figure 2.3. Bearings are fixed to the same supporting structure (not shown for clarity). A sample in the form of a disc (or a stack of discs) is kept in a sample holder attached to a freely rotating shaft with a pointer. Apart from the sample under test (and magnetising poles), all other parts should be non-magnetic.

One end of the shaft is connected to a torsion spring, for example, metallic wire or elastic thread (Tiunov, 2012). The other end of the spring is also equipped with a pointer and can freely rotate in yet another bearing. The magnetising field can be generated by a permanent magnet, electromagnet or a set of the Helmholtz coils. This depends on what control is required over the level and rotation of the magnetising field.

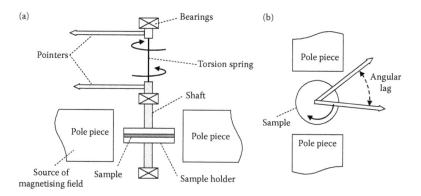

FIGURE 2.3   Torque magnetometer with a disc sample, side view (a) and top view (b).

There are two main approaches to this technique. In one, the source of the magnetising field is fixed in the position, and the sample is rotated. This requires the top bearing to be present (Figure 2.3a). The top of the torsion spring, with the pointer, is rotated in one direction. The spring passes the force and causes the sample to follow. Owing to the rotation, there are magnetic losses inside the sample, which exert an opposing force (compare with Figure 2.2). The driving and opposing forces cause the ends of the spring to twist, creating 'angular lag' between the pointers (Figure 2.3b).

The torsion spring is there to act as a part of the torque transducer, and therefore could be replaced by a different type of spring or suitable torque sensor. But in the setup from Figure 2.3, the twist of the spring (wire) is directly proportional to the torque, and it can be calibrated before the measurement.

For such apparatus, the magnetic energy or power loss is calculated from Equation 2.3 or 2.5, respectively. However, the total value is related to the volume or mass of the sample, so a larger sample yields higher absolute loss values under the same level of excitation. Thus, in order for the results to be comparable between various samples, they must be scaled appropriately to specific values, for instance, to be expressed per unit mass (W/kg) or per unit volume (J/m^3), or similar. This is calculated by knowing precisely the mass or volume of the sample under test, as required.

PRACTICAL COMMENT

The energy lost for overcoming the unwanted mechanical friction of bearings must be taken into account during this measurement. This is usually verified by carrying out rotation without the sample present or without the magnetic field applied, or both. For the correct operation of the system, the angular lag under such condition should be negligible. This is especially important with higher speed of rotation, where the mechanical balance of the rotating system also becomes crucial. Otherwise, the vibration can prevent accurate measurements. Other sources of errors might need to be taken into account as well (e.g. friction in air).

For sensitive measurements, very sophisticated apparatus can be used. For instance, Harrison et al. employed bearings based on pressurised air, and the value of torque (and hence loss) was detected by measuring the time it took to decay the rotational speed from maximum to stationary. The sample was reported to weigh just 1.2 mg ($1.2 \cdot 10^{-6}$ kg) and to be cut out from an audio tape (Harrison et al., 1999). So in such a setup, the lag angle was not measured at all, but the energy loss was measured indirectly through the decay of the rotation of the sample. However, the measurement was still based on mechanical forces acting on the sample under test.

In the other more popular approach, the magnetising field is rotated, instead of the sample. This can be implemented by physically rotating an appropriately shaped electromagnet (or magnet) around the sample. Alternatively, a set of multi-phase Helmholtz coils allows the generation of a rotating magnetic field with relative ease, also at higher frequencies.

With such an arrangement, one end of the torsion spring is fixed. Then the torque acts on the sample, which follows the rotation of the electromagnet until it is balanced by the force of the spring. Of course, the sample cannot then freely rotate around 360°, but its angular position is always dictated by the acting force due to the magnetic loss and the balancing force of the spring.

With this approach, the mechanical balance is even more important because the mass of the rotating electromagnet is much greater than that of the sample. Avoiding vibrations is more difficult, and frequencies higher than 50 Hz become impractical, especially for larger devices. Hence, the rotating field produced by a set of electro-magnets or a set of the Helmholtz coils is just easier to implement as there are no moving parts. The field rotates as it would in a stator of two-phase or three-phase motor. Measurements under dynamic rotation are described in the following sections.

The same equation is used for calculating the loss value. In order to eliminate any directional influences, the rotation should be carried out in both clockwise and anticlockwise directions, and both values should be averaged to produce the final reading, at a given excitation and frequency.

Of course, the diagram shown in Figure 2.3 is only a simplification. The axis of rotation can be horizontal, the sample can have multiple laminations and the spring and shaft arrangement can take a different form, but all the underlying principles follow those described above. For instance, it is almost always assumed that the dis-tribution of flux density in the sample is uniform because the flat disc is an approxi-mation of a 3D round ellipsoid. This might not be the case for multi-layer samples, unless each of the laminations has different diameters so that the shape of the overall stack resembles ellipsoid. And even then there could be significant shape anisotropy due to gaps between the laminations.

The torque magnetometer technique was first used over a century ago (Baily, 1896; Beattie, 1901) and it is still in use now (Soinski, 2001; Anuszczyk et al., 2009). Under moderate frequency of rotation, the instantaneous torque variations are aver-aged out by the inertia of the sensing equipment, so that a stable torque reading could be taken.

2.1.2 TORQUE MAGNETOMETER – ANISOTROPIC SAMPLE

The magnetometer described in the previous section was suitable only for isotropic materials. During the rotation, the torque is generated only due to the power loss inside of the sample and the B vector lags the H vector with the same constant angle. If small anisotropy is present, the deflection is recorded at various degrees of rota-tion and the average value represents the average loss per cycle (Brailsford, 1948; Zbroszczyk et al., 1981; Soinski, 2001).

Much higher anisotropy of magnetic properties came with the dawn of cold-rolled electrical steel in the 1930s. Significant anisotropy could cause large torque pulsa-tions, which could not be averaged out by the inertia of the system, especially at low or very low frequency of rotation.

The point-by-point averaging during slow rotation was simply not feasible any more. The crystallographic anisotropy could contribute to the torque in such a way that at some instants of time, the 'crystallographic torque' was stronger than the 'power loss torque'. In effect, the instantaneous resultant torque could be negative – for a small angle, the sample could rotate in a direction opposite to the rotation of the magnetising field.

Some researchers were still able to make measurements by calculating the difference in torque between the clockwise and anticlockwise rotation. However, this was possible only because of sophisticated measurement techniques that allowed achieving a resolution of torque detection better than 0.1% (Zbroszczyk et al., 1981).

To better understand the measurement difficulties, let us digress again to the phenomenon of anisotropy.

As it was shown in Chapter 1 (Figures 1.21 and 1.22), there are some preferred directions within a crystal. In the grain-oriented electrical steel, most of the volume is occupied by crystallites aligned such that their (110) plane coincides with the plane of the sheet, and the 'easy' direction *[100]* with the rolling direction.

When the magnetising field is applied to such a material within the plane of the sheet, then the *B* vector will tend to align to one of the preferred crystallographic directions, for example, *[100]*, in order to minimise the magnetic energy. The *B* vector will be deflected from an arbitrary direction of the *H* vector and the angular difference (the 'lag') will depend very strongly on the crystallographic structure. Such an anisotropic sample will tend to align itself in the external field as to minimise the total energy. So, even without any losses, there will be torque exerted on the sample. This naturally makes the measurement of the 'loss torque' more difficult.

It can be shown theoretically that for a single crystal of Fe–Si 3% within the (110) plane, the torque curve will be similar to that shown in Figure 2.4. These are theoretical values calculated from the so-called magnetocrystalline anisotropy constants that would occur under saturation conditions.

The curve can be theoretically calculated if the crystallographic configuration and anisotropy constants are known. For the (110) plane of Fe–Si 3%, the energy can be calculated as (Tumanski, 2011)

$$E = K_0 + K_1 \cdot (0.25 \cdot \sin^4 \theta + \sin^2 \theta \cdot \cos^2 \theta) + 0.25 \cdot K_2 \cdot \sin^4 \theta \cdot \cos^2 \theta \qquad (\text{J/m}^3) \quad (2.6)$$

where θ is the angle of rotation.

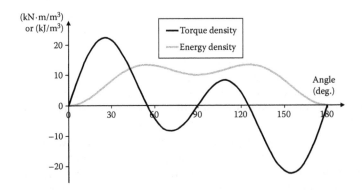

FIGURE 2.4 Torque curve for single crystal of Fe-Si 3% caused by crystallographic energy.

The point-by-point acting torque can be calculated as the point-by-point derivative of the energy from Equation 2.6. Or conversely, the energy can be calculated by the integration of the torque curve. Figure 2.4 shows both curves for comparison. For such materials, the experimental data match theoretical calculations very well (Anuszczyk et al., 2009).

However, the torque related to magnetocrystalline energy is applicable only at magnetic saturation. At lower excitation, the curve would be different and the transitions from positive to negative values would have greater slope and different shape. This is because other effects would also contribute to the torque, for example, energy of the domain walls (Zbroszczyk et al., 1981).

Nevertheless, it can be seen from Figure 2.4 that the torque can vary from highly positive to highly negative values, whereas the average would be a much smaller value. Interestingly, even for the so-called 'isotropic' non-oriented electrical steel, such torque curves can be detected (Fiorillo, 2004). This is because there is always some anisotropy for commercially available non-oriented electrical steel due to manufacturing technology.

The torque variations due to anisotropy might reach values hundreds of times greater than the torque caused by rotational loss (Brailsford, 1938). This can be overcome by measuring the torque with extremely high precision, for instance, better than 0.1% (Zbroszczyk et al., 1981).

But the more common approach is to stack several samples at staggered angles, so that the torque variations compensate out for the whole stack (Brailsford, 1938; Boon et al., 1964; Zhu et al., 1997a). With such a multi-lamination sample, the torque becomes much more stable even up to a high-magnetisation level (Figure 2.5).

2.1.3 Torque Magnetometer – Dynamic Rotation

The curve of rotational losses versus circular B has such a peculiar shape due to the hysteresis loss, which is caused by domain wall generation and annihilation. At magnetic saturation, there is only one rotating domain, hence there are no domain walls and the hysteresis component of the rotational loss vanishes. But the loss component generated by eddy current increases with both amplitude of magnetisation and frequency. At higher frequency, the peak of the loss curve might not occur at all because the eddy current loss completely overshadows the hysteretic component (Zurek et al., 2006c).

So in order to study the hysteresis loss component, measurements are carried out under so-called quasi-static conditions that are at such low frequencies at which the eddy current losses are negligible (see also Figure 1.27). Because eddy currents depend on the thickness and resistivity of the sample under test, the actual frequency will be different for different materials (Gozdur, 2004).

As mentioned above, Francis Baily was the first researcher to investigate rotational losses. He used a type of torque magnetometer (Figure 2.6), and carried out the measurements up to quite a high frequency of around 65 Hz (Figure 2.7). The apparatus employed a stationary sample with the electromagnet rotating around it.

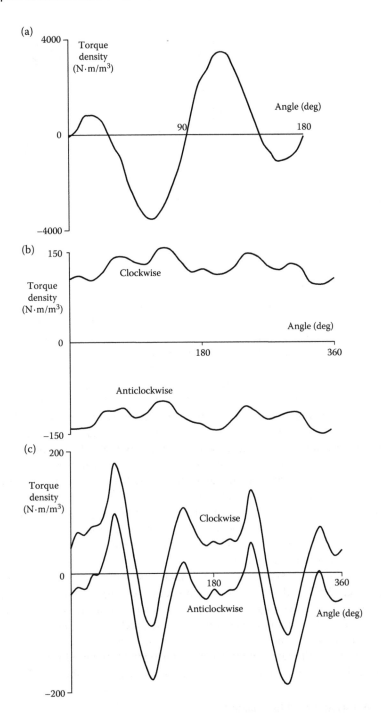

FIGURE 2.5 Torque variations for hot-rolled steel for (a) a single-lamination sample at 0.24 T, (b) a multi-lamination sample at 1.70 T and (c) a multi-lamination sample at 2.26 T. (Adapted from Brailsford F., *Journal IEE*, 83, 1938, 566.)

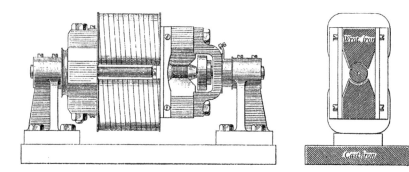

FIGURE 2.6   Baily's apparatus. (Copyright © 1896, The Royal Society. Reproduced with permission from Baily F.G., *Philosophical Transactions of Royal Society of London, Series A*, 187, 1896, 715.)

Commonly used modern non-oriented electrical steel can be 0.5 mm thick, and at 1.5 T and 50 Hz, the quasi-static hysteresis losses (rotational) are found to be comparable on the order of magnitude to eddy current loss (Zurek, 2009b).

The peak values of rotational losses per cycle measured by Baily were 14,500–15,000 ergs per cubic centimetre per cycle. Converting these to SI units (assuming material density of approximately 7800 kg/m³ for low-carbon steel) gives values on the order of 9.3–9.6 W/kg at 50 Hz, whereas for modern non-oriented electrical steel, this could be considerably lower at 4.5 W/kg (Zurek, 2009b), even though there is a significant contribution of eddy current loss.

The eddy currents estimated by Baily did not exceed 4% of the total loss. This was because the laminations used by Baily were relatively thin, between 0.081 mm and 0.141 mm, so the eddy currents were significantly reduced (as compared to typical commercial electrical steels used nowadays). But also the sample was made from 'charcoal iron', so the hysteresis losses were much greater in the first place, due to possible higher carbon content and uncontrolled crystallographic texture – as compared to modern non-oriented electrical steel.

In real-life applications, in rotating machines, the total losses are important because they decide the thermal operating conditions. So, besides the pure hysteresis component, the eddy currents (and additional losses) are also of interest, at least from engineering and design viewpoints.

Fully dynamic measurements were made as well by other researchers. For instance, Brailsford modified his apparatus to employ three pairs of the Helmholtz coils in order to generate a rotating magnetic field, also at mains frequency (Figure 2.8). This eliminated most of the mechanically rotating parts and made the measurements more accurate. The Helmholtz coils were designed such that each pair produced the same amplitude of magnetic field strength for the same driving current. Typical measured results obtained by Brailsford are shown in Figure 2.9.

2.1.4   Anisometer – Disc Sample

As the name suggests, an anisometer is an apparatus that is used for studying the magnetic anisotropy of disc samples (Figure 2.10).

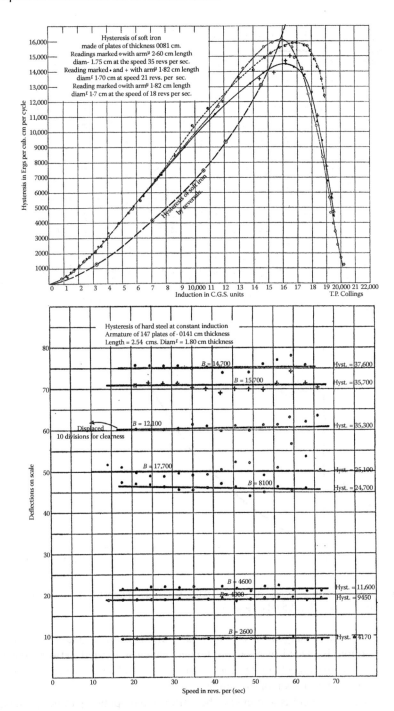

FIGURE 2.7 Example of Baily's results. (Copyright © 1896, The Royal Society. Reproduced with permission from Baily F.G., *Philosophical Transactions of Royal Society of London, Series A*, 187, 1896, 715.)

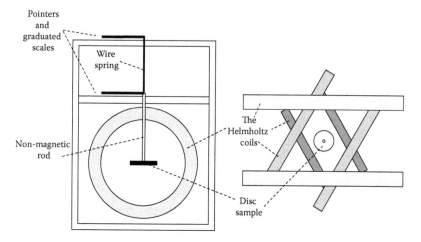

FIGURE 2.8 Dynamic torque magnetometer used by Brailsford. The drawing on the left shows only one Helmholtz coil for clarity. The drawing on the right shows a top view, with three pairs of intersecting coils displaced by 120°. (Adapted from Brailsford F., *Journal IEE*, 95 (II), 1948, 38.)

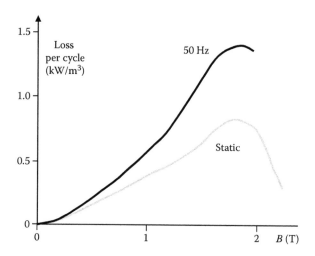

FIGURE 2.9 Dynamic losses measured with a torque magnetometer. (Adapted from Brailsford F., *Journal IEE*, 95 (II), 1948, 38.)

The mechanical arrangement is somewhat similar to the torque magnetometer. The disc sample is placed in a sample holder, which is located between the pole pieces of an electromagnet (Pluta, 2001). The sample holder is driven by a geared mechanism to allow rotation of the sample.

The magnetising field is assumed to be uniform and applied parallel to the axis of the magnetising coils. In such an arrangement, the sample will be magnetised in a direction parallel to the axis between the pole pieces of the electromagnet. The actual vector of flux density B and hence also polarisation J in the sample might

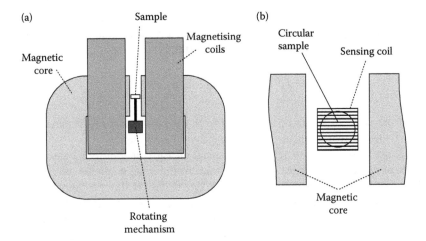

FIGURE 2.10 Anisometer: side view of the electromagnet (a) and a top view of the circular sample between the pole pieces of the electromagnet (b); see also Figure 6.11.

lie at a different angle, which is caused by the magnetocrystalline anisotropy (for anisotropic materials). This is illustrated in Figure 2.11.

As explained in the previous section, some torque will act on the sample. If the J vector is exactly parallel to the field, then no torque will be present. This can be seen, for instance, in Figure 2.4 where the torque curve passes through zero several times. At those points, the J vector is parallel to the applied H vector.

However, at any other angular position of the sample, the torque does develop because the J and H vectors are not parallel. So, we can analyse the J vector in terms of its two orthogonal components: parallel $J_\parallel$ and perpendicular $J_\perp$ to the applied excitation (Figure 2.11).

As mentioned above, $J_\parallel$ does not contribute to the torque, so there is no need to detect it, and indeed most such apparatus are not equipped with any sensor for the detection of $J_\parallel$.

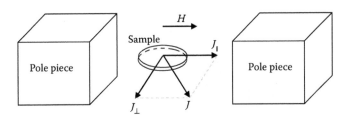

FIGURE 2.11 Definition of $J_\parallel$ and $J_\perp$ in an anisometer. (Adapted from Soinski M., *IEEE Transactions on Magnetics*, 20 (1), 1984a, 172 and Soinski M., *Magnetic materials in technology* (in Polish: *Materialy magnetyczne w technice*), Biblioteka Centralnego Osrodka Szkolenia i Wydawnictw SEP, 2001, ISBN 83-915103-5-2.)

The sample holder has a fixed search coil, whose axis is *perpendicular* to the direction of the applied magnetic field, so as to detect the $J_\perp$ component. The torque arising from a sample volume V_s is proportional to

$$\tau = J_\perp \cdot V_s \cdot H \quad (\text{T} \cdot \text{m}^3 \cdot \text{A/m}) = (\text{N} \cdot \text{m}) \tag{2.7}$$

And the value of H can be calculated from the parameters of the electromagnet (magnetising current, number of turns, length of the air gap).

So it can be seen that no actual mechanical measurement of torque needs to take place to derive characteristics similar to those from torque magnetometers.

Of course, the search coil can detect a signal *only* if the $J_\perp$ component varies (according to Faraday's law of induction). This is achieved either by the rotation of the sample or by using an alternating magnetising field. For this reason, the method is used for studying the crystallographic texture and anisotropy (Matheisel, 1975) rather than losses related to rotational magnetisation. Also, the measurements are not absolute because supporting x-ray crystallographic measurements are required, but the technique is relatively easy to use and useful for the quick assessment of anisotropy in physical experiments (Matheisel, 1975).

The measurement of the perpendicular component can also be carried out with the use of a Hall effect sensor. This method was used for obtaining the results, for instance, in Flanders (1985) and Tan et al. (1985).

The principle is the same because rotational loss is measured by means of detecting the perpendicular component of flux density. However, the detection of magnetic flux itself is based on the field lines that 'fringe out' from the sample under test due to the demagnetisation effect (Figure 2.12, see also Figure 1.19).

The measurements are made only for a 'non-oriented SiFe' sample, and it is stated in Flanders (1985) that the rotational loss W_R is calculated from the following equations[*]:

$$W_R = \int_0^{2\pi} L \cdot d\theta = \int_0^{2\pi} M \cdot H \cdot \sin\alpha \cdot d\theta \quad (\text{W}) \tag{2.8}$$

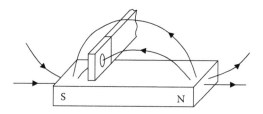

FIGURE 2.12   Detecting flux density as a 'fringing out' component. (Adapted from Flanders P.J., *IEEE Transactions on Magnetics*, 21 (5), 1985, 1584.)

---

[*] The unit of 'watt' is presumed for Equations 2.8 and 2.9 because no units are given in the paper.

But because W_R is a function of signals from Hall effect sensors ($V_{H\parallel}$ – parallel and $V_{H\perp}$ – perpendicular), Equation 2.8 is transformed into (M_S – saturation magnetisation, $_{SAT}$ – at saturation)

$$W_R = \left(\frac{H \cdot M_S}{V_{H\parallel/SAT}} \right) \cdot \int_0^{2\pi} V_{H\perp} \cdot d\theta \quad \text{(W)} \quad\quad (2.9)$$

The paper does not give any information on the accuracy of the method. Also, the graphs are plotted not in units directly related to (W) or (J) but as a function of $V_{H\perp}$ versus $V_{H\parallel}$, so direct comparison to other methods is not possible. Although the results for 'non-oriented Si–Fe disc' are shown, the method seems to be devised and used for semi-hard materials (recording tape) rather than magnetically soft (electrical steels) (Flanders, 1985).

Yet another apparatus was used with a modified approach of parallel/perpendicular signals (Grimwood et al., 1978). A flat disc sample was magnetised between pole pieces of an electromagnet, somewhat similar to the setup shown in Figure 2.10. Two pairs of coils were used to detect the parallel and perpendicular components of flux density. Because the sample was rotated very slowly (e.g. one revolution in 2 hours), the coils were vibrated at a suitable frequency and amplitude in order to induce measurable signals. The sample was put inside of a known uniform field (solenoid) to measure its 'true' dipole moment, which was then used to correct the signals from the coils during the actual measurement. The value of perpendicular component of the corrected flux density was used to calculate the torque acting on the sample in order to be able to use Equation 2.7.

Also, the sample could be vibrated instead of the coils. Such a system was described, for instance, in Perov et al. (2000). The setup was used for measurements of anisotropy constants, but the underlying principle was the same – detection of the perpendicular component. The authors referred to the apparatus as *vibrating sample anisometer* (VSA).

2.1.5 TORQUE MAGNETOMETER – RING SAMPLE

A rotating magnetic field in a sample can also be generated in a different way, somewhat similar to the construction of an actual rotating machine. The sample takes a form of a toroid stacked from several ring-like laminations (Figure 2.13). It is crucial that the sample is made as a stack of laminations because the magnetic field is applied within the plane of such laminations. If the sample was made as a toroid wound in a spiral manner from one long strip of lamination (as it is the case for commercially available toroidal transformers), then the magnetic field would rotate out-of-plane and the measurement would not be valid, at least not in the same sense as other methods described so far.

An example of such an apparatus is shown in Figure 2.13. The magnetising coils (2) and pole pieces magnetise the sample (6) across its diameter, within the plane of the laminations. The shaft (3) is driven by the motor (4), whose speed is controlled to the required frequency of rotation.

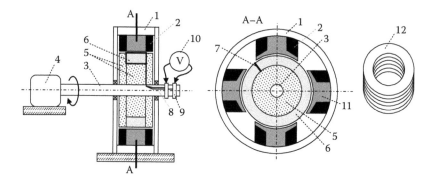

FIGURE 2.13   Apparatus for measurement of rotational losses in toroidal samples: 1 – case, 2 – magnetising coil with DC current, 3 – shaft, 4 – driving motor, 5 – non-magnetic sample holder, 6 – toroidal sample under test, 7 – measuring coil on the sample, 8 – wires positioned in the channel in the shaft, 9 – slip rings, 10 – voltmeter, 11 – magnetic poles, 12 – stack of ring-like laminations (not to scale). (Adapted from Anuszczyk J., *Method and apparatus for measurement of power loss in ferromagnetics under rotational magnetisation* (in Polish: *Sposob i urzadzenie do pomiaru strat mocy w ferromagnetykach przemagnesowanych obrotowo*), Patent application, PRL, Poland, P127–489, 1984.)

The voltmeter (10) is used for the detection of the level of flux density in the sample. The losses are measured as the total power taken by the driving motor, which is affected by the extra torque exerted on the shaft due to the additional loss. Two measurements need to be made – one with and one without the sample and the difference between the readings gives the total loss of the sample. The no-sample measurement is required to subtract all losses not related to the presence of the sample: mechanical friction, windage, etc. (Anuszczyk, 1984; Anuszczyk et al., 2009). Of course, the mass or volume of the sample must be known to convert the losses into specific values (per kg, or per m³). Similar systems were also described in Reisinger (1987) and Aouli et al. (2000). An approach with a different shape of the sample, although still based on torque measurement, was presented in Dami et al. (1996).

The method does not offer a direct means of controlling the 'circularity' of rotational magnetisation, which might be problematic for more accurate measurements. This is especially important for measurements on highly grain-oriented electrical steels, when even small errors in control of waveshape lead to large changes of measured losses (Zurek, 2005a; Zurek et al., 2006a,b). These implications are explained in more detail in Chapters 4 and 5.

2.2   THERMOMETRIC (RATE OF TEMPERATURE RISE)

Losses dissipated in the sample under test are ultimately converted to heat. Therefore, they could be detected almost 'directly' by measuring the actual temperature increase of the sample under test.

Very precise measurements of heat can be carried out by the calorimetric technique. This method is commonly used in many physics experiments, for example, for studying the so-called magnetocaloric effect (Lin et al., 2006; Wang, 2012; Kuepferling et al., 2013).

However, the ordinary calorimetric method does not lend itself well for measurements of rotational losses for two reasons. First, for correct measurement, the sample must be isolated thermally from the rest of the circuit. This is cumbersome, but possible to achieve, because the sample can be enclosed in a thermally isolated container, and the magnetising field can be applied through its walls (Ragusa et al., 2008).

Usually, there are other components that need to be placed together with the sample (e.g. *B* and *H* sensors). These add thermal inertia and have some influence on the measurement.

However, more importantly, for accurate measurement, the whole volume of the sample would have to be magnetised in a uniform way. Otherwise, there would be different amplitudes of losses at different parts of the sample. Since losses depend in a non-linear way on the excitation level, it would become impossible to measure the actual value in a controlled way.

In many measurement systems, the uniformity of excitation can be ensured only in a relatively small central part of the sample (Alinejad-Beromi, 1992; Sievert et al., 1995). In such cases, the calorimetric method cannot be used, at least not with acceptable precision. This is especially true for systems in which the sample cannot be thermally isolated (as, for instance, shown in Figure 2.24).

Nevertheless, Gorican et al. showed that for appropriately constructed magnetising yoke and round samples, uniformity of the excitation is much better, especially for non-oriented electrical steel (Gorican et al., 2003). Also, larger air gap between the magnetising yoke and the sample usually improves the magnetisation uniformity.

If the losses are assumed to be generated uniformly in the sample, then a thermometric measurement can be performed.

One of the first such measurements under elliptical rotational magnetisation was described in Strattan et al. (1962). A disc sample (silicon iron) was magnetised at 60 Hz in a set of two-phase Helmholtz coils. Copper–constantan thermocouple was used, with the claimed accuracy of temperature measurement at 0.001 K.

A special construction of the sample simulating an ellipsoidal shape was used by some researchers to further improve the uniformity of magnetisation. The apparatus used by Fiorillo and Rietto (Fiorillo et al., 1992a) (also later in Ragusa et al., 2008) is described below as an example. The technique is sometimes also called the *rate of temperature rise* because of the shape of the measured curves and the means of processing the data. The power loss is calculated from the curves, in a simpler approach by detecting the initial slope of the curve (Hollitscher, 1969; Derebasi et al., 1992) and in a more comprehensive approach by curve fitting (Ragusa et al., 2008).

Ideally, the sample would be placed in a fully adiabatic environment so that there is no heat exchange. Under such conditions, any power loss dissipated in the sample would correspond to its temperature rise, and if the dissipated power is constant, then the temperature rise would be linear. In reality, the temperature increase is not linear because completely adiabatic conditions cannot be attained. This is because at the very least heat will be lost due to radiation, so the temperature increase will

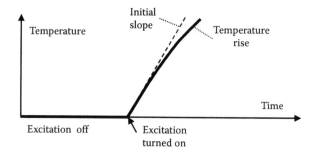

FIGURE 2.14   Measurement of power loss by detection of rate of temperature rise.

slow down and eventually plateau. So only the initial part of the curve is pertinent (Figure 2.14).

An example of such a system is shown in Figure 2.15. The sample is placed inside a cylindrical vacuum chamber, which is fitted inside the stator core of an electric motor (for the generation of rotating magnetic field). The pressure in the chamber is lowered to 0.002 Pa to give an approximation of vacuum (Ragusa et al., 2008), which is required to achieve quasi-adiabatic conditions (so that heat transfer through any surrounding gas is limited).

The sample has one or more thermocouples connected to its surface (Figure 2.16). These allow the measurement of the rise in temperature of the sample. There are also search coils for detecting the level of magnetisation of the sample. Dummy thermocouples not thermally connected to the sample can be used for cold-junction compensation (Figure 2.16).

The conditions are not ideally adiabatic, so the equation of energy balance must take into account that some energy dissipated by rotational loss heats up the sample, but some of it is exchanged with the ambient.

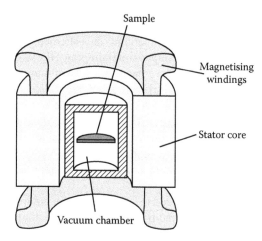

FIGURE 2.15   Stator-like core of a three-phase induction motor can be used as a magnetising yoke to generate the required 2D field in the plane of a lamination placed in a vacuum chamber; see also Figures 6.6 and 6.7.

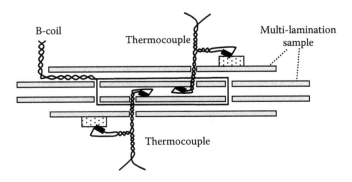

FIGURE 2.16 An example of thermometric arrangement showing a cross-section view through a sample set with four laminations. (Adapted from Fiorillo F. et al., The measurement of rotational losses at I.E.N.: Use of the thermometric method, *Proceedings of 1st 1&2DM Workshop*, Braunschweig, Germany, 1992a, p. 162.)

An assumption is also made that at the moment of starting the excitation, all components (the sample and the reservoir) are at the same ambient temperature T_0. It can therefore be written that

$$P = c \cdot \frac{d(T_m - T_0)}{dt} + K_t \cdot (T_m - T_0) \quad \text{(W/kg)} \tag{2.10}$$

where c is the specific heat of the material under test (J/(kg·K)), K_t is the heat transfer coefficient (W/(kg·K)) and T_m is the measured temperature (K).

The increase of temperature after starting the excitation is measured by the thermocouples. The experimental data follow exponential curves, which can be fitted by solving Equation 2.10 so that

$$T_m - T_0 = \frac{E}{K_t} \cdot \left(1 - e^{-K_t \cdot \frac{t - t_{on}}{c}} \right) \quad \text{(K)} \tag{2.11}$$

where t is the time (s) and t_{on} is the time instant of starting the excitation (s).

For example, non-oriented electrical steel is an alloy, so the specific heat coefficient can be found by applying the Kopp–Neumann law, which is a weighted average of constituent heat coefficients of all alloy components. In particular, for the non-oriented samples used in the experiment, it was calculated that $c = 461$ J/(kg·K) (Ragusa et al., 2008).

Another value that must be known is the Seebeck coefficient of the thermocouples. In this case, they were copper–constantan with a value of $40.41 \cdot 10^{-6}$ V/K. At and around room temperature, a coefficient of around 0.2%/K should also be taken into account. A nanovoltmeter was used for the detection of thermocouple voltage. The heat transfer coefficient K_t was found by fitting the curve into the experimental data, an example of which is shown in Figure 2.17.

When the excitation is switched off, the sample begins to cool down, a process that can be approximated with the following equation (T_1 is the temperature

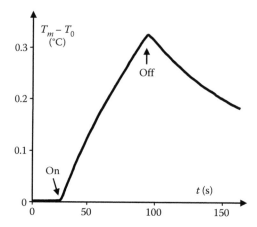

FIGURE 2.17   Temperature rise versus time under quasi-adiabatic conditions and circular B (1.5 T, 50 Hz) for a non-oriented electrical steel sample tested at 50 Hz, at INRiM laboratory. (Adapted from Ragusa C. et al., *Journal of Magnetism and Magnetic Materials*, 320 (20), 2008, e623.)

at the instant of stopping the excitation [K], t_{off} is the time instant of stopping the excitation [s]):

$$T_m - T_0 = T_1 - T_0 \cdot e^{-K_t \cdot \frac{t - t_{off}}{c}} \qquad \text{(K)} \qquad\qquad (2.12)$$

For the data shown in Figure 2.17, the heat transfer coefficient found by curve fitting was 4.4 W/(kg · K).

The accuracy of the setup was verified by carrying out one-directional excitation (i.e. not rotational) and comparing the results with the data measured in Epstein frame. The agreement was better than 2% up to 1.5 T and 50 Hz. At other frequencies or magnetisation level, the uncertainty of the method was up to 5%.

The same sample was also measured by a fieldmetric technique (described later in this chapter) and the agreement between the two completely different methods at two different laboratories was quite satisfactory, not exceeding a difference of 10% for intermediate levels of flux density.

The agreement between thermometric and fieldmetric measurements within the same laboratory can be very good, as, for example, demonstrated in Appino et al. (2009).

PRACTICAL COMMENT

As can be seen in Figure 2.17, the total temperature rise is rather small – less than 0.35 K for the whole measurement. It is quite difficult to measure such small temperature changes with high *absolute* precision and accuracy. This is the reason why thermocouples were used because they do offer sufficient resolution for performing such experiments. But even then the output signal is very small and not easy to measure.

Additionally, long intervals are required between each measurement point. The sample must be cooled down sufficiently, to permit another cycle of heating. We can see from Figure 2.17 that several minutes are required in order to measure a single point of power loss at the given level of flux density.

This is why the method cannot be used in practice at very low frequencies and low flux density values. The losses simply become too small to sufficiently heat up the sample under test. For example, loss at 0.15 T and 1 Hz would be around 500 times lower as compared to 1.5 T at 50 Hz. Therefore, over the same timescale, the temperature would have risen by less than 0.7 mK, which would be certainly very difficult to detect at all, let alone accurately.

The *rate-of-temperature-change* method can employ any temperature sensors, provided that they offer sufficient resolution for the measured signal. Thermocouples are just easy and fairly inexpensive sensors to be used for such purpose.

Interestingly, even the best thermal cameras in the market are not very suitable – the high-thermal resolution cannot be easily utilised when looking through the glass walls of the vacuum vessel (Cooper et al., 1998). Some researchers attempted using thermal imaging cameras, but the measurements were of limited accuracy, and the reported results are rather scarce. This is despite the fact that under special conditions, some thermal imaging cameras offer theoretical resolution down to 0.1 mK (Nakata et al., 1992; FLIR, 2017).

If the uniformity of magnetisation can be assured, then the thermal technique is the absolute method of loss measurement that can be used as a type of 'calibration for other methods'. The 10% agreement quoted above might not sound very 'high-precision', but comparison of rotational measurements at various laboratories showed a notorious difficulty even for the same type of measurement. The international intercomparison of results just for the fieldmetric technique gave discrepancies exceeding 50% in some cases (Sievert et al., 1995), so the agreement shown in Figure 2.18 is rather encouraging from the viewpoint of rotational measurements. It is in fact a very good example showing that the different measurement methods are complementary, rather than competing.

However, the thermal measurements cannot be treated as the absolute calibration technique unless the conditions are controlled very carefully. As mentioned above, it is difficult to achieve adiabatic conditions and uniform rotational magnetisation *for the whole sample* at the same time. If this is not the case, then an unknown level of measurement uncertainty will be introduced.

The name 'rate-of-temperature-rise' applies because the initial slope of the temperature rise curve is directly proportional to the energy loss. At the beginning of the measurement, there are no temperature differences between the sample and the ambient, so there is no heat exchange, and under ideal adiabatic conditions, the rise would be linear and infinite. In practice, it is not

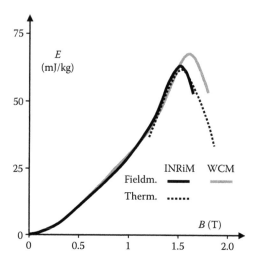

FIGURE 2.18 Comparison of results measured by a thermometric (Therm.) and field-metric (Fieldm.) methods in two laboratories: National Institute for Metrological Research, (INRiM), Italy, and Wolfson Centre for Magnetics (WCM), Cardiff University, Cardiff, the United Kingdom. (Adapted from Ragusa C. et al., *Journal of Magnetism and Magnetic Materials*, 320 (20), 2008, e623.)

because the increasing temperature difference makes heat dissipation faster and the slope slows down. However, the detection could be made by plotting a tangential line at the beginning of the rise (hence *rate of rise*), which with an appropriate calibration would give a direct measure of the energy loss, without the need for the full curve fitting procedure.

2.3 WATTMETRIC

The name 'wattmetric method' appears to be quite widely used also when referring to the fieldmetric method (Sasaki et al., 1985; Fiorillo et al., 1988; Pfutzner, 1994; Anuszczyk et al., 2009).

However, there is a subtle difference between a wattmetric method as defined, for instance, in the IEC standards (toroidal sample, Epstein frame and SST) and the commonly used fieldmetric method.

In this book, the author decided to make a clear distinction between the 'classical' wattmetric method using the magnetising currents for loss measurement and the somewhat newer fieldmetric in which the H can be measured directly by a suitable sensor (Sievert, 1992a). The main difference is that the classical wattmetric measurement relies directly on the magnetising current and does not allow localised measurements.

In uni-directional measurements, like those performed in Epstein frame or toroidal samples, the measurements are based on the principle of an unloaded transformer. A wattmeter is connected into the circuit, such that the primary current and secondary voltage are used as its input signals. This disregards any power losses in the primary winding (because the voltage drop on this winding is not measured), and the secondary winding should work under no-load conditions, thus also producing the negligible loss.

As a result, the wattmeter would measure only the power loss dissipated in the magnetic core (the primary-to-secondary turns ratio must be taken into account separately). In order to achieve the best possible accuracy, additional losses caused by the load of the wattmeter are also subtracted (IEC, 2003).

2.3.1 TOROIDAL SAMPLE

In the circuit shown in Figure 2.19, the total power loss dissipated is measured by an actual *wattmeter* for the whole sample under test – hence the name 'wattmetric' method. It should be noted that no dedicated 'sensors' are used for making such measurement. All signals come from connecting the tested sample (the unloaded transformer) directly to the wattmeter.

The toroidal sample is not a separate entity from the magnetising yoke, which was the case for all the methods described above. Also, the measurement of H or B (or conversion of signals from voltages and currents to H and B) is not actually carried out specifically, but the loss value is measured directly by the wattmeter.

This method is well established, widely used and it is standardised by international standards (IEC, 2003; ASTM, 2011).

PRACTICAL COMMENT

Toroidal samples have the advantage of being a very 'compact' magnetic circuit. The demagnetising field (see Figure 1.19) is practically eliminated and high-accuracy measurements are possible. However, usually the windings must be made manually, which can be time consuming and prone to errors with counting the number of turns, especially for larger numbers.

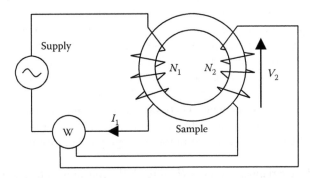

FIGURE 2.19 A simplified wattmetric method applied for the measurement of losses in toroidal sample.

Also, typically, it is not possible to compensate the $\mu_0 \cdot H$ contribution and B must be measured instead of J. This problem is discussed in detail in Chapter 4.

The greatest advantage of the toroidal sample method is that the magnetic path length is well defined and easy to calculate accurately, if the ratio of outer to inner diameter is close to 1.

Therefore, if the magnetic path length, magnetising current and number of turns of the primary winding are known, then Equation 1.3 can be used to calculate the value of H directly. More details on this are given in Chapter 3.

In order for the results of the measurements to be comparable between different laboratories, they should be carried out under the same magnetising conditions. The electricity supply is based on AC voltage, which is sinusoidal in nature. Therefore, the magnetic cores of transformers, motors and generators are also excited with conditions that resemble sinusoidal voltage.

For this reason, it was decided in the standards that the magnetisation regime should be controlled to be pure sine. Such conditions could be easily reproduced between different laboratories and the conditions could be controlled so that higher harmonics were excluded by means of feedback or low-impedance supply (discussed in Chapter 4).

2.3.2 EPSTEIN FRAME

Epstein frame (Figure 2.20) uses a similar principle as the toroidal sample. Closed magnetic circuit is created from a number of *strips* that are stacked together so that they create a square 'frame', hence the name.

The mechanical construction is such that the locations of the strips are well defined (Figure 2.20d). Each strip extends across the whole area of the corner, so that the strips are doubly overlapped, which minimises the magnetic reluctance between the strips (Figure 2.21). At the time of writing this book, the latest version of the standard (IEC, 2008; ASTM, 2014) defines that the side of the square should be 25 cm and the strips should be 30 mm wide and between 280 mm and 320 mm long. The minimum number of strips is not strictly defined but the mass of the whole sample should be at least 240 g for 280 mm strips (IEC, 2008). Obviously, in order to keep the structure symmetrical, the number of strips in the sample must be a multiple of 4. For example, if each strip weighs 18.08 g (see the value written on the strip in Figure 2.20d), then a minimum of 16 strips would be required to constitute a valid sample.

However, the stack of strips will never fill the area available inside the coil (Figure 2.20d) and a large volume of air would contribute to significant contribution from the $\mu_0 \cdot H$ component. Therefore, there air flux compensation is executed by means of a mutual inductor, which is typically cylindrical and placed at the centre of the frame (not shown in the photographs because it is underside the structure).

The doubly overlapped corners (Figure 2.21) constitute greater active cross-sectional area than in the rest of the magnetic circuit. This arguably has an influence on the definition of magnetic path length, which is arbitrarily set as 0.94 m.

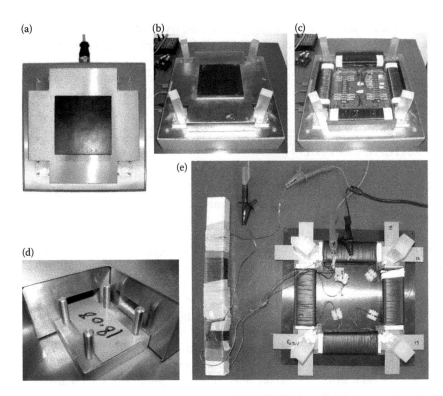

FIGURE 2.20 Epstein frame: (a) without sample, (b) with sample and the non-magnetic weights in each corner producing 1 N holding force, (c) without the top cover, (d) close-up view of one corner with a single strip and (e) non-standard experimental frame (used for the experiments presented in Marketos et al., 2007) with the mutual inductor visible on the left.

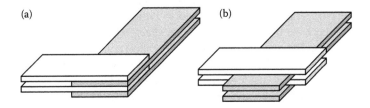

FIGURE 2.21 Double overlapping in the corners required by the Epstein frame standards for shorter (a) or longer samples (b).

Indeed, experiments carried out on Epstein frame with different dimensions (Figure 2.20e) indicate that the magnetic path length can vary depending on the type of material of the sample, magnetising frequency, etc. The apparent value of the magnetic path can change as much as from 0.88 m to 0.96 m (Dieterly, 1948; Marketos et al., 2007).

Nonetheless, these effects are neglected in the standard, the value 0.94 m is assumed constant and H is also calculated directly from the magnetising current, by means of Equation 1.3.

The output terminals of the primary and secondary windings are connected in the same way as in Figure 2.19, but are split into four equal parts, each part enwrapping one side of the square.

The method is widely accepted in the electrical steel industry and became the basic method by which electrical steels are classified and traded between manufacturers and the end users.

PRACTICAL COMMENT

Historically, the Epstein frame had a size of 50 cm and required a sample weight of 5–10 kg (Dieterly, 1948; Henney, 1950). The continuing progress in the magnetic measurements led to the 'baby' Epstein frame sized at 25 cm with the mass of the sample reduced to 0.25 kg, as used presently. Therefore, from a purely practical viewpoint, the apparatus could be loaded up in a much shorter time and much less expensive samples could be used. The mechanical construction is also fairly uncomplicated, which allowed the widespread use of this method.

The Epstein frame is flux air compensated so that the J–H characteristics of the specimen are measured, rather than B–H as in the toroidal sample method. However, the measurement itself is performed in a very similar way because the power loss is effectively also measured with a wattmeter and the magnetising conditions require setting up sinusoidal-induced voltage.

The individual strips are usually referred to as 'Epstein strips', and are sometimes used in other measurement systems.

2.3.3   SINGLE-SHEET TESTER

The third popular standardised method is the single-sheet tester (SST) (ASTM, 2009; IEC, 2010). The method is designed to be less time consuming in use than Epstein frame because the sample is a single *sheet*, which can be quickly inserted into the apparatus. The concept of the method is shown in Figure 2.22.

The primary and secondary windings are permanently placed on a suitable former. The former has a channel inside so that the sample can be slid into it. The top yoke is supported on a mechanism and can be lifted up, creating a gap through which the

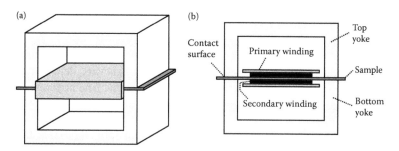

FIGURE 2.22   Single-sheet tester (a) and its cross section (b); drawings not to scale.

sample is inserted. For measurements, the top yoke is lowered in order to complete the magnetic circuit.

The secondary winding is wound closer to the sample in order to reduce the amount of air flux. In any case, the air flux is fully compensated with a mutual inductor, hence the measurement returns J rather than B, similarly as in Epstein frame. The primary and secondary windings are connected similarly as shown in Figure 2.19. Sinusoidal shape of the secondary voltage is required.

The operation of an SST is based on the assumption that magnetic reluctance of the yoke is negligibly small as compared to the reluctance of the sample. This is met by ensuring that the effective cross-sectional area of the yoke is much greater than that of the sample. The exact dimensions are specified in the standards.

PRACTICAL COMMENT

Equation 1.3 is valid directly only for a magnetic circuit with a constant cross-sectional area and constant properties. This is not the case for the SST because the contribution of the yoke would have to be taken into account separately such that $N \cdot I = H_{sample} \cdot I_{sample} + H_{yoke} \cdot I_{yoke}$.

The length of the yoke is on the same order of magnitude as the length of the sample so it cannot be neglected. However, the drop of magnetomotive force on the magnetic reluctance of the yoke is much smaller than that of the sample. The yoke should have a combined cross section of $2 \cdot 25$ mm $\cdot$ 500 mm $= 25,000$ mm^2. On the other hand, a typical sample would be, for instance, 0.5 mm $\cdot$ 500 mm $= 250$ mm^2, which is a value 100 times smaller. Therefore, in the first approximation, if the same flux is assumed to penetrate the yoke and the sample, then because of the difference in cross-sectional areas, it is $B_{yoke} \ll B_{sample}$ and thus also $H_{yoke} \ll H_{sample}$. As a result, it can be assumed that $H_{yoke} \approx 0$ and thus $N \cdot I \approx H_{sample} \cdot I_{sample}$, which approximates Equation 1.3 with sufficient practical accuracy. The magnetic path length is measured between the inner sides of the yoke.

If the sample and the yoke were made of the same material, then an error of around 1% could be expected as far as H is concerned. The standard states that reproducibility of the method is at the level of 3% for the measured power loss (IEC, 2010).

The main difficulty of the SST is that the contact surfaces of the yoke (see Figure 2.22b) must be very well made so that no additional air gaps are created. This is difficult to be achieved in practice and special polishing procedures should be employed (Beckley, 2000a).

Also, the top yoke is very heavy and cannot be allowed to rest freely on the sample. There must be a suspension mechanism that ensures that the yoke exerts only a force between 100 N and 200 N on the sample. This makes the SST more expensive to make than Epstein frame.

However, in terms of operation, the SST is faster, less expensive and more convenient than Epstein frame. There is a need only to cut a single sheet, which can be inserted in seconds into the yoke.

2.3.4 SINGLE-STRIP TESTER

Single-strip tester (SsT) is basically the same as SST, but with non-standard dimensions. In order to distinguish these two methods, the first one is sometimes abbreviated with a lowercase 's' (SsT) in order to make it different from the standardised SST.

An example of a non-standard SsT is shown in Figure 2.23. The construction follows the same principles as shown in Figure 2.22, and the measurement method is also identical, including the mutual inductor for air flux compensation.

PRACTICAL COMMENT

The non-standard SsTs are used by researchers for a variety of reasons. In many cases, the percentage change of a given trend is the subject of the investigation and an SsT method is then suitable because the results do not have to be reported with absolute accuracy. And some experiments cannot be conducted with the use of toroidal sample, Epstein frame or SST so other approaches have to be employed.

However, the non-standard SsT yokes have similar tolerances but with smaller surface area of contact. This makes them more prone to problems caused by the additional reluctance because of the air gap between the sample and the yoke, as well as with the precise definition of magnetic path length. As a result, the accuracy of the H signal measurement might be affected by errors. An example of difference between $J–H$ magnetisation curves for the same material but measured with two different yokes is shown in Figure 7.5. It is quite clear that the differences are far from being negligible.

2.3.5 OTHER WATTMETRIC METHODS

A variation of the wattmetric principle can be applied to the measurement of power loss with the help of a balanced or semi-balanced bridge circuit (Majocha et al.,

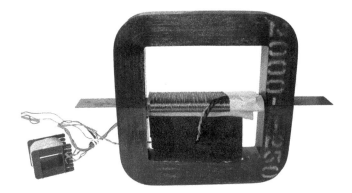

FIGURE 2.23 Experimental (non-standard) single-strip tester for performing measurements on a single strip from Epstein frame. The mutual inductor is visible in the bottom left corner.

2010; Gozdur et al., 2013). The circuit is capable of measuring the power loss from the primary side, which also includes the loss in the primary winding. However, this loss is compensated out with the bridge circuit so that only imbalanced loss is measured and it was shown mathematically and experimentally that the loss measured as the out-of-balance signal of the bridge is directly proportional to the loss of the magnetic sample.

However, there are no reports of using such methods for measurements of 2D or rotational power loss and so they will not be discussed in this book.

2.3.6 ROTATIONAL WATTMETRIC METHOD

Following the wattmetric principle, first Kaplan (Kaplan, 1961) and later Moses and Thomas (Moses et al., 1973) used an apparatus that allowed the generation of rotational magnetisation in a magnetically closed* circuit (Figure 2.24).

The sample/yoke comprised two intersecting paths for the magnetic fluxes. If the two magnetising coils (Figure 2.24) were excited with sinusoidally and cosinusoidally varying currents, then rotational magnetisation was set up at the centre of the intersection.

In such a magnetically closed circuit, the amplitudes of magnetising currents were directly proportional to the amplitudes of H in each 'branch'. Hence, the analogy to the 'classical' wattmetric method shown in Figure 2.19.

The rotating flux was set up only in a relatively small area at the intersection, so a pair of search coils were used to detect the orthogonal x and y components of the flux density. The signals were measured with a frequency analyser and the calculations of losses were based on the specific harmonics (Moses et al., 1973).

PRACTICAL COMMENT

Unfortunately, such arrangement did not allow achieving high values of rotational flux density in the area of interest, and for anisotropic samples, as little as 0.5 T was the upper limit in order to keep the x and y components sinusoidal (Moses et al., 1973).

Also, the uniformity of the magnetisation in the area of interest was questionable. Nevertheless, this experiment showed that the rotational experiments can be approached in yet another way and that the measured values match those obtained from completely different methods.

2.4 FIELDMETRIC

The fieldmetric method became widely used especially with the advent of computer technology. The computing power combined with the ease of acquisition and the processing of several signals even with small amplitudes meant that measurements

---

* It should be noted that all the rotational methods described so far used a setup in which the sample was always separated by an air gap from the magnetising yoke.

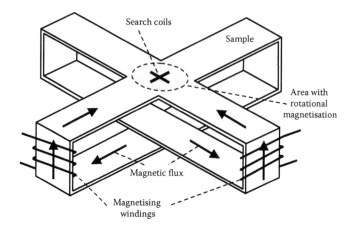

FIGURE 2.24   Apparatus used for wattmetric measurement of rotational loss. (Adapted from Moses A.J. et al., *IEEE Transactions on Magnetics*, 9 (4), 1973, 651.)

of complex waveforms became possible with less difficulty. Also, digital computing allowed performing digital integration of signals thus simplifying analogue and electronic circuitry required for operation.

All mechanical methods required high precision of machining, which is time consuming, expensive and complicates the design and processing. Moreover, moving parts such as bearings wear out with time and controlling all the sources of errors is challenging.

The thermal methods are capable of measuring the dissipated energy in a form of temperature of the sample. But again, the required apparatus is quite cumbersome in use and does not lend itself to quick measurements because of the long time required between each measurement for the whole setup to cool down to ambient temperature. The cooling phase can be several times slower than the heating, so the overall measurement time is rather lengthy.

The fieldmetric method uses no moving parts, mechanical requirements are less difficult to fulfil, the time required for setting up the system for a new sample is much shorted, and the measurements can be processed almost in real time, at least when compared to any of the two previous methods.

But perhaps more importantly, the fieldmetric method when driven by a computer program offers the possibility of precise control of the rotational flux density. This allows achieving much better 'circularity' of rotation, which is not always possible in other methods, at least not for a sample of arbitrary shape.

In the fieldmetric method, the excitation process is decoupled from the measurement of signals. It is true that the measured signals are used for controlling the waveshape of the driving currents, but the actual measurement is, or at least it certainly can be, independent of the excitation system as such.

Boon and Thompson developed an apparatus in which a square sample was used, and the yoke consisted of two paths magnetised in quadrature (e.g. sine in x and cosine in the y channel). The actual measurement was thermometric, but the concept of a two-phase yoke was introduced (Boon et al., 1965).

The wattmetric setup used by Moses and Thomas showed that it was possible to use small search coils for reliable localised measurement of power loss (Moses et al., 1973). It was also evident that both flux density and magnetic field strength can be measured by other sensors, rather than derived from the value of the magnetising current (Ewing, 1900; Brix et al., 1982; Flanders, 1985).

It was almost a natural consequence therefore to combine all these approaches and to produce a 'fieldmetric' system. The sample can have almost any shape; to give a few examples: square (Enokizono et al., 1990a), rectangular (Sasaki et al., 1985), hexagonal (Hasenzagl et al., 1996), circular (Gorican et al., 2000a) and cross-shaped (Nakata et al., 1992).

The general concept is depicted in Figure 2.25. A medium-sized sample (typically sized between 20 mm and 300 mm) is placed in a magnetising yoke, which provides a low-reluctance path for the magnetising field. The magnetising coils can be placed either on the branches of the yoke, or around the sample itself.

Typically, the rotational field is generated by means of a two-phase excitation, but a three-phase generation can also be used (Hasenzagl et al., 1996).

The important part of the fieldmetric method is the system of B and H sensors. These are placed in the region where uniform magnetisation can be achieved, at the centre of the sample.

Typically, there are two pairs of sensors, which detect the following components: B_x, B_y, H_x, H_y, each of which is a function of time. There is a great flexibility in terms of the sensors used.

B is detected by search coils or needle probes, and H by Rogowski–Chattock potentiometers, H-coils, Hall effect sensors or other integrated sensors of magnetic field. The sensors will be discussed in more detail in Chapter 3.

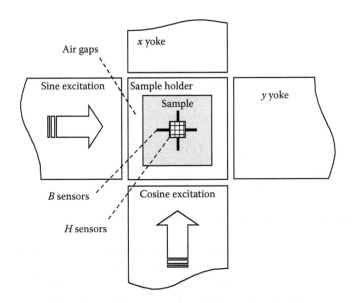

FIGURE 2.25 Fieldmetric method; drawing not to scale.

The rotational loss can be calculated by using the following formula:

$$P_{rot} = \frac{f}{D} \int_0^T \left[\frac{dB_x(t)}{dt} \cdot H_x(t) + \frac{dB_y(t)}{dt} \cdot H_y(t) \right] dt \qquad \text{(W/kg)} \qquad (2.13)$$

We can see from Figure 2.25 that the B and H components are measured locally, for instance, only at the centre of the sample. On the one hand, this allows performing more accurate measurements focused only on the area of interest and where the magnetisation is the most uniform, which is a desirable attribute.

On the other hand, the locality can also be a disadvantage. In many experimental apparatus, the area under test can be comparable to the grain size in grain-oriented electrical steels (around 20 mm, see also Figure 1.23), even if the sample itself is larger. This leads to lack of averaging over many grains, which is an inherent feature of the standardised methods such as Epstein frame (IEC 60404-2) and ring sample (IEC 60404-6).

Such lack of volume averaging is suspected to be a source of additional measurement errors (Zurek et al., 2006b) and still makes the rotational measurements not reproducible enough to be suitable for international standardisation (Sievert et al., 1995).

PRACTICAL COMMENT

There is another version of equation for the fieldmetric method. For purely circular rotating flux density (and only for this condition), it can be shown that the losses could be calculated as

$$P_{rot} = \frac{2 \cdot \pi \cdot f^2}{D} \int_0^T [B_x(t) \cdot H_y(t) - B_y(t) \cdot H_x(t)] dt \qquad \text{(W/kg)} \qquad (2.14)$$

Also, using simple geometry, in any case the instantaneous angle between the B and H vectors (purely circular or not) can be calculated as (Gumaidh et al., 1993a; Hadoud et al., 1997)

$$\phi(t) = \arctan\left[\frac{B_x(t) \cdot H_y(t) - B_y(t) \cdot H_x(t)}{B_x(t) \cdot H_x(t) + B_y(t) \cdot H_y(t)} \right] \qquad \text{(rad)} \qquad (2.15)$$

If digital technology is used, the detection of signals is carried out by a data acquisition device ('analogue inputs' in Figure 2.26), which is also used for synthesising the signals used for magnetic field generation ('analogue outputs'). These signals are fed into a power amplifier, which drives the magnetising coils, which in turn can be wrapped directly on the sample, or on the magnetising yoke.

The whole system can be connected to a computer, which is used for the measurement of signals and calculation of the final result, as well as waveshape

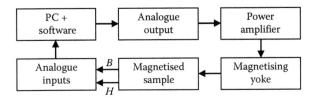

FIGURE 2.26   Block diagram of a typical fieldmetric system.

control. The control can also be based on several different mathematical approaches (as explained in Chapter 4).

The configuration of a fieldmetric system is very flexible. For this reason, a system in each laboratory looks and performs differently, and it would be impossible to describe here all the permutations currently in use by many laboratories. For instance, the sensors do not have to be positioned at x and y directions (0° and 90°), but at some other arbitrary angles. The computer is then used to re-calculate the signals into Cartesian system for further calculations according to Equation 2.13.

Also, post-processing of already-measured data can be applied for the correction of angular errors or other type of data analysis. In any case, computer control allows performing a whole set of measurements, for example, sweeping through a range of flux density amplitudes or frequencies.

But we can see that the underlying principle will be similar for all such systems. At least four signals are measured, and they are used for computing a value of rotational power loss according to Equation 2.13, or similar. It is also worth noting that Equation 2.13 requires the information about B and H values. Hence, there is full flexibility in the type of sensors – provided that the results can be calculated into SI units and used in that equation.

Typical results measured with the fieldmetric method are shown in Figure 2.27. The B loci were controlled to be circular with the characteristic 'butterfly' shapes of H loci. In general, there is some asymmetry present in the H loci, which changes with the direction of rotation (CW or ACW).

Quite high excitation can be achieved under rotational magnetisation. Approaching saturation can be recognised in the H loci by the characteristic 'swelling' of the curves (the B loci are controlled to be circular in this particular case, as in Figure 2.27a). If very high excitation is applied, then the easy direction (in anisotropic electrical steels) can also be saturated so that the whole curve 'swells' ending up somewhat circular (and definitely not 'butterfly' with a waist as in Figure 2.27), as presented in Figure 2.28.

PRACTICAL COMMENT

The sample used to measure the data in Figure 2.28 was a 3% silicon–iron conventional grain-oriented electrical steel and the value of H required to achieve this condition was exceeding 30 kA/m. At such excitation level, the hardest magnetisation direction required lower H than the hard direction. It should be noted that this is in agreement with the curves shown in Figure 1.24. The saturation also seemed to occur around 30 kA/m

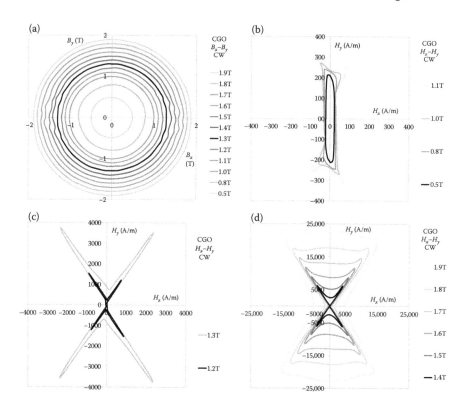

FIGURE 2.27 Typical results for conventional grain-oriented electrical steel, controlled circular B, 50 Hz, CW rotation: (a) circular B loci, and corresponding H loci at: (b) low excitation, (c) medium and (d) high.

(Bozorth, 2003), and for the area marked with an arrow in Figure 1.24, the 'hardest' direction *[111]* requires lower H than the 'hard' direction *[110]*.

More importantly, the H loci in Figures 2.27d and 2.28 exhibit the characteristic 'butterfly' shapes. This can be compared with the hypothetical shapes illustrated in Figure 1.22b – right, which also exhibit 'easy', 'hard' and 'hardest' directions.

The square-like pattern in the Figure 1.22b – left is also experimentally observed for materials with different crystallographic orientation. Some measurement examples for such materials are shown in Chapter 7.

Figure 2.29 shows examples of typical power loss curves measured with the fieldmetric technique (Zurek, 2005a). Comparison of power loss measured with fieldmetric and thermal methods is also shown in Figure 2.18.

The results shown in Figure 2.30 are for the same material as in Figure 2.29, but magnetised under controlled circular H. Interestingly, the peak value of power loss is significantly greater than it is the case for the circular B (compare Figure 2.30 with Figure 2.29).

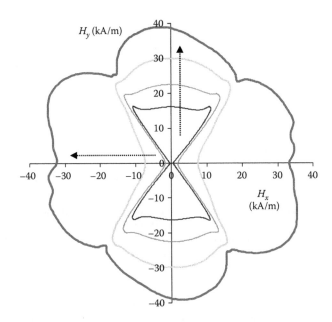

FIGURE 2.28 'Swelling' of *H* loci indicates approaching saturation, measured for conventional grain-oriented electrical steel, under controlled circular *B* up to 2.0 T, at 50 Hz, the arrows indicate directions in which the shape 'swells' the most. (Copyright © 2017 *Przeglad Elektrotechniczny (Electrical Review)*. Reproduced with permission from Zurek S., *Przeglad Elektrotechniczny (Electrical Review)*, R. 93, NR 7/2017, 2017b, p. 13.)

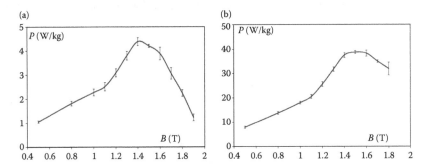

FIGURE 2.29 Rotational power loss measured under controlled circular *B* on sample of conventional grain-oriented electrical steel at 50 Hz (a) and 250 Hz (b); the error bars show reproducibility of the results within the same measurement system. (Adapted from Zurek S., *Two-Dimensional Magnetisation Problems in Electrical Steels*, PhD thesis, Wolfson Centre for Magnetics Technology, Cardiff University, Cardiff, the United Kingdom, 2005a.)

An important disadvantage of the fieldmetric technique is that in the majority of the apparatus devised so far, the area of uniform magnetisation is quite small. A typical size of *B* sensors is on the order of 20 mm (Sievert et al., 1995). This is comparable with a size of a single grain in grain-oriented electrical steels. Larger sensing areas used by the uniformity are questionable if the sample size is not increased at the same time.

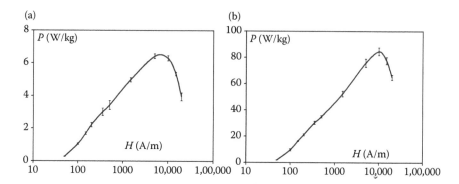

FIGURE 2.30   Rotational power loss measured under controlled circular H on a sample of conventional grain-oriented electrical steel at 50 Hz (a) and 250 Hz (b). (Adapted from Zurek S., *Two-Dimensional Magnetisation Problems in Electrical Steels*, PhD thesis, Wolfson Centre for Magnetics Technology, Cardiff University, Cardiff, the United Kingdom 2005a.)

Therefore, the measurement is quite local and does not allow for the averaging of properties over a large volume of material, which makes the reproducibility of results quite low as compared to the standardised magnetic measurements.

The small measurement area leads to additional errors, namely, that the value of power loss calculated from CW and ACW rotations can differ significantly. It was proved that a large part of this error is caused by angular misalignment between the B and H sensors (Sievert, 1990; Salz et al., 1992; Zurek et al. 2006a,b).

Even a small angular misalignment of less than 0.5° can lead to large negative values (Figure 2.31). The 'true' rotational loss is then found by averaging the CW

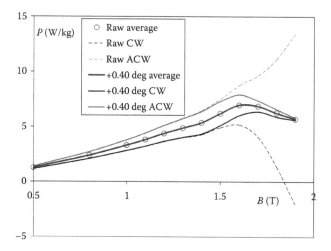

FIGURE 2.31   Rotational power loss measured on a sample of non-oriented electrical steel, without and with angular error. (Adapted from Zurek S., *Two-Dimensional Magnetisation Problems in Electrical Steels*, PhD thesis, Wolfson Centre for Magnetics Technology, Cardiff University, Cardiff, the United Kingdom, 2005a.)

and ACW measurements, which greatly reduces the errors caused by the angular misalignment.

But the errors are not eliminated completely. Intriguingly, even for non-oriented electrical steel, there remain significant differences between CW and ACW rotation (Figure 2.31). This cannot be explained on the ground of grain size, since in the non-oriented steel, the grain size usually does not exceed 0.15 mm, so averaging takes place over hundreds of grains. Therefore, there must be another mechanism for such behaviour. Latest studies by the author indicate that some of these effects could be caused by *H*-coils (Zurek, 2017e). More details are given in Chapter 3.

On the other hand, the fieldmetric technique allows more detailed analysis of the magnetisation processes because much more data can be recorded practically on the fly, under almost arbitrary quasi-static and dynamic conditions.

The fieldmetric technique is very flexible and robust, but still there are factors that limit its absolute accuracy. For instance, thermal measurements have no discernible difference between the CW and ACW measurements (Ragusa, 2013). This is why there is no competition between the various measurement methods, but they can be used to verify results from each other, as shown in Figure 2.18.

Because there is a *B–H* loop for uni-directional magnetisation in the same way, we can also plot B_x–H_x and B_y–H_y loops (Zurek, 2014a). They have rather peculiar shapes (Figure 2.32) that do not lend themselves for straightforward explanation in the same sense as the *B–H* loops for the alternating excitation (see also Figure 1.18).

The rotational *B–H* loops shown in Figure 2.32 do not offer any immediate physical meaning (Hempel, 1992). By definition, the area of the loop is *mathematically* proportional to the value of power loss component calculated from it. But otherwise, the loops change their shapes significantly depending on the level of excitation. The rotational loops can cross themselves multiple times, resulting in 'positive' and 'negative' parts (Figure 2.32). Some authors reported even loops whose areas were completely negative (Sasaki et al., 1985). This is very different from the classical hysteresis loops known for alternating magnetisation.

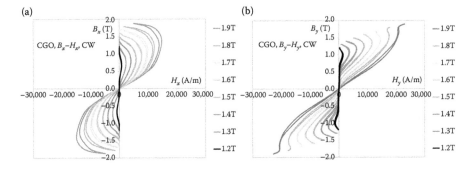

FIGURE 2.32 Rotational results presented as B_x–H_x (a) and B_y–H_y loops (b) for conventional grain-oriented electrical steel at 50 Hz. (Adapted from Zurek S., *Transformers Magazine*, 2 (2), 2015d, 44.)

However, it can be shown that Equation 2.13 can be mathematically split into two components – for directions x and y. Indeed, sometimes total loss is reported (Sasaki et al., 1985; Zurek, 2014), which is by definition equivalent to Equation 2.13 as

$$P_{rot} = P_x + P_y \quad \text{(W/kg)} \tag{2.16}$$

where

$$P_x = \frac{f}{D} \cdot \int_0^T \left(\frac{dB_x}{dt} \cdot H_x \right) dt \quad \text{(W/kg)} \tag{2.17}$$

$$P_y = \frac{f}{D} \cdot \int_0^T \left(\frac{dB_y}{dt} \cdot H_y \right) dt \quad \text{(W/kg)} \tag{2.18}$$

The split P_x and P_y components can be then averaged from CW and ACW measurements (Zurek, 2014), in order to eliminate the angular errors mentioned above:

$$P_{x,avg} = \frac{1}{2} \cdot (P_{x,CW} + P_{x,ACW}) \quad \text{(W/kg)} \tag{2.19}$$

$$P_{y,avg} = \frac{1}{2} \cdot (P_{y,CW} + P_{y,ACW}) \quad \text{(W/kg)} \tag{2.20}$$

A very interesting picture emerges with the losses plotted as $P_y = f(P_x)$. Figure 2.33 shows an example for conventional grain-oriented electrical steel, but similar shapes

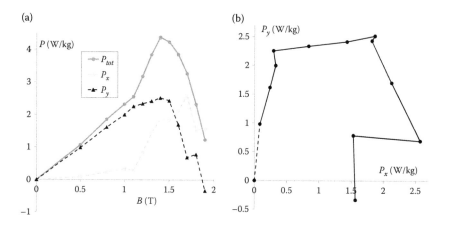

FIGURE 2.33 Conventional grain-oriented electrical steel, $P_{tot} = P_x + P_y$ measured under controlled circular B: (a) $P = f(B)$, (b) $P_y = f(P_x)$. (Adapted from Zurek S., *IEEE Transactions on Magnetics*, 50 (4), 2014, 1.)

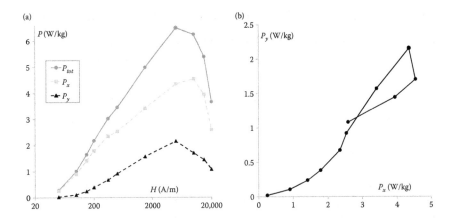

FIGURE 2.34 Conventional grain-oriented electrical steel, $P_{tot} = P_x + P_y$ measured under controlled circular H: (a) $P = f(B)$, (b) $P_y = f(P_x)$. (Adapted from Zurek S., *IEEE Transactions on Magnetics*, 50 (4), 2014, 1.)

are obtained for other anisotropic electrical steels (Zurek, 2014) magnetised with circular B. Figure 2.34 shows results measured under controlled circular H.

Interestingly, on a qualitative level, the $P_y = f(P_x)$ curves for various electrical steels seem to take only one of two types, either 'I' (slanted) or 'O' (looped), as depicted in Figure 2.35. The looped curves occur for anisotropic electrical steels (conventional grain-oriented, high-permeability grain-oriented, double grain-oriented) when magnetised under controlled circular B. However, when the same materials are excited by circular H, the curve takes the shape of the 'I' type, which is also the case for isotropic steels, regardless of the magnetisation regime.

The type 'O' curve seems to have four very distinct regions. The transition from one region to another coincides with the changes of the peculiar B–H rotational loops (Zurek, 2014). Therefore, splitting the loss in the P_x and P_y components gives the opportunity for deeper study of the phenomenon of rotational power loss. Such in-depth analysis is not possible with the other two techniques: mechanical or thermal, both of which measure the total loss rather than its components.

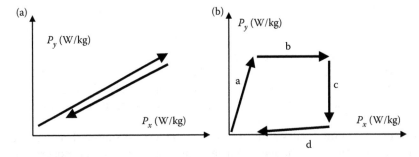

FIGURE 2.35 Two types of $P_y = f(P_x)$ curves: (a) – 'I' (slanted), (b) – 'O' (looped). (Adapted from Zurek S., *IEEE Transactions on Magnetics*, 50 (4), 2014, 1.)

2.4.1 NON-CIRCULAR AND NON-ROTATIONAL 2D FIELDMETRIC MEASUREMENTS

Because of the robustness of the fieldmetric apparatus, many researchers studied 2D properties not only in non-circular but also in non-rotating, arbitrary 2D magnetising conditions. For instance, in order to produce a rotating field, the magnetising coils in the x and y directions would be fed with currents shifted by 90° (e.g. sine and cosine). But if both magnetising coils are energised by currents that are in phase, then the resulting field does not rotate, but rather alternates in a direction dictated by the amplitudes of both currents. The control of the waveshape, signal processing and calculation of power loss remains practically the same as for the ordinary rotational fieldmetric method.

Therefore, such a setup allows quick and inexpensive investigation of directional properties – something that previously had to be done by either using the torque magnetometer or by cutting many samples at specific angles, which is quite time consuming and rather expensive (Soinski, 1984a, 1987; Emura et al., 2001; Zurek et al., 2009a).

Such results were reported, for instance, at the first 2DM workshop in 1992, measured under rotating elliptical (Enokizono, 1992) and non-rotating conditions (Ishihara, 1992).

The main difficulty with such investigations is the large number of possible permutations of 2D excitation. These are often referred to as 'vector properties'. There is a plethora of obtainable results because the excitation can be applied with various degrees of ellipticity, with non-uniform speed of rotation, with arbitrary variation of magnitude, alternating in any direction, a combination of all of the above with or without DC bias and so on (Sakata et al., 2003; Matsuo et al., 2009).

The measurements with a DC bias are especially difficult to interpret because all B sensors relying on the law of induction are intrinsically incapable of detecting the DC bias in B (see Equation 1.14). Such bias is therefore much more difficult to measure.

Some authors also investigated 'vector properties' at liquid nitrogen temperatures (Miyamoto et al., 2011). Hence, any coherent and holistic interpretation of such results is not possible, or at least it has not been carried out to date. However, there are a few exceptions.

Figure 2.36 shows a curve demonstrating how the values of power loss increase from alternating magnetisation ($B_y/B_x = 0$) to rotational circular ($B_y/B_x = 1$), and beyond. The increase is around 2.5 times, and is typical for other reported materials, for example, electrical steels (Zhu et al., 1997b; Zurek, 2000).

More directional measurements were reported, for instance, in Ishihara (1992) and Gorican et al. (2002), as well as by the author (Zurek, 2005a; Zurek et al., 2009a). Such results seemed suggesting that there could be significant asymmetry of anisotropy because of the visible differences between positive and negative angles (Figure 2.37 shows typical curves).

However, careful analysis of the systematic errors shows that such large asymmetry is caused by the measurement setup, and is not a real property of the material under test.

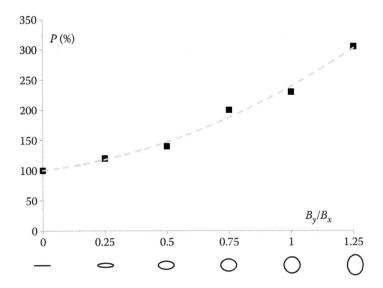

FIGURE 2.36 Changes of relative power loss P with degree of ellipticity (ratio B_y/B_x) measured at 50 Hz on a sample of Megaperm (40% Ni–Fe alloy). (Adapted from Zurek S., *Rotational Magnetisation in Flat Soft Magnetic Materials (in Polish: Przemagnesowanie obrotowe w materialach magnetycznie miekkich plaskich)*, MSc thesis, Czestochowa University of Technology, Poland, 2000.)

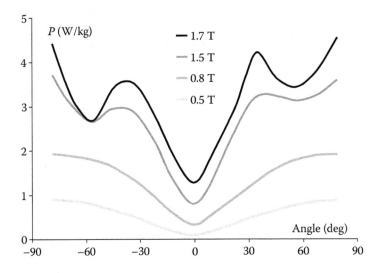

FIGURE 2.37 Anisotropy of power loss, alternating magnetisation in a round 2D tester, 50 Hz, conventional grain-oriented steel, asymmetry reaches 30% at higher excitation. (Adapted from Zurek S. et al., *Przeglad Elektrotechniczny (Electrical Review)*, R. 85 (1/2009), 2009a, 16.)

Under rotational magnetisation, the results of the power loss value are averaged from CW and ACW rotation. It was conclusively shown that such averaging compensates out most of the errors introduced by angular misalignment of the sensors. But with directional alternating measurement (i.e. non-rotating), there is only one result at a given angle and averaging cannot be carried out. Worse still, by definition, the measurement is quite local and volume averaging from multiple grains also does not take place. As a result, the asymmetry can reach 30% or more (Zurek et al., 2009a; Gorican et al., 2002).

However, this is misleading because the material does not actually exhibit such high magnitude of asymmetry. This was verified by the author by using a fairly expensive and very time-consuming approach of cutting many samples in all directions, including 'positive' and 'negative' inclination angles (Figure 2.38a). Each sample had dimensions of a single strip compatible with the Epstein frame method, that is, 30 × 305 mm. The magnetic properties were measured in a non-standard SsT (see also Figure 2.23). There were at least three samples for each direction, so the final result was averaged from three measurements.

The repeatability of the method is demonstrated in Figure 2.38b, for the 1.3 T (top curve). The solid line shows the averaged values, whereas the circular full symbols show the actual measured values for three samples. The influence of the hardest magnetising direction (in this case cut at 55°, see also Figure 1.22) is visible as a 'kink' in almost all curves. But the values for all other angles are much more symmetrical than suggested by the fieldmetric method (Figure 2.37). At increased frequencies, the curves became even more symmetrical (Zurek et al., 2009a).

The results proved conclusively that the material is symmetrical to within repeatability of the method, and certainly as high differences as 30% do not take place. Hence, such results should be treated with caution and at the very least measurements for 'positive' inclination angles should be averaged with those from 'negative' angles.

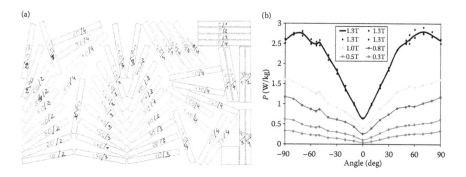

FIGURE 2.38   Verification of asymmetry of power loss anisotropy on single-strip test measurements: drawing of sample strips laser cut out from a large sheet (a) and measurement results where each point is averaged from three samples (b); conventional grain-oriented electrical steel, controlled sine *B*, 50 Hz. (Copyright © 2009 *Przeglad Elektrotechniczny (Electrical Review)*. Reproduced with permission from Zurek S. et al., *Przeglad Elektrotechniczny (Electrical Review)*, R. 85 (1/2009), 2009a, 16.)

2.4.2 THREE-DIMENSIONAL FIELDMETRIC MEASUREMENTS

Magnetic cores of rotating machines are usually made of laminations. Therefore, the magnetisation is limited mostly into in-plane 2D components, which are easier to study and comprehend. However, even the relatively well-known effects do not lend themselves for easy inclusion into the design process of magnetic cores of motors, generators and transformers.

Three-dimensional measurements became more important with the advent of soft magnetic composites. The magnetic cores could be manufactured as whole blocks, without any laminations. This allows for the flux to flow in any direction, as it results from the local magnetising conditions.

Still, there are limited publications referring to such measurements. They do not seem to be very popular and their incorporation into the design of magnetic cores is even more difficult than it is the case for the 2D data.

Hadoud used 2D rotational system, with an extra yoke to apply a DC excitation in the third direction (Hadoud et al., 1997; Hadoud, 1998). The approach was therefore essentially the same as that shown in Figure 2.25, but with one more yoke for the independent DC magnetisation.

The results implied that the rotational loss decreases with the DC field applied in such a way. However, the uncertainty of the method was rather large so it is difficult to judge if the actual reduction effect has taken place. No further research was published so far with a similar experimental setup to verify the findings.

A lot of work on 3D magnetisation was carried out by Zhu and his colleagues (Zhu et al., 2003, 2009; Zhong et al., 2005; Li et al., 2010, 2011; Zhang et al., 2016).

Again, the magnetising apparatus had similar features as for 2D fieldmetric measurements. The sample was cubic, and the magnetising yoke provided a low-reluctance path in each direction. The magnetising coils were wound on the yoke, as shown in Figure 2.39. If the part of the yoke responsible for magnetising in the third dimension is removed (or just not excited), then the setup is similar to a 2D setup shown in Figure 2.25.

Also, the same sensors were employed: search coils (B-coils) for B- and H-coils for H detection. However, owing to the significant volume of the sample, there were 12 H-coils in order to measure the field on each side of the cubic sample and for two orthogonal directions, and for each component, the H waveform was averaged from both sides of the sample.

PRACTICAL COMMENT

It should be noted that in the case of samples made from ferrite or soft magnetic composites, it is not possible to use the needle probes. Such materials have very high-electrical resistivity and thus the needle probe method would simply not work.

Of course, in the 3D setup, the use of needle probes would be very cumbersome mechanically, but this is a different problem from the fact that the needles simply would not function correctly.

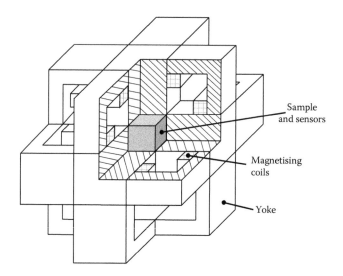

FIGURE 2.39 An example of a 3D magnetising yoke. (Adapted from Zhu J.G. et al., *IEEE Transactions on Magnetics*, 39 (5), 2003, 3429.)

A similar equation was used for power loss calculation, but naturally all three components were employed: *x*, *y* and *z* (Zhu et al., 2003), rather than just *x* and *y* as in Equation 2.13. The equation therefore becomes (Li et al., 2011)

$$P_{rot} = \frac{f}{D} \cdot \int_0^T \left(\frac{dB_x}{dt} \cdot H_x + \frac{dB_y}{dt} \cdot H_y + \frac{dB_z}{dt} \cdot H_z \right) dt \quad \text{(W/kg)} \quad (2.21)$$

The setup was controlled by a computer and the generation of very complex arbitrary 3D magnetisation waveforms was possible. The difficulty then becomes how to interpret the data and make it useful for modelling, loss prediction or core design.

The signal processing itself is quite complex because as many as 12 sensors might be required (6 for each *B* and *H*) for optimum capture of the required information (Li et al., 2010). All these signals have to be measured simultaneously.

3D measurements are capable and do generate even more data than it is the case for 2D as shown in Figures 2.40 and 2.41. Researchers still struggle with the complete understanding of 2D measurements and analysis, so stepping the analysis up to three dimensions is even more difficult.

2.5 OTHER MEASUREMENTS RELATED TO ROTATIONAL MAGNETISATION

Besides saturation flux density and permeability, rotational loss is also an important parameter from a practical viewpoint. After all, reduction of losses leads to designing more efficient rotating machines.

However, there are some other properties related to rotational magnetisation: magnetic domain observation, magnetostriction and Barkhausen noise, to name

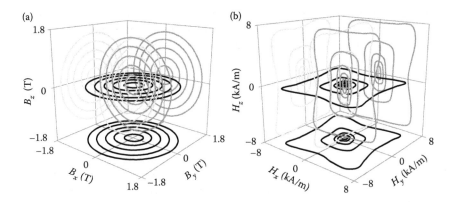

FIGURE 2.40 Typical controlled circular *B* loci (a) and corresponding *H* loci (b) under 3D magnetisation for a soft magnetic composite material Somaloy 500, at 50 Hz. (Adapted from Li Y. et al., *IEEE Transactions on Magnetics*, 47 (10), 2011, 3520.)

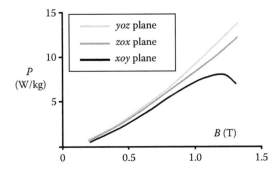

FIGURE 2.41 Rotational power loss for all 2D planes for Somaloy 500 at 50 Hz. (Adapted from Li Y. et al., *IEEE Transactions on Magnetics*, 47 (10), 2011, 3520.)

just three. Also, as described above, the measurement of 'vector properties' was attempted by some researchers.

The link between such additional measurements and practical applications is perhaps weaker, although magnetostriction is important for acoustic noise, and vector properties can be useful for computer modelling of rotational and 2D magnetisation. Nevertheless, all such studies enrich the whole picture and help in a better understanding of all interconnected phenomena occurring during rotational magnetisation. And better understanding of all involved factors can lead to designing more efficient magnetic materials.

These auxiliary measurements will be treated here only briefly, especially that there are far fewer publications on these topics pertaining directly to rotational loss.

2.5.1 Magnetic Domain Observation

Magnetic domain studies were undertaken by several researchers. However, it is difficult to briefly digest the results because naturally most of the input data are based on images, which need a great deal of interpretation for drawing any conclusions. And even if very good images are produced, great care is still required for analysis because correct conclusions can be sometimes counterintuitive (Hubert et al., 1998; Xu et al., 2011).

Such studies are often more qualitative rather than quantitative, although sometimes there are attempts to calculate values measured in numbers or units. Therefore, they usually referred to as 'domain observations' or 'domain studies' rather than 'domain measurements'.

Domain observations are often carried out under simplified or idealised conditions, as, for instance, for single-crystal samples, for which logical interpretation of the results is easier. For commercially available materials, the domain patterns are sometimes so complex (see Figure 1.9) that visual interpretation becomes almost impossible, certainly not at a *quantitative* level. On the other hand, in highly anisotropic materials, the domain patterns differ significantly depending on the plane in which the magnetisation is carried out. Numerous papers were written by physicists, which prepared the samples by cutting them in appropriate plane from a large single crystal. Such studies are of course fundamental from a physical point of view, but linking them with what happens in commercial materials is usually not straightforward.

Some techniques for domain observations require meticulous sample preparation because the surface must be polished to a high degree of flatness (mirror-like finish). This might significantly change the distribution of mechanical stresses, even if annealing is carried out (because the outer coating will be removed by definition). The observation of magnetic domains under rotational magnetisation uses the same techniques as under alternating magnetisation.

Nevertheless, domain observations do offer yet another angle of analysis of the complex problem of rotational magnetisation. The interpretation of domain images can be linked to the shape of the rotational $B–H$ loops, whose peculiar shapes can be attempted to be explained on the basis of possible domain images (Fiorillo, 1992b).

However, the actual research of rotational domain patterns is more difficult. Under rotational magnetisation, the domains can form very complex structures, which are not easy to show as images in the available scientific publication formats, nor are they easy to interpret and understand. Even such prosaic factors as resolution of the printed pictures can have degrading influence on the legibility of the domain images, especially for older publications (Phillips et al., 1974; Narita et al., 1974). Therefore, comparison of the obtained results by other researchers is less feasible than, for instance, in the case of power loss measurement for which numerical data are much easier to be analysed and verified.

The papers published so far offer multiple explanations, not always forming a coherent overall picture, and often without definite succinct conclusions – apart from a few studies that work on simplified cases or idealised materials.

Some of the first theoretical studies on magnetic domains were published by Ewing. His experimental setup is shown in Figure 1.36 and some of the results in Figure 1.35 (Ewing, 1900). His work was carried out only under uni-directional magnetisation. However, using Ewing's model of 'molecular magnets', Swinburne showed that rotational loss should decrease to zero under high-rotational excitation

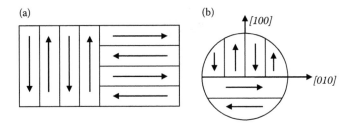

FIGURE 2.42 Simplified domain models. ((a) Adapted from Kornetzki M. et al., On the theory of hysteresis loss in rotating magnetic field (in German: *Zur Theorie der Hystereseverluste im magnetischen Drehfeld*), *Zeitschrift fur Physik, Bd.*, 142, 1955, 70. (b) Adapted from Zbroszczyk J. et al., *IEEE Transactions on Magnetics* 17 (3), 1981, 1275.)

(Ewing, 1900). This was therefore quite an important finding, which allowed further progress in the rotational studies.

Later, a simple model of domain structure was proposed, comprising two crystallites with 90° and 180° domains (Figure 2.42). Kornetzki and Lucas showed that under low-magnetic field, the rotational losses are around two times higher than the alternating losses (Kornetzki et al., 1955). However, even then it was recognised that this simple interpretation was expected to be incomplete. A different study used a similar model, again suggesting that it is correct only when reversible wall motions take place (i.e. at low excitation) (Zbroszczyk et al., 1981). Nevertheless, the 'factor of 2' increase at low excitation was confirmed experimentally (Boon et al., 1964; Fiorillo, 1992b).

Carrying out actual magnetic domain observations is difficult in practice due to the required painstaking sample preparation, or cost of the materials (e.g. monocrystals). Real-time domain observations under uni-directional magnetisation was reported only fairly recently (Moses et al., 2005), and at the time of writing this book, no real-time dynamic rotational images were published (apart from the stroboscopic technique).

The whole volume of ferromagnetic material is occupied by domains, which are separated by domain walls (see also Figure 1.8). The total energy of the system is minimised by the presence of domain structure so that almost all magnetic energy is contained inside of the material.

However, the domain walls represent magnetic discontinuities, and magnetocrystalline energy of the atomic structure might be forcing the main domains to be not exactly parallel with the surface, for instance, when the grains are misoriented. For these reasons, there can be some small leakage field on the surface, and such leakage creates field gradient.

Static domain images can be observed by using a colloid of magnetic particles or the so-called Bitter pattern method (Boon et al., 1964; Hubert et al., 1998). The Bitter colloid is a special suspension of small ferromagnetic particles in a liquid – the particles must have the right size, the liquid must be of appropriate density, drying out must be prevented, etc. It is also possible to use the technique in a form of 'domain viewer', in which the liquid is enclosed between thin and transparent membranes (Brockhaus Measurements, 2017). An external magnetic field can be

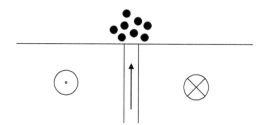

FIGURE 2.43   Leakage field at the surface attracts particles in the Bitter method. (Adapted from Hubert A. et al., *Magnetic Domains: The Analysis of Magnetic Microstructures*, Springer, 1998, ISBN 9783540641087.)

applied (perpendicular to the investigated surface) in order to increase the leakage field, thus enhancing the contrast (Xu et al., 2011).

If such a colloid is applied to the surface of a ferromagnetic material, then the particles are attracted towards a higher gradient of magnetic field. A schematic example is shown in Figure 2.43.

The actual patterns observed by this technique depend on several factors. The attraction of the particles is created by the leakage field and not by the presence of domains or domain walls as such. Therefore, it is possible to see the concentration just at the domain walls (Zurek et al., 2009c), as well as over the whole domains (Xu et al., 2011).

For this reason, the interpretation of obtained domain images should be done carefully. For instance, as seen in Figure 2.44, the contrast of the pattern swaps in many places when crossing a grain boundary, which could be perceived that domains of opposing polarities are present on both sides of such a grain boundary. This might not be the case necessarily, and observations with a different technique reveal that domains are perfectly capable of crossing through a grain boundary almost completely unimpeded (Hubert et al., 1998; Xu et al., 2011).

Of course, the particles would also be attracted to any other source of leakage field. The surface of commercially available grain-oriented electrical steels is

FIGURE 2.44   Picture from a domain viewer; in this particular case, the visible lines are more akin to domain walls rather than domains. (Copyright © 2009 *Przeglad Elektrotechniczny (Electrical Review)*. Reproduced with permission from Zurek S. et al., *Przeglad Elektrotechniczny (Electrical Review)*, R. 85 (NR 1/2009), 2009c, 111.)

(a)   (b)

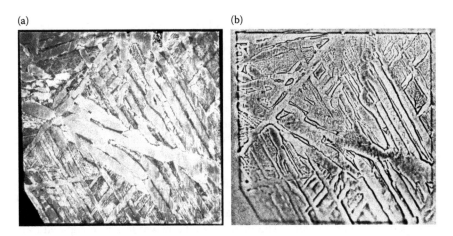

FIGURE 2.45 Etched surface of a sample of Muonionalusta meteorite (a) and patterns created in the domain viewer (b) the patterns reflect only roughness of the surface rather than the magnetic domains.

usually smooth enough so that the domains or at least domain walls can be detected with a suitable viewing system.

However, if the surface is too rough, then this technique is unsuitable. An example is shown in Figure 2.45. Samples of the meteorite* Muonionalusta are often cut and etched to show the Widmanstatten pattern, either for exhibition or decoration purposes. There are multiple very long but narrow grains (up to 50 mm and longer). The material is ferromagnetic due to the high content of iron (>90%) and nickel (>8%). However, the high-surface roughness (easily felt with a fingertip) due to etching leads to much stronger field leakage than any effects associated with the presence of magnetic domains. As a result, only the surface roughness is visible, and not the domains.

Notwithstanding, the Bitter technique is very useful and was successfully deployed by many researchers. The interpretation of typical results for (100) plane of single crystal of Si–Fe is shown in Figure 2.46.

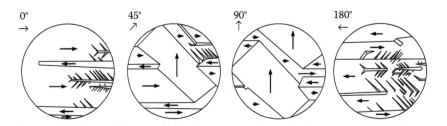

FIGURE 2.46 Domain images with a constant magnetic field applied at various directions in the (100) crystallographic plane. (Adapted from Boon C.R. et al., *Proceedings of IEE*, 111 (3), 1964, 605.)

---

* Indeed, this example is not very relevant for electrical steels but it illustrates the effect and the problems caused by surface roughness in such studies.

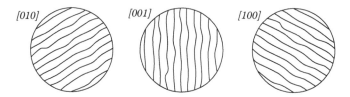

FIGURE 2.47   Domain images with a constant magnetic field applied at various directions in the (111) crystallographic plane. (Adapted from Boon C.R. et al., *Proceedings of IEE*, 111 (3), 1964, 605.)

Interestingly, such domain images are not really dissimilar from those proposed in the very simplistic 90°/180° models (e.g. compare the images for 45° and 90° from Figure 2.46 to that in Figure 2.42). In all cases, there is a significant amount of parallel and perpendicular domains. The increase in the power loss as compared to the alternating magnetisation was experimentally confirmed to be around the factor of 2 (at low excitation).

On the other hand, quite intriguing were the domain patterns for the (111) plane. Owing to magnetocrystalline anisotropy, the domain patterns took only one of the three possible configurations (Figure 2.47). The transition between each of the pattern was described as a 'jump' towards any of the easy directions *[100], [010]* or *[001]*. No intermediate patterns were reported.

Such domain patterns were very regular and equally spaced in all three positions. An estimation of wall energy was made and assuming that each re-arrangement caused the dissipation of the whole stored energy, a calculation was made, which gave as result a factor of around 8, as compared to alternating power loss. Again, this was in agreement with actual measurements taken on these samples (Boon et al., 1964).

Figure 2.48 presents domain patterns published in Zbroszczyk et al. (1981). When the excitation is parallel to the easy direction, the domains resemble bar domains, which are well known from studies of alternating magnetisation. But when the excitation is applied at a different angle, a much more complex pattern emerges. As confirmed experimentally by the torque method, this coincides with the increase of loss (Zbroszczyk et al., 1981).

Another important technique for domain observation under rotational magnetisation is the longitudinal Kerr effect (Radley, 1992). A beam of polarised light is reflected from a magnetised surface. If the surface is magnetised, then the polarisation of the light is twisted proportionally to the intensity of the field and the direction of magnetisation (Figure 2.49). If the magnetisation is reversed (e.g. in the neighbouring domain), then the twisting occurs in the opposite angle.

The light is passed through a series of various optical devices and filters in order to improve the contrast, which is very poor to start with. The images are further processed digitally (Moses et al., 2005), especially for more 'challenging' materials.

But before the advent of computer technology, only 'analogue' images could be taken. They had to be reproduced for publication usually from hard copies. In order to facilitate better clarity, the images were often shown in two versions – the original image (e.g. a photograph) and the schematic drawing of what the image shows.

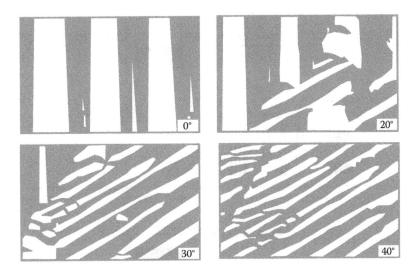

FIGURE 2.48 Domain patterns in Fe–Si3.25% single crystal, in (011) plane, with external $H = 15.84$ kA/m applied at various angles to the *[100]* direction. (Adapted from Zbroszczyk J. et al., *IEEE Transactions on Magnetics* 17 (3), 1981, 1275.)

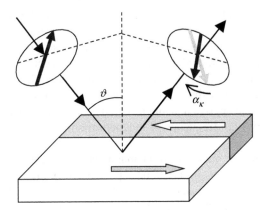

FIGURE 2.49 In the longitudinal Kerr effect, polarisation of the beam of light is rotated by an angle α_k after reflection from a magnetised surface. (Adapted from Hubert A. et al., *Magnetic Domains: The Analysis of Magnetic Microstructures*, Springer, 1998, ISBN 9783540641087.)

This practice is also used to date with more complex images, or when the authors attempt 3D analysis of the investigated domain structure.

In this method, the surface must be meticulously polished to a high degree of smoothness (referred to as 'mirror finish') in order to reduce scattering of the light and improve contrast. The angle by which the beam is twisted is very small and resolution as high as angular seconds (0.001°) is required to guarantee appropriate contrast (Neonark, 2013). In order to attain such selectivity, a quite complex optical system is required (see Figure 2.50).

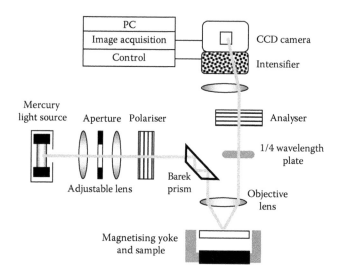

FIGURE 2.50 An example of a Kerr effect observation system. (Adapted from Moses A.J. et al., *IEEE Transactions on Magnetics*, 41 (10), 2005, 3736.)

Kerr microscopy can be used to observe static as well as dynamic images. Real-time observation is certainly the most difficult and technologically became available only recently, so some researchers deployed stroboscopic imaging in order to capture a given domain wall configuration under dynamic conditions (Phillips et al., 1974; McQuade et al., 1995). Qualitatively, the actual dynamic results did not seem to differ significantly from those obtained at static or quasi-static magnetisation. Using the stroboscopic approach, some research succeeded in domain observations and analysis at very high frequency of magnetisation, for example, up to 100 kHz (Magni, 2014).

For highly anisotropic materials, if the magnetisation direction coincides with the easy direction, then bar domains are to be expected because the conditions are identical with alternating magnetisation, especially for low-rotational frequency. Also, with no applied excitation, the domain walls will be oriented along certain preferred crystallographic directions, in order to minimise the total energy of the system.

However, increasingly complex domain configurations are created by rotating or applying the magnetising field away from the easy direction. All this depends on the crystallographic structure of the sample and the direction in which the sample was cut. Several authors investigated anisotropic samples cut at various angles from a large single crystal of silicon iron. Thus, the magnetisation could be applied in a plane of choice, rather than being limited to, for instance, the (110) plane.

For commercial grain-oriented steels, the pictures obtained by several researchers show a certain amount of similarity. When there is no excitation or the vector of applied magnetising field follows the rolling direction (0° inclination), then bar domains are indeed observed (Figure 2.51). When excitation is applied along a different direction, there will be some smaller 'transverse' domains appearing, especially around grain boundaries and other pinning sites.

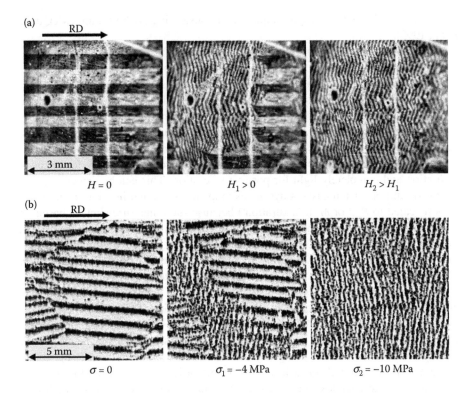

FIGURE 2.51 Static domain structures on a sample of grain-oriented electrical steel under (a) DC magnetisation along transverse direction and (b) compressive stress along the rolling direction. (Courtesy of Piotr Klimczyk, formerly of Wolfson Centre of Magnetics, Cardiff University, Cardiff, the United Kingdom. Copyright © 2009 *Przeglad Elektrotechniczny (Electrical Review)*. Reproduced with permission from Klimczyk P. et al., *Przeglad Elektrotechniczny (Electrical Review)*, R. 87 (No. 9b/2011), 2011, 33.)

An example for excitation applied to a sample of grain-oriented steel along a transverse direction (i.e. 90° to the 'easy' direction, vertically in Figure 2.51) is shown in Figure 2.51a. The surface of the sample was prepared by a magnetic disc polisher (around 1 μm roughness) and the domains were observed with the magneto-optic microscope described in Moses et al. (2006a).

Without the applied field, there are 'bar domains', and the domain walls follow the easy direction. With increasing amplitude of excitation, more volume of the sample is occupied by domains and domain walls, which coincide with the direction of the applied magnetising field (Klimczyk et al., 2011; Klimczyk, 2012).

Similar results were reported by other researchers using the Kerr technique. There were different angles at which the 'transverse' domains were appearing, but certainly the results were comparable at the qualitative level (Radley, 1992). For instance, excitation applied at various angles but with the same amplitude can show that bar domains remain clearly visible up to around 25°, and then the amount of the 'transverse' domains grows rapidly, to completely take over the whole surface at 40°, so the changeover is quite rapid when considering 360° rotation (McQuade et al., 1995).

Of course, there are many other domain observation techniques (Hubert et al., 1998; Chikazumi, 2002), but these are less applicable to studies of rotational magnetisation in electrical steels or similar commercial materials; hence there are practically no publications giving details of such research connected to rotational magnetisation.

Under alternating magnetisation, the investigated domain image might have only well-defined bar domains (e.g. Figure 2.51a for $H = 0$). During magnetisation, the domain walls move and the domain widths change, so that the 'dark' and 'light' areas have different balance, for instance, becoming completely dark for negative saturation and completely light for positive saturation. This can be detected *quantitatively*, and on such a basis, a hysteresis loop can be plotted in arbitrary units (or at best scaled to B, J or M at 'saturation') because such 'measurement' is based on detection of areas. Owing to its relative simplicity, such observation technique can be used for investigation of many materials, from nanowires to data storage surfaces (Konishi et al., 1970; Kryder et al., 1990; Kottler et al., 1998; Qiu et al., 1999; Morales et al., 2002). The technique is described in more detail in Section 3.1.9.

However, the simplicity is also quite limiting. Most apparatuses using the Kerr effect rely on a single beam of light, shone in one direction. The hypothetical condition of full saturation leads to a single domain rotating under rotational magnetisation. So the magneto-optical image would just vary in a sinusoidal way between dark and light extremes, without giving any information as to the direction of the single domain. This would call for at least two beams of light, which would make such setup much more complex (see Figure 2.50 for just a single beam). The situation is even worse at intermediate levels of excitation. As shown above, the domain patterns under rotational magnetisation can be very complex, and the attainable contrast levels are not always conducive to quantitative image analysis, even with the help of the best digital processing techniques. For these reasons, such quantitative studies are very difficult to carry out in practice for rotational magnetisation, and hence there are no published results in the literature.

PRACTICAL COMMENT

The stroboscopic technique relies on capturing images at the same instance of time for each subsequent magnetisation cycle. This allows averaging from many frames and thus improving the quality and resolution of the captured image. However, it is obvious that the cycle-to-cycle variations cannot be analysed with such a method. This is why the proper real-time domain analysis could be conducted only recently because in the past there was no equipment capable of gathering the data with sufficiently good signal-to-noise ratio.

2.5.2 MAGNETOSTRICTION AND MECHANICAL STRESSES

The domain images shown in Figure 2.51b are obtained under compressive stress, but without any magnetising field applied. However, by comparing the images with those from Figure 2.51a, it can be seen that there is a large degree of similarity between stress and applied magnetisation. On the other hand, it is known that magnetisation of a sample causes magnetostriction, but only if the so-called 90° domains are involved.

The analysis of magnetostriction can be carried out with connection to the presence of mechanical stresses within the body of the sample under test. In a classical mechanical problem, the behaviour of the system can be described by Hooke's law, which states that the strain ε (i.e. relative deformation) is directly and linearly proportional to the applied stress (compressive or tensile):

$$\varepsilon = \frac{\sigma}{E_Y} \quad \text{(unitless)} \tag{2.22}$$

where E_Y is Young's modulus of elasticity (Pa) and σ is the applied stress (Pa).

So knowing Young's modulus of the sample (material constant) and measuring the *strain*, it is possible to calculate the actual *stresses* in the sample, assuming that no *plastic* deformation takes place.

The stress can be caused by external factors, for instance, during the assembly of a transformer core. It can also be caused by internal forces arising inside the material due to the presence of magnetic domains.

Magnetostriction is a phenomenon in which the physical dimensions (e.g. length) change due to the applied magnetisation, which causes internal stress. Magnetostriction λ is usually measured as the ratio of the change of the given dimension Δl to the actual dimension l of the sample:

$$\lambda = \frac{\Delta l}{l} \quad \text{(m/m) or (unitless)} \tag{2.23}$$

For electrical steels and most practical magnetic materials, the values are very small, typically on the order of 50 µm/m ($50 \cdot 10^{-6}$ m/m) or less.[*] This means that for an Epstein frame strip, which is typically 300 mm long, the expected change in length would not exceed 0.015 mm.

It is generally accepted that the magnetostriction is 'positive' if the material elongates in the same direction as the applied magnetisation. Conversely, if the material shortens in the given direction, the magnetostriction is 'negative' (Figure 2.52). The assumption is that the volume magnetostriction is negligibly small, which allows a considerable simplification of measurements and analysis (Bozorth, 2003). This means that some dimensions can change, but the overall volume of the material remains practically constant. Volume magnetostriction does actually occur, but for most materials, it is usually much smaller than 'linear' magnetostriction and indeed in most cases, the assumption is fairly 'safe'.

---

[*] There are exceptions, for instance, Terfenol D used in magnetostrictive actuators exhibits the so-called 'giant' magnetostriction and might change its dimensions up to $2000 \cdot 10^{-6}$ m/m = 2 mm/m.

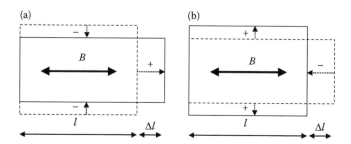

FIGURE 2.52   Positive (a) and negative (b) magnetostriction.

PRACTICAL COMMENT

The value of strain or magnetostriction is relative. The 'units' can be expressed in multiple ways by different authors. For example, all numbers listed below denote exactly the same numerical value:

- $1 \cdot 10^{-6}$ (without units) (Fiorillo, 2004)
- $1 \cdot 10^{-6}$ m/m (Ebrahimi et al., 2013)
- $1\ \mu\text{m/m}$ (Somkun, 2010)
- 1 ppm (parts per million) (Pfutzner et al., 2011)
- 1 microstrain (Beckley, 2000a)
- $1\ \mu\varepsilon$ (microstrain) (Phway et al., 2008)

The magnetostriction is caused by changes in magnetic domains, whose spontaneous magnetisation is not parallel with the applied magnetisation, and which undergo a rotation of internal magnetisation during excitation. This induces internal stresses, which are responsible for the correlated strain (deformation).

Magnetostriction depends on the direction of the magnetisation vector, and not its sense, so a change of length in the 'positive saturation' is the same as that in the 'negative saturation'. Thus, the frequency of magnetostrictive vibrations is twice the magnetising frequency.

Each magnetic domain is magnetised internally to its saturation. Therefore, if the parallel domains grow only at the expense of the anti-parallel domains, then magnetostriction (i.e. change of dimensions) does not occur – because in both cases the magnetisation is applied in the same direction (but opposite sense).

This happens especially at low excitation applied along the easy magnetisation direction. The main change is that the large bar domains (see also Figures 1.9 and 3.40) grow and shrink and the perpendicular domains remain mostly unchanged. But the strain is the same for positive or negative magnetisation (Figure 2.52), so that no changes of dimensions take place.

The domains whose directions differ from the applied magnetic field are affected more at higher excitation levels. The flux density vector in each domain will be forced to rotate towards the direction of applied field (Figure 1.12) and this will cause changes in dimensions.

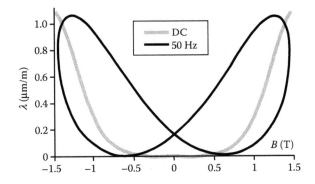

FIGURE 2.53 Surface magnetostriction measured with Epstein frame as a function of average flux density B and frequency (DC and 50 Hz) for 0.5-mm-thick sample. (Adapted from Rasilo P. et al., *IEEE Transactions on Magnetics*, 49 (5), 2013, 2041.)

It is well known that the magnetisation process is affected by dynamic phenomena. With increasing magnetising frequency, the contribution of eddy currents increases (Equation 1.36) and the number of domain walls changes (Sasaki, 1980). At even higher frequency, the skin effect becomes significant, so that the instantaneous B at the surface of the sheet might have a significantly different value than that in the middle. Magnetostriction is caused by changes in B of each domain. So if there are differences in B through the thickness of the sheet, this will have some effect on the behaviour of magnetostriction.

Under quasi-static excitation (DC), the magnetostriction curve can have very small hysteresis.* But under dynamic excitation (e.g. 50 Hz), the magnetostrictive hysteresis can be significant (Figure 2.53).

A deformation can be caused only if there is a stress acting on the body of the material. In a general case, the stress is a 3D quantity and could be tensile (pull apart) or compressive (push together). It can therefore represent a linear vector, which is perpendicular (normal) to a given analysed surface, thus resulting in a linear deformation along the same direction.

As stated above, it is necessary to measure mechanical deformation in order to measure magnetostriction. Under uni-directional magnetisation, this can be achieved by fixing one end of the sample (e.g. an Epstein strip) and measuring the displacement of the free end. The actual measurement method or type of sensor is irrelevant as long as the appropriate values (in appropriate units) are detected.

Under uni-directional magnetisation, several researchers used laser technology, for instance, single-point laser vibrometer (Phway et al., 2004). The laser beam is shone at the free end of the sample; the beam reflects and is measured back by the device. The apparatus had resolution up to 2 nm, which was sufficient for a 30-mm-long sample. Conceptual drawing of the experimental setup is shown in Figure 2.54.

---

* The term 'hysteresis' in general (and as referred to Figure 2.53) denotes the fact that the ascending branch of any curve is different from its descending branch. Such behaviour cannot be described by a single-value function and is often referred to as 'history dependent'. It should be noted that the behaviour shown in Figure 2.53 is a completely different type of hysteresis from the B–H loop described in Chapter 1.

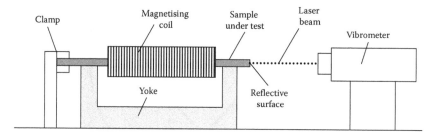

FIGURE 2.54 Measurement of magnetostriction with laser vibrometer (side view). (Adapted from Phway P.P.T. et al., *Journal of Electrical Engineering*, 55 (10/S), 2004, 7.)

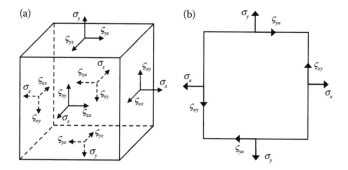

FIGURE 2.55 Stress components of a cubic element (a) and of a planar element (b). (Adapted from Somkun S., *Magnetostriction and Magnetic Anisotropy in Non-Oriented Electrical Steels and Stator Core Laminations*, PhD thesis, Wolfson Centre for Magnetics, Cardiff University, Cardiff, the United Kingdom, 2010a.)

However, there can also be a tangential component of stress* (ς in Figure 2.55), which is referred to as shear, and causes angular deformation. These stresses cause deformation, hence magnetostriction, in the appropriate direction. The stress can be caused by magnetisation, but also by external factors (e.g. assembly of the magnetic core) so their analysis is important for real-life applications.

Because of the multiplicity of stress components, the situation complicates when the strain is to be measured under rotational magnetisation or at various directions (Soinski, 1989). Nevertheless, it is possible to apply some simplifications. The first one is the assumption of negligible volume magnetostriction, which is quite valid for electrical steels.

Another assumption is the fact that the electrical steel laminations are thin, namely, the length of a given sample is much greater than its thickness. The measurements are thus confined mostly to the two co-planar surfaces, as the changes of dimensions will be much greater than through the thickness.

---

* The normal stress is denoted by the Greek letter σ (sigma). The tangential stress is denoted by many authors with the Greek letter τ (tau). However, in this book, τ was used to represent torque. In order to avoid confusion, it was therefore chosen to represent tangential stress with ς (also sigma, but an alternative character).

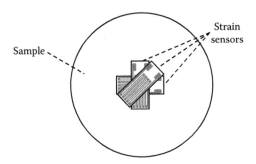

FIGURE 2.56 Typical strain gauge arrangement on a circular sample – the sensors are positioned at 0°, 45° and 90°. (Adapted from Somkun S., *Magnetostriction and Magnetic Anisotropy in Non-Oriented Electrical Steels and Stator Core Laminations*, PhD thesis, Wolfson Centre for Magnetics, Cardiff University, Cardiff, the United Kingdom, 2010a and Krell C. et al., *Przeglad Elektrotechniczny (Electrical Review)*, R. 81 (Nr 5/2005), 2005, 47.)

But the presence of the tangential stress means that just two sensors (x, y) are *insufficient* to capture the full picture of magnetostriction under 2D excitation (Krell et al., 2005). At least *three* sensors are required, and they are typically used in the so-called *rosette* configuration, in which the angular positions of each sensor are, for instance, 0°, 45° and 90° (Figure 2.56). Strain gauge is a typical type of sensor for such measurements (Enokizono et al., 1990b; Krell et al., 2005; Somkun, 2010; Kai et al., 2011a; Klimczyk, 2012).

The strain caused by magnetostriction λ in an arbitrary direction α and arbitrary instant of time t, hence $\lambda(\alpha,t)$, can be calculated from the signals of the three sensors in the rosette configuration (Krell et al., 2005):

$$\lambda(\alpha,t) = \cos(\alpha) \cdot [\cos(\alpha) - \sin(\alpha)] \cdot \lambda_{0°,t} +$$
$$+ 2 \cdot \cos(\alpha) \cdot \sin(\alpha) \cdot \lambda_{45°,t} - \sin(\alpha) \cdot [\cos(\alpha) - \sin(\alpha)] \cdot \lambda_{90°,t} \quad \text{(m/m)} \quad (2.24)$$

where $\lambda_{0°,t}$ is the strain signal from the 0° sensor (m/m), $\lambda_{45°,t}$ is the strain signal from the 45° sensor (m/m) and $\lambda_{90°,t}$ is the strain signal from the 90° sensor (m/m).

As evident from Equation 2.24, the result of a magnetostriction measurement is a function of the direction on the 2D plane (surface of the sample), as well as time. Moreover, the actual values also depend on the level of excitation (e.g. expressed as B). Showing all these information in a single graphical way is difficult and the authors use various ways – for instance, either as a function of B or α. The function of t is usually included indirectly because for each case, a full magnetisation cycle is shown. If magnetostriction is shown versus B, then the graph takes the shape of magnetostrictive hysteresis (Figure 2.53). Alternatively, the distribution of magnetostriction can be shown (Figure 2.57).

Magnetostriction is associated with the presence of mechanical stresses. It can therefore be expected that the additional mechanical stress applied to the sample under test should affect magnetostriction. This is indeed the case and several such studies were carried out.

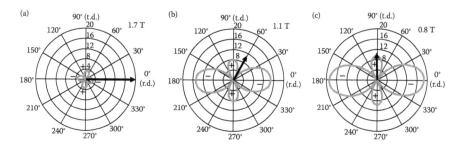

FIGURE 2.57 Typical distribution of magnetostriction (in ppm) for high-permeability grain-oriented electrical steel under elliptical magnetisation at given instants of time, where the *B* vector is positioned at 0° (a), 60° (b) and 90° (c). (Adapted from Pfutzner H. et al., *IEEE Transactions on Magnetics*, 47 (11), 2011, 4523.)

In uni-directional measurements, the stress is usually applied in the same direction as the magnetisation. The application of tensile stress is fairly easy – a force is applied so as to stretch the sample. But compression is much more difficult because thin laminations are very prone to 'buckling' or bending out of the plane of lamination, as well as any rotation or twisting. For this reason, quite sophisticated clamping arrangement has to be used (Klimczyk, 2012).

The force can be applied with a mechanical actuator, for instance, pneumatic cylinder (Klimczyk et al., 2011) or simply a screw on a suitable thread (Permiakov et al., 2005). The displacement is measured by suitable sensors, for example, accelerometers, whose outputs are double-integrated to convert the signal from acceleration to displacement. A load cell can be used in order to facilitate an accurate measurement of the applied force. A second displacement sensor can be attached to the fixed end of the sample in order to compensate for the vibration of the whole system. The major components of such a system are shown in Figure 2.58.

Compressive stress that is applied along the rolling direction of grain-oriented electrical steel is capable of increasing both the magnetostriction and power loss

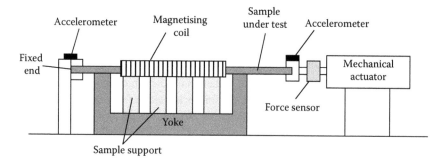

FIGURE 2.58 Typical system for measuring the magnetostriction under external stress. (Adapted from Klimczyk P. et al., *Przeglad Elektrotechniczny (Electrical Review)*, R. 87 (No. 9b/2011), 2011, 33.)

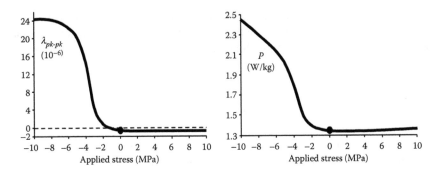

FIGURE 2.59 Typical curves of magnetostriction and power loss under external stress for grain-oriented electrical steel, measured along rolling direction, the symbol • denotes the measurement under reference conditions (no stress). (Adapted from Klimczyk P. et al., *Przeglad Elektrotechniczny (Electrical Review)*, R. 87 (No. 9b/2011), 2011, 33.)

(Figure 2.59). This is strongly connected to the changes of the magnetic domain structure.

The application of compressive stress creates somewhat similar domain patterns as if the excitation was applied along the transverse direction, that is, perpendicular to the rolling direction (Klimczyk et al., 2011). In such a configuration, the majority of domains are rotated by 90°, and this must lead to increased magnetostriction – when compared to the 'no stress' reference condition (Figure 2.51).

Several authors attempted investigation of magnetostriction and domain patterns with or without the presence of uni-axial external stress (Imamura et al., 1981; Masui, 1995; Hashi et al., 1996; Klimczyk, 2012). It is difficult to distil the general conclusions, partly because the domain-related studies are not measurements in the absolute sense.

The application of 2D stress (also sometimes referred to as 'biaxial') calls for more complicated mechanical arrangement. Several authors used cross-shaped sample, as shown in Figure 2.60 (Basak et al., 1978; Ng et al., 1992; Hubert et al., 2005;

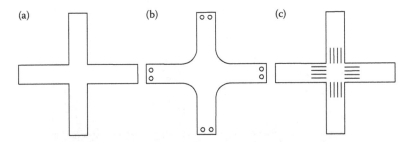

FIGURE 2.60 Examples of cross-shaped samples used for measurement under mechanical stress. (a) Simple cross-shaped sample to be clamped in vice-like jaws. (Adapted from Basak et al., *Proc. IEE*, 152 (2), 1978, 165.) (b) Sample with rounded inner corners. (Adapted from Hubert O. et al., *Przeglad Elektrotechniczny (Electrical Review)*, R. 81 (Nr 5/2005), 2005, 19.) (c) Sample with slits. (Adapted from Kai Y. et al., *IEEE Transactions on Magnetics*, 47 (10), 2011a, 4344.)

Kai et al., 2011a). Such approach is very similar to the two-phase magnetising yoke, where a combination of excitations in the x and y directions yields a resultant excitation in an arbitrary direction, depending on the amplitude and phase of the components. The same technique can therefore be used for the application of mechanical stress.

Each arm of the cross is attached (by vice-like jaws or screws) to a mechanical actuator. For the application of compressive stress, the sample must be protected against buckling out. This can be done either by appropriate sample holder (Kai et al., 2011a) or even by gluing inside of a 'sandwich' dummy sample, whose mechanical properties are similar, in order to achieve low stress incompatibility (Hubert et al., 2005). The area of interest is located at the centre of the cross, where other sensors can be attached: B needles, H-coils, strain gauges, etc. (compare with Figure 2.25). For accompanying magnetic measurements, usually the fieldmetric method is used.

The sample from Figure 2.60c has additional slits around the area under test. These slits are made to improve the uniformity of the applied stress (Kai et al., 2011a). The same authors used even an eight-arm double-cross sample (Kai et al., 2012).

Such an apparatus is capable of recording a plethora of results, depending on the many possible conditions of measurement – for example, there were over 100 curves reported in just one paper (Kai et al., 2011a). These are not easy to interpret and again it is difficult to draw generalised conclusions. The shapes of rotational B–H loops can be significantly affected by the application of stress, even to the point of 'negative' areas developing (Kai, 2011b) (see also the description of the fieldmetric method above). Interestingly, one of the expressed conclusions is that compressive stress can lead to reduction of power loss.

Yet another way to apply mechanical stress is when a sample is compressed between two rigid formers; an example is shown in Figure 2.61 (see also Figure 6.19) (Yamamoto et al., 2011, 2014).

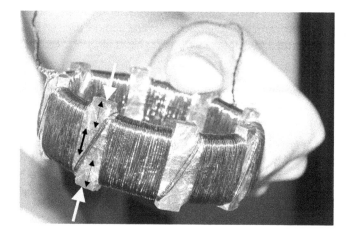

FIGURE 2.61   Ring sample used for measurement with axial stress, solid black arrow shows magnetic sample, dashed black arrows show the thickness of the mechanical structure used for applying the axial stress (white arrows). (Courtesy of Jeanete Leicht, formerly of Institute for Technological Research (IPT), Brazil.)

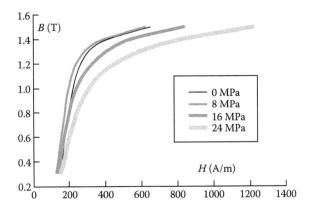

FIGURE 2.62 *B–H* curves for a toroidal sample of non-oriented electrical steel subjected to axial compressive stress, magnetised at 50 Hz. (Adapted from Yamamoto K.-I. et al., *Przeglad Elektrotechniczny (Electrical Review)*, R. 87 (No. 9b/2011), 2011, 97.)

A toroidal sample is sandwiched between two rings with multiple mechanical supports around the perimeters (Yamamoto et al., 2011). The primary and secondary windings are wound around the whole structure. This leads to substantial air flux contributing to the voltage induced in the secondary winding, but if the dimensions are known, this can be compensated numerically. Otherwise the measurement is carried out similarly to the standardised method (IEC, 2003).

Interestingly, measurements carried out on a sample of non-oriented electrical steel, subjected to compressive axial stress (normal to the surface of the lamination), show that under low-compressive stress (e.g. 8 MPa), the permeability increases and correspondingly power loss decreases, as compared to the no-stress state (Figure 2.62). High values of stress (up to 24 MPa) invariably significantly lower the permeability and increase the power loss (Yamamoto et al., 2011).

2.5.3 BARKHAUSEN NOISE

The measurements of magnetostriction described in the previous section are 'proper' measurements, which can be quantified with defined accuracy (albeit limited in practical applications). Strain sensors can be calibrated and results can be reported in numerical values with appropriate units.

Similarly to the domain observations, the Barkhausen noise 'measurements' are not quantitative. The results depend strongly on the instrumentation involved, type of sensors, filtering, etc. Calibration is not possible, at least not in an absolute sense. But qualitative studies were carried out by many researchers, especially under uni-axial magnetisation.

Barkhausen noise is generated by abrupt un-pinning of the domain walls from pinning sites inside the volume of the sample. The pinning sites can be introduced by many causes such as chemical impurities and precipitations, mechanical defects and so on. For these reasons, the technique is applied for non-destructive testing of materials. For instance, micro-cracks developing due to mechanical fatigue or internal

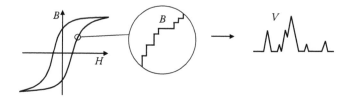

FIGURE 2.63   Sudden changes can be detected on the B–H characteristics.

stresses introduced from thermal surface hardening cause significant changes to the distribution of pinning sites – a condition that strongly affects the intensity of Barkhausen noise.

Magnetically, Barkhausen noise manifests itself as sudden changes of the B–H curve (Figure 2.63). Under uni-directional magnetisation, the B signal is most frequently detected with a search coil (B-coil), or secondary winding. Because the induced voltage is a derivative of B (see Equation 1.14), any sudden change appears as a sharp spike in the voltage signal.

Under normal conditions, the induced 'global' voltage is much greater in amplitude than the local Barkhausen noise, so any minute impulse changes are over-shadowed. This is especially the case if the voltage is detected with an analogue-to-digital converter because often the voltage resolution is too coarse to detect the small spikes of voltage. Several techniques are used in order to facilitate the detection of the Barkhausen noise.

The most obvious one is to suppress the changes of the relatively high-global voltage. This can be achieved by using two identical search coils (e.g. 500 turns each), which are connected in series opposition (Figure 2.64a). Their global voltages will be very similar because the global B changes will be the same for both coils. However, they will be physically placed at slightly different locations on the sample under test. The sudden voltage spikes are stochastic (random), so their signals will *not* balance out from the two coils (Patel et al., 2006). If the two coils are placed too

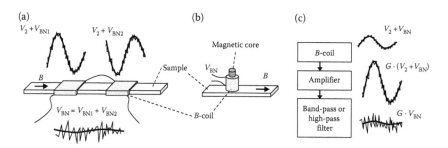

FIGURE 2.64   Typical techniques for Barkhausen noise measurement: two coils in series opposition (a), surface sensor (b), amplification with gain G and filtering for extracting the V_B signal from the induced secondary voltage V_2 (c). (Adapted from Patel H.V. et al., *Sensors and Actuators A*, 129, 2006, 112; Augutis V. et al., *Materials Science*, 12 (1), 2006, 84; Chukwuchekwa N., *Investigation of Magnetic Properties and Barkhausen Noise of Electrical Steel*, PhD thesis, Wolfson Centre for Magnetics, Cardiff University, Cardiff, the United Kingdom, 2011.)

close, they tend to suppress signal from each other (see also Figure 2.66) (Moses et al., 2006b).

The Barkhausen jumps are caused by sudden changes of domain wall positions. At the instant of such an event, some electromagnetic field will be emitted from the sample. This field can be detected by a suitable induction sensor. Several studies used a small magnetic core (e.g. soft ferrite, 3 mm diameter), with a coil (e.g. 1000 turns), placed perpendicular to the surface of the sample (Augutis et al., 2006; Patel et al., 2007; Wilson et al., 2009). This method has the advantage that the high-global voltage is inherently absent. The drawback is that the Barkhausen signal is also smaller and hence more difficult to detect.

A positive feature is that the measurement is much more localised (than the global search coils), and this by definition facilitates local measurements, and even scanning the surface for Barkhausen noise activity and correlation with other magnetic properties and parameters (Zurek et al., 2009c).

Both these techniques can be, and usually have to be, combined with additional signal processing. Two major operations involve wideband analogue signal amplification, and analogue or digital high-pass or band-pass filtering (typically at various bands between 300 Hz and 300 kHz [Parakka et al., 1997; Moses et al., 2006b; Stupakov et al., 2008; Prabhu et al., 2012]). Especially, the latter step contributes to the lack of 'absoluteness' of the method. Different researchers use various filter settings for low and high cut-off frequencies, so the measured values cannot be compared directly but only on a qualitative level.

Additional modification of the noise signal is sometimes performed by setting a threshold (for instance, 4 μV), below which all values are rejected (Patel et al., 2006). This helps in suppressing other sources of continuous small noise (e.g. thermal noise caused by the resistance of the B-coil), but introduces further 'personalised' modifications to the measurement procedure, so that absolute measurements cannot be carried out.

When Barkhausen noise is isolated from the global voltage, it becomes obvious that its maximum occurs at the maximum of induced voltage and therefore at the greatest rate of change of B (Figure 2.65). Once the B waveform reaches its peak for a brief instant, nothing changes ($dB/dt \approx 0$, so $V \approx 0$), no domain walls move, so any Barkhausen noise activity also ceases. This can be even better illustrated with trapezoidal magnetisation, where the 'flat top' of B waveform is much longer, so the Barkhausen noise disappears. On the other hand, for a triangular B waveform, the changes are forced to be practically constant over the whole excitation cycle, hence as a result, the Barkhausen noise activity is almost constant, without any 'quiet' intervals (Patel et al., 2006).

Several means of quantifying the signals were applied in the literature. These are usually based on statistical methods because the Barkhausen noise itself is stochastic in nature. The most well known are RMS value, total sum of all amplitudes (TSA), total number of peaks (TNP), power spectrum (PS) and kurtosis (Zhu et al., 2001; Hartmann, 2003; Patel et al., 2007).

RMS value is the root mean square value of the full Barkhausen noise signal (BN). Any amplification is taken into account, so that the result is scaled to the units of the raw measured voltage. For this reason, the absolute calculated value is usually quite small, for instance, below 1 mV at sinusoidal excitation in electrical steels, for

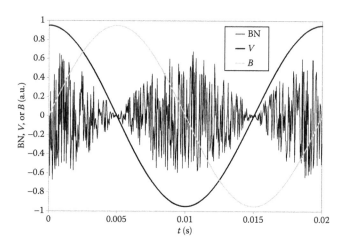

FIGURE 2.65 Typical Barkhausen noise (BN) waveforms under controlled alternating sinusoidal B and its phase relationship to the global secondary voltage (V) and flux density (B); all waveforms are scaled so that they have comparable amplitudes (hence arbitrary units 'a.u.' are shown on the vertical axis). (Adapted from Patel H.V. et al., *Sensors and Actuators A*, 129, 2006, 112.)

measurements on single Epstein strips, as shown in Figure 2.66 (Moses et al., 2006b; Patel et al., 2006; Chukwuchekwa, 2011).

The RMS value of BN signal is calculated using a similar equation as for any other RMS value of a signal:

$$V_{BN,RMS} = \sqrt{\frac{1}{T} \cdot \int_0^T \left(V_{BN}\right)^2 dt} \quad (V) \qquad (2.25)$$

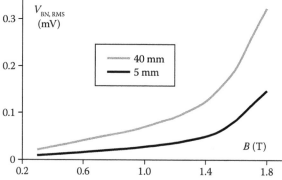

FIGURE 2.66 Typical RMS curve, at 50 Hz for a sample of grain-oriented steel (single Epstein strip) and varying distance between the two opposing B-coils. (Adapted from Moses A.J. et al., *Journal of Electrical Engineering*, 57 (8/S), 2006b, 3.)

However, if a signal is digitally processed, then the whole information is contained in a set of numbers (e.g. a 1D array), rather than continuous function, so the equation actually becomes ($V_{BN,i}$ is the ith voltage value in the digitised BN sequence, N_{BN} is the number of data points of BN)

$$V_{BN,RMS} = \sqrt{\frac{1}{N_{BN}} \cdot \sum_{i=0}^{N_{BN}-1} \left(V_{BN,i}\right)^2} \quad (V) \qquad (2.26)$$

Total sum of all amplitudes (TSA) of BN is also convenient but does not count simultaneously occurring peaks, which can occur quite frequently, especially for 'larger' events. The calculation is carried out by adding all absolute values of all the acquired data points (Chukwuchekwa, 2011), which is synonymous to

$$V_{BN,TSA} = \sum_{i=0}^{N_{BN}-1} \left|V_{BN,i}\right| \quad (V) \qquad (2.27)$$

The resulting value of TSA is much higher (e.g. at the level of 1 V) than RMS because all the data are summed together. Figure 2.67 shows typical graph of TSA.

Total number of peaks (TNP) of BN counts the number of detected peaks, in integer numbers (hence it is unitless). For instance, a peak detection library function can be used as supplied in the commercially available LabVIEW programming environment (Chukwuchekwa, 2011) (the actual settings used for the function were not specified). The procedure is therefore sensitive to the used peak detection method (and its parameters). Figure 2.68 shows a typical TNP curve.

Power spectrum (PS) of BN is calculated as a Fourier transform of the BN signal, and the curve is plotted like any other PS versus frequency (Figure 2.69) (Augutis et al., 2006). Such curve illustrates that the lower-frequency events contain more

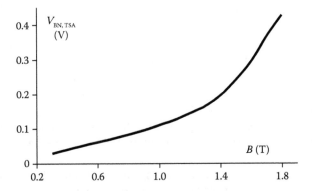

FIGURE 2.67 Typical TSA curve, at 50 Hz for a sample of grain-oriented steel (single Epstein strip) measured with two opposing *B*-coils. (Adapted from Moses A.J. et al., *Journal of Electrical Engineering*, 57 (8/S), 2006b, 3.)

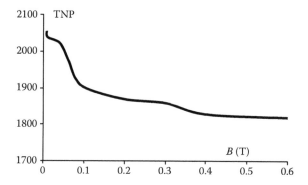

FIGURE 2.68 Typical TNP curve, at 50 Hz for a sample of non-oriented steel (single Epstein strip) measured with two opposing B-coils. (Adapted from Chukwuchekwa N., *Investigation of Magnetic Properties and Barkhausen Noise of Electrical Steel*, PhD thesis, Wolfson Centre for Magnetics, Cardiff University, Cardiff, the United Kingdom, 2011.)

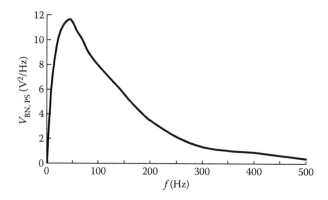

FIGURE 2.69 Typical power spectrum curve, at 50 Hz for a sample of grain-oriented steel (single Epstein strip) measured with two opposing B-coils. (Adapted from Augutis V. et al., *Materials Science*, 12 (1), 2006, 84.)

energy than the higher-frequency ones. Many of the BN events occur in an avalanche fashion, when several BN peaks occur almost at the same time, greatly overlapping and thus appearing like a high-amplitude long-lasting peak ('large event'). By calculating the PS, it is possible to show that the longer peaks have lower frequency assigned to them ($f = 1$/time), so that more energy is involved at lower frequency. The previously described coefficients (RMS, TSA, TNP) are not capable of capturing this information because by definition all the BN events are lumped into a single number.

However, the sensor and signal processing equipment have their own frequency-dependent characteristics and appropriate numerical compensation has to be carried out in order to arrive at the 'true' BN PS. The procedure is not straightforward, so there can be significant sources of discrepancies between various measurement apparatus.

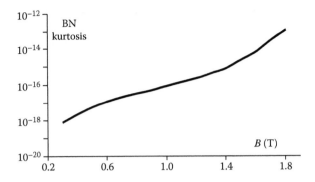

FIGURE 2.70 Typical kurtosis curve, at 50 Hz for a sample of grain-oriented steel (single Epstein strip) measured with two opposing *B*-coils. (Adapted from Moses A.J. et al., *Journal of Electrical Engineering*, 57 (8/S), 2006b, 3.)

Other authors might use a different definition of 'power spectrum'. For example, the curves can be plotted not versus frequency, but versus flux density, so the total power of the BN signal has to be calculated (Moses et al., 2006b).

Kurtosis is a coefficient that describes the statistical 'peakedness' of the BN signal (Moses et al., 2006b). The 'peakedness' can be conceptually explained as the measure of how 'narrow' (i.e. 'peaky') or 'wide' (i.e. 'flat') is the bell-shape curve of normal distribution. The kurtosis can be calculated* as

$$\text{Kurtosis} = \frac{1}{N_{BN}} \cdot \sum_{i=0}^{N_{BN}-1} (V_{BN,i} - V_{BN,\text{mean}})^4 \qquad (2.28)$$

where $V_{BN,i}$ is the subsequent (*i*th) value in the sequence of BN and $V_{BN,\text{mean}}$ is the mean value of the sequence (Moses et al., 2006b).

A typical curve of the BN kurtosis versus peak value of sinusoidal flux density is shown in Figure 2.70.

The same sensing techniques, signal processing and calculations can be used for the analysis of Barkhausen noise under rotational magnetisation. Of course, this can be combined with various methods such as mechanical (Malkinski et al., 1993) or fieldmetric (Enokizono et al., 1999).

However, as explained above, the Barkhausen noise 'measurement' is not a measurement in an absolute sense. Even for alternating magnetisation, the conclusions can be drawn practically only at a qualitative level.

The very presentation of results from measurements of rotational Barkhausen noise poses a challenge. Various authors used different figures, from visually illus-

---

* It is unclear from Moses et al. (2006b) as to what SI unit (if any) should be used, although as judging by Equation 2.28, the units should be (V⁴). According to Cox et al. (2010), the first moment of probability distribution function is the *mean* (average, a measure of location or expectation), the second moment is *variance* (a measure of data dispersion), the third moment is *skewness* (a measure of asymmetry about the expectation), and the fourth is *kurtosis* (a measure of heaviness of the 'tails' or 'peakedness' in the centre).

trating 3D surface (less useful for any numerical analysis) (Malkinski et al., 1993) to somewhat exotic 'attractors' (Enokizono et al., 1999).

Similarly to alternating excitation, the conclusions from measurements of Barkhausen noise under rotational magnetisation are mostly qualitative. But owing to the increased complexity of measurement, the statements seem to be even less defined than it was the case for alternating excitation. To give a few quotes:

[…] quantitative predictions are difficult to make […]

Grimwood et al. (1978)

[…] the Barkhausen noise and anisotropy measurements […] can be used for material testing as supplementary methods.

Malkinski et al. (1993)

The power spectrum density and the correlation function showed the Barkhausen noise was aperiodic.

Enokizono et al. (1999)

3 Sensors and Sensing Techniques

The three main measurement methods used in studies of rotational magnetisation (torque, thermal and fieldmetric) were described in the previous chapter. The underlying mathematical equations can be applied regardless of the type of sensors used, provided that the values are measured or scaled to the appropriate SI units – as required by each of the method.

Various sensors can be used to measure the same quantity, almost independently of the actual measurement method itself. The most commonly used sensors and sensing techniques are described in this chapter. Many are closely related to those used in uni-axial measurements because similar physical quantities are measured.

3.1 FLUX DENSITY SENSORS

Most of the sensing techniques are described below with the focus on the detection of flux density B, rather than polarisation J. For magnetically soft materials, it can be often assumed* in practice that $B \approx J$. This is true provided that excitation H is relatively low, that is, saturation is not approached, so that the difference arising from Equation 1.9 is indeed negligible for practical applications.

However, by no means is this always the case. The difference between B and J is proportional to $\mu_0 \cdot H$ and must be carefully evaluated in order to make sure that an appropriate level of accuracy is attained from the scientific and metrological viewpoint. The correction of the $\mu_0 \cdot H$ component can be taken in several ways. If signals are processed digitally and required dimensions are known, then, it is possible to re-calculate the results because $J = B - \mu_0 \cdot H$.

The $\mu_0 \cdot H$ component can also be eliminated by providing a mutual inductance coupled to the applied excitation of H (described in Chapter 4). However, this is only possible in magnetically closed circuits, in which the magnetising current is directly proportional to the applied H, which is not the case for most of the magnetising apparatus used in rotational magnetisation measurements. For certain configurations, a dummy sensor can be used that produces signal proportional to the $\mu_0 \cdot H$ component and can be subtracted.

In any case, researchers should be aware of the difference between measuring B and J and apply appropriate corrective measures according to the requirements in a given method.

---

* The $B \approx J$ assumption can be made in the sense that for engineering applications the power loss measured with the toroidal sample method (based on B) can be used for the design of real magnetic cores comparably to the loss measured, for instance, with the Epstein frame method (which is based on J).

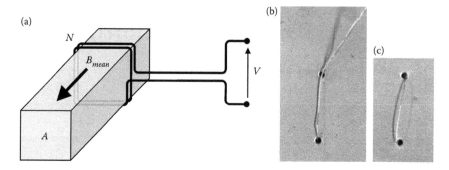

FIGURE 3.1   Induction coil – a sensor of variable flux density (a), close-up view of single-turn B-coil made on a lamination of non-oriented electrical steel, with 0.5 mm holes spaced by 5 mm with single turn of 0.1-mm-diameter enamelled wire (b) and underside of the same B-coil (c); the dimensions of the holes and their spacing were chosen in order to show more details in the photographs.

3.1.1   INDUCTION COIL (B-COIL)

Induction coil (also *B coil, B-coil, search coil, sensing coil* or *pickup coil*) is the most basic but also the most robust type of sensor (Tumanski, 2011). The concept of B-coil is shown in Figure 3.1.

In its simplest form, it is sufficient just to wind at least one turn of wire around the given magnetic circuit under test (Figure 3.1b and c). The main requirement is that the wire is electrically insulated from itself and anything else. The measurement of B is performed by detecting the induced electrical voltage V and processing it accordingly.

As already discussed in Chapter 1 (Figure 1.14), varying B induces V in the enclosing coil. The amplitude of this voltage is proportional to the rate of change of B, number of turns N of the coil and the active area A enclosed by the search coil (Figure 3.1a).

The voltage produced in the coil is proportional to B averaged* over the area A; hence it is denoted here as B_{mean}. This relationship as a function of time is known as *Faraday's law of induction*, and for a uni-directional system using scalar notation, it can be expressed by Equations 3.1 and 3.2.

PRACTICAL COMMENT

In the literature, the words *mean* and *average* can be used interchangeably. However, in electrical engineering, the term *average voltage* is used to denote the *mean* value of a *rectified* signal; hence, in this particular sense,

---

* The averaging can have important implications. For instance, B at different locations within the volume of the sample can have different values. Such a situation takes place, for example, around drilled holes (see Figure 3.18) or in measurements at elevated frequencies, where the skin effect might become significant. B-coil is incapable of detecting such conditions – it can only detect B averaged over the whole area enclosed by the coil (hence B_{mean}). The same applies if the sample is magnetised in a non-uniform way on purpose.

average $\neq$ *mean* and careful attention should be paid to the implications of each name. This is especially important when the topic of voltage measurement is discussed (at the end of this chapter).

Ordinary mathematical mean is zero for one cycle of pure sine because all the values in the positive half cycle are exactly equal to all the values in the negative half cycle. Therefore, electrical engineers use the short-hand name *average* to denote *mean value of a rectified signal*, which for a pure sine is a non-zero quantity, lower exactly by a factor of $2/\pi$ ($\approx$0.6366) as compared to the peak value.

In this book, both words *average* and *mean* (if used without any further description) refer to the ordinary mathematical mean value, without any rectification. When a rectified mean voltage is discussed below, it is specified that the rectification takes place, for each use of the term 'average'. For clarity, the symbol 'AVG' in this book always denotes 'mean value of a rectified signal'.

It should be noted that *mean* also applies to concepts such as shown in Figure 3.1 and used in Equations 3.1 and 3.2. In such case, the voltage induced in the sensor is proportional to the arithmetical mean of all points on the cross-sectional area under the coil. The amplitude of B changes with time, but the value taken into account is averaged over the area or volume, at a given instant of time.

The negative sign in Equation 3.1 means that the induced electromotive force $EMF(t)$ would produce a current with polarity opposing the changes of magnetic flux. However, the voltage *measured* across the coil as $V(t)$ is positive (Lubelski, 1982), so that the minus sign does not occur (see also description of numerical integration in Chapter 8).

$$EMF(t) = -N \cdot A \cdot \frac{dB_{mean}(t)}{dt} \quad (V) \qquad (3.1)$$

$$V(t) = N \cdot A \cdot \frac{dB_{mean}(t)}{dt} \quad (V) \qquad (3.2)$$

PRACTICAL COMMENT

The *EMF* in Equation 3.1 is a quantity that is induced *inside* the coil and cannot be measured. The voltmeter connected across a coil will measure the voltage developed *across the voltmeter*, which has a positive sign (Lubelski, 1982). Therefore, Equation 3.2 should be used in such calculations because this is the waveform that will be detected in reality by *any* voltmeter. This topic is explained in detail in Chapter 8.

An important conclusion that can be drawn from Equations 3.1 and 3.2 is that constant B does not generate any voltage because if $B = const$, then its derivative

$dB/dt = 0$. Hence, constant B or a constant offset cannot be easily* detected with a search coil.

It can also be derived from Equations 3.1 and 3.2 that for a *pure sine* there is a direct relationship between the measured RMS voltage V_{RMS} and B_p (peak value of B) (Jezierski, 1975):

$$B_p = \frac{V_p}{2 \cdot \pi \cdot f \cdot A \cdot N} \approx \frac{V_{RMS}}{4.44 \cdot f \cdot A \cdot N} \quad (\text{T}) \qquad (3.3)$$

The numerical value '4.44' is related to the ratio between the peak and the RMS value of sine, and it is precisely equal to $\pi \cdot \sqrt{2}$ ($\approx$4.44).

PRACTICAL COMMENT

In this chapter, most equations related to sensors will be given in a form that allows direct calculation of the measured quantity. Equation 3.3 is usually written in a form $V = f(B)$ like in Equations 3.1 and 3.2, but in order to calculate B, it must be rearranged anyway; hence $B = f(V)$ is more useful in practice, from the viewpoint of sensor design.

V_{RMS} in Equation 3.3 is useful because most voltmeters are calibrated to measure the RMS value, so a simple voltage measurement can be used for calculating B_p. However, it should be borne in mind that Equation 3.3 holds *only* for sinusoidal voltage and cannot be used for distorted signals. On the other hand, Equation 3.4 is applicable only to waveforms that do not contain even harmonics. For an arbitrary shape of the waveform, full integration must be employed.

It can be often assumed in magnetic measurements that the measured signals have only odd harmonics, with negligible even harmonics. The voltage can be heavily distorted with higher harmonics, but if the 'positive' half-cycle is a mirror image of the 'negative' half-cycle, then even harmonics are not present. If this is the case, then it can be derived from Equations 3.1 and 3.2 that

$$B_p = \frac{|V|_{mean}}{4 \cdot f \cdot A \cdot N} = \frac{V_{AVG}}{4 \cdot f \cdot A \cdot N} \quad (\text{T}) \qquad (3.4)$$

where $|V|_{mean} = V_{AVG}$ denotes the average value of 'rectified' voltage (V) (IEC, 2003).

PRACTICAL COMMENT

$|V|_{mean}$ in Equation 3.4 must be 'rectified', namely, absolute values must be used for the calculation of the average. Otherwise, as mentioned above, an

---

* However, with appropriate techniques, it is possible to detect constant B with a search coil. Some details are given later in this chapter.

average of purely sinusoidal signal over one cycle yields zero because the positive and negative half-cycles cancel each other out.

Equation 3.4 holds also for distorted voltage provided that there is no *even* harmonics – such that the positive and negative half-cycles are symmetrical about the horizontal axis. This is generally true for symmetrical alternating magnetisation, but it is not the case, for example, if the excitation contains some DC offset.

3.1.1.1   Analogue Signal Processing

Equations 3.3 and 3.4 give a method for converting a voltage value directly into B_p value. This can be useful in applications where the shape of the B waveform is less relevant, and only the peak or RMS value is of interest, such as, for instance, in the rotational apparatus shown in Figure 2.13, where a voltmeter is employed for this very reason.

However, if the shape of the *waveform* also needs to be known, as it would be for the fieldmetric method, then full integration must be performed:

$$B(t) = \frac{1}{N \cdot A} \cdot \int V(t)dt + C_{integral} \quad (\text{T}) \tag{3.5}$$

where $C_{integral}$ (T) is a constant value resulting from integration and initial conditions.

PRACTICAL COMMENT

$C_{integral}$ in Equation 3.5 is simply a mathematical artefact resulting from the calculation of an integral. As it was stated above, under normal conditions, the DC offset cannot be detected by the search coil, so even if B has some offset, there will be none in the induced V. In a result of Equation 3.5, all values of the B waveform should oscillate around zero. In practice, the constant is often simply subtracted because it cannot exist under alternating excitation (the B-coil cannot detect it). This $C_{integral}$ value will be omitted in the following equations, for the sake of clarity.

Such constant might denote, for example, the initial state of the material, which could be magnetised to some unknown value of B. Therefore, the constant must have the same units of (T) as the other components of the equation.

The integration as required from Equation 3.5 can be performed in several ways. One of the first invented methods was the mechanical inertia of moving parts in a galvanometer (Figure 3.2). Such technique was used in the so-called *ballistic* measurements.

A galvanometer is a device used mostly for detecting zero voltage, but it is in fact a very sensitive voltmeter. If there is any voltage present, then a current flows through the movable (rotating) coil, and the pointer will be deflected appropriately – for instance, to the right for a positive voltage polarity, and to the left for negative.

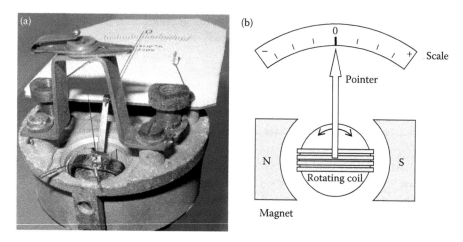

FIGURE 3.2 Galvanometer: (a) Photograph of the inner structure. (Photograph by Wojciech Pysz, CC-BY-SA-2.5, Wikimedia Commons. Adapted from Pysz W., *Galvanometer*, Wikimedia Commons, 2017, accessed online February 17, 2017 http://commons.wikimedia.org/wiki/file:galvanometer_late_20cent.jpg.) (b) Diagram explaining the functions of components.

The movable coil is suspended on calibrated springs, so the amount of deflection is directly proportional to the amplitude of the current.

The electrical resistance of the coil had to be chosen appropriately as not to overload the sensing circuit. If the mechanical inertia of the moving parts and damping friction were set carefully, then a total deflection of the needle was proportional to the integral of the voltage applied over a certain time interval.

Interestingly, the ballistic method is suitable for detecting a constant value of B. The sensor must be made in such a way as to allow a rapid withdrawal from the place where B is to be measured, to a location where B is negligible, which is usually sufficiently far away from the excitation. As a result, the B penetrating the search coil changes rapidly from some fixed value (maximum) to zero. This change induces a transient impulse of voltage (and hence current), which pushes the needle accordingly. It is therefore possible to measure B by knowing the proportionality factor (damping, mechanical inertia of the pointer system, etc.).

Of course, this is a mechanical system, where mechanical factors such as friction can have a considerable influence and significantly reduce accuracy. Moreover, such technique is only suitable for measuring constant or at best very slowly changing B. At 50 Hz, any changes are too quick to be visually observed in this way, especially that they will be impeded by mechanical inertia of the moving system.

Nevertheless, such ingenious techniques were used over a century ago for important work both in the field of magnetic sensors and magnetic measurements (Chattock, 1887; Ewing, 1900; Rogowski et al., 1912; Rogowski, 1916).

Tumanski gives many examples of practical electronic circuits for analogue integration of voltage (Tumanski, 2011). Some of these circuits are useful only for specific types of sensors or for high-frequency measurements. For instance, passive integration is possible with just a single resistor and single capacitor, which together act as a low-pass RC filter. However, for such a circuit to operate correctly, the inductive reactance of the B-coil must be much greater than the impedance of the RC filter, so the applicability of this method is limited (practically used B-coils usually have quite small impedance). Also, such passive integration significantly attenuates the output signal.

In practical applications, the integration function is often reliably achieved with a circuit comprising integrating operational amplifier. Such circuits are capable of simultaneously performing integration as well as amplification of the signal of interest. The electronic circuit shown in Figure 3.3 performs analogue integration, whose result* is

$$V_{out}(t) = -\frac{1}{R \cdot C} \cdot \int V_{in}(t)dt + V_0 \qquad (V) \qquad\qquad (3.6)$$

where C is the capacitance of the feedback capacitor (F), R is the total input resistance (Ω) and V_0 is the initial voltage (V).

The total input resistance is the sum of the input resistor R_{in} (a deliberately used component) and the self-resistance of the B-coil (not shown in Figure 3.3), which might be a significant value, especially for a B-coil with many turns of fine wire. V_0 is the initial voltage across the terminals of the amplifier at the moment of switch on.

In order to reduce errors resulting from voltage offsets, the circuit can be modified, for instance, by adding the resistor R_0, which limits the low-frequency bandwidth (Tumanski, 2011). With such modification, higher-frequency measurements (AC) can be carried out with relative ease and satisfactory accuracy.

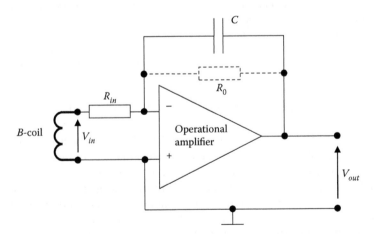

FIGURE 3.3 Integrating operational amplifier with typical values $R_{in} = 10 \text{ k}\Omega$, $C = 10 \text{ μF}$. (Adapted from Tumanski S., *Handbook of Magnetic Measurements*, CRC Press, 2011, ISBN 9781439829516.)

---

* The minus in Equation 3.6 is a consequence of using inverting implementation of an operating amplifier, as shown in Figure 3.3.

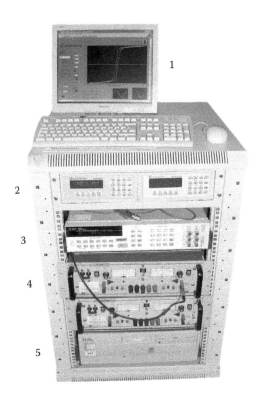

FIGURE 3.4 LakeShore Fluxmeter 480 used in quasi-DC permeameter (*B–H* loop tracer); the magnetic yoke and sample are not shown because the setup can be used with various systems: Epstein frame, toroidal sample, custom yokes, etc. The marked items are: 1 – measured and displayed *B–H* loop, 2 – LakeShore Fluxmeter 480 (*B* measurement), 3 – voltmeter (*H* measurement), 4 – Kepco programmable power supply, 5 – computer. Equipment courtesy of Wolfson Centre for Magnetics, Cardiff University, Cardiff, the United Kingdom.

For instance, a Lakeshore Fluxmeter 480 uses a similar technique (Figure 3.4). The AC measurements can be performed from 10 Hz to 10 kHz with 1% accuracy, and from 2 Hz to 50 kHz with 5% accuracy. The voltage input range can vary from millivolts to tens of volts, and the *B*-coil parameters (number of turns *N* and area *A*) can be programmed in a very wide range (Lakeshore, 2004).

Very low-frequency measurements are more difficult to carry out. One of the largest problems is the integrator drift, which cannot be as easily eliminated as in the AC mode. For this particular device, the DC accuracy on its own is quite good – at the level of 0.25%. The DC drift is 0.0004% of full scale per minute of operation. For an ideal instrument with the input of the integrator short-circuited, the output should remain at zero. But in a real instrument, the output slowly drifts and indicates some erroneous value, usually increasing with time.

The input ranges are specified usually in volt-seconds (V · s). The lowest DC range for the same instrument is 3 mV · s, the internal offset is 10 μV · s and the integrator drift is 1 μV · s/min. So after 1 minute of operation, the total error can be up to 2% of the full scale, and it will increase with time, generally in an unpredictable and non-linear way (Scholes, 1970; Garcia et al., 2006).

PRACTICAL COMMENT

The practical significance of the integrator drift is that quasi-static hysteresis loss (see Figure 1.27) needs to be measured at very low frequency, for some cases well below 0.1 Hz. This leads to a situation where a single loop requires many minutes to trace one full cycle. But during this time, the integrator might drift by such a large amount that the measurement becomes difficult to carry out accurately.

For instance, if the measurement would have to be carried out at 0.004 Hz (250 s), then the drift could cause error comparable to 10% of the full scale. As a result, the starting and finishing point of a $B-H$ loop might not meet so that the loop would not close (Garcia et al., 2006; Mimura et al., 2012). There are mathematical corrections that can be applied, but the drift is non-linear, which makes such methods impractical for intervals longer than a few minutes. It will be impossible to evaluate the real errors even if it is possible to post-process the values such that the loop will become closed.

The input ranges are specified in volt-seconds ($V \cdot s$) because the integrated signal is proportional not only to the input voltage, but also to its duration. This is explained below in more detail.

An analogue electronic integrator will also introduce some phase shift in the signal. A value of 0.1° of phase error was reported in the literature (Sievert, 1990).

3.1.1.2   Digital Signal Processing

The advent of computer technology allowed measurements of induced voltage with digital data acquisition devices. As a result, all the necessary signal processing could be carried out within appropriate software, rather than with analogue circuits.

Sampling frequency of the device dictates the number of samples per second that can be acquired. According to the Nyquist–Shannon theorem, the signal must be sampled with a frequency *at least* twice of the highest harmonic.

For example, a distorted 50 Hz signal with 100 significant harmonics requires a *minimum* sampling frequency of 50 Hz $\cdot$ 100 harmonics $\cdot$ 2 = 10 kHz or 10 kS/s (10,000 samples per second). If the sampling frequency is not sufficiently high, then a phenomenon called aliasing can cause for 'ghost' harmonics to be apparently detected, as shown in Figure 3.5.

The resulting frequency of such ghost harmonic is the difference between the real frequency of the waveform and the sampling frequency: $f_{alias} = f_{real} - f_{sampling}$. Once the signals are converted into the digital domain, it is not possible to recover the information about the original frequency of the signal. For this reason, aliasing should be avoided, either by appropriately high-sampling frequency or by analogue filtering of the input signal, or a combination of both.

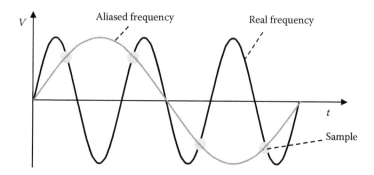

FIGURE 3.5   Aliasing occurs if the sampling frequency is less than twice the frequency of the measured signal.

Also, if the digital signal is used for active feedback (waveshape control), then better accuracy of phase information might be required, which again calls for increased sampling frequency. In the author's practical experience, at least 10 points per cycle of given harmonic are required to ensure a satisfactory stability of digital feedback. So, in the example given above, the factor would increase from 2 to 10, and for 50 Hz and 100 harmonics, the minimum sampling frequency becomes 50 kS/s.

The voltage resolution of an analogue-to-digital converter dictates the smallest detectable voltage change. The theoretical limit is a full scale of the given range divided by the number of available steps. For a typical device, the highest range is ±10 V. So, with a 12 bit resolution, the smallest detectable voltage change is +10 V−(−10 V)/(2^{12}−1) = 20 V/4095 ≈ 5 mV.

On the other hand, the lowest voltage range could be ±0.2 V. Hence, for typical 12 bit, 16 bit, 18 bit and 24 bit data acquisition devices, the theoretical resolution would be 100 µV, 6 µV, 1.5 µV and 24 nV, respectively.

However, the actual measurement accuracy is affected by many factors such as internal noise, offset and temperature influence. The absolute accuracy can be on the order of magnitude worse, and, for instance, for a 16 bit device (resolution of 6 µV at the ±0.2 V range), the specified absolute uncertainty can be as high as 90 µV, and it usually worsens with temperature changes. An increase of 10°C above the calibration temperature can cause an additional offset error of over 200 µV (NI, 2013a), which is 0.1% of the full range (vs. 0.003% of the theoretical resolution). This is especially important if the data acquisition device is mounted inside the computer (Figure 3.6c), where the temperature can be easily 30°C or even more above ambient – even in air-conditioned laboratories.

PRACTICAL COMMENT

Both sampling frequency and voltage resolution are especially important, for instance, in Barkhausen noise measurements (see Section 2.5.3), where even at a slow excitation (1 Hz), the investigated signal can have significant frequency spectrum up to 300 kHz but the RMS voltage of only 300 µV (Figure 2.66).

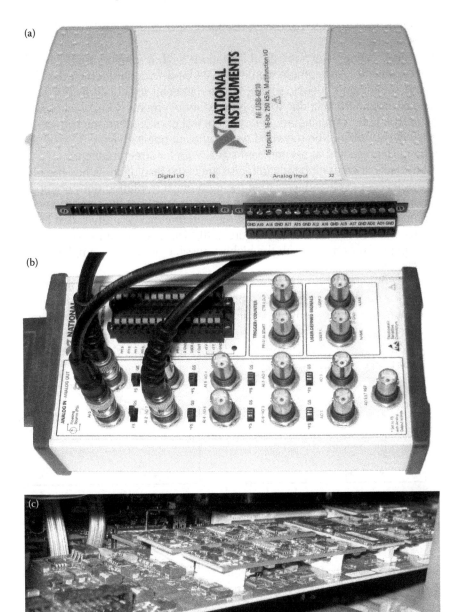

FIGURE 3.6 Example of data acquisition devices: (a) External USB-powered NI USB-6210, 16 bit, 250 kS/s, non-simultaneous sampling, around $700. (NI, 2013a) (b) An external box with BNC connectors for connection of signals to analogue inputs and outputs. (c) Internal PCI-powered NI PCI-6120, 16 bit, 1 MS/s, simultaneous sampling, around $5500. (NI, 2013b) Equipment courtesy of Megger Instruments Ltd, the United Kingdom.

For this reason, high speed and high resolution are required and even 24 bit devices are used.

However, owing to technological limitations, high resolution usually limits the bandwidth. An example would be a 24 bit data acquisition card with 200 kHz sampling frequency, so that the fundamental signal bandwidth was limited to around 90 kHz due to the Nyquist requirements (Chukwuchekwa et al., 2010). If the sampling frequency is not fast enough, then the high-frequency events will not be recorded with sufficient fidelity (Figure 3.7a), or they will be misrepresented as a significant signal with incorrect frequency.

Conversely, increasing bandwidth also limits the resolution. When the number of points per cycle reduces, then the theoretical resolution does not apply anymore because there are not enough points to represent the small changes. Such reduction of performance can be measured with a quantity called ENOB (effective number of bits, see also Figure 4.48). This is yet another reason for performing measurements with the highest possible number of sampled points per cycle (Platil et al., 1999).

Many commercial devices are available, and they are commonly referred to as ADC or A/D (analogue-to-digital converter), DAQ (data acquisition device), DAS (data acquisition system), AI (analogue input), etc. They differ greatly in

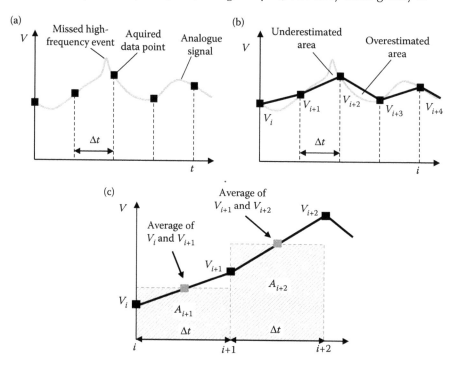

FIGURE 3.7   Digital integration with the trapezoidal rule: (a) digital signal represented by a sequence of equally spaced instantaneous values, (b) approximation of area (digital integration) with a series of trapezoids, (c) area of trapezoids is equal to the area of rectangles whose height is calculated as an average of starting and ending values for each pair of points.

price, with the least expensive at around $100 (as of 2017). Better-performance instruments (higher resolution and sampling frequency) are more expensive and a good-quality device (e.g. 16 bits, 1 MS/s, simultaneous sampling) can cost several thousands of dollars (Figure 3.6c).

The analogue voltage signal is digitised by the device. The digital signal is then represented as a series of numerical values, typically as a one-dimensional (1D) sequential array, as shown in Figure 3.7.

Such digital signal can be processed in the digital domain similarly to an analogue signal in analogue electronics. For instance, the analytical integration by means of Equations 3.5 can be approximated by summation with rectangular, trapezoidal or Simpson rules, depending on the required precision of approximation (NI, 2011).

An integral is mathematically synonymous with the area under the curve. This can be approximated, for instance, with a series of trapezoids, as shown in Figure 3.7b. It should be noted that this is only an approximation with limited accuracy because the assumed partial areas are underestimated in some places, and overestimated in others.

The area A_i under the curve of the ith trapezoid can be calculated numerically as (Figure 3.7b and c)

$$A_i = \frac{V_{i-1} + V_i}{2} \cdot \Delta t \qquad (V \cdot s) \qquad (3.7)$$

or using alternative, but mathematically equivalent notation:

$$A_{i+1} = \frac{V_i + V_{i+1}}{2} \cdot \Delta t \qquad (V \cdot s) \qquad (3.8)$$

This is effectively an area of a rectangle, whose length of one side is equal to the average between the two points (Figure 3.7c).

Commonly, the acquired points are spaced by a fixed interval of time Δt, effectively a constant for a given digital signal that simplifies calculations. In order to find the 'indefinite' integral at a given point (as when converting from $V(t)$ to $B(t)$ values but for the whole *waveforms*), it is necessary to compute a sum of all the preceding values from the beginning of the waveform (current partial sum) (NI, 2011; MathWorks, 2013):

$$Integral_i = \Delta t \cdot \sum_{i=0} \left(\frac{V_{i-1} + V_i}{2} \right) \qquad (V \cdot s) \equiv (T \cdot m^2) \qquad (3.9)$$

which is synonymous with

$$Integral_i = Integral_{i-1} + \Delta t \cdot \frac{V_{i-1} + V_i}{2} \qquad (V \cdot s) \equiv (T \cdot m^2) \qquad (3.10)$$

PRACTICAL COMMENT

For a signal with N points, there will be only $N-1$ trapezoidal areas. For instance, in Figure 3.7c, three data points are shown but only two corresponding trapezoids. So a voltage signal with N points would seemingly translate into a flux density signal with only $N-1$ points. However, the calculation of the first partial area requires a value before the first point at $i-1$, according to Equation 3.10. If this value is not provided as one of the initial conditions, then the first partial area is assumed as zero, and therefore the signal becomes effectively 'shifted' by some offset value (see also Figure 3.8c where all values are positive). If the signal can be assumed symmetrical (e.g. pure sine), then this integration offset can be subtracted in the post-processing, but this is not always possible.

Often, in a typical practical measurement, the signal is periodic and symmetrical. According to Equations 3.1 and 3.2, there should be no offsets in the voltage waveform. So if *exactly* one full cycle (or an integer number of them) of induced voltage is analysed, then the processing can be applied as follows: 1 – remove any DC offset by subtracting the average value from the voltage waveform, 2 – perform the digital integration, 3 – again remove any DC offset by subtracting the average.

Also, it should be noted that the units resulting from Equation 3.9 or 3.10 are $(V \cdot s)$ synonymous with $(T \cdot m^2)$, which is the product of flux

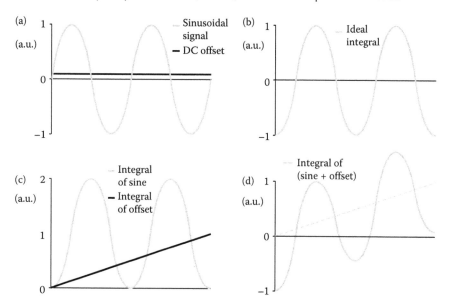

FIGURE 3.8 Normalised waveforms with arbitrary units for clearer illustration: (a) separate waveforms showing sinusoidal input voltage and 10% offset, (b) ideal integral of sine with the initial conditions taken into account, (c) actual integrals of pure sine (without initial conditions) and integral of offset, (d) 'ramped' output resulting from the integral of the sum of sine + offset, even with correctly applied initial conditions; two cycles are shown for better illustration of the ramping effect.

density B (T) and area A (m²), and also number of turns (unitless). The integral is divided by the product of $N \cdot A$ as in Equation 3.5 in order to scale the result with the actual parameters of the B-coil used. This is the primary reason for the input ranges of analogue fluxmeters to be specified in (V · s), and also why the cross section and the number of turns must be entered before the measurement can be carried out – so that the device can return the answer in the correct units of (T). This is not possible if the parameters of the B-coil are unknown.

Obviously, the accuracy of A (or N) directly impacts on the absolute accuracy of detected B and hence also on the accuracy of the measured P.

The information about the area A and number of turns N of the search coil can be used for calculation of the actual B waveform composed from a series of B values. We thus get for each subsequent point of the waveform:

$$B_i = B_{i-1} + \Delta t \cdot \frac{V_{i-1} + V_i}{2} \cdot \frac{1}{N \cdot A} \quad \text{(T)} \qquad (3.11)$$

The calculation of the first point dictates the mathematically introduced offset of the whole waveform. If the initial point is assumed zero, then the whole waveform is biased accordingly, for instance, as shown in Figure 3.8c (all values are positive). It should be noted that an integral of a sine waveform is a negative cosine, which should not have any DC offset (Figure 3.8b). But as shown in Figure 3.8c, the integrated waveform will start at zero, so that a large DC offset is introduced.

In many cases, it is not possible to know in advance the value of the initial condition – in general because the signal is being processed, so the values are unknown yet. For this reason, the initial value can be set to zero, and if the signal is symmetrical, then the resulting bias (Figure 3.8c) can be subtracted in the post-processing (after the integral was calculated).

However, if the input voltage does contain an erroneous offset or a DC component, then this value will also be integrated, resulting with a 'ramped' signal as shown in Figure 3.8d – even if all initial conditions and integration offsets are applied correctly. Such additional erroneous offset can be added, for instance, by the data acquisition device, whose internal offset depends on the operating temperature.

A B-coil made as a single turn wrapped around a single Epstein frame strip (30 mm wide, 0.3 mm thick), which is magnetised sinusoidally at 1 T and 50 Hz, according to Equation 3.3 will output RMS voltage of around 2 mV.

As discussed in the example above, the data acquisition device can introduce its own offset error, which can reach 0.2 mV – or in this example case up to 10% of the B-coil signal. Of course, this offset is not caused by the sensor, but is an artefact of the measurement device. If this offset is not removed (e.g. by numerical calculations), it will be integrated and the calculated B waveform will be severely affected, as illustrated in Figure 3.8d.

This effect is similar to the influence of drift in fluxmeters, as described above. It can be seen from Figure 3.8c and d that the final value of the integrated offset differs significantly from the initial value, even though the two should be equal, as it is the case for the ideal integrated waveform shown in Figure 3.8b. If a *B–H* loop was plotted, then the end point would not meet the start, and the loop would not 'close' properly (Mimura et al., 2012).

Integration with the so-called *Simpson rule* is performed similarly to the trapezoidal method. The difference is that the curve is approximated not with a straight line between the two points, but a polynomial interpolated over at least three points. The mathematical formulae are somewhat more complex, but allow achieving smaller integration errors. However, if pre-defined digital integration functions are used (LabVIEW®, MATLAB®), then more attention is required to supply the correct initial or boundary conditions for correct calculations, which can be more troublesome for some measurements. The apparent benefits of using Simpson rule tend to vanish if higher sampling frequency is used, or if the signal-to-noise ratio is poor because interpolation between points strongly affected by random noise cannot improve the accuracy of the measurement. Rectangular integration is even simpler, producing larger inaccuracies. For this reason, the trapezoidal integration is usually a good compromise between accuracy, ease of implementation and speed of calculation.

The digitised signals are represented as a series of numerical values,[*] usually equally spaced in time. These values are used on a point-by-point basis. For instance, it is necessary to multiply values of *dB/dt* and *H* for each point to numerically compute Equation 2.13. It is therefore clear that in order to be able to do so, the number of points in the *dB/dt* and *H* waveforms must be explicitly equal, and they must be sampled at the same instants of time. This translates into a requirement that sampling frequency for all concerned signals must be the same. Such procedure will result in a new set of values, a new waveform, whose subsequent values are proportional to instantaneous power loss.

This curve must be integrated in a similar way as the integration for *B* is performed, for example, by means of Equation 3.11. Finally, the results are 'scaled' to appropriate values by multiplying by frequency and sample density (Equation 2.13).

If the calculation was performed for a uni-directional magnetisation, then Equation 2.13 would mean calculating the area enclosed by the *B–H* loop. Some instantaneous values would be positive (energy stored) and some negative (energy returned). This is illustrated in Figure 1.18a. The area in question is divided into a number of small areas (e.g. trapezoidal), as many as there as points in the digitised waveforms (precisely *N*–1). The digital integration will follow the rules of Equation 3.11, as illustrated in Figure 3.7.

PRACTICAL COMMENT

Equation 2.13 requires the input variables $dB_x(t)/dt$ and $dB_y(t)/dt$. Obviously, numerical differentiation could be used and calculate these point-by-point

---

[*] Such series of values can be referred to by several names. For instance, in MATLAB (MathWorks, 2014) and Mathcad (Mathcad, 2014), a series of numbers can be a 'vector', which has a different meaning from a vector in physics (an entity that has direction, magnitude and sense). On the other hand, in LabVIEW (NI, 2014a), names such as '1D array' or 'analog waveform' can be used. However, all such sets of subsequent values are numerically synonymous.

waveforms from the $B_x(t)$ and $B_y(t)$ signals. However, as evident from Equation 3.5, $B(t)$ is first calculated from $V(t)$ by means of integration.

Therefore, it is actually better to use the $V(t)$ information directly because no additional errors will be introduced by performing unnecessary in this case integration and differentiation. By rearranging Equation 3.2, we can see that $dB(t)/dt = V(t)/(N \cdot A)$. And since $(N \cdot A)$ is a constant, then it can be taken before the integral.

$H(t)$ must be integrated normally (e.g. if detected by an H-coil). And obviously, the integral of $V(t)$ will still have to be calculated in order to compute the waveform $B(t)$, but this would be a *separate* action from the power loss calculation.

Detailed step-by-step examples of numerical integration for both a waveform and power loss calculation are given in Chapter 8.

3.1.1.3   Practical Implementation

For studying rotational magnetisation, many researchers used B-coils whose width was around 20 mm (Sievert et al., 1995), as shown in Figure 3.9. A single turn is the smallest possible number. For example, the frequency could be varied from 1 Hz and B from 0.5 T, with the sample thickness of 0.27 mm. So the lowest expected voltage can be around 12 μV. This is just about detectable as the least significant change with a 16 bit device, whose resolution is 6 μV at ±0.2 V range. But such change is well below the stated absolute uncertainty of such device, which is usually much wider than the resolution.

The apparent resolution can be improved by a factor of $\sqrt{N}$ (where N is the number of readings) by employing the technique called dithering and waveform averaging (Zurek, 2005a; Zurek et al., 2007a). For instance, an averaging from 100 readings gives a theoretical improvement of resolution by a factor of 10, or from 6 μV to 600 nV. However, it should be noted that the absolute accuracy is *not* improved, and

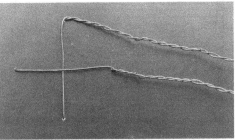

FIGURE 3.9 A typical sample for rotational measurements (H-coils were placed underneath the sample) and single-turn 20-mm-wide B_x and B_y orthogonally positioned search coils at the centre of the sample. (Courtesy of Joanna Kaczmarzyk, CC-BY-3.0, *Encyclopedia Magnetica*.)

can even worsen if the processed signals are not triggered in a precise way (Zurek et al., 2005c). This is described in more detail in Chapter 4.

Three main approaches can be used for boosting of the *B*-coil signal: amplification, increasing cross-sectional area, or increasing the number of turns.

Some measurement systems are equipped with signal pre-amplifiers (Bajorek et al., 1999). The amplitude of the processed signal can be increased so that the sensitivity of the data acquisition device is no longer an immediate problem. However, pre-amplifiers introduce their own errors, which can depend on both the amplitude and frequency of the processed signal. An additional problem is the fact that a signal amplifier will introduce its own noise, offset and phase shift (Scholes, 1970). Such factors are quite difficult to quantify in practice and most researchers tend not to use such pre-amplifiers, unless it cannot be avoided. But if these factors can be quantified, controlled or eliminated (e.g. through a calibration procedure), then amplification can be used. After all, data acquisition devices employ internal amplification in order to achieve functionality of multiple voltage ranges.

The *B*-coil signal can also be improved by using a greater cross-sectional area of the coil. Some researchers use search coils wrapped around the whole available width of the sample. However, it should be borne in mind that the sensor detects *B*, which is averaged over the whole area in question (see B_{mean} in Figure 3.1a). So this method can be used *only* if the area in question is magnetised in a uniform way. This is *very rarely*, if ever, the case. This conclusion applies, for example, to most of rotational magnetising systems, in which the magnetising yokes have two distinct magnetic channels (*x–y*). This was found both by simulations and experiments, with some results illustrated in Figure 3.10 (Boon et al., 1965; Moses et al., 1973; Enokizono et al., 1990a; Alinejad-Beromi, 1992; Enokizono, 1992; Nencib et al., 1995).

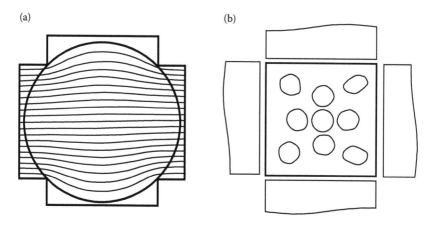

(a)   (b)

FIGURE 3.10   Non-uniformity in two-channel rotational magnetisation setup: (a) As modelled in FEM. (Adapted from Nencib N. et al., *IEEE Transactions on Magnetics*, 31 (6), 1995, 3388.) (b) Measured on a real sample. (Adapted from Alinejad-Beromi Y., *Rotational Power Loss Measurement System under Controlled Magnetization*, PhD thesis, University of Wales, Cardiff, the United Kingdom, 1992; Enokizono M., Studies on two-dimensional magnetic measurement and properties of electrical steel sheets at Oita University, *Proceedings of 1st 1&2DM Workshop*, Braunschweig, Germany, 1992, p. 82.)

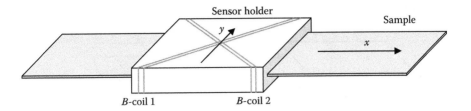

FIGURE 3.11 Two-dimensional measurement system with perpendicular coils placed diagonally (magnetising yokes not shown for clarity). (Adapted from Sasaki T. et al., *IEEE Transactions on Magnetics*, 21 (5), 1985, 1918.)

For this reason, the measurements are usually limited to a small area at the centre of the sample, typically 20 mm wide (Sievert, 1990; Sievert et al., 1995, 1996).

However, some researchers did use *B*-coils wound around the whole width of the sample. Such method was described, for instance, in Sasaki et al. (1985), Enokizono et al. (2003) and Kimura et al. (2016), where the authors decided to place the *B*-coils at 45 degrees to the detection axis (Figure 3.11). The voltage signals can then be mathematically processed to derive the orthogonal B_x and B_y components.

If such coils are exactly orthogonal and placed at $+45°$ and $-45°$ with respect to the reference direction (e.g. x), then appropriate components can be obtained, for instance, just by adding (for x direction) and subtracting (for y direction) (Sasaki et al., 1985).

Later, it was found that magnetising yokes with rotational symmetry offer better homogeneity, so, for example, a 60-mm-wide *B*-coil could be used for an 80-mm-wide sample (Gorican et al., 2003). Also, in 3D magnetisation, it is impractical to use smaller *B*-coils. They have to be positioned at the faces of the sample cube or wound around the whole volume of the sample (see also Figure 2.39) (Zhu et al., 2003).

To be able to make a *B*-coil narrower than the sample, it is necessary to drill holes, which can then be used to thread the wires through them (Figure 3.12).

It is possible to make such *B*-coil with more than one turn. This increases the output voltage and makes it easier to detect. However, from a practical viewpoint, it is very difficult to use a wire thinner than 0.1 mm. The electrical insulation on commonly used enamelled copper wire is easily damaged, and thicker insulation cannot be easily used, for two main reasons.

First, a thicker coil would require larger diameter holes, which would adversely affect the local magnetic properties of the sample. Because the method is destructive, it is generally accepted that the holes should be as small as possible (Loisos et al., 2001). The diameter of the holes impacts directly on the systematic measurement errors, as described in more detail below (Figure 3.18) (Zurek, in press).

The second difficulty is that if flat *H*-coils (described in the following sections) are to be used, then they must be placed as close as possible to the surface of the sample. The two *B*-coils crossing each other at the centre of the sample create a certain local thickness, which significantly increases the minimum distance between the *H* sensors and the sample surface (distance *l* in Figure 3.12c and d).

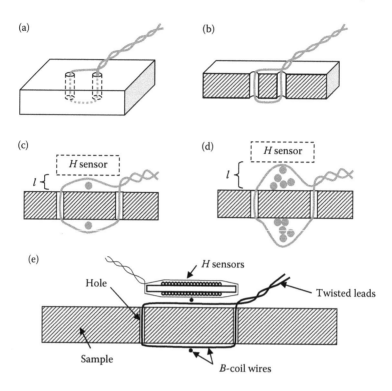

FIGURE 3.12 *B*-coil is made by threading the wire through small holes drilled in the sample (drawings not to scale): (a) overview, (b) cross-sectional overview, (c) cross-sectional view of two crossing coils, (d) distance *l* increases with number of turns, (e) example of positioning of single-turn *B*-coils and corresponding two-axis *H*-coil.

For these reasons, the author decided to use holes with 0.3 mm diameter (see also Figure 3.13) and the enamelled copper wire was 0.1 mm thick with a single turn each, even though their width was just 20 mm (Zurek, 2005a). The detection of the small signals was realised directly by using a 16 bit data acquisition device. No pre-amplification was used. The signals were digitally averaged (which gave significant improvement in the apparent sensitivity) and digitally integrated by means of Equation 3.11, which yielded the flux density waveform for each *B*-coil separately.

3.1.1.4 Practical Difficulties and Measurement Errors

There are several factors that can contribute to errors of *B* measurement. The most obvious one is incorrectly specified number of turns of a *B*-coil. However, typical *B*-coils have few turns (e.g. one or five [Sievert et al., 1995]), so this is rarely a source of error, unless human mistake is made when the value is input into calculations.

Another is the positioning of the drilled holes. Even if they are drilled with a precise machine, their positions will never be ideal. This immediately contributes to errors due to the resulting cross-sectional area of the holes caused by the error Δl as defined in Figure 3.14 caused by the geometrical offset between the ideal and the real position of the holes.

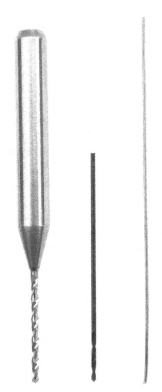

FIGURE 3.13 Drill bits with 'active' diameter of 0.5 mm (left) and 0.3 mm (middle) and enamelled copper wire 0.2 mm (right) suitable for making local B-coils (see also Figure 4.14). (Courtesy of Joanna Kaczmarzyk, CC-BY-3.0, *Encyclopedia Magnetica*.)

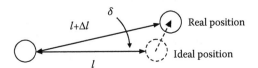

FIGURE 3.14 Actual location of a B-coil always differs from the ideal.

Additionally, a real B-coil is never placed at the ideal angle, so the geometrical offset can cause an additional angular misalignment δ (Figure 3.14). If such a B-coil were to detect a rotating vector, then an additional phase error will be introduced, due to the angular misalignment δ. Of course, the same applies problem for both B-coils but with different values, for x and y directions.

In a general case, it is not possible to know exactly the contribution of such errors. For this reason, the researchers 'take great care' during the preparation of the sensors to minimise all such contributions, so that they can be assumed to be 'negligible' in actual measurements. Estimation of resulting uncertainties can be attempted, but these are open to subjective interpretations and can be significantly underestimated (Sievert et al., 1995), especially that their exact contribution (for instance, towards the measured value of power loss) is complex and unknown in a general case, because it will also depend on the properties of the sample. This problem is discussed at more length in Chapter 5.

Angular misalignment and phase errors can be additionally increased by the fact that wire diameter must be smaller than the hole diameter (Figure 3.15, also compare with Figures 3.1b, c and 3.9) so as to be able to thread the wire through the hole. In the first approximation, it can be assumed that the holes are small enough so they do not disturb the global distribution of flux density (but even this is not true, as discussed below). Hence, the position of the wires in the holes leads to similar errors as the misalignment of the holes themselves (as in Figure 3.14). However, in the worst case, these two effects can add, further increasing the angular misalignment and phase errors.

It should also be remembered that phase errors can be introduced by signal processing. For instance, a multiplexed data acquisition device (i.e. non-simultaneous sampling) introduces small phase shifts between subsequent measured signals, and for the fieldmetric method, at least four signals must be measured: B_x, B_y, H_x, H_y. The partial phase errors can have a very significant influence on the rotational power loss, especially at a higher excitation level (Figure 3.16).

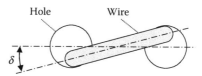

FIGURE 3.15   Possible real positioning of a wire inside the holes (compare with Figures 3.1b,c and 3.9).

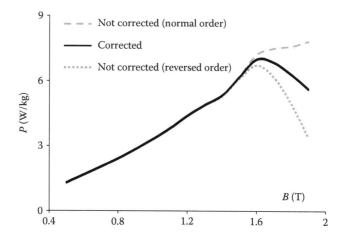

FIGURE 3.16   Rotational power loss errors caused by non-simultaneous digital sampling which introduced a phase shift of 0.06° (360° divided by 1024 points and 6 channels). (Adapted from Zurek S., *Two-Dimensional Magnetisation Problems in Electrical Steels*, PhD thesis, Wolfson Centre for Magnetics Technology, Cardiff University, Cardiff, the United Kingdom, 2005a.)

For this reason, it is best to use simultaneous sampling devices, but in the past, this was not always possible due to cost implications. Such errors can be corrected to a large degree if the characteristics of the data acquisition device are known (Zurek et al., 2005c). Fortunately, digital technology continues to improve and less expensive devices with simultaneous sampling are presently more available and affordable.

B-coils can encircle not only the sample cross section but also a certain amount of air flux (see the empty area inside the coils in Figure 3.12c and d). This is important if the sample is very thin, as, for instance, in the case of amorphous ribbon. Because, in general, the shape of the wire-wound coil cannot be determined with precision, it is therefore not possible to compensate for this air flux, as it is done, for instance, in Epstein frame.

However, it is possible to create *B*-coils by depositing a thin layer of conductive material directly on the surface of the sample under test (Figure 3.17). There are multiple methods that can be employed for such a purpose: electro-deposition, cathodic sputtering in low-pressure discharge, vacuum evaporation of metals, chemical vapour deposition, etc.

Such sensors can be extremely thin, even less than 0.1 μm, while the wire is around 0.1 mm thick. It is also possible to make the deposition through a drilled hole so that a continuous coil encloses the area under test with an absolute minimum of 'air flux' (Basak et al., 1997). An additional difficulty is that the inside surface of the drilled holes must be first insulated before depositing a conductive material.

For this reason, the deposition technique is easier to use with the so-called 'needle probe' approach. The same technique can be used for making a thermocouple. Both types of such sensors are described in the following sections.

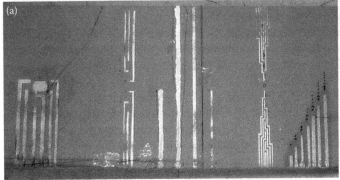

FIGURE 3.17 Sensors made by depositing a thin conductive layer on the surface of grain-oriented electrical steel. (Adapted from Mazurek R., *Effects of Burrs on a Three Phase Transformer Core Including Local Loss, Total Loss and Flux Distribution*, PhD thesis, Cardiff University, Cardiff, the United Kingdom, 2012.) (a) Courtesy of Rafal Mazurek and vacuum chamber for physical vapour deposition and (b) equipment courtesy of Wolfson Centre for Magnetics, Cardiff University, Cardiff, the United Kingdom.

With such deposited coils, the air flux is minimised to a great extent, but the manufacturing procedure is more complex and more expensive. Positioning is prone to similar errors because electrical insulation must be removed in a similar fashion as for a wire-based coil. Additionally, the deposition process is usually controlled by stencils or masks, so the resulting width of the conductive strip is usually quite significant, for example, 1 mm or even wider (Figure 3.17, see also Figure 3.23).

The area of such a search coil can be very small, so the sensing can be very local. On the one hand, this is an advantage because the coils can be positioned in the uniformly magnetised area (e.g. see Figure 3.10b), or within a single grain of grain-oriented electrical steel if such a grain is to be investigated (see also Figure 1.23).

But on the other hand, the 'locality' can also be a disadvantage. Measuring within a single grain would not provide sufficient averaging of the global properties of the investigated grain-oriented electrical steel (see also results shown in Chapter 7). Moreover, the drilling operation does introduce mechanical stresses, which inevitably affect to some degree the local magnetic properties of the sample. Ideally, the sample should be annealed to recover its properties (Loisos et al., 2001; Loisos, 2002). However, annealing cannot be used for domain-refined electrical steel or amorphous materials because the global magnetic properties would be significantly affected by the annealing (Zurek, 2005a).

Last but not least, a drilled hole introduces a significant discontinuity for the magnetic flux, albeit on a small scale as compared to the size of the sample (Tumanski, 2000a; Tamaki et al., 2009; Borg Bartolo, 2015; Wanjiku et al., 2015a; Zurek, in press). Also, the distribution of H is perturbed, capable of creating a change in local H reaching tens of A/m.

The effects of drilled holes are investigated, for instance, in Zurek (in press). As shown in Figure 3.18a, there is 'flux crowding' around the holes (darker means higher amplitude). A B-coil detects B in the direction perpendicular to the wire. As a result, the B-coil encloses the area with elevated B, whose localised amplitude can reach, for example, 1.35 T if the sample was magnetised at 1.0 T as measured far away from such a hole (in the uniform region).

The elevated amplitude reduces with the distance from the hole. However, the B-coil extends to another hole, which has similar elevation. Therefore, for such a B-coil, measuring B under uni-directional magnetisation will always detect values higher because the mean value will always include the elevated edges around the holes. For this reason, the 'B_{mean}' value is used in Equations 3.1 and 3.2.

As evident from Figure 3.18b, the elevated values extend over different distances, depending on the hole size. Therefore, for the same B-coil width (the same distance between the holes), the detected mean value of B will be a function of the hole diameter (or more precisely the ratio between the hole diameter and the coil width) as well as the amplitude of B. The discrepancy can reach 10% for a B-coil with 1 mm holes and 10 mm width (i.e. 10% error for 10% ratio). It should be noted that, although for different reasons, this error is of similar order as it is the case for needle probe technique (Loisos, 2002) described in the next section.

In order to ensure discrepancy smaller than 1%, the hole diameter would have to be less than 0.2 mm for 20-mm-wide B-coils (Zurek, in press). This might be impractical because of the minimum practical hole size, which can be made with a

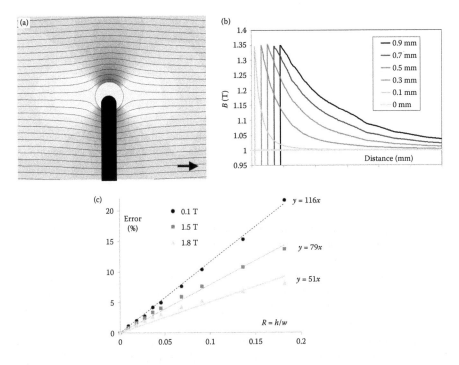

FIGURE 3.18 Distribution of B around holes (more figures are also shown in Chapter 7): (a) qualitative grey-scale colour map of amplitude (black vertical line shows the position of the B-coil wire), (b) distribution of B for different hole diameters at 1 T for non-oriented electrical steel, (c) systematic error versus ratio R (h/w = hole diameter/B-coil width). (Copyright © 2017 *Przeglad Elektrotechniczny* (*Electrical Review*). Reproduced with permission from Zurek S., *Przeglad Elektrotechniczny* (*Electrical Review*), in press.)

drill (see Figure 3.13). Therefore, the B-coil width might be dictated by the hole size in order to ensure sufficiently low level of measurement error.

PRACTICAL COMMENT

The voltage induced in a small B-coil can have a very small amplitude. The coil is also usually positioned at a relatively large distance from the voltmeter circuit. Therefore, there is a certain area encircled between the connecting leads and any variable magnetic flux penetrating this area will induce additional voltage, which will be unwanted and erroneous. The simplest way to minimise this effect is to tightly twist the wires together, as shown in Figure 3.12e, which can lower the unwanted signal significantly, for example, factor of three as compared to loose twisting. Even greater improvement is achieved by using smaller-diameter wires. For instance, reducing the wire diameter from 0.1 to 0.05 mm can reduce it as much as fivefold because tight twisting produces smaller effective loops (Zurek et al., 2008a) (see also Figures 4.17 and 4.19).

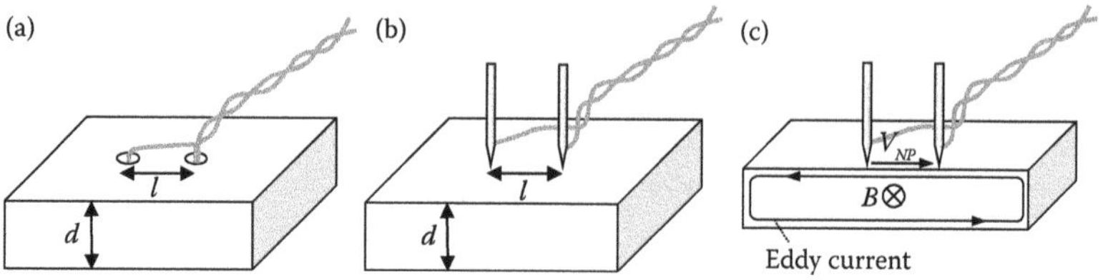

FIGURE 3.19   The concept of the needle probe technique: (a) single-turn B-coil with l and d dimensions, (b) single pair of needles with identical location and (c) cross section through the sample; drawings not to scale.

3.1.2 NEEDLE PROBES

The *needle probe technique* (also referred to as *NP* or *NPT*, *B-probes* or *stylus probes*) was probably described for the first time in an Austrian patent published in 1955 (Czeija et al., 1955).

The technique can be used only with electrically conducting samples because it relies on the presence of eddy currents and voltage drop due to the resistivity of the sample. The method is based on Faraday's law of induction, with similar equations as the B-coil described in the previous section.

If the sample is electrically homogeneous, then its surface can be treated as a single closed turn of an induction coil. Global eddy currents flow within the sample due to varying B. The sample is resistive, so the current flow produces a proportional voltage drop on the surface of the sample (Figure 3.19c). This voltage can be detected by making electrical contacts with the surface of the sample.

For larger sensing area, it is often assumed in practice that the thickness of the sample is significantly smaller than the distance between the needles so that $d \ll l$. For a single pair of needles (Figure 3.19b and c), the voltage V_{NP} can be therefore approximated as (Loisos, 2002)

$$V(t) \approx \frac{1}{2} \cdot d \cdot l \cdot \frac{dB(t)}{dt} \quad \text{(V)} \qquad (3.12)$$

Equation 3.12 is not strict because effectively an assumption is made that the sample is infinitely thin. This introduces an error, which is directly proportional to the ratio d/l. For instance, if the sample has a thickness of 0.5 mm and the distance between the needles is 5 mm, then the expected theoretical error would be 9.1% (Loisos, 2002). Such error arises because the voltage that would be induced along the thickness of the sample is not detected (as it would be with a comparable B-coil). This can lead to indicating non-physical values, for instance, in excess of 2 T even if the actual value present in the sample was $B = 1.4$ T (Loisos et al., 2001).

PRACTICAL COMMENT

The above mentioned assumption ($d \ll l$) should be compared with the form of Equations 1.21 to 1.23, where a similar approach was taken.

The relationship between the electric field E and flux density B is visualised in Figure 1.26. E has the unit of (V/m), whereas B is measured in (T) $\equiv$ (s · V/m^2). Only change of B in time t contributes to the electric field, so the unit of dB/dt is (T/s) $\equiv$ (V/m^2).

Therefore, a sensor such as B-coil effectively integrates the electric field E (V/m) around the area with dimensions of length l and thickness d, so the total length of the perimeter of 2 · ($l + d$) with the units of (m). The number of turns N of a B-coil would contribute to how many times this integration would be performed around a given area. Time t (s) is involved because only the change of B in time contributes to the electric field so that dB/dt has the unit of (T/s).

Therefore, the voltage induced in a B-coil is $V = E \cdot 2 \cdot (l + d) \cdot N$ with the units of (V/m) · (m) = (V). For the needle probe approach, the voltage between the needles is proportional just to the length between the needles, which is $V = E \cdot l$ (Figure 1.26). Hence, for a B-coil, the proportionality is 2 · ($l + d$) · N, whereas for the needles, it is just l. The ratio of these two factors is $l/[2 \cdot (l + d) \cdot N]$, so when compared to a single-turn B-coil, the voltage induced between the needles will be $l/[2 \cdot (l + d)]$. If the assumption $d \ll l$ can be made, then the factor becomes $l/(2 \cdot l) = 0.5$, which is why the factor ½ is used in Equation 3.12.

So, strictly speaking, as derived above, the factor $l/[2 \cdot (l + d)]$ should be used instead of ½. For the hypothetical dimensions mentioned above of $l = 5$ mm and $d = 0.5$ mm, the factor becomes $5/[2 \cdot (5 + 0.5)] \approx 0.4545$, which differs from 0.5 by the value of 9.1% mentioned above.

However, it should also be remembered that there can be some additional effects that can affect the induced voltage. Non-uniform magnetisation was mentioned above but also large grains (in electrical steels) can distort the local distribution of B and thus the induced voltage. Large increase as well as decrease of local B can be experienced for grain-oriented electrical steel. The difference can be so large that even B-coil voltage can be easily influenced (see also Figure 7.31b), so the effect on the voltage in the needle probe technique can be expected to be of even large magnitude.

A modification of the method is required in order to measure these values with greater accuracy. The concept of the modified needle probe is illustrated in Figure 3.20.

Figure 3.20a shows a configuration with two B-coils. Two holes are drilled and wires are threaded through them. The wires run to the edge of the sheet, rather than being connected above the sample, as it is usually the case for the B-coils (Figure 3.19a).

As a result, the two induced voltages V_{A2} and V_{A1+A2} are directly proportional to the areas A_2 and the sum of $A_1 + A_2$, respectively.

The two B-coils can be replaced with two pairs of needles located at the same points as the holes. Neglecting positioning errors and air flux, the induced voltages are analogous as for the configuration with the B-coils. Therefore, in order to measure B in the area A_1, the two voltages would need to be subtracted:

$$V_{A1} = V_{A1+A2} - V_{A2} \qquad \text{(V)} \qquad (3.13)$$

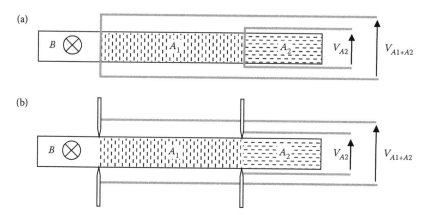

FIGURE 3.20   Illustration of the 'modified' needle probe technique: (a) configuration with two B-coils and (b) corresponding configuration with two pairs of needles. (Adapted from Loisos G., *Novel Flux Density Measurement Methods of Examining the Influence of Cutting on Magnetic Properties of Electrical Steels*, PhD thesis, Wolfson Centre for Magnetics, Cardiff University, Cardiff, the United Kingdom, 2002.)

The subtraction can be performed numerically, or, for instance, by the use of a bridge circuit (Werner, 1957; Loisos, 2002).

The greatest advantage of NPT is the fact that the probes can be moved on the surface of the sample, thus providing a localised measurement at a chosen area of interest. This cannot be done with B-coils because localised measurements can be made only by drilling appropriate number of holes in the sample under test, and the more holes are drilled, the more the structure of the sample is affected. For measurements of *very small* areas, the B-coil method cannot be used as it is too destructive (Loisos et al., 2001; Crevecoeur et al., 2008; Zurek, in press).

Unfortunately, the 'modified' NPT is difficult to use in practice because it would require running the wires to the side of the sample. The measurement can be made, but in general, free movement is not possible. If multi-position measurements have to be made, then multiple pairs of needles can also be employed (Crevecoeur et al., 2008). Hence, the most common approach is to use the 'simple' NPT as shown in Figure 3.19b and c.

For ease of use, spring-loaded probes are often employed, usually around 25 mm long (Figure 3.21). These are widely used in the testing of electronic circuits, in the so-called 'bed-of-nails' method (Peak Group, 2014). The probes have thicker outer part so that the thinner tip can slide inside, with a telescopic action, by compressing the inner spring. There are many types of needles, but the most useful ones are those with sharp tips such as conical or multi-sided (Peak Group, 2014). The spring helps in holding a steady force keeping the needle tip against the surface of the sample.

One of the applications of the needle probes for 2D measurements is shown in Figure 3.22 (see also Figure 4.17). An electrical steel sheet sample is used with removed coating so that the needles can make an electrical contact with its conductive surface. The needles are mounted on a suitable head, which is attached to a computer-controlled 3D positioning system (Konadu, 2006).

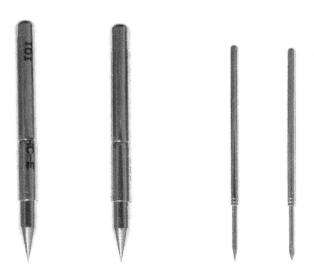

FIGURE 3.21   Typical spring-loaded needle probes usable for NPT, with conical tips on the left and multi-face on the right (see also Figure 6.22); length of the needles is around 25 mm. (Courtesy of Joanna Kaczmarzyk, CC-BY-3.0, *Encyclopedia Magnetica*.)

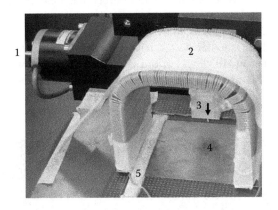

FIGURE 3.22   An example of a setup with movable measurement head (the arrow indicates a needle probe protruding from the measurement head): 1 – 3D positioning system, 2 – magnetising yoke, 3 – measurement head with sensors, 4 – sample, 5 – global *B*-coil. (Konadu, 2006) Equipment courtesy of Wolfson Centre for Magnetics, Cardiff University, Cardiff, the United Kingdom.

The measurement is carried out by lowering the head to the surface of the sample. The needles make contact with the sample. The induced signals are measured with a data acquisition device. Two pairs of needles are used, for B_x and B_y components. The H_x and H_y signals are measured by a 3D Hall effect sensor (described in the following sections) placed between the needles (Konadu, 2006; Moses et al., 2008).

After the signals are measured at a given location, the head is lifted clear off the surface, moved to a new position and lowered again for the next measurement.

The method can be used for the measurement of selected areas, as well as for scanning the whole surface, location after location.

The amplitude of localised *B* can change from one position to the next. For this reason, a global *B*-coil was used in order to control the excitation of the sample with a constant amplitude.

The 'needle probes' can also be made by the deposition of conductive tracks (Figure 3.23). The surface of electrical steel is covered with an electrically insulating layer. The insulation is removed in pre-defined locations. Conductive tracks are deposited, so that electrical contacts are made with the body of the sample through the spots of removed insulation. The tracks are guided to the side, where terminating wires are connected, for signal measurement.

The main advantage of such a solution is the very small thickness of the overall setup, around two orders of magnitude thinner than if needles or wires were used (Mazurek, 2012). Of course, with such an approach, it is not possible to move the locations because by definition the sensors are fixed to the surface. Another advantage is the elimination of air flux influence, which is not a trivial problem when needles are used.

An interesting theoretical analysis of another modification of the needle probe method was given in Loisos et al. (2001) and Loisos (2002). The mathematical derivation is too long to be repeated here in its entirety, but it can be illustrated with the schematic drawing in Figure 3.24.

The main idea is to use five needles, one of which is movable. The areas A_1 (large grey triangle) and A_2 (smaller dotted triangle) can be calculated if the distances between the needles and thickness of the sample are known. Owing to the similarity of the angles, it can be seen that the area A_2 can be subtracted from A_1, giving a quadrilateral figure (whose area in Figure 3.24 is grey but not dotted).

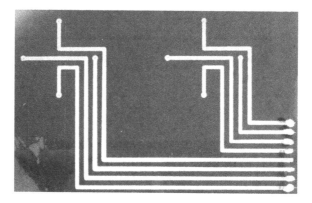

FIGURE 3.23   NPT-type sensors made by depositing a thin conductive layer on the surface of grain-oriented electrical steel (the image shows 160 mm wide area). (Adapted from Mazurek R., *Effects of Burrs on a Three Phase Transformer Core Including Local Loss, Total Loss and Flux Distribution*, PhD thesis, Cardiff University, Cardiff, the United Kingdom, 2012. Photograph courtesy of Rafal Mazurek, formerly Wolfson Centre for Magnetics, Cardiff University, Cardiff, the United Kingdom.)

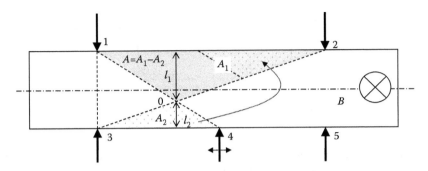

FIGURE 3.24 Diagram of the method proposed by Loisos. (Adapted from Loisos G., *Novel Flux Density Measurement Methods of Examining the Influence of Cutting on Magnetic Properties of Electrical Steels*, PhD thesis, Wolfson Centre for Magnetics, Cardiff University, Cardiff, the United Kingdom, 2002.)

Voltages of all needles are measured, and they can be added and subtracted as required. It can then be shown mathematically that the difference between voltages $V_{2,3,5}$ (triangle 2–3–5) and $V_{1,3,4}$ (triangle 1–3–4) is proportional to the difference of areas $A = A_1 - A_2$.

As a result, the measurement can be limited to the differential area $A = A_1 - A_2$, and the shape of this area can be changed by moving the central needle 4. If needle 4 moves towards 5, then point 0 moves towards the centre of the sample, area A_1–A_2 reduces, but it has similar width for all depths of the sample.

On the other hand, if needle 4 moves towards 3, then point 0 also moves towards 3 and the differential area becomes more triangular, thus presenting the varying width for different depth levels.

Therefore, such method could be potentially useful for measuring B distribution through the thickness of the sample. This could be important in the context of Equation 1.33, which would require the knowledge of distribution of flux density through the thickness of the sample.

Unfortunately, no such measurements were presented to date. The practical limitations of the needle probe method (described in the next section) would have to be controlled to a large degree before such measurement can be successfully carried out with sufficient resolution and accuracy.

Another hypothetical implementation of NPT was described in Zurek (2005a) and Zurek et al. (2006d). The underlying concept is the same as for the regular use of the method. However, the electrical contact between the needles and the sample could be replaced by capacitors formed by conductive pads placed or deposited directly onto the insulating layer (Figure 3.25).

The capacitance of each such capacitor is dictated by the area A_C of its conductive pad, the thickness of the insulating layer, as well as the electrical permittivity of the insulator (Figure 3.25).

However, the pad area A_C would be limited in a practical setup. Hence, the resulting capacitance would be small (typically significantly less than 1 nF), and its impedance around power frequency (50 Hz) would exceed 1 MΩ. For correct measurement, the internal impedance of the voltmeter (Figure 3.25) would have to be

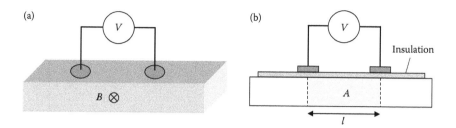

FIGURE 3.25 Capacitive effect used as means of detecting the needle probe method (compare with Figure 3.19b): (a) general view and (b) cross-sectional view. (Adapted from Zurek S., *Two-Dimensional Magnetisation Problems in Electrical Steels*, PhD thesis, Wolfson Centre for Magnetics Technology, Cardiff University, Cardiff, the United Kingdom, 2005a.)

significantly larger (perhaps TΩ) to allow accurate measurement of such voltage. As a result, such measurements could not be carried out with a voltmeter, but it would require a more sophisticated device, such as an electrometer.

In order to avoid the influence of air flux, the connecting leads would need to be twisted, and the length of such twisted wires could contribute a significant parasitic capacitance (tens of pF).

As stated in one of the previous sections, a single-turn B-coil wrapped around an Epstein strip, 30 mm wide and 0.3 mm thick and magnetised sinusoidally at 1 T outputs a voltage of around 2 mV. According to Equation 3.12, a pair of needle probes would therefore produce around half of that voltage, so 1 mV. Typically, excitation of the sample requires several volts (or more) to drive the magnetising coil with sufficient current. The capacitance between the magnetising coil and the yoke, as well as the sample, would be much greater than the capacitance of the capacitive pads from Figure 3.25. As a result, the body of the sample would be floating at a common-mode voltage, whose amplitude is several orders of magnitude greater than the signal to be measured (volts versus millivolts).

The rejection of such high common-mode signal in connection with very high-internal impedance of the voltmeter is outside of the technical capabilities of commonly used conventional voltmeters and data acquisition devices. Increasing the magnetising frequency reduces the capacitive impedance of the sensor, but equally reduces any other parasitic capacitive impedance such as the twisted wires and the coil–yoke–sample system. The author made a prototype of such a sensor, but due to the abovementioned difficulties, it was not possible to perform a successful measurement of B (Zurek, 2005a). Much more controlled environment and/or better measurement equipment would have to be used.

3.1.2.1 Signal Processing

The level of output voltage of the needle probe technique is similar to a B-coil sensor of comparable size. The voltage waveform is proportional to the derivative of the flux density waveform. The amplitude is lower, but in the first approximation is on the same order of magnitude as for a single-turn coil.

Therefore, all aspects of signal processing described above for B-coils are directly applicable also the needle probe method. The same analogue or digital signal-processing equipment can be used for both methods, so the measurement results can be directly compared.

The calculation of B value is performed by using integration, but of course Equation 3.12 should be used in combination with Equation 3.5, hence effectively assuming that $N \approx 0.5$ (for the 'simple' configuration of NPT).

3.1.2.2 Practical Difficulties

Needle probe technique offers the possibility of a measurement that is practically *almost* non-destructive, localised and movable to a different position on the sample. The sensors can be made so small and thin that a hand-held measurement head can be inserted even between laminations of a transformer core (Krismanic et al., 2000).

However, there are several practical aspects that need to be taken into account in order to perform accurate measurements. Some of such practical difficulties are described below, but this is not meant to be an exhaustive list, but rather an indication of the likely problems to be encountered by researchers.

Precise positioning of the needles on the surface of the sample requires mechanical precision. Commonly used spring-loaded needles (Figure 3.21) have their own mechanical tolerances. The tip is retractable, so, apart from the axial movement, due to the lack of absolute rigidity, there can be additional relative radial motion of the sharp tip with respect to the outer body of the probe. If the spacing between the needles in a pair is 5 mm, then movement of a tip by 0.1 mm represents a potential error of $\pm 2\%$.

The non-movable body of the probe must be held by some other mechanical arrangement, whose precision would be prone to similar misalignment than that described for B-coils (Figure 3.14).

For just two needles, the *absolute* accuracy of placement is less relevant because the relative distance between the needles and their angular alignment with respect to the sample is much more critical.

However, for the 'modified' technique, four needles are required – two on each side of the sample (Figure 3.20). In such a configuration, the precise location of the 'top' and the corresponding 'bottom' needles must be controlled *at the same time*. Depending on the mechanical configuration, from a practical viewpoint, this could mean that precision in absolute sense would have to be required, and this is usually more difficult to attain.

The problem of relative or absolute positioning becomes even more pronounced if the sample is not magnetised uniformly. The apparatus shown in Figure 3.22 is far from ideal because the magnetising C-yoke is placed only on one side of the sample. At the interface between the sample and the yoke, there is a large component of magnetic flux, normal to the surface of the sample. This can generate significant planar eddy currents, which are capable of locally affecting the symmetry of magnetisation within the thickness of the sample.

A pair of C-yokes, one on each side of the sample, offers a much better symmetry (Crevecoeur et al., 2008), and it is one of the reasons for such a configuration recommended in the single-sheet tester method (Figure 2.22) described by the international standard (IEC, 2010), rather than just a single-sided yoke.

The symmetry and uniformity of magnetisation is critical for the correct measurement of flux density with the needle probe method (Loisos et al., 2001; Oledzki, 2003; Pfutzner et al., 2004). Otherwise, non-physical amplitudes can be measured, which can not only approach the saturation flux density of the material under test (>2 T [Loisos et al., 2001]) but can even reach completely non-physical values (>4 T [Pfutzner et al., 2004]), which are obviously erroneous.

Another important source of error can be the so-called 'air flux'. Spring-loaded needles (Figure 3.21) have retractable tips. From a mechanical viewpoint, it would be easier to attach the wires to the non-retractable body. But this would lead to enclosing a large area through air, between the needles, the connecting wires and the voltage-measuring instrument – similarly as in Figure 3.25. Any B in air (i.e. the 'air flux') would induce additional voltage and thus introduce an error in the measurement. This must be avoided, for instance, by attaching the wires to the *retractable tips*, close to the sample, and not to the top of the needles (see Figure 3.19b and c as well as Figure. 4.17). At the very least, the wires should be attached at the lowest point of the non-movable parts as visible in Figure 6.22. Some researchers used copper foil positioned close to the sample surface in order to minimise this effect (Sievert, 1990). This problem is also described in more detail in Section 4.4.3 (Figure 4.72c).

The voltage signal induced between the needles is relatively small, and wires should be guided in such a way as to reduce the distance between the wires and the body of the sample (Figure 3.20b). For movable or scanning systems, this is difficult (Figure 3.22), but the deposited thin-film sensors easily fulfil this condition (Figure 3.23). Another way would be, for example, to use PCB technology to mount the needles. The electrical connections could be then made as conducting tracks on the surface of the PCB so that they would be displaced only by the thickness of the coating on the PCB (see e.g. Figure. 6.24). Such PCB could also comprise integrated H-coils as well as miniature signal amplification mounted directly where the sensors are placed (Zurek, 2017e).

On the one hand, needle probes allow easier access to smaller area under test than it would be possible with B-coils. But the thickness of the needles is finite, and this dictates physically the smallest distance between them. Equation 3.12 is only approximate and eventually the error resulting from neglecting the thickness of the sample becomes significant and comparable to the measured signal (Loisos, 2002).

Also, for the fieldmetric method, an H sensor should be positioned *between* the needles (Figure 3.26), and this is one of the greatest advantages of the NPT. B and H sensors can be integrated into one 'sensing head', which can be freely moved

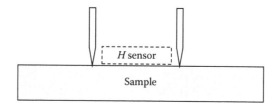

FIGURE 3.26 H sensor can be placed between the needles, very close to the surface of the sample (see also Figure 3.12).

around. Probably one of the first uses of such integrated head was described in Stauffer (1958), which was later published also in Tompkins et al. (1958) (see also Figure 3.55).

The presence of an H sensor poses another limit on the proximity of the needles. For rotational measurements, larger distances are advisable (e.g. >20 mm), but smaller spacing can also be used, for example, 6.3 mm in Moses et al. (2008) or even less (Aihara et al., 2011).

To overcome the linear resolution, some researchers used a row of several closely spaced needles, which were positioned at 0.16 mm from each other, but the voltages were measured with respect to a much more distant needle (15 mm away). Such modifications are useful for investigating the degradation of magnetic material at the edge of the sample due to the employed cutting technology (Crevecoeur et al., 2008).

For a B-coil, it could be possible to boost the voltage by increasing the number of turns (see Figure 3.12), but for needle probes, there are no 'turns'. So the only way to increase the voltage is to increase the distance between the needles, according to Equation 3.12, or to use the 'modified' technique, which encloses a greater cross-sectional area of the sample. Alternatively, signal amplification could also be used.

Another point to consider is that the needle probe method requires electrical contact with the conductive body of the sample. So the measurement can be performed at most on a *single* lamination – a restriction that does not apply to the B-coil method. For NPT, the single lamination can be a part of a larger core, but the measurement would be limited to the single lamination in electrical contact with the needles.

For the laminated core to perform well magnetically, the separate laminations must be isolated from each other electrically. In electrical steels, this is commonly achieved by appropriate insulating coating, which is quite hard mechanically. The needles must be either exceptionally sharp or be pressed with a large force to puncture through such insulation. Alternatively, the sample can be used in an uncoated version (Figure 3.22), either by the coating not applied in the first place, or removed later. In any of such situations, the non-destructiveness of the method is highly questionable because the introduction of mechanical pressure is capable of changing the domain structure (Figure 2.51), and absence of coating can completely change the magnetic properties of the sample. In such case, the decoating of the sample could affect the measurement in a greater degree than the abovementioned errors of the needle probe technique.

On the other hand, the coating on electrical steels is often very hard. If the needles are required to puncture through the coating, they get dull quickly and must be replaced frequently. This can further contribute to the unknown positional misalignment of the tips, and calibration of distances of moving tips is quite difficult, if possible at all.

Last but not least, the NPT method relies on the resistivity of the sample under test. It was shown in the literature that there can be a certain amount of anisotropy of resistivity in electrical steels (Soinski, 1985). There is no information in the literature on rotational measurements how the anisotropy of resistivity could impact on measurement accuracy of the needle probe technique. It would be interesting to investigate such effects.

3.1.3 INDIRECT B-COIL

The B-coil method described in the previous section is a direct measurement, in the sense that the sensor measures flux density in the position where it is placed. There is also a technique that allows an *indirect* measurement, albeit with reduced absolute accuracy. The principle of such indirect measurement is depicted in Figure 3.27.

A magnetising coil is wound on a yoke (magnetic core), which is placed directly on the sample (assuming no yoke–sample air gap). The method relies on the assumption that the amplitude of magnetic flux Φ is constant throughout the magnetic circuit, hence

$$\Phi_{yoke} \approx \Phi_{sample} \quad (Wb) \tag{3.14}$$

where Φ_{yoke} is the magnetic flux in the yoke (Wb) and Φ_{sample} is the magnetic flux in the sample (Wb).

However, the amplitude of flux density is not constant due to different cross-sectional area of the yoke and the sample. So, from $\Phi = B \cdot A$, it follows that

$$\Phi_{yoke} = B_{yoke} \cdot A_{yoke} \approx B_{sample} \cdot A_{sample} \quad (Wb) \tag{3.15}$$

where B_{yoke} is the flux density in the yoke (T), A_{yoke} is the cross-sectional area of the yoke (m²), B_{sample} is the flux density in the sample (T) and A_{sample} is the cross-sectional area of the sample (m²).

An 'indirect' B-coil is wrapped around the yoke as close as possible to the surface of the sample, as shown in Figure 3.27 (Oka et al., 1998a). The induced voltage $V_{indirect}$ is proportional to the number of turns (N) of the indirect B-coil, as well as the magnetic flux in the yoke:

$$V_{indirect} = N \cdot A_{yoke} \frac{dB_{yoke}}{dt} \quad (V) \tag{3.16}$$

Hence, from Equations 3.15, 3.16 and 3.5, it can be derived that

$$B_{sample} \approx \frac{1}{N \cdot A_{sample}} \cdot \int V_{indirect} dt \quad (T) \tag{3.17}$$

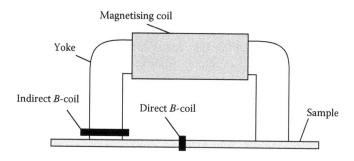

FIGURE 3.27 Concept of an indirect B-coil.

There are at least two important additional effects that must be taken into account, namely, if the sample is wider than the yoke (Figure 3.28), then the magnetic flux spreads out (Wood et al., 2009), and the magnetised cross-sectional area of the sample is therefore difficult to estimate; hence Equation 3.17 cannot be used without additional correction factors, or using dummy yokes to improve the flux distribution.

The other factor is the fact that the magnetic flux in the sample is also assumed to be uniform. This might not be the case for a 2D measurement. For this reason, this method does not seem to be reported in the literature for rotational measurements (the yokes might be used for excitation, but not for indirect measurement). However, it can be used for localised measurements, also with automatic scanning of the surface (Moghadam et al., 1993).

Nevertheless, the technique was employed in commercial testers for electrical steels (Hall, 2001; Soken, 2014) and even for low-permeability (non-ferromagnetic) materials (Bajorek et al., 2000a). A commercial example is shown in Figure 3.29. The device is capable of measuring the specific power loss of electrical steels at 50 and 60 Hz under uni-directional excitation, with specified 5% accuracy as compared to the Epstein frame method.

However, precise knowledge of the thickness of the sample (to within 0.01 mm) is required and must be manually entered (which is a consequence of the approximate Equation 3.17).

3.1.4 NON-ENWRAPPING *B*-COIL

The *non-enwrapping method* is an indirect measurement, in some sense similar to the Soken tester described above. An induction coil is used to detect *B*, but the coil can be applied in a 'non-enwrapping' way. Such a method is useful especially in the electrical steel industry for on-line measurements, where enwrapping coils cannot be used due to technological restrictions (Khanlou et al., 1992; Beckley, 2000b; Beckley et al., 2000c; Passadis et al., 2003).

Of course, the excitation must also be applied in a non-enwrapping way. An example is shown in Figure 3.30, where a single layer of conductors effectively creates a 'current sheet'. The current is returned in 'remote bundles'.

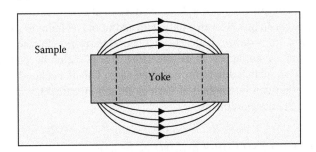

FIGURE 3.28 Top view showing flux spreading in a sample wider than the yoke (see also Figure 3.27).

(a)

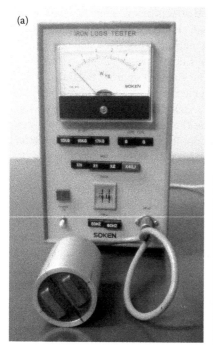

(b)

FIGURE 3.29   Soken tester uses indirect B-coil for flux density measurement: the whole device (a) and the measurement head (b). Equipment courtesy of Wolfson Centre for Magnetics, Cardiff University, Cardiff, the United Kingdom.

Underneath the 'current sheet', the sample is magnetised to the maximum flux density. However, there is no defined returning path, so the flux eventually must leave the sample sheet and return back through the surrounding air. A sensing coil can be positioned in a similar non-enwrapping way so that it senses the returning flux (Figure 3.30c). Knowing the dimensions involved, it is possible to define and calibrate performance so that B in the sample under test is calculated from the measured signals.

However, the method is more applicable to large-area measurements because the relationship between the induced signals and the measured flux density is neither direct nor linear. For this reason, it will not be described here in detail. Measurements (with 5% accuracy) can be performed only with additional techniques such as non-linear compensation, for instance, with artificial neural networks (Passadis et al., 2003).

Although there are no reports in the literature of application in rotational studies, the description was included here for completeness. This is because with this technique the excitation can be applied at various directions (Beckley, 2000a), thus in theory allowing 2D measurements.

3.1.5   HALL EFFECT AND OTHER ELECTRONIC SENSORS

There are several sensors in which the proportionality between the detected B and the output voltage V of the sensor is direct. Therefore, there is no need for integra-

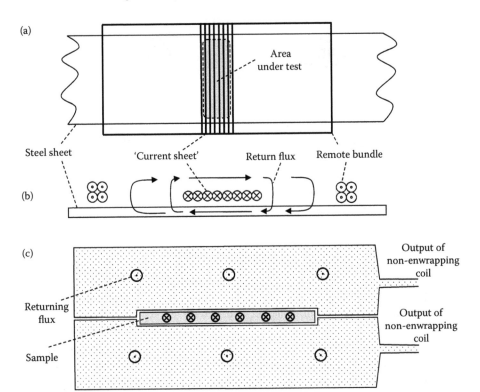

FIGURE 3.30 Non-enwrapping tester: (a) top view, (b) side view, (c) cross section of the sheet and the non-enwrapping B-coils (drawing not to scale). (Adapted from Beckley P., *Electrical Steels, A Handbook for Producers and Users*, European Electrical Steels, 2000a, ISBN 0-9540039-0-X.)

tion by means of Equation 3.5, but the instantaneous B value can be calculated by means of a proportionality factor C_x (T/V), also referred to as *gain*, *constant* or *sensitivity*:

$$B(t) = C_x \cdot V(t) \quad (\text{T}) \tag{3.18}$$

Hall effect sensor (Figure 3.31) is a popular type of such transducer. There are two main types of 'Hall sensors', the so-called switch type and linear.

When a *switch*-type sensor (Figure 3.30b) is exposed to a sufficiently strong source of magnetic field, then the output changes state to 'on', and when the field is removed, it switches to 'off' (or vice versa). Such 'on/off' characteristics is used, for instance, in proximity switches, and cannot be used for sensing waveforms, but it is shown here just for illustration.

The output signal V_H of a *linear* Hall effect sensor (Figure 3.30a) changes proportionally to the amplitude of B, which penetrates the sensor, so that Equation 3.18 is fulfilled within given limits.

(a)   (b)

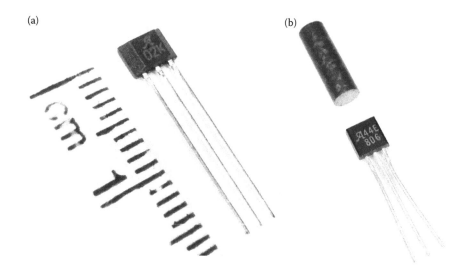

FIGURE 3.31   Hall effect sensors: (a) Linear A1302. (Allegro, 2013a) (b) Switch-type A3144 with an activating permanent magnet. (Allegro, 2013b)

The operating principle of a Hall effect sensor is illustrated in Figure 3.32. A current I is driven along a semiconducting plate. When there is no external magnetic field, then the electrons flow straight along the plate and there is no voltage difference across the plate.

However, when flux density B is applied perpendicularly to the surface of the plate, then the electrons are deflected from their paths (see the curved arrow in Figure 3.32). This leads to a concentration of charges on one edge, which can be measured as a voltage difference (the so-called Hall voltage V_H). The effect is directly proportional and if the direction of B is reversed, the voltage V_H also changes its polarity.

The sensitivity is also proportional to the amplitude of I, but this cannot be too high because the plate has a certain electrical resistance, which can produce

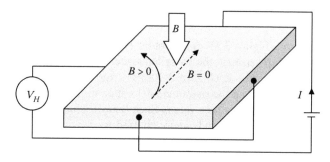

FIGURE 3.32   Hall effect sensor. (Adapted from Tumanski S., *Handbook of Magnetic Measurements*, CRC Press, 2011, ISBN 9781439829516.)

self-heating due to the applied current. A data sheet provided by the manufacturer describes all parameters and conditions for correct use of such devices.

For instance, a typical device (Allegro, 2013a) requires a nominal supply voltage of 5 V and during normal operation it consumes a maximum of 11 mA. The sensor contains an electronic circuit, with a current source, precise amplifiers, filters, trim control, etc. The output voltage without an applied magnetic field is 2.5 V (a half of the supply voltage) and increases or decreases linearly with the applied B. Therefore, all measurements must be performed by subtracting the offset, either numerically or by differential measurement to a suitable 50% voltage divider connected to the same power supply (e.g. with a bridge circuit).

For this particular sensor, the sensitivity is specified in the data sheet as '1.3 mV/G', which in SI units is 13 V/T or 13 mV/mT. So, for each 1 mT of applied B, the output voltage would change by 13 mV.

According to Equation 3.18, in order to calculate B from the output voltage V_H, the proportionality factor should be expressed in (T/V), which is a reciprocal of the sensitivity; hence, 1 V/13 T = 0.0769 T/V.

For this specific sensor, the maximum voltage change can be only 2.5 V (from the mid-point of 2.5 to 5 V or from 2.5 to 0 V). So the maximum measurable B would be ±0.19 T. Any higher B would saturate the output of the sensor. So, in order to measure higher B, a sensor with lower sensitivity would have to be used.

The sensitivity varies from sensor to sensor of the same type, even by as much as 30% and the manufacturer gives at best only a minimum, typical and maximum values (Allegro, 2013a). This means that *each* sensor should be calibrated individually.

As mentioned above, the sensor detects B. However, the sample–sensor configuration should be carefully analysed before such sensor could be used for the measurement of a particular B component.

Theoretical analysis shows that the normal component (perpendicular to the surface) of B vector does not change at the boundary between two media with different permeabilities μ_1 and μ_2, although the total amplitude or the tangential components can differ significantly as illustrated schematically in Figure 3.33 (Johnk, 1975; Singh, 2011).

For instance, it can be assumed that permeability μ_2 represents the sample under test and μ_1 the surrounding air. Therefore, the normal component B_{2n} just inside the

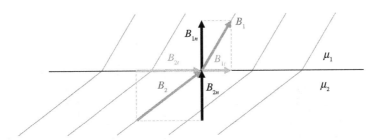

FIGURE 3.33 Normal component of flux density does not change at the boundary, so $B_{1n} = B_{2n}$. (Adapted from Johnk C.T.A., *Engineering Electromagnetic Fields & Waves*, Wiley International Edition, New York, USA, 1975, ISBN 9780471442905; Singh Y., *Electro Magnetic Field Theory*, Pearson Education India, 2011, ISBN 9788131760611.)

sample is equal to B_{1n} just outside. Hence, measuring the normal component directly at the surface of the sample should in theory be equal to that inside of the material – but *only* in the case of the normal component, and *only* very close to the surface.

The difficulty of using this method lies in the configuration of the yoke–sample system and the physical sizes involved. As can be seen from Figure 3.31, a typical Hall effect sensor is significantly larger (e.g. $1.5 \times 3 \times 4$ mm) than the thickness of typical electrical steel sheet (~ 0.5 mm).

In order to measure B within the plane of the sample, the sensor would have to be placed in an air gap, as shown in Figure 3.34. This would present a complete magnetic discontinuity at least on the order of 1.5 mm long (i.e. the thickness of the sensor). Even if the edge of the sample could touch the body of the sensor, most of sensor volume is just a non-magnetic packaging with the actual sensing area being much smaller (Allegro, 2013a). And if the sensor was placed inside of a little window (Figure 3.34, right), then the flux would flow around the discontinuity, preferring the path of lower reluctance (similarly as shown for the round hole in Figure 3.18a). Of course, the equality of normal components according to Figure 3.33 would be preserved as such, but the actual normal component would be significantly different due to the presence of the discontinuity, so the proportionality would be indirect, and unknown in a general case.

Nevertheless, Hall effect sensors can be and are widely used for the measurement of flux density. There are commercially available instruments called 'gaussmeters' or 'teslameters' (Figure 3.35). The probes are made as thin as possible, but this still usually means more than 0.5 mm (Lakeshore, 2009). A probe can comprise a Hall sensor for a measurement in a single direction (transverse, tangential or axial) as well as a single probe capable of measurement in all three orthogonal directions (B_x, B_y, B_z). Such three-axis sensors are also available as standalone sensors (Melexis, 2013).

Gaussmeters can be used, for instance, for the measurement of B between the magnetic poles of electromagnets, inside of Helmholtz coils, solenoids, etc. The result can be re-calculated in the units of (A/m) because such measurements are performed basically in the surrounding air, which is non-magnetic, linear and lossless. Therefore, the measurement of H can also be carried out by means of measuring B. More details are provided in the following sections.

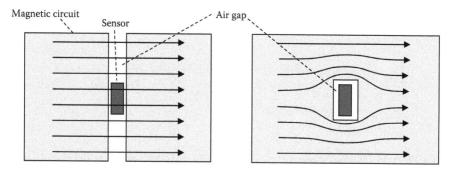

FIGURE 3.34 For proportional measurement, the sensor must be placed in a full-width air gap, not in a small 'window'.

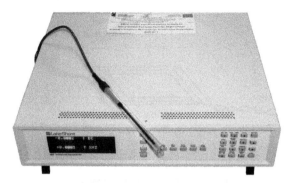

FIGURE 3.35 Gaussmeter LakeShore 460, with a probe in its protective tube. (LakeShore, 2009) (Photograph courtesy of KBR Magneto, Poland.)

Hall effect sensors could be used for the detection of flux normal to the flat surface of the sample, but this component is usually of less importance than the in-plane components. As a result, there are few publications describing the use of such sensors in 1D, 2D, 3D and rotational measurements. This is because of the relatively large size of such sensors and the fact that B measurement can be performed only under special conditions (Flanders, 1985).

A Hall effect sensor can inherently detect a DC value of B – something that cannot be achieved as easily with a B-coil. However, because of the built-in electronic circuits, there is usually an upper frequency limit, for instance, 400 Hz (LakeShore, 2009).

Additional limitations apply to other similar sensors. For example, *magnetoresistive* sensors have wide bandwidth (up to MHz) but might require set/reset function (Honeywell, 2008), which complicates the operation.

Fluxgate magnetometers, which are physically even larger (Figure 3.36) and also have limited upper frequency, for example, 1 kHz (Stefan Mayer Instruments, 2013).

FIGURE 3.36 Fluxgate sensor FLC 100 (length around 50 mm). (Stefan Mayer Instruments, 2013)

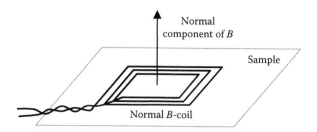

FIGURE 3.37   Flat *B*-coil for measurement of normal component of *B* (drawing not to scale).

However, such sensors as Hall effect, magnetoresistive or fluxgate could be used for the detection of the normal component of *B*, similarly to the normal *B*-coil described in the next section (Figure 3.37).

3.1.6   NORMAL *B*-COIL

In some applications, there is a need to measure *B* perpendicular (normal) to the sample surface. As illustrated in Figure 3.33, the normal component of *B* does not change at the interface between two materials. Therefore, a flat 'pancake' coil placed on the sample surface will detect the normal component. The coil shown in Figure 3.37 is rectangular but of course any suitable shape can be employed (e.g. circular) (Oka et al., 1998b; Li et al., 2010, 2014a). As for an ordinary *B*-coil, the *B* measurement requires the knowledge of number of turns and active cross-sectional area, or calibrated area–turn product.

The number of turns can be counted precisely, but the active cross-sectional area is more difficult to derive. This is because in an ordinary *B*-coil, the area is usually taken as the cross-sectional area of the sample (Figure 3.1a) and any air flux is normally neglected due to its small contribution. This is no longer the case for the normal *B*-coil, which does not encircle a well-defined cross-sectional area of the sample. The turns usually spread out somewhat (Figure 3.37), so the effective area cannot be easily calculated. In the first approximation, it can be taken as the average of the areas enclosed by the inner-most and the outer-most turns. But for more precise measurement, such coil would have to be calibrated, in order to ensure that the turn–area product is precisely defined for a given coil.

For instance, such coils can be used in 3D systems, where the sample has the shape of a cube. In the absence of flux leakage, the flux entering one face of the cube can be measured as the normal component. In 3D systems, the accuracy of measurement can be affected by the fact that there is an additional leakage and not all flux detected by the coil can enter the sample. Reducing the air gaps around the sample might be beneficial in suppressing this effect (Li et al., 2014).

3.1.7   PIN SENSOR

A non-magnetic opening inside the magnetic material produces a local re-distribution of magnetic flux as shown in Figures 3.18 and 3.34. However, *some* flux will flow through the non-magnetic aperture and its amount can be detected so that an indirect detection of the main flux can be performed.

The technique was described, for instance, in Pfutzner et al. (2014) and the authors also referred to it as 'dummy' sensor because the measurement is not direct, and requires an auxiliary magnetic pin or rod on which a B-coil is wound. The sensor is designed to be used for the measurement of B distribution in larger magnetic cores rather than in single laminations (Shilyashki et al., 2015a).

The core under test must have an opening such as a cylindrical hole (Figure 3.38). Such holes are routinely used for the alignment of subsequent laminations and mechanical fixing of the core in large transformers.

A movable pin is inserted in the hole. The pin has such a diameter as to fit inside of the hole, but can be still moved freely. At one end of the rod, there is a pair of small holes, through which a B-coil is wound (Figure 3.38b). The magnetic rod is therefore a carrier of the embedded B-coil. During magnetisation of the core, some fraction of magnetic flux will flow through the magnetic rod and it can be detected by the embedded B-coil. The amount of flux flowing through the rod also depends on the amount of air gap between the core and the rod as well as magnetic permeability of the latter.

The characteristic of such a detection method is not linear and in general case cannot be specified in an analytical way. Therefore, the measurement is only relative, and must be linearised either by pre-calibration of the sensor for a given operating conditions, or by comparison to the average of all measurements, which is then referred to the global B in the core (Pfutzner et al., 2014; Shilyashki et al., 2014). This is possible because a global B-coil can be used to detect the mean flux density, averaged over the whole cross section of a given location.

It can be concluded that similar measurements can be performed by other types of sensors, not only a B-coil. For instance, a Hall effect sensor could be used in place of the embedded coil. However, such sensors would introduce additional air gaps and this would further reduce the amount of flux penetrating the magnetic rod and thus adversely affect the sensitivity of the method.

The method itself is capable of detecting the differences at the level of 0.01 T in a core magnetised to around 1.7 T (Shilyashki et al., 2014).

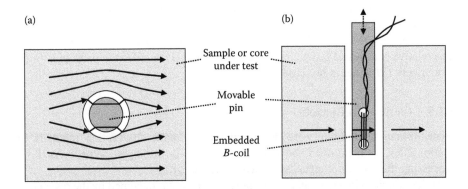

FIGURE 3.38 Pin sensor (drawing not to scale): (a) top view, (b) cross section. (Adapted from Pfutzner H. et al., Magnetic dummy sensors – A novel concept for interior flux distribution tests, *Proceedings of 13th 1&2DM Workshop*, Turin, Italy, 2014, p. 59.)

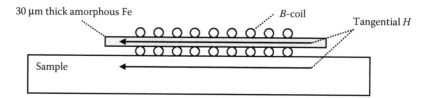

FIGURE 3.39   Tangential *B* sensor (drawing not to scale). (Adapted from Pfutzner H. et al., *Magnetic dummy sensors – A novel concept for interior flux distribution tests*, *Proceedings of 13th 1&2DM Workshop*, Turin, Italy, 2014, p. 59.)

3.1.8   Tangential *B* Sensor

Another indirect method reported for the measurement of *B* employs a tangential *B* sensor (Pfutzner et al., 2014; Shilyashki et al., 2015b). A multi-turn *B*-coil is wound onto a thin magnetic core (the coil carrier), for instance, a single strip of Fe-based amorphous ribbon. Such a sensor is very thin and can be placed directly onto the surface of the sample under test (Figure 3.39).

The measurement actually relies on the equality of tangential *H*, as it is described below for the *H* sensors (see also Figure 3.44). As with the embedded sensor, the detection relies on the permeability of the magnetic carrier so the detection is indirect and non-linear. A calibration must be performed, for instance, by placing the sensor on top of the material under test, for example, inside Epstein frame, and excited to such values of *B* as it would be expected on the real sample or core.

The main difficulty lies in the fact that it is only the tangential *H* that is equal inside and outside of the sample (Figure 3.44). For the same *B*, the *H* will be different for samples of various grades of electrical steels, or even various batches of the same material. So, for more precise measurement, the calibration would have to be carried out with the exact material for which the measurement is expected to be carried out, in order to make sure that the mutual relationship between the permeability of the magnetic carrier and the permeability of the sample under test is captured and calibrated. In the general case, this is not possible due to the unavailability of calibration samples. Therefore, the *absolute* accuracy of the method cannot be guaranteed for a general case.

Nevertheless, the method would be sufficient to detect *relative* changes of *H*, and thus relative changes of *B* (by employing Equation 1.11) on different parts of a transformer core (Pfutzner et al., 2014).

3.1.9   Optical Techniques

Some *optical techniques* were described with domain observation methods in Section 2.5.1. As mentioned above, flux density can be attempted to be 'sensed' by processing the domain images, as shown in Figure 3.40.

If the sample is in a demagnetised state, then the areas occupied by the respective 'dark' and 'light' domains will be the same. And, for example, if there are all 'light' domains, then the sample is magnetised towards the positive saturation, and 'dark' domains indicate negative saturation.

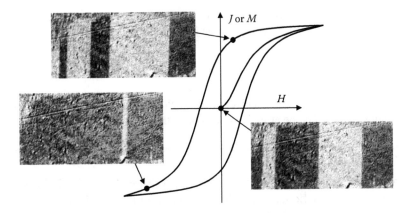

FIGURE 3.40 Detection of 'magnetisation' through the correlation of image processing and applied excitation.

It should be noted that the value of excitation can be known quite precisely, for instance, because the sample was magnetised in a magnetically closed circuit, so H could be calculated from the magnetising current. Of course, there would have to be a precise time-frame correlation between the applied excitation and the acquired frame-by-frame images.

This technique can be applied only on a qualitative level because absolute values of B (or J, or M) cannot be measured through image processing. The investigated area is usually small. Moreover, only the surface can be observed.

For the experiments in which a hysteresis loop is constructed, the vertical axis is typically expressed in values relative or normalised to the saturation value (e.g. M/M_s) (Gubbiotti et al., 2005; Westphalen et al., 2007). Some authors use even less 'magnetic' units such as 'degrees of rotation' of the light beam (Suzuki et al., 1990) or 'grey levels' (Ionita et al., 2008).

3.2 MAGNETIC FIELD STRENGTH SENSORS

There are several sensing techniques that allow the detection and measurement of magnetic field strength H. Two main approaches are used in practice – based on magnetising current or on the tangential component of the magnetic field strength.

3.2.1 MAGNETISING CURRENT SENSORS

In a magnetically closed circuit (see, for instance, Figures 2.19 and 2.24), the magnetising coil is wrapped directly around a magnetic circuit. The 'sample' is either the magnetic circuit itself or is an inherent part of the magnetic circuit. Crucially, there is no air gap between the 'sample' and the 'magnetising yoke', and additionally the reluctance of the magnetising yoke is negligible with the reluctance of the sample. The cross section of the sample is assumed constant throughout its entire length.

Under such conditions, the magnetic field strength H generated by the magnetising coil is directly proportional to the current I applied to the coil, which is a consequence of Ampere's circuital law. For such simple magnetic circuit, we can therefore apply Equation 1.3, which can be further simplified to

$$H(t) = C_x \cdot I(t) \quad (\text{A/m}) \tag{3.19}$$

where $C_x = N/l$ is the constant of a given setup, such that N is the number of turns of the magnetising coil and l is the effective magnetic path length under consideration (m).

PRACTICAL COMMENT

The relationship of Equation 3.19 is directly proportional and also holds for instantaneous values; hence it is a direct function of time (t). This means that, for instance, for sinusoidal or triangular magnetising current, the peak value of I is synonymous with the peak value of H, and when I is crossing zero, H is also zero, etc. However, if the current is measured with an AC ammeter, detecting the RMS value, then the value of H calculated with Equation 3.19 would also be RMS.

As evident from Equation 3.19, in such a method, the measurement of H reduces to the measurement of current I (and performing very simple calculation) as shown in Figure 3.41.

There are several ways in which the current can be detected and theoretically almost all suitable ammeter-like devices could be used for such a purpose. However, in practice, only a few approaches are used.

3.2.1.1   Shunt Resistor

One of the simplest methods is to use a non-inductive *shunt resistor*, which is connected in series with the magnetising coil. This current-sensing technique is actually a basis for many analogue and digital ammeters, which employ an internal low-resistance shunt resistor (Figure 3.41b). In such ammeter A, actually an inter-

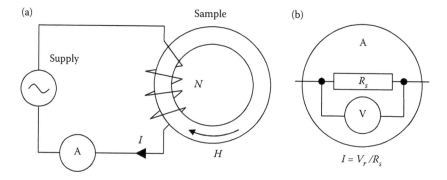

FIGURE 3.41   Principle of the measurement of H through the measurement of I in magnetically closed circuits (a) and an ammeter A employing a shunt resistor R_s and a voltmeter V (b).

nal voltmeter is used to measure the voltage drop V_r across the shunt resistor and the result is scaled accordingly by the resistance value. The value is converted to current using Ohm's law: $I = V_r/R_s$ (Figure 3.41b).

An example of a shunt resistor used by the author for some of his measurements is shown in Figure 3.42 (see also Figure 4.39). This particular component had a nominal resistance of 0.47 Ω and a parasitic inductance of 40 nH (Tyco, 2006). For instance, even at 1 kHz, the inductive reactance is only 0.25 mΩ, which is less than 0.055% of the resistance value. Therefore, in most practical measurements of rotational losses, other sources of errors are much more significant. But of course even such a small error should be taken into account in the estimation of total measurement uncertainty of the given system.

A shunt resistor must be chosen to meet at least two criteria: power dissipation and signal level. So the resistor must be capable of dissipating maximum power safely, without overheating, which might cause inaccuracies (temperature drift of the resistance value), destruction of the component or even injury of the operator (hot surfaces).

For this reason, the resistor in Figure 3.42 is designed to be attached to a suitable heat sink (see also Figure 4.39), which facilitates cooling, thereby lowering the operating temperature. The maximum power rating for the resistor in Figure 3.42 is 250 W, which means that with 0.47 Ω, the maximum current to be safely used is 23 A. Without a heat sink, the rating is much lower, and could be as little as 10% of the maximum rating, or for this case only around 7 A of RMS current. A physically smaller component would have much lower power rating, which might be unsuitable from this viewpoint.

For the same current, the dissipated power can be lowered by using a resistor with a lower resistance value. However, this adversely affects the signal-to-noise ratio. If the value was too small (e.g. 1 mΩ), then the voltage developed across such a resistance would also be very low and it would be more difficult to measure accurately.

FIGURE 3.42 High-power non-inductive thick-film shunt resistor 0.47 Ω with four terminals (size around 60 mm). (Tyco, 2006) Equipment courtesy of Megger Instruments Ltd, the United Kingdom.

Therefore, the choice of a shunt resistor requires a reasoned compromise. It might be required to use two or more shunt resistor values to cover the whole expected measurement range (see Figure 4.39).

The resistance value itself does not influence the measurement accuracy directly. If the conditions are controlled carefully (power dissipation, non-inductive device, temperature stability, etc.), then the power loss accuracy is not impacted adversely, and the value can be, for instance, anywhere between 0.1 and 10 Ω, and even outside of those values when required (Zurek, 2015b).

PRACTICAL COMMENT

The compromise needs to include other factors as well. In rotational measurements, the excitation currents can be heavily distorted and the peak voltage across the shunt resistor can be significantly higher than the RMS value. Even for sinusoidal excitation, the voltage drop across 0.47 Ω with 23 A would be over 15 V peak, which is *above* the safe specification of typical data acquisition devices (NI, 2013a). A higher value of resistance will exacerbate the problem and can lead to permanent damage to sensitive equipment due to the high-voltage levels.

Another important choice is the connection of the shunt resistor into the circuit. This is important if the power supply is not galvanically isolated, and also if a non-isolated voltmeter device is used. For instance, a typical non-isolated mains-powered oscilloscope can be used as a data acquisition device (voltmeter). The power supply and the voltmeter must be connected in such a way that their grounds are connected together (Figure 3.43). Otherwise, significant current could flow through the ground of the oscilloscope and damage it. Of course, such measurement would be inaccurate and thus worthless, but the damage to the expensive equipment and risk of electrocution is an immediate safety problem.

Therefore, using a separating/isolating transformer between the power amplifier and the magnetising yoke gives additional advantage of full galvanic

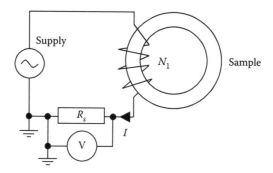

FIGURE 3.43 Circuit with a non-isolated supply and non-isolated voltmeter (e.g. oscilloscope); the shunt resistor must be connected to the ground of the power supply if the voltmeter is grounded as well.

separation between grounds. Such a transformer is discussed in complete detail in Chapter 4 (Figures 4.36 and 4.39).

Commonly available non-inductive resistors are not precision components in the absolute sense. The resistance value can differ by up to 10% (Tyco, 2006), which is an industry standard tolerance, even though the stability of the value can be very good (250 ppm/°C or better).

Therefore, there is a requirement for calibration by measuring the resistance with a high-precision ohmmeter. For a low-value resistance (e.g. below 1 Ω), precise measurement requires a four-terminal connection, so that the current terminals are separated from the voltage terminals (see Figure 3.42).

3.2.1.2 Other Current Sensors

One of the important advantages of a non-inductive shunt resistor is that the wave-form phase shift introduced by the sensor can be very low. It is also fairly easy to change 'input range' simply by using a resistor of a different value. Amplification of the signal is also possible, but attention must be paid to any additional phase shifts introduced by the amplifying circuit. Under typical operating conditions, no ampli-fication is usually needed.

There are many types of current sensors (also referred to as *transducers*) that could be used for current measurement. For instance, *current transformers* offer a kind of 'wireless' measurement because the current path does not have to be broken for the sensor to detect the current (as compared with the shunt resistor approach). There are also other transducers: Hall-effect-based or fluxgate-based current sen-sors. Full galvanic isolation from the measured circuit is achieved almost by defini-tion, due to electromagnetic coupling.

However, in practice, there are several difficulties. The first one is that the phase shift is usually much larger in such sensors, as compared to the shunt resistors. And in rotational measurements, even the smallest phase shifts (e.g. 0.05°) can be devas-tating as demonstrated in Figure 3.16 (Zurek, 2005a; Zurek et al., 2005c, 2006a,b).

Another factor is the linearity of the transducers. The so-called closed-loop sen-sors exhibit much better linearity than the open-loop version, and the values could be at the 0.1% and 1% levels, respectively (Telcon, 2014a, b). Such sensors have a magnetic core so there could be an additional error caused by internal hysteresis, and because the signal processing involves the transducer and the associated electronic circuit, there are voltage offsets, temperature drifts and so on.

These additional errors can add up to significant values. For instance, the closed-loop sensor HTP25NP (Telcon, 2014a) has a nominal output current of 25 mA, and an offset of up to 0.25 mA, which is 1% of the full scale. The bandwidth is from DC to 150 kHz, where the stated error is −1 dB (which is equivalent to around 13%). For open-loop sensors, the errors are usually greater.

For the particular shunt resistor discussed above, such error would occur theoreti-cally only above 2 MHz, which is obviously a much greater operating bandwidth.

AC signals can be measured with current transformers, but these are incapable of measuring DC or very low-frequency signals. Achieving the required low-value phase shift would also be a great challenge because even 'precision current transformers' can have phase errors at the level of 10 angular minutes, which is 0.16° (Yokogawa, 2011).

The low-cost fluxgate sensors suffer from similar error levels (LEM, 2014a). On the other hand, there are fluxgate-based sensors that offer excellent accuracy and stability (also for DC currents), but these are very expensive (LEM, 2014b), as compared to more traditional sensors.

There are also other types of current sensors. In general, it is irrelevant which sensor technology is actually used. But any choice must be carefully analysed from the viewpoint of the associated measurement errors. The phase shift becomes critical in the fieldmetric technique if higher B values (e.g. above 1.5 T for electrical steels) are attempted – as evident from Figure 3.16.

3.2.2 Tangential Sensors

Theoretical analysis shows that at the boundary between two media, the tangential component of H does not change as illustrated in Figure 3.44 (Johnk, 1975; Singh, 2011).

For instance, let us assume that permeability μ_2 represents the sample under test and μ_1 the surrounding air. Therefore, the tangential component H_{2t} just inside the sample is equal to H_{1t} just at the surface outside (Figure 3.44). Hence, in theory, the tangential component measured directly at the surface of the sample should be equal to that inside of the material – but *only* in the case of the tangential component.*

Therefore, the measurement of H inside of the material can be simplified to the measurement of H at the surface of the sample under test. In theory, any type of an H sensor can be used for this purpose, but in practice some are more suitable than others.

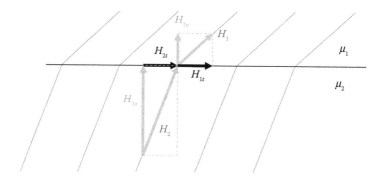

FIGURE 3.44   Tangential component of H does not change at the boundary, so $H_{1t} = H_{2t}$ in the absence of surface currents. (Adapted from Johnk C.T.A., *Engineering Electromagnetic Fields & Waves*, Wiley International Edition, New York, USA, 1975, ISBN 9780471442905; Singh Y., *Electro Magnetic Field Theory*, Pearson Education India, 2011, ISBN 9788131760611.)

---

* There are several assumptions made for this derivation. One of them is the lack of surface currents, which could flow, for instance, if the analysed material is a superconductor, or even a simple wire with current (e.g. part of the magnetising coil) placed directly on the surface. However, in most of the rotational measurements, the analysed materials are not superconductors, so the equality of tangential components does apply. Nonetheless, the limitations of this approached should be borne in mind.

3.2.2.1 *H*-Coil

A very popular type of sensor is the so-called *H*-coil shown in Figure 3.45 (Brix et al., 1982; Enokizono et al., 1990a; Moses, 1992; Hasenzagl et al., 1996; Bajorek et al., 1999; Gorican et al., 2002; Zurek, 2005a).

A flat coil is wound with a fine insulated wire on a non-conductive and non-magnetic (paramagnetic or diamagnetic) former. According to Figure 3.44, the coil should detect *H* as close as possible to the surface of the sample. For this reason, the supporting former is a flat and thin sheet.

The wire used for the coil should also be as thin as possible, for at least two reasons. First, a thick wire would increase the distance *l* (Figure 3.45a) between the centre of the coil and the surface. This is especially important for a 'two-axis' *H*-coil, in which one coil is wound on top of the other (Figure 3.45b–d), so that the distance *l* is increased. Such double *H*-coil uses the available space in a more efficient way, also allowing the detection of H_x and H_y components from the same localised area.

Second, more turns of a thinner wire can be wound within the same length of the former, which improves the signal-to-noise ratio.

An example of such a double- or two-axis *H*-coil is shown in Figure 3.45d. A former was cut by precision machining (0.01 mm nominal tolerance) from a plastic sheet 0.25 mm thick. The width and length of each coil was 20 mm, and with a 0.05 mm wire, there were 230 turns wound as a single layer, in order to minimise the distance *l* by avoiding an increase of the overall coil thickness (Zurek, 2005a).

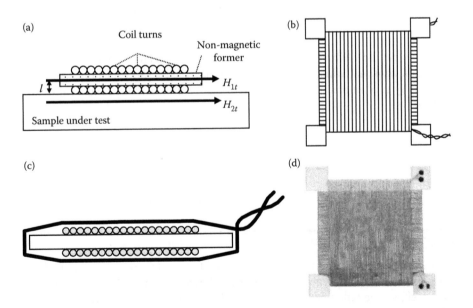

FIGURE 3.45 Tangential *H*-coil (drawings not to scale): (a) cross section of a single-axis *H*-coil for which $H_{1t} \approx H_{2t}$ (see also Figure 3.44), (b) top view of a two-axis *H*-coil, (c) cross section of a two-axis *H*-coil, (d) close-up photograph of a prototype of double *H*-coil sensor 0.25 × 20 × 20 mm; the *H*-coils were wound by Richard Wakely at Wolfson Centre for Magnetics, Cardiff University, Cardiff, the United Kingdom (see also Figure 4.18).

Such a sensor was placed on top of the B-coils so that all four sensors would measure signals from within the same local area on the sample (see also Figure 3.12e).

The underlying physical operating principle of an H-coil is the same as that of the induction coil described in the previous sections. The voltage induced in the coil can be described by Equations 3.1 and 3.2 because it is related to the *flux density* penetrating the coil. However, the coil is wound on a non-magnetic former, whose permeability can be assumed in practice to be equal to 1 (see also Chapter 1), because of the linear relationship between B and H for a non-magnetic material. Therefore, substituting Equation 1.7 into Equation 3.2 gives

$$V(t) = N \cdot A \cdot \mu_0 \cdot \frac{dH_{mean}(t)}{dt} \quad (\text{V}) \tag{3.20}$$

where N is the number of turns (unitless), A is the active cross-sectional area of the coil (m²), μ_0 is the permeability of free space (H/m) and $H_{mean}(t)$ is the mean magnetic field strength penetrating the active volume inside the coil (A/m) and t is the time (s).

The signal processing is analogous to that of a B-coil. The induced voltage is usually of low amplitude and also requires integration so procedures described in Sections 3.1.1.1 and 3.1.1.2 are equally applicable. The H waveform is therefore derived as

$$H_{mean}(t) = \frac{1}{N \cdot A \cdot \mu_0} \int V(t)dt \quad (\text{A/m}) \tag{3.21}$$

H-coils can be calibrated (see Figure 3.46) so that each coil will have its own calibration constant, which can also include the μ_0 term. So Equation 3.21 would be normally used in practice as a product of the integrated voltage and a calibration constant:

$$H_{mean}(t) = C_x \cdot \int V(t)dt \quad (\text{A/m}) \tag{3.22}$$

where C_x is the calibration constant $\left(\dfrac{1}{H \cdot m} \right)$, found for each coil separately.

PRACTICAL COMMENT

The calibration constant in Equation 3.22 can be employed in a number of ways. For instance, the value of μ_0 does not have to be included in the constant itself, so the value and the unit would be appropriately different. This is simply the consequence of the given approach and does not change the performance of the sensor or the method in any way.

For most H-coils, the number of turns can be known or counted precisely (e.g. under a microscope), so the only unknown value is the effective cross-sectional area A. As discussed below, this cannot be calculated with sufficient absolute accuracy and the only way to precisely learn its value is to perform calibration. However, in order to exclude any errors in the counting of the number of turns, the product of these two values $N \cdot A$ can be assumed to be unknown, so that it can be found precisely and directly from the calibration procedure, which excludes any human error during counting of turns.

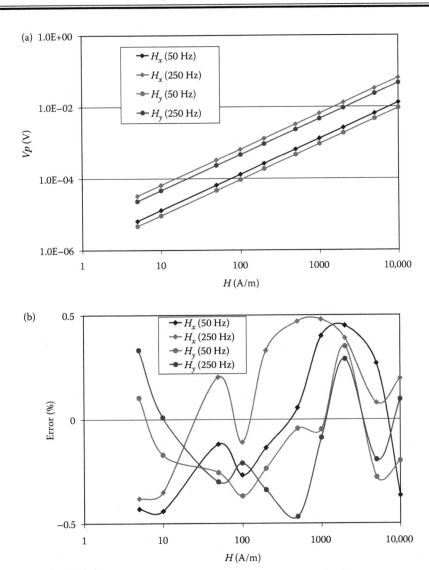

FIGURE 3.46 Calibration of H-coils: (a) measured curves for various frequencies and H values for both H-coils from Figure 3.45d and (b) linearity error over whole measurement range does not exceed $\pm 0.5\%$ for either H-coil. (Adapted from Zurek S., *Two-Dimensional Magnetisation Problems in Electrical Steels*, PhD thesis, Wolfson Centre for Magnetics Technology, Cardiff University, Cardiff, the United Kingdom, 2005a.)

3.2.2.1.1 Practical Difficulties

The various parameters of an H-coil require a compromise. For the best accuracy of the tangential measurement (Figure 3.44), the coil must be as flat and thin as possible. However, this reduces the active cross-sectional area, which leads to a worsened signal-to-noise ratio.

For instance, if the sensor (Figure 3.45d) was to detect $H = 100$ A/m at 50 Hz, then the induced voltage would be less than 1 mV (Figure 3.46a). However, such and even lower voltages can be successfully measured directly by data acquisition devices with sufficient resolution, for example, 16 bits for 200 mV input range (Zurek, 2005a).

The winding wire must be thin in order to be able to wind a sufficient number of turns to increase the voltage and compensate for the low cross-sectional area. But winding a very thin wire is difficult because of its vulnerability to mechanical damage.

Without specialised winding machines, the thinnest practical wire is around 0.05 mm (although some researchers successfully used even finer wires, for instance, see Figure 6.21). Such a wire is still strong enough to be wound on a flat former, and it is possible to be terminated by ordinary soldering, for instance, to some thicker leads.

Thinner wire has proportionally lower tensile strength and can break during winding, especially due to the shape of the former (winding on a flat former is much more difficult than on a cylinder). Also, if its electrical insulation is not protected, it can be very easily damaged by scratching during normal use. For this reason, the H-coils used by the author and shown in Figure 3.45d were covered with a thin layer of varnish after manufacturing.

On the other hand, thicker wire occupies more space (width per turn) and fewer turns can be wound on the same length, so the output signal would be lower for comparable dimensions. Winding more than one layer of coil is less than optimal. It increases the overall thickness of the coil, but even more importantly there is less control over the positioning of single turns for the upper layers. If the neighbouring wires cross themselves, this increases the thickness even further, as well as makes the coil non-uniform, which impacts on the measurement precision as well.

An additional problem with a multi-layer winding is that it significantly increases the parasitic self-capacitance of the coil. Such capacitance might impact measurement accuracy at higher frequencies by introducing unwanted phase shifts in the output voltage. For this reason, such H-coils should also be calibrated at the highest expected frequency in order to investigate any potential problems.

In practice, the size of an H-coil depends on the area of interest. If the sample under test is larger and the area of 'uniform magnetisation' is larger, then the size of an H-coil can also be greater, thus offering better signal-to-noise ratio. Some researchers used H-coils $1 \times 50 \times 50$ mm (Hasenzagl et al., 1996), but a commonly used size for rotational measurements is around 20×20 mm (Sievert et al., 1995).

Advances in precision machining allow making much smaller H-coils, for instance, as small as 4×4 mm, 0.6 mm thick, and wire as thin as 0.015 mm (Aihara et al., 2011) (see also Figure 6.21).

The flat former usually has a precisely known (or at least easily measurable) thickness and width and the diameter of the wire is also known. Therefore, it should be possible to calculate the effective cross-sectional area. However, as can be seen from

Figure 3.45c, only one of the orthogonal coils follows the contour of the former, whereas the other is 'slanted' towards the edges. This increases the cross-sectional area of such a coil and complicates calculations.

For instance, for the coil shown in Figure 3.45d, the effective constant calculated from the theoretical area–turn product of the inner coil would be 0.25 mm · 20.00 mm · 230 turns = 1150.00 mm^2 if the diameter of the wire was completely excluded (just the cross-sectional area of the former), or 0.30 mm · 20.05 mm · 230 turns = 1383.45 mm^2 if the diameter of the bare wire was included, or 0.317 mm · 20.067 mm · 230 turns = 1463.08 mm^2 if the maximum outer diameter of the enamelled wire was included with the insulation. The actual calibrated constant was 2363.72 mm^2 (Zurek, 2005a), so even for the largest theoretical value, the error is still around 40%. Similar discrepancies would be for the other coil, which has even greater effective area (see Figure 3.45c), so the calibration must be carried out in any case.

This underestimation is more severe for thinner flat formers because any non-perfections in the windings have a greater impact on the effective area. For thicker formers, the influence of the wire diameter and its insulation is less significant, but of course, for precise measurements, calibration must be undertaken, for example, in a long solenoid, as shown in Figure 3.51b (Zurek, 2005a), or a Helmholtz coil (Chen et al., 2017).

An additional problem is the field non-uniformity. A widely used assumption is that the field at the centre of the sample, or the area of interest, is uniform. However, in a general case, the measured field could contain not only tangential but also some normal component. The magnetic field lines would penetrate the H-coil at some angle, hence with some normal component (Figure 3.47, see also Figure 4.4) and this could cause additional unknown errors as discussed in the next section.

This can be improved by employing a shielding lamination (e.g. made of the same material as the sample under test), which reduces the normal component and equalises the magnetic field strength (Zhu et al., 1993; Makaveev et al., 2000a). However,

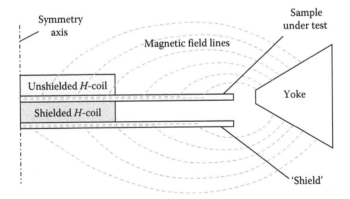

FIGURE 3.47 H-coil exposed to non-uniform field with normal component. (Adapted from Makaveev D. et al., Measurement system for 2D magnetic properties of electrical steel sheets: Design and performance, *Proceedings of 6th 1&2DM Workshop*, Bad Gastein, Austria, 2000a, p. 48.)

such technique introduces unknown contribution of the demagnetising field of the shield as well as the complication of the mechanical setup. Some implications are discussed in more detail in Chapter 4.

The field non-uniformity also poses a different problem, namely, the H at the surface of the sample might have a different magnitude than at some distance from the sample. But the sensor has a finite thickness and measures H, which is somewhat different from that at the surface. As shown in Figure 3.47, this effect can be reduced (but not completely eliminated) by using shielding or 'sandwiching'.

It was also shown, for instance, in Nakata et al. (1987) that the H at the surface can be obtained by extrapolation from more than one sensor, so that the errors are reduced. The values of tangential H measured at increasing distances from the sample surface do not change rapidly, and in the first approximation, a straight line can be used for extrapolation at small distances (Figure 3.48c, see also Figure 3.54). Some authors used second-order polynomials for extrapolation in order to improve the accuracy (Hall et al., 2009).

However, such extrapolation complicates the measurement procedure. The H is measured as a waveform, so that the data from each sensor are a function of time. So if there are data from two sensors, then the values must be extrapolated for each time instant separately. For rotational magnetisation, this would require using at least four sensors of H (Figure 3.48b). The mechanical arrangement therefore becomes quite complex, and the inaccuracies resulting from any angular misalignment could be significantly exacerbated. The problems become even greater if the multiple H-coils are not placed

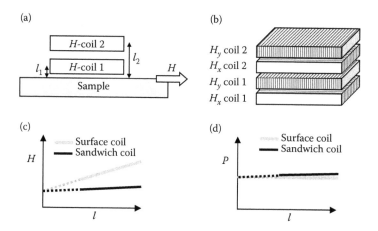

FIGURE 3.48   Extrapolation of H at the surface of the sample from two sensors: (a) Uni-axial arrangement of the sensors. (Adapted from Nakata T. et al., *IEEE Transactions on Magnetics*, 23 (5), 1987, 2596.) (b) Two-dimensional arrangement. (Adapted from Tumanski S., Investigations of 2D parameters of electrical steels, *Proceedings of 6th 1&2DM Workshop*, Bad Gastein, 2000b, p. 185.) (c) Variation of H with distance of H-coils from sample surface. (d) Variation of P with distance of H-coils from sample surface. (Adapted from Zhu J.G. et al., Measurement and modelling of losses under two dimensional excitation in rotating electrical machines, *Proceedings of 5th International Workshop on Two-Dimensional Magnetization Problems*, Grenoble, France, 1997b, p. 63.)

on the same side, but on opposite sides of the sample (Chen et al., 2017), in an attempt to decrease the effective distance l for each coil (Tumanski, 2000b).

It is therefore not surprising that there are limited examples of such extrapolations for 2D measurements reported in the literature, Nakata et al. (1987) and Hall et al. (2009) being examples for a uni-directional measurement, and Zhu et al. (1993) and Makaveev et al. (2000a) for 'sandwiching' or shielding. The extrapolation technique must be used if precise measurement of H is to be carried out. Otherwise, the demagnetising field effect falsifies the results, which cannot be compared, for instance, to those from single-sheet testers.

Nevertheless, the extrapolation is carried out in some modern commercial rotational testers, for example, in the rotational system made by Brockhaus Measurements (Brockhaus, 2014). An example of practical implementation is shown in Figure 4.15.

For uni-axial magnetisation, the method can be applied with very good results. For instance, it was shown in Stupakov (2012) that the H value can be extrapolated to the sample surface through tangential measurement with an array of three Hall effect sensors (Figure 6.17). The $B–H$ loop could be measured with good repeatability even in the presence of a large air gap (up to 10 mm) between the sample and the yoke. Under such conditions, the measurement of H through a magnetising current is not possible in a general case because of the great influence of the reluctance of the air gap.

It should be noted that H can decrease (Figure 3.48c) as well as increase (Tumanski, 2000b) towards the surface of the sample (see also Figure 4.25d).

PRACTICAL COMMENT

With linear extrapolation, the following function can be applied for calculating H at the surface of the sample (Steentjes, 2017), as based on Figure 3.48a:

$$H_{surface} = \frac{l_2 \cdot H_1 - l_1 \cdot H_2}{l_2 - l_1} \quad (A/m) \qquad (3.23)$$

where l_1 is the distance of the first H-coil positioned closer to the surface (m), l_2 is the distance of the second H-coil positioned farther from the surface (m), H_1 is the magnetic field strength measured by the first H-coil (A/m) and H_2 is the magnetic field strength measured by the second H-coil (A/m).

Of course, more than two data points can be used for extrapolation (Stupakov, 2012), but so far, this was not demonstrated in rotational measurements.

3.2.2.1.2 Errors Caused by Off-Axis Field Components

Rotational power loss initially increases with B, but then reaches some maximum value and begins to decrease towards saturation, so that normally the curve should exhibit just one local maximum. However, the curves measured with the fieldmetric technique for clockwise and anticlockwise rotation can exhibit much more complex shapes. There can be very clear *multiple* local extrema in such a curve (Figure 3.49)

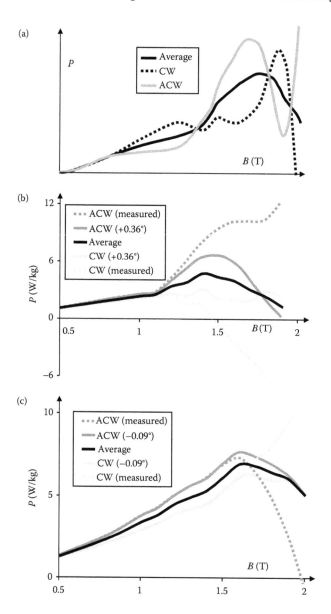

FIGURE 3.49   Measured CW and ACW power loss can differ significantly from the average: (a) Several minima and maxima without phase shift correction. (Adapted from Gorican V. et al., Performance of round rotational single sheet tester (RRSST) at higher flux densities in the case of GO materials, *Proceedings of 7th 1&2DM Workshop, Ludenscheid*, Germany, 2002, p. 143.) (b) Several maxima and minima with phase shift correction. (c) Persistent difference between CW and ACW curves. (Adapted from Zurek S., *Two-Dimensional Magnetisation Problems in Electrical Steels*, PhD thesis, Wolfson Centre for Magnetics Technology, Cardiff University, Cardiff, the United Kingdom, 2005a; Zurek S. et al., Errors in the power loss measured in clockwise and anticlockwise rotational magnetisation. Part 2: Physical phenomena, *IEE Proceedings, Science, Measurement and Technology*, 153 (4), 2006b, 152.)

for a given direction of rotation (Gorican et al., 2002; Zurek, 2006b), even though the average loss follows the expected one-maximum shape.

The large positive/negative values were proved to be caused by phase shifts between the signals due to angular misalignment of the sensors. This can be corrected by recalculation of the values, as discussed in Chapter 5. However, there are some differences (Figure 3.49c) or distortions that cannot be corrected in this way.

There are some suggestions in the literature as for the sources of such behaviour. They could be caused by local non-uniformity of magnetisation, large grains, problems with feedback, wire twisting (similar as for B-coils) and so on. However, these are mostly just some unverified ideas because the effect was not sufficiently studied experimentally and explained (so far).

For instance, in some measurements on non-oriented electrical steels, there can be a persistent difference between CW and ACW measurements even at lower excitation (Figure 3.49c) (Zurek, 2005a; Brockhaus, 2014). They occur despite the fact that the sample has low anisotropy, the sample has no large grains, it is *not* difficult for the feedback to achieve purely circular magnetisation, the sample is fairly thick (0.5 mm) so the signal-to-noise ratio is large, and so on. More importantly, the differences occur very clearly even below 1 T where the non-linearity of magnetisation is low and the uniformity of magnetisation is very good.

A study carried out by the author reveals that the accuracy of H-coil measurement could be affected by the off-axis components of the H vector (Zurek, 2017c,e).

An ideal H-coil would detect only the component of the vector that is parallel to the main H-coil axis. Such H-coil should be completely immune, for instance, to the normal H component, which is perpendicular to the large surface of the flat former onto which an H-coil is wound. Interestingly, the analysis shows that the way the single return wire is positioned can also have an impact on the immunity of the coil against the off-axis components, with the phase of the unwanted signals significantly affected.

This effect is illustrated in Figure 3.50. The drawing shows simplistically just two turns in order to better illustrate the involved 'partial' areas. But it was shown that the total areas are actually independent of the number of turns (Figure 3.50g) (Zurek, 2017c).

Figure 3.50a shows a configuration with the return wire positioned at the side of the H-coil. This produces an area equal to 50% of the flat square of the former. If such a configuration is exposed to a normal H component (H_z), then some erroneous signal will be induced, proportionally to the size of the area and the distribution of H_z through that area.

The significance is that any single lamination in an open magnetic circuit is exposed to a large normal component of H, as shown schematically in Figure 3.47 (this is further discussed in Chapter 4). The distribution is such that at the centre of the system the perpendicular component is zero, but it is negative on one side and positive on the other, which results directly from the magnetic symmetry of the system.

The partial areas in Figure 3.50a are 'wider' on one side, so the negative and positive components will induce different signals that will *not* cancel each other out (Zurek, 2017c,d).

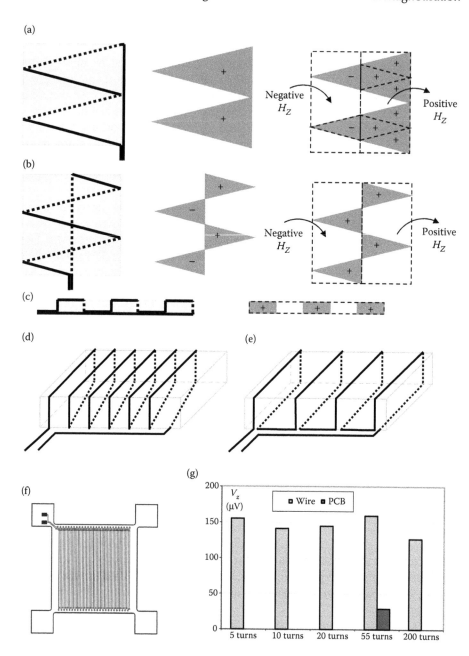

FIGURE 3.50   Sensitivity of an *H*-coil to off-axis components: (a) top view – side return wire creates large partial areas for H_z, (b) top view – central return wire creates differently shaped partial areas for H_z, (c) side view – partial areas for H_y might also have some influence, (d) normal wire-wound *H*-coil, (e) concept of an 'ideal' *H*-coil with improved immunity to off-axis components of *H*, (f) PCB layout of an 'ideal' *H*-coil, (g) comparison of absolute sensitivity to H_z of wire-wound and PCB-based 'ideal' *H*-coil.

The situation does *not* improve significantly even if the return wire is positioned through the centre of the *H*-coil. This time the partial areas are similar and due to the way the wire is looped, they produce 'negative' loops on one side and 'positive' loops on the other (due to the fact the wire would encircle the local area in a clockwise, or anticlockwise, direction). But unfortunately, the H_z can also be negative on one side and positive on the other. As a result, all the areas can contribute towards some erroneous signal and there is no cancellation for some angles (Figure 3.50b). The total active area can be up to 25% of the area of the flat former.

Similar unwanted sensitivity applies if the *H*-coil is analysed from the side view (that is looking at the edge of the former), for its immunity to the H_y component (Figure 3.50c). There can be areas (around 50% of the corresponding area of the flat former), which all contribute with the same sign, so again some erroneous signal is induced (as shown in Figure 3.50c).

Poor immunity to H_z and H_y components can in theory lead to significant errors, even though there is only a 'single turn', as compared to the H_x axis, which would normally have tens or hundreds of turns.

Under rotational magnetisation in the first approximation, the H_y component has a comparable magnitude to the H_x component. These are normally detected by two separate *H*-coils (Figure 3.45) but of course each coil is always penetrated with both components at different instants of time. So the H_x coil *should* have perfect immunity against the H_y component, and vice versa. If this is not the case, some errors are inevitably introduced.

Let us assume parameters of an *H*-coil with 20 mm width, 0.5 mm thickness and 200 turns. It can be shown that the side-view area–turn product can be up by 0.25% as compared to the main axis sensitivity, which would be 100%. Such a small value does not seem significant, but in the worst case, it can contribute to a phase error of arctan(0.25/100) = 0.14° (Zurek, 2017c). A phase shift of 0.14° is definitely not negligible, as can be seen from Figure 3.16 or Figure 3.49c.

The situation is even worse if a large H_z component is applied in the analysis because the area–turn product can be as high as 5% of the main axis sensitivity (Zurek, 2017c). In the worst case, this could lead to a phase shift of arctan(5/100) = 2.9°, which can introduce even larger errors. This is especially applicable because, as shown in Figure 4.28a for a single lamination sample, the amplitude of the H_z component can be several times *higher* than the H_x component (Zurek, 2017c).

Such large phase errors could easily cause the multiple local minima and maxima in the CW and ACW curves (Figure 3.49a and b) and could also explain the persistent difference between CW and ACW curves at lower excitation (Figure 3.49c). Theoretical calculations show that the errors caused by the H_z component are indeed capable of introducing such CW–ACW differences at the level exceeding 5% (Zurek, 2017d).

The performance of an 'ideal' *H*-coil was analysed by the author (Zurek, 2017e). The concept of the 'ideal' configuration is shown in Figure 3.50e. With the 'wires' guided in a special way, the unwanted partial areas are practically eliminated and the immunity to off-axis components of *H* is greatly improved, as shown by the analysis presented in Zurek (2017e). Such a coil can be produced by using PCB technology with the layout for a single coil shown in Figure 3.50f. The sensitivity to the H_z component is greatly reduced as compared to the wire-wound *H*-coil of the same size and number of turns, as clearly evident from Figure 3.50g. Photographs of such PCB-based *H*-coil prototypes are shown in Figure 6.24.

3.2.2.1.3 Angular Misalignment and Phase Shifts

Several researchers investigated the possible influence of angular misalignment between the orthogonal H_x and H_y coils (Sievert, 1992a; Zurek et al., 2006a; Aihara et al., 2011). The non-magnetic former on which the coils are wound is usually precisely machined, so the 'mechanical' dimensions and angles are rather well defined.

However, the 'magnetic angle' between the two orthogonal *H*-coils can differ from the 'mechanical angle', for instance, due to the way the wire of the coil was wound. This angle can exceed 0.6° (Aihara et al., 2011) or even 1.8° due to manufacturing variations (Sievert, 1992a). It should be stressed here that it is the wire winding procedure that is mostly responsible for the magnetic angle and not the shape of the former (Aihara et al., 2011).

The measurement of such an angular misalignment can be carried out in a solenoid. This requires a non-magnetic turntable to be placed in a solenoid (Figure 3.51). The detection of the magnetic angle can be achieved by measuring when the signal

(a)

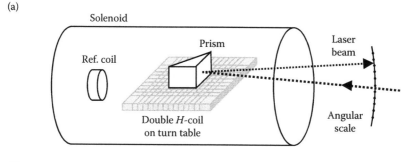

(b)

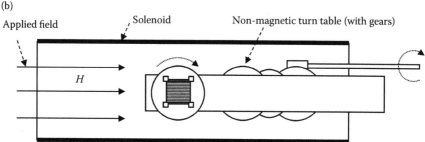

FIGURE 3.51 Measurement of the 'magnetic' angle between H_x and H_y coils: (a) With optical detection of angular position. (Adapted from Sievert J., Studies on the measurement of two-dimensional magnetic phenomena in electrical sheet steel at PTB, *Proceedings of 1st 1&2DM Workshop*, Braunschweig, Germany, 1992a, p. 102; Aihara S. et al., Characteristic evaluation of 4 mm-square-sized double *H*-coil, *Przeglad Elektrotechniczny (Electrical Review)*, R. 87 (9b/2011), 2011, 73.) (b) With electromechanical detection of angular position. (Adapted from Zurek S., *Two-Dimensional Magnetisation Problems in Electrical Steels*, PhD thesis, Wolfson Centre for Magnetics Technology, Cardiff University, Cardiff, the United Kingdom, 2005a.)

in one coil (e.g. H_y) is at zero, or rather the best obtainable minimum (because the RMS value will always differ from zero, e.g. due to the unwanted capacitive coupling between the H_x and H_y coils). After rotating by 90°, the other coil (e.g. H_x) should exhibit a minimum in output signal.

A correction for such angular misalignment can be introduced by re-calculating the non-orthogonal components into orthogonal ones (see also Figure 5.11) (Zurek et al., 2006a).

Also, it is very important that the H-coils are positioned at the same angle as the B-coils. For symmetrical rotational measurements, the CW–ACW errors are greatly reduced by calculating the average from both directions of rotation (Sievert, 1990; Zurek et al., 2006a). Of course, all care should be taken in order to align the B and H sensors, and correction can be added for such misalignment as well (Maeda et al., 2007; Yanase et al., 2007). This can be more difficult for round samples because positioning of the rolling direction lacks a clear reference angle, whereas in square, rectangular or hexagonal samples, one of the edges can easily serve such a purpose. This problem is discussed in more detail in Chapter 4.

However, if non-symmetrical or non-rotational measurements are performed, then the CW-ACW averaging cannot be implemented and this can cause significant errors in the detection of B or H components, and thus also power loss and other magnetic properties. Some authors suggested that the angular phase shift between the sensors cannot be responsible for such great errors and for the resulting asymmetry in uni-axial power loss measured at various inclination angles (Gorican et al., 2002). Nevertheless, a careful analysis of the involved effects, and measurements on samples cut at various degrees show that it is the sensor misalignment that can be and indeed must be responsible for such results, mostly due to lack of clockwise–anticlockwise averaging (Zurek et al., 2009a).

There could also be some additional phase shifts caused by parasitic self-capacitance of the coil. For a single-layer H-coil, the parasitic values would be much lower, but multi-layer windings would be undoubtedly more affected and the measurement accuracy could be affected, especially at higher frequencies. However, this effect is difficult to quantify for a general case, as it depends on the geometry, frequency and distortion of the signals (content of higher harmonics).

Many of these effects could be probably reduced if the PCB-based sensors were utilised (Figure 6.24). The orthogonality of such sensors would be significantly improved so that the mechanical and magnetic angles were much better aligned with each other. Needle probes could be mounted directly in such PCB, thus further reducing angular misalignment between the B and H sensors.

3.2.2.2 Rogowski–Chattock Potentiometer

Rogowski–Chattock potentiometer (RCP), also referred to as *Chattock coil*, or *Chattock–Rogowski coil*, similarly to the *H*-coil, relies on the tangential field equality as illustrated in Figure 3.44. The sensor is essentially a coil wound on a former. Its underlying principle is therefore similar to any other induction coil – magnetic flux penetrating the coil induces voltage signal in it. The sensing technique based on a flexible coil was first described in Chattock (1887) and then seemingly independently in Rogowski et al. (1912), who also patented the idea (Rogowski, 1916).

Therefore, it would be more correct to refer to the sensor as 'Chattock–Rogowski', but the 'Rogowski–Chattock' name appears to be used more frequently in the literature.

In the case of the flat H-coil, the whole sensor is exposed to an assumed uniform magnetic field because it is placed in the area of the sample where the field is most uniform, by design of the experiment. As a consequence, each turn has almost the same voltage induced in it, and the measured signal is proportional only to the number of turns (and of course the cross-sectional area of the coil).

Owing to the magnetic reluctance of the sample, there exists a magnetic potential V_M along the surface sample. The RCP performs a line integral of H over the distance between its two ends 'A' and 'B' (Figure 3.52), thus measuring the value of this magnetic potential difference, or the difference in magnetomotive force (Salz et al., 1992).

The value of the corresponding integral does not depend on the path taken for integration. Therefore, the integration over path $l_{surface}$ is the same as over path l_{coil} (Figure 3.52) (Salz et al., 1992), so it can be written that the magnetic potential $V_{M(A,B)}$ between the points 'A' and 'B' is

$$V_{M(A,B)} = V_{M(B)} - V_{M(A)} = \int_A^B \left(\vec{H}\right)d\vec{l} = \int_{l_{surface}} \left(\vec{H}_{surface}\right)d\vec{l} = \int_{l_{coil}} \left(\vec{H}_{coil}\right)d\vec{l} \quad \text{(A)} \quad (3.24)$$

The critical parameter of any RCP is the uniformity of its winding, and this constraint is much more stringent than it is the case for an H-coil. If an H-coil is exposed to a uniform field, then the non-uniformity of winding is almost irrelevant, as long as it does not influence the 'magnetic angle'.

This is not the case for the RCP, which in general is exposed to much greater field variation along its size. For instance, the field at the surface of the sample can be significantly different than 5 mm above the sample (see Figures 3.48c, 3.54b and 4.4).

An RCP is exposed to such spatial variations and the uniformity of winding is critical for performing a correct line integral according to Equation 3.24. Magnetic flux linking to an infinitesimal part of the coil length, for a uniformly wound coil, can be described as

$$d\Phi_{coil} = A \cdot \mu_0 \cdot \vec{H} \cdot \frac{N}{l_{coil}} \cdot d\vec{l} \quad \text{(Wb)} \quad (3.25)$$

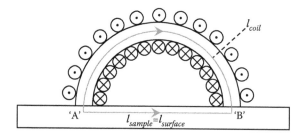

FIGURE 3.52 Operating principle of a Rogowski–Chattock potentiometer.

where $\dfrac{N}{l_{coil}} \cdot \vec{dl}$ is the number of turns for the infinitesimal length dl, due to the *uniform* turn density defined as the number of total turns N divided by the total length of the coil l_{coil}. The more the turns there are for a given length, the more the flux couples with the coil, so the sensitivity is improved. This is synonymous with using a greater number of total turns for an *H*-coil.

Therefore, the magnetic flux in an RCP is a line integral of Equation 3.25 (Fiorillo, 2004):

$$\Phi_{coil} = A \cdot \mu_0 \cdot \frac{N}{l_{coil}} \cdot \int\limits_{l_{coil}} \left(\vec{H}\right) \vec{dl} \qquad (\text{Wb}) \tag{3.26}$$

The voltage induced in an RCP is directly proportional to the time derivative of the magnetic flux (see also Equations 3.1 and 3.2), which is a function of the magnetic potential, which in turn is 'line integrated' by the coil itself along the way in which the RCP coil is 'bent'. Hence

$$V(t) = A \cdot \mu_0 \cdot \frac{N}{l_{coil}} \cdot \frac{d\left(\int\limits_{l_{coil}} \left(\vec{H}\right) \vec{dl} \right)}{dt} = A \cdot \mu_0 \cdot \frac{N}{l_{coil}} \cdot \frac{d\left(V_{M(A,B)} \right)}{dt} \qquad (\text{V}) \tag{3.27}$$

So, in order to measure the magnetic potential (magnetomotive force), the signal must be integrated in a similar way as it was for *B*-coil or *H*-coil, but with a different proportionality coefficient:

$$\int\limits_{l_{coil}} \left(\vec{H}\right) \vec{dl} = \frac{l_{coil}}{A \cdot \mu_0 \cdot N} \cdot \int V(t) dt \qquad (\text{A}) \tag{3.28}$$

If the magnetic field on the surface of the sample is uniform and the RCP does not enclose any other sources of magnetising field (like an excitation winding), then it can be written that

$$\int\limits_{l_{coil}} \left(\vec{H}\right) \vec{dl} = H \cdot l_{surface} \qquad (\text{A}) \tag{3.29}$$

Hence, H can be calculated by dividing the magnetomotive force by the length of the path:

$$H = \frac{\int\limits_{l_{coil}} (\vec{H}) \vec{dl}}{l_{surface}} \qquad (\text{A/m}) \tag{3.30}$$

So, in order to calculate the magnetic field strength from the induced voltage, the following equation should be used:

$$H(t) = \frac{l_{coil}}{A \cdot \mu_0 \cdot N \cdot l_{surface}} \cdot \int V(t) dt \quad \text{(A/m)} \tag{3.31}$$

which with the use of an appropriate calibration constant C_x becomes

$$H(t) = C_x \cdot \int V(t) dt \quad \text{(A/m)} \tag{3.32}$$

PRACTICAL COMMENT

If the surface path is taken in Equation 3.24, then it can be seen that the case becomes equivalent to the flat H-coil described in the previous sections. So, both sensors should measure the same quantity (compare Equations 3.32 and 3.22). In practice, this might not be the case in the absolute sense, due to practical difficulties, as discussed below. However, the accuracy of both sensors should be comparable, if they have similar dimensions and are made with similar care, for example, the windings are made with a similar wire and are wound uniformly.

The comment before Equation 3.29 states that the field should be uniform. This means that the measurement should be performed over an area for which the sample under test is magnetised uniformly. Otherwise, the RCP would still measure a correct value, but the measured result would be questionable from an experimental point of view due to non-uniform magnetisation of the sample under test.

3.2.2.2.1 Practical Implementation

The RCP sensing coil itself can be made in several ways. In his original paper, Chattock described a flexible rubber rod, 37 cm long, with a coil wound carefully around it, with the wire interspaced with cotton thread to ensure uniformity (Chattock, 1887). Such a sensor could be bent into various shapes so that its ends could be positioned at desirable points in space, in an arbitrary way (Figure 3.53a). This was especially useful for ballistic measurements with a galvanometer (see also Figure 3.2).

On the other hand, in his patent, Rogowski described a flattened cross section, more akin to a belt (Figure 3.53b) (Rogowski, 1916).

In rotational measurements, some researchers used the configuration shown in Figure 3.53c (see also more examples in Chapter 6) (Xu, 1995). The structure is rigid, which helps in retaining a constant distance between the coil ends, which is one of the factors affecting the calibration constant (see Equation 3.31). The cross-sectional area is rectangular and thin.

An interesting variation of the sensor is shown in Figure 3.53d. The magnitude of H eventually reduces to zero at a sufficiently large distance from the sample.

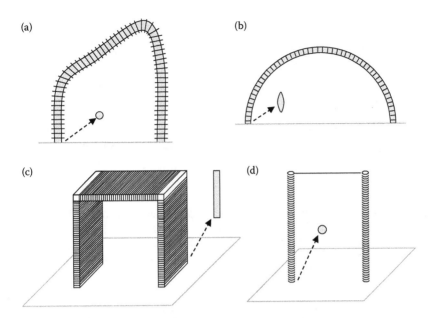

FIGURE 3.53 Examples of Rogowski–Chattock coils (drawings not to scale): (a) Original Chattock's coil wound on flexible rubber rod. (Drawing based on a description given in Chattock A.P., *Philosophical Magazine, Ser. 5*, 24 (146), 1887, 23.) (b) Rogowski's coil. (Adapted from Rogowski W., *Measuring magnetic properties of materials and apparatus therefor*, US Patent 1,204,489, 1916.) (c) RCP sensor. (Adapted from Xu J., Recent experiments on rotational power loss measurement at PTB, *Proceedings of 4th 2DM Workshop*, Cardiff, 1995, p. 9.) (d) RCP sensor with non-continuous coil. (Adapted from Hempel K.A. et al., A phenomenological description of the anisotropic and non-linear properties of electrical sheet under general two-dimensional magnetic excitation by means of reluctance tensor, *Proceedings of 5th 2DM Workshop*, Grenoble, 1997, p. 93.)

Therefore, the remote part of the RCP would contribute a negligible voltage, and the sensor could be simplified to two straight long coils, electrically connected in series (Hempel et al., 1997). It is unclear what error can result with such simplification and how long the coil should be so that the far ends are placed in 'zero field'. This approach is questionable due to the presence of a strong normal component of H extending to a considerable distance above the sample (as shown in Figure 3.47 and discussed in Chapter 4).

A direct comparison of performances of an H-coil and RCP was carried out in Xu (1995). A measurement head comprised two pairs of needle probes, an orthogonal double-axis H-coil and two overarching RCPs (see also Figure 6.22). Typical results are shown in Figure 3.54.

The measurement of power loss is in fairly good agreement, better than 5% (as can be judged visually from the published graphs, numerical data were not provided). The measurement of H amplitude differs to a greater extent, perhaps exceeding 10% in the worst case. However, such agreement is still significantly better than as compared between the *same* technique, that is, H-coils but used by different laboratories (Sievert et al., 1995).

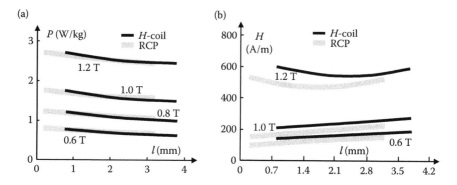

FIGURE 3.54 Comparison between rotational properties measured with H-coils and Rogowski–Chattock potentiometers. (a) Rotational power loss. (b) H values (only H_y values are shown here for clarity; H_x values had similar discrepancy). (Adapted from Xu J., Recent experiments on rotational power loss measurement at PTB, *Proceedings of 4th 2DM Workshop*, Cardiff, 1995, p. 9.)

An interesting variant of the implementation was described in Stauffer (1958) and Tompkins et al. (1958), and recently also in Gaworska-Koniarek et al. (2016), namely, a 'modified Chattock coil' was integrated with a pair of needle probes (Figure 3.55).

The modification was such that the 'modified Chattock coil' did not touch the surface of the sample, but was placed horizontally above the sample and small magnetic yokes extended from the coil to the sample. As with any other yoke, it can be approximated that the drop of magnetomotive force in the yoke is negligible (as compared to other components in such magnetic circuit), so that the magnetic potential measured

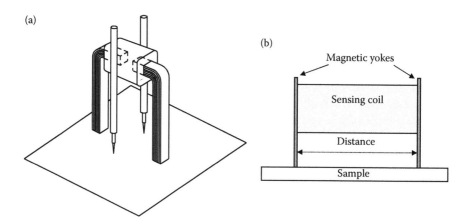

FIGURE 3.55 Modified RCP sensors: (a) Integrated head for measuring the H signal with a coil and two yokes and the B signal with a pair of needle probes. (Adapted from Stauffer L.H., *Methods of and devises for determining the magnetic properties of specimens of magnetic material*, 1958, Patent US 2828467.) (b) H sensor with magnetic yokes. (Adapted from Gaworska-Koniarek D. et al., Magnetic field strength sensor, *Symposium of Magnetic Measurements and Modeling*, Czestochowa-Siewierz, Poland, 2016.)

across the distance between the mini-yokes (Figure 3.55a) was the same as the magnetic potential difference at the sample surface. Therefore, it would be more beneficial to add small magnetic yokes joining the ends of the two partial coils in Figure 3.53d to achieve a possible improvement of the performance of such two-part RCP.

Therefore, it can be seen that in some sense, the sensor shown in Figure 3.55 is a combination of an H-coil and RCP. On the one hand, the measured quantity is proportional to the magnetic potential on the surface of the sample. However, on the other hand, the sensing coil is wound on a rigid former and the uniformity of winding is less important – similarly to an H-coil.

The implications on the measurement accuracy of such an approach were not reported so far. The sensor itself might offer reasonable performance (Gaworska-Koniarek et al., 2016), but the presence of magnetic yokes above the surface of the sample might introduce an unknown contribution to the distribution of the local magnetic field in magnetically open circuits. As can be seen in Figure 3.47 (and Figure 4.4), there can be a significant component of H perpendicular to the surface of the sample. Such component would find favourable path through the mini-yokes in the modified RCP sensor (Figure 3.55), thus injecting the additional flux in/out of the sample. For this reason, further studies of such sensors would be required before they could be used with guaranteed performance in a system with significant air gaps (which is usually the case for measurements of rotational power loss).

3.2.2.2.2 Practical Difficulties

The RCP coil must be wound in such a way as to have a well-defined and constant number of turns per unit length, or in other words – the turns must be distributed very uniformly and the active area of each turn must be identical for each turn. Otherwise, the 'turn density' becomes a variable function of the position along the length, and cannot be treated as constant in Equation 3.26. This function would have to be then defined under the integral, and calculated accordingly in an analytical or numerical way. Theoretically, this can be done, but in practice, it is not possible because the non-uniformity of 'turn density' cannot be defined or measured with sufficient precision.

Hence, such mathematical function would be unknown in a general case and thus the integral could not be evaluated correctly. As a result, the induced voltage would not be directly proportional to the derivative of the magnetomotive force and the measurement would be incorrect, with an unknown level of uncertainty.

The cross-sectional area as well as the length of the supporting 'former' must also be uniform over the whole length of the structure. This is especially important for bendable sensors (e.g. Figure 3.53a), which could distort while bent, and thus have effective local non-uniformities. For such coils, the wires of the neighbouring turns must not touch because they would restrict free bending, leading even to a change in the total length of the coil (Chattock, 1887).

The H at the surface of the sample is different than at a certain distance from it. An RCP has a non-negligible height and therefore in a general case, it has to be assumed that the H penetrating the RCP is non-uniform. This is especially more problematic for sensors, which are not made on a 'smooth' former. For instance, the sensor shown in Figure 3.53c has sharp corners, over which it is not possible in practice to maintain uniformity of the winding. The sensor shown in Figure 3.53d dispenses altogether with the top part, which can create even greater errors, so the usage must be carefully justified.

Usually, the calibration procedure would need to be performed as well. As with the *H*-coil, it is difficult to precisely know the active cross-sectional area, length of the coil and length between the ends. And as with the *H*-coil, these are the 'magnetically active' values, not those which can be measured on a mechanical former. Therefore, the signal processing would be the same as that for the *H*-coil (Equations 3.22 and 3.32), once the calibration constant is known. The calibration constant would encompass all those values, and give direct proportionality to the integrated voltage signal.

Another difficulty is that the winding of the coil must be wound is such a way that during the measurement the first and last turns are positioned as close as possible to the surface of the sample. As can be seen in Figure 3.54, the RCP is claimed to be positioned closer to the surface of the sample than the *H*-coil. However, the shapes of some of the curves in Figure 3.54 (for both *P* and *H*) seem to suggest that perhaps the distance from the sample was slightly larger than assumed – if the RCP curves were shifted to right (or *H*-coil curves shifted to left), the agreement would have been better. Nonetheless, the minimum distance from the surface is a recognised practical problem and it should be taken into account during preparation of such sensors.

In theory, the extrapolation of values towards the surface can also be carried out in the same way as for *H*-coil – by using more than one sensor for each direction, each positioned farther away from the sample. However, to date, there were no reports in the literature of such experiments.

As with the *B*-coils, two RCP sensors can also be placed at different angles (e.g. +/−45°) for the detection of orthogonal components of *H*, as shown in Figure 6.23.

Of course, PCB technology could also be used for making RCPs and therefore very good uniformity, positioning and mechanical stability of windings could be achieved. However, reduced sensitivity is to be expected because the PCB tracks cannot be positioned as tightly as thin wires (Zurek, 2017e).

As with any induction coil, the DC component cannot be measured, unless a ballistic method is employed (or the movement is initiated by vibrating the coil or the sample).

3.2.2.3   Hall Effect Sensor

The Hall effect sensor was described in Section 3.1.5 (see Figure 3.31), where its potential use was discussed for the measurement of *B*. Of course, the operating principle and signal processing remain exactly the same and only the positioning of the sensor would be different. Most of the comments given in Section 3.1.5 remain true also for the detection of *H*.

Once again, the measurement of *H* would rely on the tangential field approximation (Figure 3.44). Therefore, the sensor must be placed as close as possible to the surface of the sample. Hence, the thickness of the sensor and positioning of the internal sensing element inside of the given electronic package are important parameters.

A Hall effect sensor always detects *B* penetrating it, according to Equation 3.18. Therefore, the results should be recalculated by scaling it with the μ_0 constant because the sensor would be positioned in air, and is itself non-magnetic. This can be included in the main constant C_x with units $\left(\dfrac{\mathrm{A}}{\mathrm{m} \cdot \mathrm{V}}\right)$, so that the measured instantaneous *H(t)* is

$$H(t) = C_x \cdot V(t) \quad \text{(A/m)} \tag{3.33}$$

The abovementioned typical sensor A1302 has its sensitivity given in the data sheet as 0.0769 T/V. This value should be divided by μ_0 in order to satisfy the units of C_x in Equation 3.33, which results with $7.958 \cdot 10^5 \dfrac{A}{m \cdot V}$ or $795.8 \dfrac{A}{m} \cdot \dfrac{1}{mV}$ so every 1 mV of output voltage corresponds to a change of 798.5 A/m of input H. Hence, $H = 100$ A/m would produce only 126 µV.

Such sensitivity poses two problems. First, the sensitivity itself is quite low as compared to other sensors. An example of an H-coil given above produced around 1 mV per 100 A/m, which is almost on the order of magnitude better. So, the measurement would require a voltmeter with not only much better resolution, but also capable of measuring much lower signals with the same accuracy to give comparable results. So, for H measurements, the sensor should have as high sensitivity as possible (and a 'low-noise' version would be desirable too). On the other hand, extremely high values of H can be measured without saturation because for this sensor, even $H = 800$ kA/m would produce only around 1 V, so the sensor would still not saturate (but neither would an H-coil).

The second and possibly more important difficulty is that the output voltage is *directly* proportional to the applied H, not its derivative. It is true that no need for integration simplifies signal processing. But signal integration (be it with analogue electronics or numerical processing) also acts as a very effective low-pass filter, which quite significantly reduces high-frequency noise in the signal.

The internal noise of the abovementioned Hall effect sensor can be as high as 150 µV (Allegro, 2013b). This is therefore higher than the useful signal produced for a field of 100 A/m, which would be around 125 µV. The implications are illustrated in Figure 3.56.

The grey signal in Figure 3.56a is a simulated waveform comprising pure sine with amplitude 125 µV with 150 µV of superimposed random noise. The resulting peaks are therefore up to 275 µV. Therefore, such voltage would be produced as the output signal of the Hall effect sensor, so the calculated H waveform would have *identical* shape, with the same high-frequency noise content (Figure 3.56b), just scaled by some constant as defined in Equation 3.33. An ordinary B–H loop plotted with such waveform would look extremely noisy in the H coordinates and the precision of B–H loop area calculation would be significantly impacted.

The situation is very different for an H-coil. Let us assume that identical voltage signal is produced by the H-coil (grey waveform in Figure 3.56a). The noise is the same, but the signal must be first integrated by means of Equation 3.22 and scaled by the calibration constant to produce the same $H_{mean}(t)$.

However, the integration procedure acts as a low-pass filter and suppresses the high-frequency noise significantly, so that the real signal is represented as being much smoother (black waveform in Figure 3.56b). An area of B–H loop plotted with such waveform will be calculated with a greater degree of precision (albeit still somewhat noisy in this particular case). In addition, as shown above, a comparable H-coil has greater sensitivity, so the produced underlying output signal is significantly greater than for the Hall effect sensor. Moreover, a Hall effect sensor

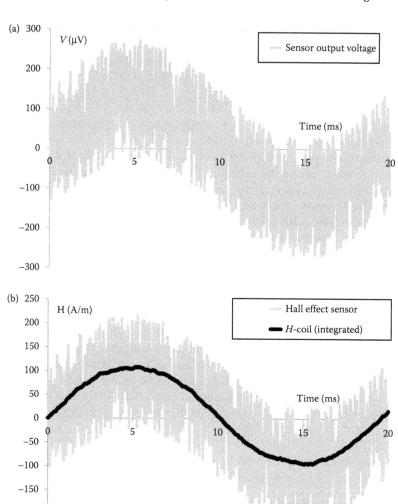

FIGURE 3.56 Comparison of simulated noisy signals: (a) noisy output voltage $V(t)$ of a hypothetical sensor, (b) calculated $H(t)$ for a Hall effect sensor (grey – direct proportionality) and an H-coil (black – integrated); for better illustration, the amplitude and phase of the integrated signal were normalised to be the same as the input signal but it can be seen that proportionally the high-frequency content was suppressed.

requires additional circuitry (supply voltage or current source), which can introduce additional noise. The sensors and/or the additional circuitry can also be susceptible to quite significant temperature drifts (Sentron, 2000; Melexis, 2013).

Of course, the signal can be filtered by some passive or active low-pass filter. However, such filters can introduce phase or gain errors (Jacob, 2004), and with the phase sensitivity of the rotational measurement, the results could be grossly erroneous.

Nevertheless, Hall effect sensors were sometimes employed for rotational (Moses, 1992) and 2D measurements (Moses et al., 2008). An advantage of a Hall effect sensor is that it can be manufactured in a fairly small package (e.g. 3 × 5 × 5 mm), and there are sensors capable of detecting all three orthogonal components simultaneously: H_x, H_y and H_z (Melexis, 2013).

For a single Hall effect sensor, the typical complication is that the sensing direction is usually perpendicular to the largest surface of the electronic package because this is the easiest configuration in which the flat plate can be built into an electronic chip (e.g. see Figures 3.31 and 3.32). Therefore, in order to detect the tangential component, such chip would have to be positioned vertically, which would result in the sensing point being located much farther away from the sample surface, which of course should be avoided (Figure 3.44). This is why the multi-axis sensors are useful because the sensing point can be put closer to the sample surface even if the electronic chip is positioned horizontally, namely, parallel to the sample surface.

The small dimensions allow carrying out much more localised measurements and even scanning the sample of the surface with point-by-point measurements (Konadu, 2006). Typical results are shown in Figure 3.57, and the measurement system in Figure 3.22.

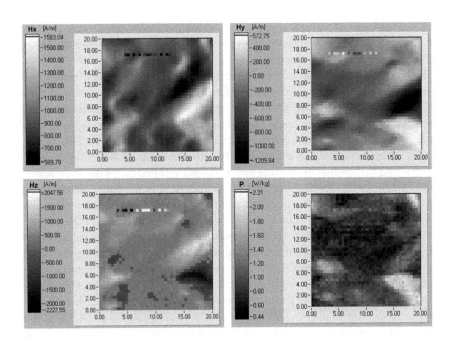

FIGURE 3.57 Results of scanned H_x, H_y, H_z and P on 20 × 20 mm surface of the sample magnetised at 50 Hz under uni-directional conditions; the system employed a three-axis Hall effect sensor Sentron 3D-H-30. ((Sentron, 2000). Images courtesy of Sam Konadu, formerly at Wolfson Centre for Magnetics, Cardiff University, Cardiff, the United Kingdom.)

3.2.2.4   Other Sensors of Magnetic Field

There exist a number of other sensors capable of detecting B or H (Ripka et al., 2003): fluxgate, based on magnetoresistance (also with their anisotropic AMR, giant GMR and tunnel versions TMR), magnetoimpedance, vibrating coil magnetometer, etc.

Some of these are physically too large to be considered for the measurement of H with the tangential approach. This is usually the case for the abovementioned fluxgate sensor (Figure 3.36). Also, the vibrating coil magnetometers require rather sophisticated mechanical driving mechanism. Usually, the vibrating frequency (or excitation frequency in fluxgates) limits the upper frequency at which a measurement can be made.

The ordinary magnetoresistive effect (MR) is usually too weak to be used in linear sensors (Jander et al., 2005). And the magnetoimpedance seems to be less popular in commercial applications.

The AMR, GMR and TMR technologies are capable of attaining very high sensitivity to the magnetic field. The sensors can be made as integrated circuits, which are dimensionally small – comparable in size to Hall effect devices (Mason, 2003), and also very similar in physical appearance (see also Figures 3.31, 6.15 and 6.17).

For instance, the three-axis sensors HMC10xx have a typical sensitivity specified as '3.2 mV/V/gauss' (Honeywell, 2006). This means that the sensitivity is a function of the supply voltage. With the typical value of 5 V, the sensitivity is therefore 16 mV/gauss, or 160 mV/mT, which translates to around 0.2 mV/(A/m). So a change of 100 A/m will produce a voltage of 20 mV, which is significantly higher than H-coil, RCP or Hall effect sensors discussed above.

However, ferromagnetic materials are an integral part of such sensors. The required operating point is set by using, for instance, an internal biasing magnetic field. Exposing the sensor to a too strong external field can create an irreversible offset, which can be nulled only through a 'resetting' procedure (Mason, 2003), usually with some additional resetting circuitry (Honeywell, 2006). This is one of the very undesirable features because the resetting current can be as high as several amperes, and the 'damaging' field as low as 0.5 mT (400 A/m) for the more sensitive devices (Honeywell, 2008). Even more importantly, it is not immediately obvious that the sensor could require resetting and thus erroneous measurements could be performed, without the operator realising that there is a problem with the sensor.

Nevertheless, such sensors are fairly inexpensive and readily available commercially. Some researchers used them instead of Hall effect sensors, in similar applications. For instance, an array of such sensors can be mounted on the surface of the sample in a uni-directional single-strip tester, and H is measured by averaging the signal from all the sensors in the array (see also Figure 6.15) (Tumanski, 2011).

Also, as with the Hall effect sensors, the output voltage signal of an AMR sensor is directly proportional to the detected H. Therefore, any noise will not be reduced by integration and it will be shown directly in the $H(t)$ waveform, similarly as shown in Figure 3.56b.

The small size can be an advantage if the sensor is to be used for the detection of a very localised field, as in Figure 3.57. However, with the measurement of the global properties of a given sample, the small size of the sensor can be a strong disadvantage. The sensor is too small to perform spatial averaging of H, and even sensors as large as H-coils can struggle with large-grain materials such as grain-oriented electrical steel. For this reason, averaging from an array of sensors might give more repeatable measurement, perhaps comparable with averaging obtained by the magnetising current method (Figure 6.15) (Tumanski, 2011).

As mentioned above, two- and three-axis AMR sensors are also available (Honeywell, 2006, 2008) and in theory could be employed for rotational measurements. Still, most of the literature on rotational measurements seems to focus on H-coils or RCPs.

3.3   TEMPERATURE SENSORS

As discussed in the previous chapter, the thermometric method requires measuring the temperature at least with very high resolution, better still with very high absolute accuracy.

A thermocouple is a commonly used temperature sensor. It is also possible to use a thermistor, or some researchers employed even thermal imaging camera, for contactless sensing. All thermometric techniques rely on the temperature rise of the sample under test. This is usually a very slow process (as compared to other techniques) so that a single measurement point (e.g. at a fixed B value) takes minutes, which is the biggest disadvantage of such methods.

3.3.1   THERMOCOUPLE

The operation of a thermocouple is based on the Seebeck effect. A conductor exposed to a temperature gradient develops a potential difference between the ends placed at different temperatures. The amount of voltage is proportional to the value of the temperature gradient and for a given material it can be expressed as the *Seebeck coefficient*. Different materials have different values of the coefficient, so if two different materials are connected at one end and exposed to the same temperature gradient, there will be a measurable voltage between the other ends of the two wires, placed at different temperatures.

Therefore, a thermocouple is usually made as a pair of different wires, connected together at one and, by fusing, spot welding or a similar technique – this connection is often referred to as *junction* (the fused droplet of metal is visible at the end of wires in Figure 3.58a). The wires are isolated from each other over the whole length, apart from the fused junction (it can be seen that the top wire has an additional thin insulation in Figure 3.58a). The other end is connected to a suitable voltage-measuring circuit, capable of rescaling the signal into temperature (e.g. multimeter with a display).

Each junction contributes to the total voltage difference (Ripka et al., 2003), including the connections to the voltmeter. In ideal measurement systems, two junctions would be used, with one placed at a reference temperature, for instance, in an ice bath, so that a 'calibrated' 0°C can be obtained, and in this way, thermocouples are used in sensitive laboratory temperature measurements.

However, under normal operating conditions, this is cumbersome and many devices employ *cold junction compensation*, for instance, by using additional

(a)

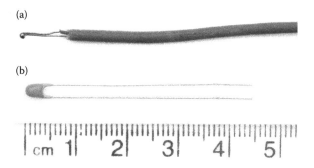

(b)

FIGURE 3.58 Temperature sensors: K-type thermocouple (a) and thermistor (b). (Courtesy of Joanna Kaczmarzyk, CC-BY-3.0, *Encyclopedia Magnetica*.)

electronics (Maxim, 2007). A further complication is that the Seebeck coefficient is not constant but varies somewhat with temperature and for more accurate measurements these variations must be taken into account, for instance, by using look-up tables (Ripka et al., 2003).

A popular, inexpensive and widely available thermocouple is the K-type* (pair of alloys chromel–alumel), which has a sensitivity (Seebeck coefficient) of around 40 µV/K (Ripka et al., 2003), so a change of 1 K or 1°C will produce a voltage change of around 40 µV. Therefore, if a temperature difference of 25 K is to be measured (e.g. the difference between the room temperature of 25°C and the reference temperature of melting ice 0°C), then the measured voltage would be around 40 µV/K · 25 K ≈ 1 mV.

However, there are also additional non-linearities, which cannot be included in a constant value. Some researchers approximate the changes with an additional temperature coefficient, for instance, 0.2%/K (Ragusa et al., 2008). However, in a general case, a simple linear equation cannot be used and it is best to refer to a look-up table, as these are available for temperatures over the whole range from −260°C to 1370°C (NIST, 2014). These are usually given with the cold junction reference at 0°C.

Such non-linearities can be approximated with polynomials, and these are also available as look-up tables, including two forms as 'coefficients' for voltage as a function of temperature and 'inverse coefficients' for temperature as a function of voltage (NIST, 2014).

Of course, for temperature measurement, the 'inverse coefficients' are more useful. The equation contains up to 10 coefficients. The general equation is of the form:

$$T_m = C_0 + C_1 \cdot V + C_2 \cdot V^2 + \cdots + C_n \cdot V^n \quad (°C) \tag{3.34}$$

where C is the constant or coefficient whose denominator has units of volts raised to the power of the coefficient number, so that C_0 (°C), C_1 (°C/V), C_2 (°C/V²) and

---

* Some types of commercially available thermocouples are designed by capital letters. There are several types designed for various temperature ranges and sensitivities: B, E, J, K, N, T, etc. Type K appears to be quite popular commercially.

C_n (°C/V^n) and V is the output voltage of the sensor with reference junction at 0°C (V) so that by definition at 0°C the output voltage is 0 V.

If all coefficients are used, then the *absolute* error can be from −0.02°C to +0.04°C. For instance, for a K-type thermocouple, the coefficients* have the following values (for a range 0–500°C):

- $C_0 = 0$°C (trivial value in this case)
- $C_1 = 25.08355$°C/mV
- $C_2 = 0.07860106$°C/mV^2
- $C_3 = -0.2503131$°C/mV^3
- $C_4 = 0.08315270$°C/mV^4
- $C_5 = -0.01228034$°C/mV^5, etc.

Let us assume that the measured output voltage is exactly 1 mV (with respect to 0°C), which according to the table would correspond to a temperature of exactly 25°C. Including an increasing number of coefficients gives

- C_0 to $C_1 = 25.084$°C (error +0.084°C)
- C_0 to $C_2 = 25.162$°C (error +0.162°C)
- C_0 to $C_3 = 24.912$°C (error −0.088°C)
- C_0 to $C_4 = 24.995$°C (error −0.005°C)
- C_0 to $C_5 = 24.983$°C (error −0.017°C) and so on

Therefore, using only some coefficients might not give the required precision and using the look-up tables might give better results, rather than the polynomial approach of Equation 3.34.

Usually, simple multimeter-type devices have insufficient display resolution to carry out power loss measurements for which the required resolution would have to be 0.001°C or better. Therefore, the voltages must be measured with a suitable high-precision voltmeter and processed accordingly. For example, nanovoltmeter was used for the research described in Ragusa et al. (2008). There are also commercially available data acquisition devices that are designed to have appropriate input signal conditioning and voltage resolution to perform such task (NI, 2014b). Multiple input channels are available, with cold junction compensation to within 2°C (of the absolute value), 24 bit resolution, etc. Commercial software such as LabVIEW can be used for communication with such devices, data acquisition and automatic re-scaling of the results into appropriate units.

Typical measurement results with a thermocouple are shown in Figure 2.17. It should be noted that a total temperature change in that figure is only around 0.3°C for measurement at 50 Hz and 1.5 T, which is the reason for the very high relative resolution required for such measurements. If lower frequency and/or B is applied, then the temperature rise is even smaller and measurements can be much more noisy. However, measurements with resolution better than 0.001°C are possible, and indeed were carried out (Fiorillo et al., 1992a). And as always, it should be borne in mind that resolution does not mean accuracy. If the reference junction tempera-

---

* It should be noted that the NIST tables have the units of mV, not V.

ture changes by some amount, so does the output voltage of the sensor, even if the measured temperature does not. Hence, stability of the cold junction compensation technique is critical for both good accuracy and precision.

An easy way to improve the resolution (or to reduce the noise) of the voltage signal is to increase the time over which the reading is averaged. For instance, the device NI 9211 (NI, 2014b) is capable of processing 14 readings per second, so the time of a single temperature reading is less than 0.1 s, which is significantly shorter than the long measurement times of the rising temperature (usually >10 s).

However, the various time intervals must be taken into account also because of the thermal inertia of the sensor. The temperature of the sample will not be transferred *instantaneously* to the sensor, but it takes a certain time for the body of the sensor to respond because a certain additional mass must change its temperature, and the energy must be transferred through some thermal resistance.

As a result, when the temperature of the sample changes, the temperature changes reported by the sensor will always lag by a certain time, which could take several seconds for larger sensors. It is therefore essential that the sensor has very good *thermal connection* to the sample, for instance, by using metal-loaded paint (Ragusa et al., 2008). If more than one thermocouple is used, it will also be essential to insulate them from each other and from the sample electrically because their tips are usually bare wires and surface eddy currents can significantly influence the small voltages (see also Section 3.1.2 on the needle probe technique). In any case, it is a good idea to electrically isolate the thermocouple and the sample under test, even if just a single sensor is used. Unfortunately, any electrical insulation usually means higher thermal resistance.

The thermal inertia can be reduced by using custom-built thermocouples. For instance, some researchers resorted to depositing thermocouples as thin layers of appropriate metal strips, similar to as shown in Figure 3.17 (Mazurek, 2012). Owing to smaller metal volume and larger surfaces involved in making such thermocouples, they will respond quicker to the thermal excitation.

PRACTICAL COMMENT

The thermoelectric voltage (Seebeck effect) always occurs as a consequence of any joints of dissimilar metals. This is a useful phenomenon in thermocouples, but it is also present as an unwanted effect in *other* voltage measurements.

As discussed above, the thermoelectric contribution can be on the order of 40 μV/K, or in other words, a change of just 1°C is capable of introducing an additional and unwanted offset of 40 μV. In many cases, such values cannot be ignored when connecting many types of sensors described in this chapter. For example, an unwanted DC offset can cause:

- A 'ramp' when integrating a voltage from a *B*-coil, *H*-coil or RCP (Figure 3.8)
- Offset in readings of a shunt resistor, Hall effect or magnetoresistive sensor
- Offset in a temperature measurement by means of a thermistor
- Apparent residual strain in a strain gauge bridge and so on

The copper–copper effect is quite small, so is the copper–tin (both values are around 0.2 µV/K), but the copper to tin–lead can be around 5 µV/K (important for soldered joints). More worryingly, the copper to copper oxide can exceed 1000 µV/K (Kidd, 2012), so even a fraction of 1°C might have a strong effect on the detected voltage. Therefore, proper connections are quite important for eliminating such effects, but careful analysis of the voltage offsets in the system is also essential.

It is also quite clear that due to such effects, minute DC offsets will be omnipresent in *any* measurement apparatus even such that employs the most precise equipment.

3.3.2 THERMISTOR

A thermistor is a non-linear resistor, whose value changes with temperature. If the resistance increases with increasing temperature, then the device is said to have positive temperature coefficient (PTC). If the resistance decreases with temperature, then the device has negative temperature coefficient (NTC). The NTC-type thermistors are more frequently used for temperature detection or measurement. They are usually made as small beads of semiconductor with two terminals across which the resistance is measured (Figure 3.58b).

There are data acquisition devices that are designed specifically to be used with thermistors, for example, SCXI-1503 (NI, 2007a).

The resistance of NTC thermistor changes in a mostly exponential manner, and within some range can be approximated with the following Steinhart–Hart equation:

$$R_m = R_{ref} \cdot e^{\beta \cdot (1/T_{ref} - 1/T_m)} \quad (\Omega) \qquad (3.35)$$

where R_m is the measured resistance (Ω), R_{ref} is the resistance (Ω) at some reference temperature, for example, room temperature of 25°C, β is the beta coefficient (K) often specified by the manufacturer of the device, T_{ref} is the reference temperature (K) and T_m is the measured temperature (K).

Therefore, Equation 3.35 can be rearranged so that the measured temperature can be calculated from the reference values and the measured resistance:

$$T_m = \frac{\beta}{\dfrac{\beta}{T_{ref}} - \ln\left(\dfrac{R_m}{R_{ref}}\right)} \quad (K) \qquad (3.36)$$

It should be noted that calculations based on the β coefficient are usually carried out for units of K, and not °C or any other degrees such as °F (Trietley, 1986; Vishay, 2009). So the reference temperature value must also be converted to the appropriate units.

For instance, a thermistor could have a base value of 10 kΩ at a room temperature of 25°C (the actual device shown in Figure 3.58b). The coefficient $\beta = 3977$ K can be applied within the range 25–85°C, over which Equation 3.35 (and therefore also Equation 3.36) will approximate the actual values to within 0.75%. If greater absolute precision is required, then look-up tables are also provided, for instance, with values every 5°C, over the full range (Vishay, 2009).

As with a Hall effect sensor, self-heating is one of the limits. This is because the measurement of resistance requires for some current to be passed through the sensor, and for the abovementioned thermistor, the maximum power dissipation must be limited to 0.25 W.

However, this is only the maximum value for 'rated' power (preventing the destruction of the sensor or other *significant* effects). Even a very small current will cause *some* self-heating, which will increase the temperature of the thermistor by some value and thus lead to overestimation of the measured temperature. This is very much device-specific and manufacturers' recommendations should be followed closely. For instance, in other typical device, a power of only 7.5 mW is capable of increasing the internal temperature of the thermistor by 5°C (Murata, 2012). Therefore, for a 10 kΩ device, the recommended operating current is only 0.12 mA, which can still increase the thermistor temperature by 0.1°C above its ambient. This is significant if the total change to be measured is just 0.3°C (Figure 2.17).

The voltage drop would be around 1.2 V, which is not difficult to detect as such. However, in order to detect a change of 0.1°C, the change in voltage would be only around 0.4%, which would require a resolution of around 5 mV just to detect such a change. This is a larger voltage that it would be for a thermocouple, but a similar percentage change. Therefore a high-resolution voltage measurement would still be required.

But as can be seen in Figure 2.17, for power loss measurements, much smaller changes must be detected, so the signal generation and measurement system must be designed accordingly in order to handle the required level of precision.

The sensing element also has a certain thermal inertia, hence does not respond instantaneously to the applied temperature. The time is proportional to the size of the sensing bead, and for the thermistor shown in Figure 3.58b, the thermal time constant specified by the manufacturer is 1.2 s. So, for the sensor to equalise with the ambient, the total time can be up to 5 s (*at least* three time constants), and this is for a sensor that weighs only 0.22 g (Vishay, 2009). For faster measurements, this might be too slow, and a physically smaller or lighter sensor would have to be used. But if the measurement is fairly slow (tens or hundreds of seconds, Figure 2.17) and the characteristics have time to stabilise its slope, then such a sensor can still be suitable.

3.3.3 PYROMETRIC METHODS

Pyrometry relies on the amount of electromagnetic energy radiated from a hot body. If the temperature is hot enough, then the radiated frequency is so high that it becomes a visible light (e.g. hot iron glows visibly in red, yellow and white). For lower temperatures, the frequency is proportionally lower, with a significant part of the spectrum residing in the invisible infra-red range. By measuring the amount of radiated energy, the temperature of a hot body can also be approximated, with various degrees of accuracy (e.g. depending on measurement conditions). Therefore, such methods allow remote detection of temperature, without the need of thermal contact with the sample under test. A detailed description of the underlying physical principles is too voluminous to be included here, and the reader is advised to refer to the vast available literature on the subject, for instance, Ripka et al. (2003), Optris (2006) and FLIR (2014).

There are two types of devices using the infra-red detection, which are widely used for temperature measurement and which are described below: infra-red thermometer and thermal imaging cameras.

Nowadays, both are digitised, so normally the user has no access to the way the signals are processed. The output information is numerical, namely, the device either displays the measured temperature on its screen or makes it available for digital transmission, for example, via serial or USB interface.

3.3.3.1   Infra-Red Thermometer

An example of an infra-red thermometer is shown in Figure 3.59. It uses an electromagnetic spectral range with wavelengths 8–14 μm (Optris, 2006).

The device has an optical system optimised for infra-red range, which can focus on a given patch of a surface. This particular device has two modes, one for measuring larger areas and one for measuring small spots (e.g. 2 mm diameter). The temperature is averaged over the whole measurement area. The device is equipped with a laser guide system, which aids the aiming and focusing. It can be mounted on a tripod and programmed with a computer, via USB interface.

In theory, such device would be capable of measuring the temperature of a given spot on the sample surface, similarly to a thermocouple or a thermistor. The device can also be mounted and pointed onto a precise area of the sample.

However, there are several reasons why using such a device is difficult in practice, and it would be rather difficult to implement it for the rate of temperature rise for power loss detection.

One of the biggest complications is the fact that the measurement accuracy depends very strongly on the properties of the measured surface. The radiation of energy depends on a coefficient called *emissivity*, which can vary in a very wide range even for the same material, caused purely by the smoothness of the surface (Optris, 2006). For instance, a very smooth surface behaves like a mirror, so instead of the

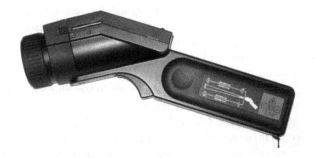

FIGURE 3.59   Hand-held infra-red thermometer. (Optris, 2006) Equipment courtesy of Megger Instruments Ltd, the United Kingdom.

temperature of the sample, the pyrometric thermometer would measure the temperature of whatever reflects in that surface – similarly to as a regular photographic camera would record reflection in the mirror, and not the surface of the mirror itself.

So, for correct measurement, the emissivity coefficient would have to be calibrated precisely, for instance, with the use of a thermocouple, and this procedure can be advised even by the manufacturer of the given infra-red thermometer (Optris, 2006). This can be overcome to some degree by obscuring the smooth metallic surface with black paint. But it would still not guarantee an absolute accuracy.

Another, perhaps more trivial difficulty can be that, for many such devices, the measurement result is reported in a pre-configured format, usually with a resolution of only 0.1°C (Optris, 2006). This would be insufficient for most measurements at lower power and/or low frequency (compare with Figure 2.17).

For more precise measurements, the sample under test would have to be placed in adiabatic conditions, as described in Chapter 2. If this is achieved with the help of vacuum, then the thermometer would also have to be placed either in the vacuum (not suitable for most devises) or outside and carry out the detection through the wall off the vacuum chamber. But for an ordinary glass wall, this is usually not possible because it blocks some part of the infra-red spectrum (Cooper et al., 1998).

PRACTICAL COMMENT

Interestingly, in some infra-red thermometers, the ultimate sensing element is still a thermocouple. The infra-red optics is used only to filter and focus the energy beam onto a thermocouple, which is calibrated accordingly to perform such measurements (Ripka et al., 2003).

3.3.3.2 Thermal Imaging Camera

In many ways, a thermal imaging camera, an example of which is shown in Figure 3.60, operates similarly to the infra-red thermometer described in the previous section, but a camera records temperatures of many 'spots' simultaneously. The sensing element is much more sophisticated and high-resolution devices can cost more than $10 k, but low-cost low-resolution devices are available for around $200. The measurement is also based on processing a specific band of infra-red radiation emitted by a hot body.

There are thermal imaging cameras designed for research applications, with a visual resolution up to 1280×1024 pixels, and a thermal resolution of 20 mK (0.02°C). Temperature differences down to 0.001 K can be processed with a special processing technique called 'lock-in' (FLIR, 2014), but this is not straightforward and cannot be applied in all cases. Nevertheless, thermal imaging measurements were reported by some researchers (Nakata et al., 1992; Shimoji et al., 2011). An example of a thermal imaging used for the detection of temperature rise of a transformer core is shown in Figure 3.61a (Marketos, 2002; Marketos et al., 2006).

However, thermal imaging suffers from the same problems as infra-red thermometry. It is even more difficult to put the whole camera in the vacuum chamber, both because of the size of the camera and the unsuitability of most cameras to operate in vacuum.

FIGURE 3.60 Thermal imaging camera X6530sc. (From FLIR, *Thermal Imaging Camera X8000sc, Data Sheet*, 2014. Photograph from of FLIR Systems. With permission.)

Pyrometric detection through an ordinary glass wall is also not possible in a general case (Cooper et al., 1998). A special window must be constructed from exotic materials such as 'sapphire glass' (Shimoji et al., 2011; Sato et al., 2014), and the authors acknowledged that *'the temperature resolution of thermography is generally 10–100 times inferior to that of temperature measurement using a thermocouple, and a longer time is needed for the measurement'*.

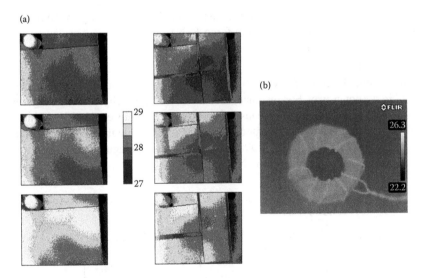

FIGURE 3.61 Thermal imaging with non-adiabatic conditions: (a) corners of two different transformer cores energised at 1.5 T and 50 Hz, with the temperature measured after 30, 60 and 106 s. (Courtesy of Filippos Marketos, formerly of Wolfson Centre of Magnetics, Cardiff University, Cardiff, the United Kingdom). (b) Toroidal sample with the primary winding heated by the magnetising current more than the magnetic material is heated by magnetic losses – the wire has lighter colour than the ambient or the core. (Courtesy of Megger Instruments Ltd, the United Kingdom). All images recorded in open air at room temperature.

Thermal imaging remains a very good method for indicating overall temperature changes in larger cores, or in systems with higher absolute temperature increase (Marketos et al., 2006). However, it is still significantly less suitable for use as a 'sensor' for rate-of-temperature-rise technique, unless the technology is improved in future to have resolution significantly better than 20 mK.

3.4   MECHANICAL SENSORS

There is a plethora of mechanical sensors and sensing techniques, which in principle can be employed for sensing stress and strain, and it would be impossible to describe them all. However, some of the most popular ones are discussed below, to show some principles and examples of such measurements.

3.4.1   FORCE, TORQUE AND STRESS SENSORS

Compressive or tensile stress is measured in the same unit as pressure (pascal), and according to the definition $Pa = N/m^2$. Therefore, if the surface area is known (e.g. cross-sectional area of the sample subjected to a given stress value), then the problem simplifies to the measurement of force in the units of newtons (N).

Similar methodology can be applied for the measurement of torque, whose unit is (N · m). The acting torque can be balanced out by a force acting on an 'arm' with such a length that the product is equal to the applied torque (see also Figure 3.63b). So, if the length l in (m) of the 'arm' is constant and known, then again the measurement simplifies to the measurement of force F (N).

3.4.1.1   Weight

In many cases, the force does not have to be *measured* as such because it can be *applied* in a known way. For instance, it is sufficient to attach a known weight with the mass m to generate force F because of gravity g (Garshelis et al., 1986; Klimczyk, 2012). Therefore, the force generated by a mass can be calculated by using Earth's gravity $g = 9.80665$ m/s^2 (NPL, 1998), so that $F = m \cdot g$ (Figure 3.62) and a weight of 1 kg produces a force of 9.80665 N.

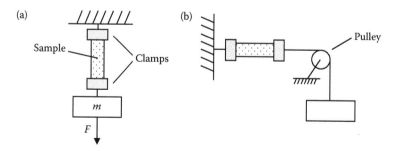

FIGURE 3.62   Force does not have to be measured if the applied value is known: (a) vertical and (b) horizontal.

PRACTICAL COMMENT

The precision of such 'measurement' will depend on the accuracy to which the values are known or estimated. For instance, g can vary between the poles and the equator on the Earth, and errors up to 0.5% can be introduced if they are not accounted or compensated for (NPL, 1998).

Of course, other parasitic effects such as friction or additional weight of any clamps should be included. For example, for extremely precise measurements, the buoyancy effect should also be taken into account, and the weight of the given object appears lower when surrounded by gas with higher pressure.

The general concept shown in Figure 3.62 is included only as an example. There are many other methods of applying force: based on mechanical screw (Kanada et al., 2011), pneumatic (Klimczyk et al., 2011), hydraulic (Yamamoto et al., 2011), etc. However, a known mass gives a very robust and inexpensive way of not only applying, but also measuring, force, as described in the next section.

3.4.1.2 Spring

One of the simplest, most versatile and robust sensor of force is a mechanical spring. The force acting on the spring causes its deflection or change of dimensions. The spring resists the force and some equilibrium is reached, so for a greater deflection, greater force must be applied.

For an 'ideal' spring obeying Hooke's law with a linear deflection,[*] the relationship is directly proportional. This is illustrated in a very simplified way in Figure 3.63 (see also Figure 2.1) and can be described as (the minus denotes that the force of the spring opposes the applied force)

$$F = -C_x \cdot l \quad \text{(N)} \tag{3.37}$$

where F is the applied force (N), C_x is the characteristic constant[†] of the 'linear' spring (N/m) and l is the deflection of the spring (m).

Of course, similar equation applies to any other spring whose deflection is not linear, but, for instance, rotational, and the name *torsion spring* is normally used. A simplified result of using a torsion spring is shown in Figure 2.2 (see also Figure 2.3), for which it would be

$$F = -C_x \cdot \phi \quad \text{(N)} \tag{3.38}$$

where F is the acting force (N), C_x is the characteristic constant of the torsion spring (N/rad) and ϕ is the angular deflection of the spring (rad).

In magnetic measurements, a torsion spring is sometimes made from a straight piece of a round wire (Figures 2.3 and 2.8). Therefore, Hooke's law can be applied

---

[*] Such versions are also referred to as *tension* or *compression springs*, depending on the design.
[†] The spring constant in Equation 3.37 is also commonly referred to by the letter 'k' in the literature.

and the string constant can be calculated from material properties. However, helical springs can also be used for the same purpose (Figure 2.6), and in a general case for precise measurements, the spring should be calibrated, by applying known torque value or force value.

The calibration can be carried out by using weights of a known value for linear deflection, or known weights hanged on a known arm for angular deflection (or torque calibration), as shown in Figure 3.63b.

A helical spring is attached to a supporting structure (Figure 3.63a). A known force F_1 is applied to the spring by attaching a known mass m_1, so the deflection is l_1 (not shown for clarity). Then a greater mass is attached such that $m_2 > m_1$ and this produces a greater force $F_2 > F_1$ and thus a greater deflection l_2.

If it is assumed that the spring behaves in an ideal linear way and that its characteristic constant C_x is indeed a constant, then the deflection is directly proportional to the applied force. According to Equation 3.37, it can be written that $l_1 = F_1/C_x$ and $l_2 = F_2/C_x$. The difference between the deflections is $\Delta l = l_2 - l_1$. Thus $\Delta l = F_2/C_x - F_1/C_x$ and the spring constant C_x can be calculated as*

$$C_x = \frac{F_2 - F_1}{\Delta l} = \frac{m_2 - m_1}{\Delta l} \cdot g \quad \text{(N/m)} \quad (3.39)$$

Therefore, calibration simplifies to knowing values of g, m_2, m_1 and measuring a linear differential deflection or distance Δl.

Similar calibration can be applied to a rotational system, when torque measurement is required. The simplified setup is shown in Figure 3.63b. A torque τ is applied to the shaft, which is mounted horizontally. There is an 'arm' with length or radius r attached to the shaft. A mass m is suspended from the arm and can be adjusted such as to equalise the torque because $\tau = F \cdot r = m \cdot g \cdot r$. There can also be a version

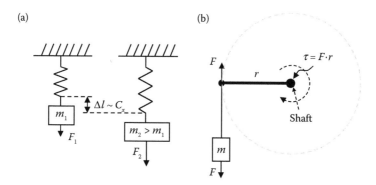

FIGURE 3.63 Measurement of force using springs and weights for a linear deflection (a) and similar setup for a measurement of torque (b).

---

* It should be noted here that the equation relies on the applied force, and not the force with which the spring reacts. Therefore, the minus sign from Equation 3.37 disappears due to the equality of the spring force and the applied force, but with opposite signs.

with a spring, whose other end is fixed, so that the deflection of the spring is a direct measure of the force, as described previously. Such approach was used, for instance, in systems shown in Figures 2.3, 2.6 and 2.8.

Therefore, the sensing of force is simplified to a measurement of deflection, linear or angular, together with information about the mass and gravity. The deflection can then be scaled in appropriate units, as required by the rest of the 'signal processing' for a given system, for example, W/kg (Baily, 1896; Brailsford, 1948).

3.4.1.3 Load Cell

There are several types of load cells, based on different operating principles. Some of them rely on the intermediate measurement of strain (see the next section) through calibrated characteristics of its inner deforming structure.

In general, the user does not have to know the details of the inner structure of such sensor, with the exception for operating limitations. For instance, the accuracy of some sensors can be significantly affected by off-axis forces, and care must be taken to avoid such conditions (Measurement Specialties, 2012). For instance, special mechanical guides must be installed to minimise the off-axis forces (Figure 3.64).

Load cells are usually calibrated by the manufacturers (or the accuracy is guaranteed otherwise), and the output voltage signal V_{out} of the sensor is directly proportional to the applied force, with given sensitivity, so the force applied to the sensor can be calculated as

$$F = C_x \cdot V_{out} \quad \text{(N)} \tag{3.40}$$

where C_x is the sensitivity of the sensor (N/V).

FIGURE 3.64   Apparatus with a load cell (white arrow) and guides (black arrows). (Adapted from Klimczyk P., *Novel Techniques for Characterisation and Control of Magnetostriction in G.O.S.S.*, PhD thesis, Wolfson Centre for Magnetics, Cardiff University, Cardiff, the United Kingdom, 2012. Image courtesy of Piotr Klimczyk, formerly of Wolfson Centre of Magnetics, Cardiff University, Cardiff, the United Kingdom.)

Sensitivity is sometimes not given directly in the data sheet, but it is evident from other values, such as the full-scale output (FSO). For instance, the cell ELHS can have a range of 1 kN, with an FSO of 200 mV. So the sensitivity is 1000 N/0.2 V = 5000 N/V. The output signal can be either positive or negative, and the accuracy can be at the level of 0.1% (Measurement Specialties, 2012). Hence, for example, the output voltage of 1 mV corresponds to an applied force of 5000 N/V · 0.001 V = 5 N.

The main advantage of a load cell is usually very small change of dimensions, practically negligible, as compared to a spring-based system. The load cell can be inserted directly between the force generator (e.g. a pneumatic actuator) and the sample under test (Klimczyk et al., 2011).

The signal processing simplifies to an accurate voltage measurement, with perhaps amplification to match the input range of the voltmeter. Low-pass filtering might also be advantageous, if there are no problems with phase-shift errors for dynamic measurements (e.g. at 50 Hz).

3.4.2 STRAIN SENSORS

Force acting on an elastic body causes strain ε, which can be measured as a relative deformation $\Delta l/l$, that is, the length of deformation in a given direction relative to the whole dimension of the sample in the same direction (see also Equation 2.23).

Therefore, the measurement of strain becomes mostly a question of measuring the deformation or change of dimensions. Such measurements can be accomplished in a number of ways, and only the most popular are discussed below.

An example of a very simple strain gauge[*] is shown in Figure 3.65. The gauge has two parts, which can move independently against each other. Each part is fixed

FIGURE 3.65 Simple mechanical strain gauge used for measuring the cracks in buildings. (Photograph by RoySmith, CC-BY-SA-2.5, Wikimedia Commons.)

---

[*] Of course, the sensor from Figure 3.65 would not be used for the measurement of strain in electrical steels, but it is used here just as an illustration of the concept of a deformation measurement (deformation = strain).

with an adhesive, with such positioning that at the moment of fixing the millimetre scales are aligned at the zero position. The gauge will then visually indicate by what amount both sides moved in relation to each other (and thus the changes in the wall crack will be measured). There are no forces as such acting on the gauge – the measurement is purely of the deformation of the underlying structure (relative movement of the opposite ends).

3.4.2.1 Resistance Strain Gauge

A resistive strain gauge is usually made as a meandered conductive track on a substrate (Figure 3.66). The multiple parallel fragments have a certain electrical resistance, which depends on the length and cross section of the given construction. When the sensor is stretched or compressed (in the direction shown by the arrow in Figure 3.66a), then its dimensions change, which is reflected in the total resistance as measured across the two terminals.

For correct operation, the sensor must be affixed to the given surface to be deformed, such that the whole surface of the sensor is attached to the corresponding surface of the sample, for example, by adhesive, glue or cement. Otherwise, the strain is not transferred accurately from the sample to the sensor (NI, 1998). Recommendations and details are usually given by the manufacturers of the sensors. Incorrectly applied adhesive can pre-stress the sensor in an unwanted way so attention must be paid to the manufacturer recommendations.

In many sensors, the change of resistance is directly proportional to the applied deformation, and the maximum allowed strain is on the order of single percents, for example, up to 3% maximum, with only special constructions allowing tens of percents (20% maximum) (Micro-Measurements, 2010). The base value of resistance (without any applied stress) depends on the size and construction of the sensor, but

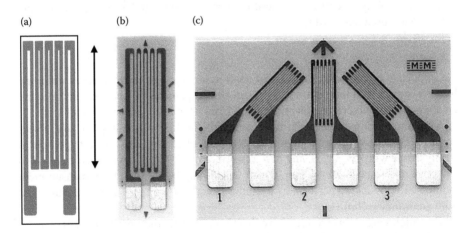

FIGURE 3.66 Resistive strain gauge with larger pads for soldering the leads: (a) schematic drawing of a sensor with the arrow showing the direction of the measured deformation, (b) photograph of a typical one-axis resistive strain gauge 500UW CEA Series, (c) rosette-type strain gauge 125UR CEA Series. (From Micro-Measurements, a brand of Vishay Precision Group, USA. With permission.)

typical values can be tens to thousands of ohms, with seemingly popular values of 120 Ω or 350 Ω (NI, 1998; Micro-Measurements, 2010; Omega, 2014).

An important feature of strain gauges is a parameter referred to as *gauge factor* (*gage factor*, *GF*, also just *G*, strain sensitivity *k*, and many other names and symbols). The value is not always specified (Micro-Measurements, 2010), and then it is probably equal to 1, but this should be checked with the manufacturer. For many sensors, the gauge factor is greater than 1, for instance, 2.1 or even up to values as large as 200 for strain gauges based on semiconductors (Klimczyk et al., 2011; Kulite, 2014). Some manufacturers give this value also in percent (e.g. 150% to denote *GF* = 1.5).

This parameter means that the given elongation corresponds to proportionally greater change in the sensor resistance, by the specified factor (NI, 1998). Therefore, a strain gauge with higher *GF* will have greater sensitivity with the same elongation, so that, in order to calculate the strain, the following equation should be used:

$$\varepsilon = \frac{\Delta R}{R \cdot GF} \quad \text{(unitless)} \tag{3.41}$$

where *R* is the base resistance of the sensor (Ω), *ΔR* is the change of resistance (Ω) and *GF* is the gauge factor (unitless).

Thus, the value of *GF* has a direct impact on measurement accuracy but its tolerance can be quite wide, for example, ±5%, even though for a single package can be better, at the level of 1% (Omega, 2014).

There are many other parameters that can affect the overall accuracy, and the choice must be carefully analysed, for instance, type of strain-sensitive alloy (resistive material), type of carrier (backing material), grid resistance, gauge pattern, self-temperature compensation, gauge dimensions (active length), quality of adhesive, long-term stability, operating temperature, test duration, number of load cycles, etc. (Micro-Measurements, 2014).

3.4.2.1.1 Signal Processing

An important problem that can exhibit itself as an apparent strain is variation of the base resistance with temperature. This can be compensated to some extent by using an appropriate temperature-invariant alloy from which the sensor is made.

Maximum strain in electrical steels is usually at the level of 20 ppm, so the sensor must be chosen accordingly for the detection of such values. For a hypothetical strain sensor with *GF* = 2.0, and base resistance of 120 Ω, this would correspond to a change in resistance of 40 ppm, which would only be 4.8 mΩ. Such measurement would be impossible to perform by a simple direct two-wire resistance measurement with an ordinary ohmmeter.

By employing a bridge-type circuit (Figure 3.67), the small changes can be detected with less difficulty (but still far from being easy). Additional temperature compensation can be achieved with a dummy strain sensor exposed to the same temperature, but not exposed to strain (e.g. a separate sample that is not magnetised).

There are many circuits that can be employed for such measurements: quarter bridge, half bridge, full bridge, with two- or three-wire connections, etc. (NI, 1998).

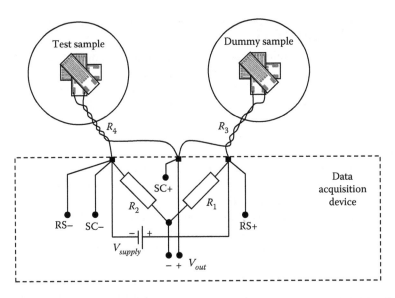

FIGURE 3.67 Simplified half-bridge circuit used for the measurement of rotational magnetostriction (only one channel shown for clarity) connected to a data acquisition device, which comprises the internal balancing resistors and power supply. (Adapted from Somkun S., *Magnetostriction and Magnetic Anisotropy in Non-Oriented Electrical Steels and Stator Core Laminations*, PhD thesis, Wolfson Centre for Magnetics, Cardiff University, Cardiff, the United Kingdom, 2010a; NI, *National Instruments, NI USB-9237, User Guide and Specifications*, 2007b.)

In practice, half-bridge circuits are usually satisfactory for the measurement of magnetostriction in electrical steels (Somkun et al., 2010; Klimczyk et al., 2011).

There are dedicated measurement devices (e.g. data acquisition device DAQ NI 9237), which comprise built-in bridge circuits, with compensating resistors and power supplies, so that the whole measurement is simplified. They can offer up to 24 bit resolution, so the smallest detectable strain can be theoretically as small as 0.006 ppm (Somkun, 2010). For comparison – the magnetostriction in non-oriented electrical steel can be up to 20 ppm, so there is more than enough resolution for most such magnetostriction measurements.

Even with a bridge configuration, there will be some current flowing through the strain gauge, so self-heating can also be a problem, as it is for thermistors. This will be mostly compensated in a bridge circuit, but for precise measurements, it is better to avoid the effects or at least minimise them to a negligible level. More details of strain gauge measurements can be found in numerous publications, such as, for instance, NI (2012).

As discussed in Chapter 2, a rosette-type configuration of strain sensors is required for the measurement of 2D magnetostriction. This requires three uni-axial sensors to be placed at various angles, for instance, all three sensors are stacked together at angles 0°, 45° and 90° (see also Figure 2.56) (Krell et al., 2005; Somkun, 2010). In order to minimise any angular errors with placing single sensors, it is better to use a pre-configured rosette sensor (e.g. Figure 3.66c), available from many suppliers (Micro-Measurements, 2014; Omega, 2014; Tokyo Sokki Kenkyujo, 2014).

PRACTICAL COMMENT

Ordinary commercial strain gauge sensors have full-scale deformation aimed at greater values (e.g. 1% = 10,000 ppm) than commonly encountered in magnetostriction measurements (e.g. less than 100 ppm). Therefore, the measurement at the low end of the sensors is quite demanding even if achieving the desired resolution is possible.

Temperature-related issues are important problems to be solved, and some of them can be avoided simply by using an appropriate type of sensor (based on temperature-compensating alloy) and by using a dummy sample.

However, self-heating of the sensor must also be taken into account because changing the temperature will affect the base resistance of the sensor. For example, if 2.5 V is applied to a 120 Ω sensor, the current will be 21 mA, and the associated loss at the level of 52 mW might be too much to be acceptable. This is despite the fact that the dummy sensor will also help in compensating such changes to a large amount.

The self-heating effects can be investigated by running the sensor without any strain applied. If the reading keeps changing shortly after powering up (without the strain being applied), then the power must be reduced. This can be achieved simply by using a sensor with higher base resistance, for instance, 350 Ω with the same supply voltage. The power will be reduced for a similar size of sensor and thus self-heating should also be reduced.

Moreover, in prolonged measurements at high excitation and frequency, the sample under test will increase its temperature – which is after all the basis for the thermometric method. It is difficult to expose the dummy sample to the same excitation, so there will be a temperature difference between the two sensors, which is more difficult to compensate. The easiest way to reduce this effect is to limit the time over which the excitation is applied (so that the differences are minimised).

The signal level is very low and the signal-to-noise ratio is generally poor. Good electrostatic shielding of all connections is important, as well as good signal conditioning (amplification and filtering). A half-bridge circuit is usually sufficient as such, but in order to achieve the required level of resolution and precision, some additional wires can be required, as shown in Figure 3.67.

The supply voltage V_{supply} can be different at the bridge connections due to additional voltage drop across the connecting wires. This can be detected using the 'remote sensing' connections (RS+ and RS−). Another pair is used for shunt calibration (SC+ and SC−), which allows the correction of sensor resistance, gain errors, etc. (NI, 2007b). Such techniques are normally required because the terminating wires to a strain gauge can be quite long, and neglecting their resistance would cause significant errors. A similar solution can be applied for the removal of any residual strain, which might be present after the gluing operation.

Additionally, copper changes its resistance by a significant amount of around 0.4%/°C, so if the total lead resistance is 1 Ω, then the temperature change of just 1°C (e.g. from 25°C to 26°C) would cause a change of resistance by 10 mΩ. This is significant because for a sensor with a base resistance of 120 Ω, with gauge factor 2.0, detecting a maximum strain of 20 ppm, this would mean a 'useful' resistance change of only around 5 mΩ. Therefore, an uncompensated change of lead resistance would be greater than the value to be measured. This change manifests itself as an offset, and can be eliminated, for instance, by recalibration of readings before each measurement. This is possible because the measured strain is usually dynamic (e.g. 50 Hz) so that any constant offsets can be eliminated either by recalibration, or numerically.

Another potentially significant problem is the fact that a flat strain sensor encloses a certain area with its conductive track, and any variable magnetic flux penetrating this area will induce additional voltage. The effect is similar as for the H-coils, as shown in Figure 3.50 (see also B-coil in Figure 3.37), especially that the normal H component will be typically present for a single lamination.

The connecting wires should be carefully twisted, in a similar way as described for B-coils and H-coils in previous sections (Somkun, 2010). But any flux normal to the surface of the sample will still couple into the strain gauge and induce voltage, which will contain frequency harmonics equal or comparable to the measured strain. As evident from Figure 3.66, there will be a certain amount of area encircled just by the internal wiring of the sensor, even if the connecting leads are twisted perfectly.

Such unwanted voltage cannot be eliminated by recalibration. One way of reducing the errors is by using special gauges that are 'folded' in half, so that the induced voltage is eliminated to a large degree. Another method can be employed if the flux does not have even harmonics because the magnetostriction usually occurs at twice the magnetising frequency. Therefore, a Fourier analysis can be employed to filter and isolate out the even harmonics (signal due to magnetostriction) from the odd harmonics (signal due to magnetisation).

The physical size (length) of a strain gauge is also important, especially for the measurement of magnetostriction in grain-oriented electrical steel, in which the grains can be very large. Otherwise, the strain is not averaged over several grains and the measurement repeatability can be adversely affected (Somkun, 2014).

Last but not least is the required number of signal inputs. If a rotational measurement is carried out by means of the fieldmetric method, then there will be at least four signals related to magnetisation, that is, B_x, B_y, H_x and H_y. But if the rotational magnetostriction is to be measured as well, then this requires the use of additional three channels for the strain gauges (as described in Chapter 2). Because of the specific requirements for signal conditioning and processing, the magnetisation signals would be commonly measured by a separate device than the magnetostriction signals. Thus, there arises a need for synchronisation of acquisition of all the signals. In some cases, the multiple devices can be triggered simultaneously, but if this is not possible then the same signal can be supplied to all the devices and all can trigger independently on the same signal (Somkun, 2014).

3.4.2.2   Laser Vibrometer

An example of a laser vibrometer is shown in Figure 3.68 and the concept of a magnetic measuring system in Figure 2.54.

Such a 'sensor' is capable of achieving very high absolute linear resolution, for example, on the order of 2 nm (Phway et al., 2004). However, one beam of laser is capable of measuring the displacement of only a single point. But in order to measure strain/deformation, the measurement should detect a change of distance between at least two points simultaneously.

Therefore, the measurement with this technique can be performed only under special conditions, for instance, if one end of the sample is fixed (so its position is 'known') and the movement of the other end is detected (Phway, 2007).

The measurement of magnetostriction at the centre of the sample (e.g. as it is the case with a strain gauge described above) is much more difficult to perform, and with a single 'sensor head' would not be possible in a general case.

Such laser vibrometer equipment is rather expensive (even significantly more than $10 k, depending on the configuration) because besides the sensor head, it also requires a dedicated controller unit (Polytec, 2014).

The signal processing is dictated by the controller unit, and the exact details are usually given in the operating manual. The measured values and signals can be fed from the controller unit, for instance, to a digital oscilloscope, which can be used to store and export the data in a digital form, for further processing or spreadsheet calculations.

The absolute resolution of 2 nm puts additional requirements on the measurement setup. The system should be mechanically isolated from other sources of vibrations, as these can cause similar displacements and thus adversely affect the measurement precision. Loud acoustic noise could introduce additional vibrations and this is partly the reason why such equipment would be used inside a soundproof room (Figure 3.68b). Vibrations of the magnetising yoke or the coils can also be detrimental.

If magnetostriction is to be measured on samples comparable to a single Epstein strip, then 20 ppm magnetostriction in 30 mm sample means deformation by 6 μm, which should be measurable even without such additional sophisticated conditions, provided of course that a single-point measurement would be sufficient (one end of the sample mechanically fixed).

(a)   (b)

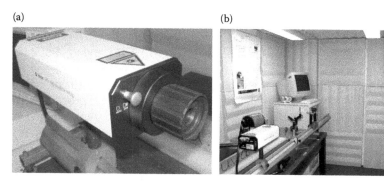

FIGURE 3.68   Laser vibrometer (a) in a soundproof room (b). Equipment courtesy of Wolfson Centre for Magnetics, Cardiff University, Cardiff, the United Kingdom.

Special conditions such as a soundproof room are useful if the magnetostriction measurement is to be performed together with acoustic noise detection, which is recognised as an important problem for modern power transformers (Phophongviwat, 2013).

3.4.2.3   Accelerometers

For dynamic magnetisation, the strain can also be measured by employing the accelerometers (Klimczyk et al., 2011; Klimczyk, 2012).

A time derivative of a displacement is velocity, and a time derivative of velocity is acceleration. Therefore, by performing the double integration of the signal proportional to acceleration with respect to time, it is possible to calculate values that are directly proportional to displacement. If two such sensors are used at different locations on the sample, then the difference in displacements is a measure of deformation and therefore strain (Figure 3.69). Dynamic magnetostriction λ can therefore be calculated as (Klimczyk, 2012)

$$\lambda = \frac{1}{\omega^2} \cdot C_x \cdot \int \int V \quad \text{(unitless)} \tag{3.42}$$

where $\omega = 2 \cdot \pi \cdot f$ is the pulsating frequency (rad/s), C_x is the calibration constant $\left(\dfrac{1}{V \cdot s^4}\right)$ (depending on the type of sensor and its sensitivity, gain of amplifier, length of sample, etc.) and $\int \int V$ is the double-integrated coupler output ($V \cdot s^2$).

The use of two accelerometers overcomes the limitation of such systems as single-beam lasers, which cannot perform a relative measurement. The voltage signal from the two accelerometers can be passed through a summing amplifier (in order to subtract them) and only then double-integrated by means of Equation 3.42 (Klimczyk, 2012).

The double integration must be performed carefully because each integration procedure can add a constant, similar to Equation 3.5 and illustrated in Figure 3.8. Therefore, the integration can be performed only for signals that do not contain any DC offset (e.g. under symmetrical sinusoidal excitation); otherwise, the constant cannot be distinguished from other offsets. With digital processing, a mean value of the waveform can be subtracted after each integration in order to eliminate any offsets or constants (Klimczyk, 2012).

The piezoelectric accelerometer used in the apparatus shown in Figure 3.69 had a useful operating range up to 5 kHz, and with a self-resonant frequency around 40 kHz (Dytran, 2013, 2014). The sensitivity was expressed in the data sheet as '100 mV/g' ('g' is the Earth's gravity 9.81 m/s²), so this translates into a value of around 0.01 V · s²/m. The measuring range can be, for instance, '20 g' (Dytran, 2013), which is equivalent to around 200 m/s².

The peak acceleration g_a can be calculated by double differentiation of the displacement over time, so that

$$g_a = \omega^2 \cdot l \quad \text{(m/s}^2\text{)} \tag{3.43}$$

where $\omega = 2 \cdot \pi \cdot f$ is the pulsating frequency (rad/s) and l is the displacement (m).

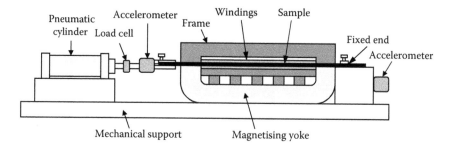

FIGURE 3.69 Apparatus with a load cell and two accelerometers for measuring the magnetostriction. (Adapted from Klimczyk P., *Novel Techniques for Characterisation and Control of Magnetostriction in G.O.S.S.*, PhD thesis, Wolfson Centre for Magnetics, Cardiff University, Cardiff, the United Kingdom, 2012.)

The output noise of this particular sensor has a value defined as *'0.0008 grms'* (Dytran, 2013), which is equivalent to around $g_a = 0.011$ m/s² peak (for a sinusoidal signal).

The smallest detectable displacement would be comparable to the noise level, and it can be calculated by re-arranging Equation 3.43. At 50 Hz magnetisation, the magnetostriction will mostly have a 100 Hz component. Therefore, for 300 mm sample, the discernible displacement would be around 30 nm (which is around 0.1 ppm).

There are accelerometers with even greater sensitivity, for instance, *'1 V/g'* (0.1 V · s²/m) with a full-scale range of *'5 g'* (Kistler, 2012). The sensitivity threshold is given as *'120 µgrms'*, which is 0.0017 V · s²/m peak and it would allow the detection of displacement at the level of 4 nm. It should be noted that this is comparable in practice to the resolution of the laser-based system described in the previous section.

Such accelerometers require careful signal amplification, conditioning and processing, for instance, by means of the so-called charge amplifiers, sometimes built into the sensor itself (Klimczyk, 2012). Piezoelectric accelerometers are often made in a cylindrical housing, with threaded connections for mechanical attachment to the vibrating mass. Their physical appearance is similar to other piezoelectric sensors, for instance, as shown in Figure 3.70.

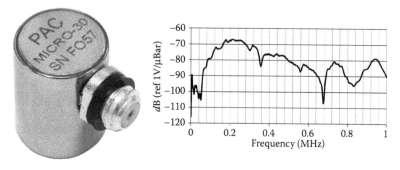

FIGURE 3.70 Piezoelectric sensor Micro30 and its frequency response. (From Mistras, *Micro30 Sensor, Product Data Sheet, Physical Acoustics Corporation*, 2011. Images courtesy of Mistras Group Ltd, the United Kingdom. With permission.)

3.5   BARKHAUSEN NOISE SENSORS

Barkhausen noise generates the *electromagnetic noise* as well as the so-called *magneto-acoustic emissions* (MAEs). Both types of sensors are discussed below.

3.5.1   MAGNETIC BARKHAUSEN NOISE SENSOR

Barkhausen noise is detected by the same type of sensors as flux density, but the induced voltage is additionally processed with some amplification and analogue filtering. This is because the high-frequency Barkhausen noise has much lower amplitude than the main flux density waveform, which must be suppressed. This can be achieved either by high-pass filtering or by two coils connected in series opposition, or with a single coil, but with more stringent high-pass filtering. Also, the output signal is normally not integrated, as it would be for a B-coil.

Other B sensors such as Hall effect of fluxgate are usually incapable of sufficient resolution, dynamic range (frequency spectrum) or signal-to-noise ratio to be useful for Barkhausen noise measurements.

Two examples of Barkhausen noise sensors are shown in Figure 2.64. In the two-coil system, the large voltage is suppressed by means of series opposition, but the Barkhausen noise is more stochastic in its nature so that it is measured locally for each partial coil and thus not suppressed.

Another approach is to use a 'perpendicular' B-coil, possibly with some magnetic core, for example, made of magnetically soft ferrite. The main component of flux density is not detected, but the sensor can be affected by the normal flux component (perpendicular to the sample surface).

Barkhausen noise is usually analysed for frequencies ranging between 300 Hz and 300 kHz. Therefore, the sensing coil (or coils) must be designed such that they are capable of processing such high frequencies. The upper spectrum could become a problem especially for physically large coils with many turns. The inter-turn and inter-layer parasitic capacitance could become so large that the self-resonance of coil could occur within the expected measurement range (e.g. below 300 kHz). There is also an additional parasitic capacitance from the coil to the sample, for instance, if the coil is to be wound directly onto an Epstein strip sample.

The self-resonance occurs due to parasitic properties and will always be present for *any* coil and cannot be avoided, although with special signal processing, it can be suppressed or corrected to some degree (Tumanski, 2007a).

Barkhausen noise sensors should be referred to as a 'detection technique' because they do not perform a 'measurement' in the strict sense of the word, namely, the 'measured' quantity cannot be calibrated to some absolute values and the results depend strongly on the type of sensor used, even if applied to the very same sample (Patel, 2009).

This is caused by the way the Barkhausen noise is quantified. For instance, the signal can be processed in such a way as to calculate the RMS value of the Barkhausen noise, so the quantity will be measured in volts. But a sensor with more turns will produce higher voltage, so signals from two different sensors will differ, possibly in a non-proportional way. Even if the result was to be scaled by the number of turns, the relationship is non-linear because of the different effects of filtering, different sensitivities due to 'noise floor' of the system, parasitic capacitance, etc. The same applies to power spectrum measured in watts. The input signal is still the same as for the RMS voltage, so any difference in RMS voltage would be even greater in the power spectrum (Patel, 2009).

Similar limitations occur for other 'measurable' quantities such as counting the number of peaks or calculating the kurtosis (Zhu et al., 2001; Hartmann, 2003; Patel, 2009). For instance, a single-coil system will detect Barkhausen jumps only local to this coil. But a two-coil sensor will detect jumps for two places, which will be summed up. So the total number of peaks will be different for the two measurements, and absolute results cannot be compared in a general case.

Nevertheless, the repeatability of results measured with the same system and the same sensor can be rather good, at the level of 1% scattering around the mean value (Patel, 2009).

Apart from amplification and band-pass or high-pass filtering, the signal processing simplifies to voltage measurement. If the data are acquired digitally then, further digital processing can be applied, depending on the type of quantity desired for the 'measurement'. The equations defining the most frequently applied quantities are given in Chapter 2, Section 2.5.3.

3.5.2 MAGNETO-ACOUSTIC SENSOR

Barkhausen noise jumps are caused by rapid local changes of domain structure. If these changes include domain rotation, then magnetostriction occurs. The mechanical vibrations caused by such jumps cause acoustic noise, which can be measured by suitable techniques. This noise is often referred to as *magneto-acoustic emissions* (MAE) and is effectively an indirect measure of the Barkhausen noise. It is claimed that the magneto-acoustic measurements contain information about the Barkhausen jumps occurring at greater depths in the sample under test (for thicker samples) than it is the case for perpendicular magnetic sensor (Wilson et al., 2009), so such properties can be used for non-destructive testing.

Acoustic vibrations are mechanical in nature so they are detected as such. Therefore, suitable sensors must be responding to mechanical vibrations, and one of the most popular types of sensors is a piezoelectric transducer (Ranjan et al., 1987; Dhar et al., 1992; Ng et al., 1994; Wilson et al., 2009).

A typical piezoelectric sensor has a wide frequency bandwidth, ranging up to 1 MHz (depending on size and mass). The response through frequency range is nonlinear, with a maximum sensitivity around the self-resonant frequency of the sensor. For instance, the sensor Micro30 has an operating range up to 400 kHz, with the resonant frequency at 125 kHz or 225 kHz (depending on the type of definition) (Mistras, 2011). The sensor is 10 mm diameter and has a height of 12 mm (Figure 3.70).

The output signal is proportional to the excitation applied to the sensor, multiplied by the sensitivity at the given frequency. For instance, the sensitivity can be expressed as $S = -67.5$ dB with the reference level[*] of 'V/μbar' (Mistras, 2011), where μbar $= 10^{-6}$ bar $= 0.1$ Pa.

---

[*] This unit can also be expressed as 'dB/V/μbar' (PUI Audio, 2008). The 'μbar' units are retained here for the benefit of the reader because most sensors appear to have their parameters expressed in this way.

The sensitivity S is defined as

$$S = 20 \cdot \log(C_x) \qquad (\text{dB ref V/}\mu\text{bar}) \tag{3.44}$$

where C_x is a ratio of the output voltage (V) to the applied pressure (μbar). The equation can be rearranged in order to derive the proportionality constant:

$$C_x = 10^{S/20} \qquad (\text{V/}\mu\text{bar}) \tag{3.45}$$

So for the example value of $S = -67.5$ dB, it can be calculated that $C_x = 422$ V/bar $= 4.22$ mV/Pa. If converted to SI units, the value $S = -67.5$ dB ref V/μbar would be equal to $S = -47.5$ dB ref V/Pa. Therefore, a pressure of 1 Pa applied to the sensor will produce an output voltage of around 4.22 mV.

However, this would be correct only for the point at which the sensitivity is defined (which is often given as 'peak sensitivity'). For other frequencies, the sensitivity changes *significantly* (20 dB change is an *order of magnitude*), as it is evident from Figure 3.70. The 'measurement' is therefore highly non-linear.

Equation 3.45 is given here only as an example, as this calculation is normally superfluous and never used. This is because the 'measured' magneto-acoustic activity is relative to some reference point: at a given excitation level, or before and after thermal annealing, etc. The magneto-acoustic activity is recorded as voltage (RMS or amplitude), as measured in the given signal chain, specific for a given experimental setup (Figure 3.71) (Ranjan et al., 1987; Dhar et al., 1992; Ng et al., 1994; Wilson et al., 2009).

The signal processing introduces additional non-linearities, for instance, by the virtue of the band-pass filtering (Figure 3.71). So the combined non-linearity of the sensor-to-output is generally not known and therefore the results from different measurement systems cannot be compared in a direct way. Therefore, as for the magnetic Barkhausen noise, it is also more a 'detection' rather than a 'measurement' technique.

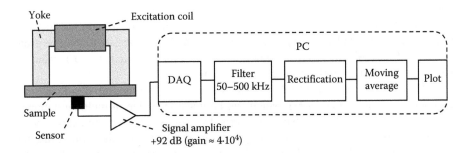

FIGURE 3.71 Simplified block diagram of signal processing for magneto-acoustic emissions (for clarity, only components related to the sensor are shown). (Adapted from Wilson J.W. et al., *IEEE Transactions on Magnetics*, 45 (1), 2009, 177.)

3.6 VOLTAGE-SENSING TECHNIQUES

Vast majority of the sensors described in this chapter transform one physical variable into a proportional output *voltage* signal. Sometimes this voltage must be amplified before it can be measured, but in many cases, the voltage measurement can be performed directly across the sensor output.

PRACTICAL COMMENT

There are also sensors and transducers in which the signal is generated as output current, not voltage. The full-scale signal can vary between standardised levels of 0 and 20 mA or depending on the standard alternatively as 4–20 mA (Tumanski, 2006). However, such output signals are more useful for industrial applications. In laboratory conditions, the sensors with output voltage are used much more commonly. And even if the output current is used, then it can be easily converted to voltage by using, for example, the shunt resistor approach (Figure 3.41b).

Therefore, the voltage measurement becomes as important part of the signal-processing chain as the sensor itself, regarding the features such as resolution, signal-to-noise ratio, absolute accuracy, range of applicability, etc.

Voltage can be measured by a number of devices, and the choice is driven by the type of quantity that should be measured. Three most frequently used devices are: voltmeter, data acquisition device and oscilloscope.

The description of 'voltmeters' given below is only for the implications of their use with magnetic sensors listed above. The exact operating principles of voltmeters will not be discussed in depth because they are provided extensively elsewhere in the literature, for instance, in Tumanski (2006).

The voltage-to-value conversion can be achieved in a number of ways (analogue and digital), but the actual physical approach is usually irrelevant from the functional viewpoint. In most cases, the signal processing is 'transparent' to the person operating the equipment. In fact, it is not possible for any non-expert to follow the operation due to sheer internal electronic complexity of modern precise voltmeters (Horowitz et al., 2015).

However, the operator *does* need to know what are the limits of applicability for each range: level of voltage, frequency, accuracy, etc. Any device should not be used outside of its specification as this might cause significant errors (e.g. frequency bandwidth) or even danger to the user (e.g. high voltage).

The underlying concept of all voltmeters is to convert a measured voltage into a numerical value (Figure 3.72), which can be made available in multiple optional ways: analogue scale, digital display, numerical storage, digital communication port, etc. In almost every case, at some stage of signal processing, the value must be converted to a 'number', which can be then used in the equation defining the mathematical function of the sensor – so that the input physical variable is quantified numerically with a given precision.

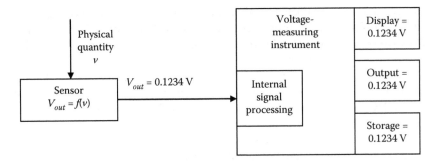

FIGURE 3.72 The concept of a voltage-measuring instrument processing an input voltage signal from a sensor into an output numerical value.

In recent years, digital data acquisition devices became very popular and ordinary voltmeters are practically not used for rotational measurements. Nevertheless, they will be briefly discussed for completeness.

For instance, let us assume that the output voltage of a given sensor is 1 mV peak at 50 Hz and the signal contains higher harmonics up to 300 kHz with amplitudes between 1 μV and 1 mV (as it is the case for Barkhausen noise measurements). It is clear that the resolution and bandwidth of a voltage-measuring equipment must be sufficient to measure with required precision over the full bandwidth from 50 Hz to 300 kHz, with all the instantaneous values appropriately sampled. Sometimes, this is not possible and additional signal amplifiers must be used in order to increase the signal level to a measurable level.

For digital devices, if the sampling frequency is insufficient, then high-frequency events can go completely undetected, as indicated in Figure 3.7a. So the choice of an instrument must also be guided by the application and the type of signal to be measured.

3.6.1 Voltmeter

A voltmeter converts continuous voltage waveform into a number (a single value) proportional to a specific quantity related to that waveform. Three most popular types of values are: DC, RMS and 'average'.

A voltmeter measuring DC value is designed with a certain low-pass filtering, which ignores frequencies above some threshold. Slow and very slow changes are still registered – after all the instrument must be capable of displaying a 'new' value if it changes. Such devices are useful for measuring the magnetic properties under DC-like conditions (DC or quasi-DC), but not for AC excitation.

The most commonly available AC voltmeters measure the RMS value. Depending on the internal signal processing (especially for the older-type analogue devices), the RMS value can be measured only for an AC component, while the DC component is at best ignored and at worst causes other errors (Tumanski, 2006). There are also digital multimeters (DMMs) that can be switched between the DC mode in which the AC component is suppressed by low-pass filtering or averaging, or the AC mode in which the DC component is eliminated by AC coupling (for instance, by connecting a capacitor in series). Therefore, the user must be aware of the limitations of a given device.

Peak value is the highest value occurring in any given waveform. This includes any distortions of a given waveform.

RMS value of a variable signal is defined as 'root mean square' and the definition is that the RMS value would be equal in terms of dissipated power (heat) on a resistor to a given DC value so that $P_{RMS}(V_{RMS}) = P_{DC}(V_{DC})$. For a pure sine, the RMS value is $1/\sqrt{2} \approx 0.707$ of the peak value.

Mean value is an arithmetic average of all values in a waveform. For a pure sine, the mean value is equal to zero because the contributions of the positive and negative half-cycles cancel each other out.

There is also a value, which is defined as an average of the rectified signal $|V|_{mean}$, which for the purpose of this book will be referred here as AVG; therefore, $V_{AVG} = |V|_{mean}$. For a pure sine with ideal full-wave rectification, AVG $= 2/\pi \approx 0.637$ of the peak value.

PRACTICAL COMMENT

It should be stressed that mean or average value is the mathematical mean of all subsequent instantaneous values over a given interval of time. For a pure sine, *mean value = average value* = zero.

However, as also discussed in the Practical comment in section 3.1.1, the average value of a rectified signal is a useful concept. This special type of average is denoted here by AVG and it can be employed, for example, for calculation of *B* in Equation 3.4.

There is also another name that is used especially in electronics. This is *peak-to-peak* value, which is the difference between the maximum and minimum of a given waveform, whereas the ordinary *peak* value can be defined as the *peak-to-zero* value, or simply put the 'maximum value'.

There is also a value specified as *true RMS* (TRMS), which also denotes RMS, but with the explicit information that the RMS is also correct for distorted signals, up to the specified *crest fact* (ratio of peak to RMS values). The distinction is made because older-type voltmeters would show the RMS value correctly only for sinusoidal, and not for distorted waveforms.

The *form factor* is defined as a ratio of RMS to AVG and for pure sinusoidal waveform it is $\pi/\sqrt{8} \approx 1.111$. This factor can be used as a means of simple identification of distortion or harmonic content of a sinusoidal waveform.

For instance, in the international standard describing measurements with a single-sheet tester, the shape of the secondary voltage should be controlled such that $FF = 1.111 \pm 1\%$. In the past, this was achieved by measuring the secondary voltage with two voltmeters: one detecting the RMS value and another detecting the AVG value, because $FF = V_{RMS}/V_{AVG}$ (IEC, 2010). Nowadays, it is easy to derive such values numerically with sufficient precision.

For a pure DC signal, it results from mathematical definitions that all values are equal, namely, peak $=$ RMS $=$ mean $=$ AVG.

On the other hand, for a sinusoidal signal (pure or with a DC offset), the values differ from each other so that peak > RMS > AVG > mean = DC offset. Moreover, if the DC offset is included or excluded, the RMS and AVG values will differ according to the ratio of magnitudes of the DC and AC components, so that exclusion of the DC component will lower both the RMS and the AVG value. This is illustrated in Figure 3.73.

Normally, the user does not have much influence as to how the 'DC' or 'AC' measurement is performed in a given voltmeter, apart from selecting one or the other option. It is thus important to study the appropriate user manuals to understand the differences, especially between 'AC mode', 'DC mode', 'AC coupling' and 'DC coupling', as they can have very different meanings (especially as compared between multimeters and oscilloscopes).

There is a plethora of various voltmeters available commercially, with various degrees of functionality and accuracy, and almost always greater accuracy means greater price. Typical examples are shown in Figure 3.74.

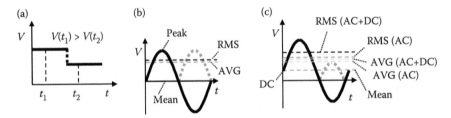

FIGURE 3.73 Examples of various combinations of DC and AC voltages: (a) two levels of DC voltage, (b) RMS and AVG (average rectified) and mean (average) voltage of symmetrical sinusoidal voltage and (c) for a sine with DC offset the RMS and AVG values change if the DC offset is included or excluded

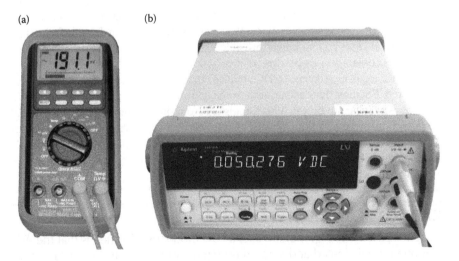

FIGURE 3.74 Typical hand-held (a) and laboratory (b) digital voltmeters. Equipment courtesy of Megger Instruments Ltd, the United Kingdom.

DMM or digital voltmeters (DVM) can offer very good accuracy for reasonable price, and nowadays in most cases they will be cheaper and perform better than equivalent analogue devices.

In recent decades, some manufacturers began providing special DMMs for laboratory and industrial measurements. These devices can output voltage measurements (such as RMS) as well as export the acquired data in a digital form. Therefore, they are in effect a cross-over between ordinary voltmeters and data acquisition devices (described in the next section) (NI, 2008a).

There can be device-specific performance that can significantly influence the measured signal. For instance, a data acquisition device might employ high-impedance (1 MΩ) inputs below 10 V and a much lower impedance voltage divider (10 kΩ) for higher voltages (NI, 2013b). Such significant change can cause significant errors in magnetic measurements as compared to other sources of inaccuracies (Zurek, 2015b).

3.6.2   DATA ACQUISITION DEVICE

The progress with computer and digital technology made possible manufacturing fairly inexpensive and good-quality data acquisition devices, some examples of which are shown in Figure 3.6. They became ubiquitous and provide measurement of instantaneous voltage waveform, as dictated by the sampling frequency. The acquired waveform is then represented as a series of numerical voltage values, each assigned to its own instant of time, as shown in Figure 3.7. Some of the properties of signal acquisition and processing were described in Section 3.1.1.2.

Once the signal is in the digital domain, the user then has access to perform whatever signal processing is required, and all the values such as peak, RMS, AVG and mean or DC offset can be calculated numerically, by using the appropriate software, or a spreadsheet application.

However, the transition from analogue to digital domain is a critical operation, which also directly impacts on measurement accuracy (among other effects). The result of the conversion is a set of numbers, and if they contain false information, then in general case, it is no longer possible to differentiate between the correct and the erroneous data.

If the data acquisition device is 'DC coupled', then both DC and AC components will be measured. But most devices introduce some unwanted apparent offset and noise (NI, 2013a, b), which cannot be easily distinguished from the real input signal. Additional effects such as thermoelectric voltages also exacerbate the problem. A calculation of RMS or AVG values from such data would result in some level of unknown error.

On the other hand, if the input is 'AC coupled', then any real DC offset will also be eliminated, which might be undesirable.

PRACTICAL COMMENT

It should be noted that in multimeters, a 'DC mode' usually means that the AC component is eliminated and ignored during measurement. But for data acquisition devices and oscilloscopes, 'DC coupling' means that all signal components, DC and AC, are acquired.

In most cases, the 'AC mode' of multimeters is comparable to the 'AC coupling' of data acquisition devices and oscilloscopes. In such case, there is some high-pass filtering so that DC components (like offsets) are neglected. However, not all devices might follow such distinction, and definitely the threshold frequency for low-pass or high-pass filtering will differ from one instrument to the next.

For this reason, the appropriate operation manuals should be studied for each instrument before it can be used with confidence. The manufacturer of the instrument should be contacted if there is any doubt about the technical specification, especially those specifications that can change without notice.

Digital processing can be applied accurately if the given operation is known to eliminate only the erroneous part of the signal. For instance, an alternating output voltage signal from a B-coil cannot contain any DC component by the virtue of Equations 3.1 and 3.2. This is true even if the input B does contain very strong DC offset because a derivative of a constant is zero. Therefore, any DC component in such voltage can be safely subtracted, as this will eliminate all unwanted offsets (input offset, thermoelectric voltage, etc.). If the offsets are not eliminated, then the integral will contain a 'ramp' as illustrated in Figure 3.8.

However, such processing can be applied only in special cases, for instance, if the acquired signal is alternating and the acquired length is equal to *exactly* integer number of cycles, so this is a very strong incentive to perform magnetic measurement in such a way (more details are given in Chapter 4). Then an arithmetic mean value is synonymous with the offset to be eliminated. In other cases, this might not be possible in such a straightforward manner, and other measures must be taken.

Of course, the input can be 'AC coupled' in hardware by using a high-pass filter in series with the sensor. However, then the DC components are eliminated and again they could be subtracted numerically. But if the sensor can produce a real DC offset proportional to the applied quantity (e.g. a Hall effect sensor), then an accurate measurement cannot be performed without further processing.

As discussed above, sampling frequency is an important parameter of digital devices. There is a continuous technological progress and ever faster devices are available commercially.

However, faster sampling frequency usually means lower resolution because it is difficult to keep up with the analogue-to-digital conversion when the sampling frequency increases. This is the main reason for which many digital oscilloscopes described in the next section have only 8 bit resolution because they are aimed usually at very high frequencies (up to GHz).

Nevertheless, there are data acquisition devices with 24 bit resolution, for DC as well as AC measurements (NI, 2007b, 2014c).

3.6.3 OSCILLOSCOPE

An oscilloscope is effectively a very fast voltmeter of instantaneous voltage, combined with some additional functionality of display, especially useful for the

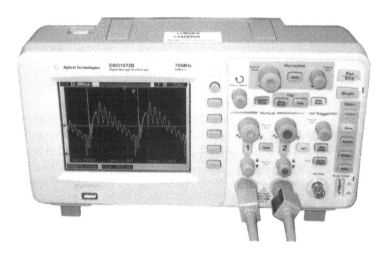

FIGURE 3.75 A typical two-channel digital oscilloscope. Equipment courtesy of Megger Instruments Ltd, the United Kingdom.

measurement of alternating or repetitive signals. An example of a typical modern oscilloscope is shown in Figure 3.75.

Before the dawn of digital technology, analogue oscilloscopes were used for displaying and measurement of hysteresis loops (Lord, 1952). For a unidirectional magnetisation, the H signal can be directly proportional to the magnetising current and the B signal can be obtained by passive integration through an RC low-pass filtering of the induced voltage. By displaying these two signals in an X–Y mode, the hysteresis loop can be suitably plotted (Goldman, 1999).

Digital oscilloscopes contain an internal data acquisition device (see the previous section), whose digital output is processed and displayed accordingly.

Most commercial oscilloscopes are capable of acquiring the signals up to very high frequencies, usually up to MHz and even up to tens of GHz. The penalty of such sampling frequency is the reduced resolution – for example, just 8 bits (Avrunin, 2013) or 10 bits (Gorican et al., 2000b). This in itself is not a problem because the resolution of the data acquisition device does not directly influence the accuracy of the measurement in the same way as other factors (Zurek et al., 2007a). But appropriate assessment of such factors should be made in order to investigate the likely consequences.

Digital oscilloscopes usually provide a means of exporting the acquired numerical data, sometimes in an automated way (e.g. through a communication port). Oscilloscopes provide accessible means of 'data acquisition', and indeed many researchers throughout the years used this approach (also with analogue oscilloscopes) (Lord, 1952; Oka et al., 1989; Khachan, 1992; Wang et al., 1999).

Indeed, the use of a high-speed oscilloscope might be the only practical means of performing measurements at very high frequencies (e.g. 1 MHz), where the ordinary data acquisition devices would be too slow (Kim et al., 2000).

In most aspects, the data are acquired similarly to other data acquisition devices, and once exported, the numerical data can be processed in the same way.

4 Measurement Apparatus

Hardware configuration of a rotational magnetising system depends heavily on the measurement method: mechanical, thermometric, wattmetric or fieldmetric (Chapter 2). The size of the sample must match the size and shape of the magnetising yoke, which also imposes requirements on the configuration of magnetising coils, power amplifiers and control of the whole system. So the choice of size/shape is one of the first to be made when a new measurement system is designed and constructed. Many other features are also a consequence of this important decision.

In recent years, majority of rotational magnetisation studies employed the fieldmetric technique. This is probably due to the relative robustness of the method which allows performing many tests with the same setup. For instance, measurements of properties under non-circular magnetisation (elliptical, DC-biased, vector properties) are much easier to obtain with the fieldmetric method. Signal analysis with this method also offers great robustness.

Magnetising systems employing the fieldmetric technique tend to share at least similar concepts of certain features. This is less of a case for other measurement techniques, for which researchers applied more diverse solutions. Therefore, it is more difficult to generalise and summarise due to the sheer number of possible combinations, especially that most of such systems are prototypes unique to a given research laboratory so there are no two identical systems. For this reason, many features described in this chapter will be relevant mostly to the fieldmetric systems. Nonetheless, some features are similar between all of the measurement techniques, which were discussed in detail in Chapter 2, and sensors in Chapter 3.

In general, smaller samples require less power to magnetise, and this makes it easier for achieving higher level of excitation. But on the other hand, the available measurement area might be too small to produce sufficient averaging of the measured quantities. For instance, in grain-oriented electrical steels, the length of a single grain can reach as much as 20 mm.

Also, for some methods, it is very difficult to apply measurements over larger areas. This is the case for domain observation. Large areas are more difficult to be polished with sufficient precision, and the observation must be confined to a limited area anyway due to optical resolution and the capabilities of the applied analysis method. After all, in practice, there are only a limited number of domain walls and their configuration, which can be tracked simultaneously with sufficient precision.

Therefore, the size of the sample (and hence also the magnetising yoke) must be selected to be fit for purpose, as dictated by the quantity to be measured and analysed. In practice, samples ranging from 15 to 500 mm were used – depending on the given study (Nakata et al., 1992; Malkinski et al., 1993; Kedous-Lebouc et al., 1995; McQuade et al., 1995; Sievert, 2011). However, a 'popular' range for the fieldmetric method is sample size of around 60–100 mm so that the area of 'uniform magnetisation' is at least 20 mm (Sievert et al., 1995).

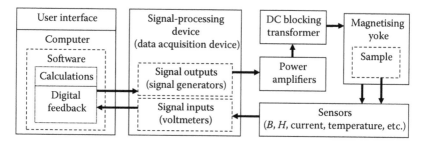

FIGURE 4.1   Block diagram of a 'typical' measurement system.

Therefore, there are many similarities between various measurement apparatus. A block diagram of a 'typical' system for rotational measurements is shown in Figure 4.1.

For explanation purposes, some topics described in this chapter are illustrated with the help of waveforms recorded for uni-directional magnetisation. This helps in visualising the discussed concepts with simpler case, and the conclusions are equally applicable to uni-directional as well as two-dimensional (2D) and rotational systems.

4.1   SAMPLE

As mentioned above, the size and shape of the sample are driven by the shape of the magnetising yoke, which could be cross-shaped, square, rectangular, hexagonal, circular and so on (Zurek, 2005; Guo et al., 2008; Sievert, 2011), and this will be discussed in more detail in the following sections. In some yoke designs, the sample does not need to have a specific shape but is simply a sheet larger than the yoke, which applies the excitation and measurement locally to some part of the sheet (see for instance Figure 3.29). There are also several specimen configurations which can be used with the same yoke. Photographs of some apparatus are also shown in Chapter 6.

A sample in an open magnetic circuit* is subjected to the so-called *demagnetising field* (Soinski, 1990; Fiorillo, 2004). The demagnetising field is created because once the sample is magnetised then there are apparent magnetic poles at the opposite ends of the sample. These poles become themselves sources of magnetic field, and some field lines will close back through the sample under test, opposing the direction of the applied field. This phenomenon was also discussed briefly in Chapter 1 (see Figure 1.19).

The demagnetising field depends on the sample shape, and for a hypothetical spherical specimen (Figure 4.2a), the demagnetising coefficient would be exactly $N_d = 1/3$ (Fiorillo, 2004).

The demagnetising coefficient reduces for the decreasing ratio of thickness to length, namely the longer and thinner is the sample, the lower the demagnetising coefficient in the 'elongated' direction. On the other hand, if a very large sheet was to be magnetised in a direction perpendicular to its surface, then the demagnetising

---

* An open magnetic circuit comprises at least one air gap with non-negligible length.

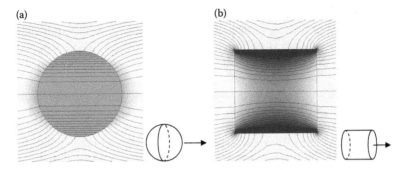

FIGURE 4.2 FEM of a spherical (a) and cylindrical (b) sample in a magnetic field; the shapes have the ratio of height/diameter = 1 to better illustrate the resulting non-uniformity.

coefficient will approach unity (Fiorillo, 2004). This means that the permeability of the *sample* will appear to be very low and comparable to free space so that $\mu_r \approx 1$, even if the *material* has very high permeability. It should be stressed here that this would be the apparent permeability of the *sample*, not *material*. The sample shape therefore can have significant effect on the measurement resulting in an 'open' magnetic circuit (Soinski, 1990).

For a cylinder magnetised along its axis (Figure 4.2b), the demagnetising factor can be calculated as (Fiorillo, 2004)

$$N_d = 1 - \left(1 + \frac{1}{C_x^2}\right)^{-0.5} \qquad \text{(unitless)} \qquad (4.1)$$

where C_x is the constant for a given ratio of the height of the cylinder to its diameter (unitless).

The demagnetising coefficients in three-dimensions (x, y, z) are such that their sum is unity (Fiorillo, 2004): $N_{d,x} + N_{d,y} + N_{d,z} = 1$. Two of these coefficients will be equal for a cylindrical shape due to rotational symmetry.

PRACTICAL COMMENT

Let us assume that there is a sample cut-out from a sheet as circular disc 0.35 mm thick and 100 mm diameter so that $C_x = 0.35/100 = 0.0035$. The demagnetising coefficient for magnetisation *perpendicular to the sheet surface* would be $N_{d,z} = 0.9965$, which means that the demagnetising field would be very strong. For a peak polarisation $J_p = 1.5$ T, the corresponding demagnetising magnetic field strength can be calculated from the following equation (the minus denotes that the demagnetising field opposes the applied field) (Soinski, 2001; Fiorillo, 2004)

$$H_d = -N_d \cdot \frac{J_p}{\mu_0} \qquad \text{(A/m)} \qquad (4.2)$$

and for the assumed sample would be at the order of $H_d = -1.2$ MA/m. For this reason, measurements of *material* permeability in direction perpendicular to the lamination surface are very difficult in practice. Firstly, large excitation must be applied for such experiment. Secondly, the results are very strongly influenced by the shape of the specimen.

The demagnetising coefficients for a cylinder magnetised *parallel to the sheet surface* will be $N_{d,x} = N_{d,y} = (1 - N_{d,z})/2$ so that $N_{d,x} = N_{d,y} = 0.00175$, so according to Equation 4.2 at the same level of 1.5 T, the demagnetising field would be around -2.1 kA/m. This is several orders of magnitude lower, but still a relatively high value, because in the easy magnetising direction for the grain-oriented electrical steel in a closed magnetic circuit in order to reach 1.5 T the required excitation would be much smaller, only at the level of 50 A/m (Cogent, 2002a). For non-oriented steel, it would be 2.5 kA/m (Cogent, 2002b), which is the same order of magnitude as the demagnetising field for such dimensions. So a quite high excitation has to be applied just to overcome the demagnetising field.

These values scale accordingly with the shape and thickness of the sample, as dictated by Equations 4.1 and 4.2. For instance, if the thickness reduces to 0.03 mm (i.e. as for amorphous ribbon) with the same diameter of 100 mm, then the corresponding demagnetising field value at 1.5 T would be around 180 A/m. The value of the demagnetising field is significantly lower, but it is still the same order of magnitude as $H = 80$ A/m required for such typical material for obtaining $B = 1.5$ T under closed magnetic circuit conditions (Metglas, 2011b).

The values scale accordingly, but this applies only if the sample was to be magnetised without any magnetising yoke, for instance in a solenoid or a Helmholtz coil (e.g. see Figure 2.8). If the sample is placed inside magnetising yoke, then its pole pieces become magnetic poles with the polarity opposing that of the sample so that the demagnetising fields of the yoke and the sample compensate each other to some degree, being a function of the length of the air gap. For vanishing air gap, the demagnetising field also reduces significantly. In a closed sample–yoke system (without any air gap), the properties very close to that of the *material* can be measured because the influence of the sample shape is reduced so significantly that it is often assumed to be negligible. This is the case for single-strip testers (Figure 2.22 and 2.23) and of course even more so for a toroidal sample (Figure 2.19), which itself creates a closed magnetic circuit – without the need for any yoke.[*]

An ellipsoidal shape exposed to otherwise uniform magnetic field exhibits uniform distribution of B within the sample volume, as shown for a sphere in Figure 4.2a (field lines inside the sample are parallel).

---

[*] Nevertheless, none of these methods are ideal; for instance, results from wound toroidal cores, Epstein frame and single-sheet testers are not completely comparable in the absolute sense, because of the differences in the measurement apparatus and procedure. They are even less comparable for magnetically open systems.

This is not the case for other shapes. For instance, Figure 4.2b shows a cylindrical object, which has the same value of B_{mean} as the sphere in Figure 4.2a (averaged over the whole volume), but the distribution is definitely non-uniform. It should be noted that such non-uniformity is caused simply by the shape of the sample and the resulting demagnetising effects – even before the influence of the magnetising yoke is considered. The magnetising yoke can introduce further problems with non-uniformity, or it can help in minimising them. Nonetheless, in any case, the most central part of the sample volume exhibits better uniformity (Figure 4.2b) as compared to the edges, which by definition represent sharp magnetic discontinuities. For this reason, the testing in such samples is usually limited to the central part of the sample where there is an assumption of 'sufficient uniformity'. Divergence of less than 1% is typically an arbitrary target.

PRACTICAL COMMENT

On a qualitative level, a cuboidal shape exhibits similar non-uniformity in its central cross section as the cylindrical shape shown in Figure 4.2b. Namely, along the axis of symmetry (horizontally in Figure 4.2b), the average value of B is lower than on the surface. A square or rectangular sample is a cuboid whose height is very small (thickness of the lamination). Therefore, on a qualitative level, similar non-uniformities will be present as shown in Figure 4.2b. This has important implications for samples made from multiple laminations.

4.1.1 SINGLE LAMINATION

A lot of measurements of 2D and rotational magnetic properties are carried out for electrical steels. These are produced in the form of sheets, with various grades and thickness usually between 0.15 and 1 mm. Therefore, the value of the abovementioned demagnetising field can differ significantly even for the same diameter of the sample if a different grade with different thickness is measured.

The simplest configuration is a single lamination cut to whatever shape is required by the magnetising yoke. The sample is positioned within the plane of the generated magnetic field, usually in such a way as to ensure symmetrical magnetisation (Figure 4.3).

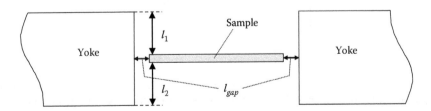

FIGURE 4.3 Schematic representation of symmetrical positioning of a single lamination sample within the magnetising yoke; $l_1 = l_2$ and l_{gap} is also the same on both sides of the sample.

Normal components of *B* or *H* are then equalised on both sides of the sample, and planar eddy currents are reduced due to the symmetry of the setup. This is especially important if needle probe sensors are used for *B* measurement (Loisos et al., 2001; Oledzki, 2003; Pfutzner et al., 2004). The sample is normally placed in a sample holder of some kind to ensure well-defined positioning of the sample and sensors.

The normal components are equalised on both sides, but they can still have very large values. As discussed in Chapter 3, in the first approximation, the tangential component of *H* increases linearly from the surface of the sample (as shown in Figure 4.4). The results shown in Figure 4.4b are calculated in 3D FEM for a circular sample 60 mm diameter and 0.5 mm thick (Zurek, 2015a). At the centre of the sample, the *H* component inside of the sample is indeed equal to that just outside of the sample – there is no step change in the *amplitude* of black curve in Figure 4.4b (only the slope changes instantaneously, but not the amplitude). This confirms the theoretical derivation for tangential sensors as illustrated in Figure 3.44.

In a completely symmetrical setup, exactly at the centre of the sample, there is no normal component so the full magnitude of the vector is numerically equal only to the tangential component.

It should be noted though that for this particular case, at the distance of 0.5 mm from the sample surface (e.g. position 0.75 mm in Figure 4.4b), the *H* value reaches 86 A/m, as compared to 46 A/m inside the sample. So if the total thickness of a single *H*-coil was 0.5 mm, then it would detect an average value of 66 A/m, which would be an error of 43% with respect to the *H* in the sample. This is the main reason why a multiple-sensor extrapolation technique should be ideally used in order to arrive at the 'real' value inside the sample (see also Figure 3.48).

This example also illustrates very well that if a single *H*-coil is used then it certainly should be positioned 'as close as possible' to the sample surface, with every

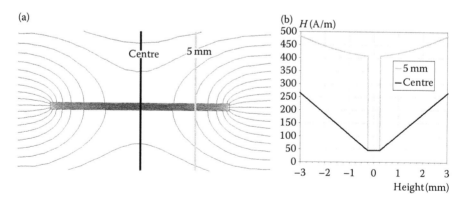

FIGURE 4.4 Non-uniformity of *H* around single lamination: (a) cross section (dimensions not to scale, for illustration only) and (b) magnitude of *H* (modulus of tangential and normal *H*) plotted along the lines centre and 5 mm (the position of 0 mm coincides with the centre of the lamination thickness). (Reproduced with permission from Zurek S., *Przeglad Elektrotechniczny [Electrical Review]*, R. 91 [Nr 12/2015], 2015a, 43. Copyright © 2015 *Przeglad Elektrotechniczny [Electrical Review]*.)

0.1 mm potentially making significant difference. Otherwise, additional measurement errors can be encountered, with an unknown magnitude, possibly highly dependent on the configuration of the magnetic circuit. On the other hand, it was shown that very good agreement between measurements in closed and open magnetic circuits can be achieved if the *H* extrapolation is carried out carefully and correctly (Hall et al., 2009).

However, even directly at the surface of the sample, but at some distance away from its centre (5 mm in this case), there is already a drastic *step change* in the *amplitude* of *H* values (grey curve in Figure 4.4b). The tangential component is almost constant, but there is a significant unwanted normal component (see also Figure 3.44).

As shown in Figure 4.4b, the values jump from 46 to 415 A/m of the vector modulus. This means that the normal component is 412 A/m, or 9 times greater in amplitude than the tangential component. If the sensor of *H* (like an *H*-coil) has some sensitivity in that direction, then additional measurement errors will occur.

Theoretical estimations show that such behaviour can lead to differences exceeding 5% for clockwise–anticlockwise rotational power loss measurement (Zurek, 2017c,d), and such differences are indeed typically present in many published results (see also Figure 3.49c).

Of course, the normal component grows gradually from the centre of the sample towards the outside, as can be judged from Figure 4.4a.

4.1.2 MULTIPLE LAMINATIONS

As discussed above, the ellipsoidal shape offers the most uniform magnetisation through the sample volume (Figure 4.2a). For this reason, some researchers chose to use a shape which approximated an ellipsoid by stacking several laminations, cut to different sizes – Figure 4.5 (see also Figure 2.16).

The uniformity of magnetisation improves, but only on the expense of increased demagnetising field, because of the increased thickness, due to the demagnetising coefficient. Therefore, such sample will be *more difficult* to magnetise to a comparable level, and more apparent power will be required.

This approach is suitable only for circular or elliptical discs, because any other shape (square, rectangular, hexagonal, etc.) would introduce additional non-uniformity due to shape anisotropy.

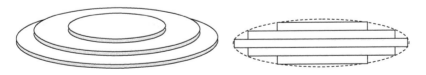

FIGURE 4.5 The shape of ellipsoid can be approximated by a stack of several circular laminations. (Adapted from Fiorillo F. et al., The measurement of rotational losses at I.E.N.: Use of the thermometric method, *Proceedings of 1st 1&2DM Workshop*, Braunschweig, Germany, 1992a, p. 162; Fulmek P.L. et al., Energetic model of ferromagnetic hysteresis: Magnetization of grain oriented steel sheets in asymmetric directions, *Proceedings for 5th 1&2DM Workshop*, Grenoble, France, 1997, p. 17.)

Also, in the case of fieldmetric measurements on anisotropic materials, all the laminations should be aligned along the same direction, so additional demagnetising field is not introduced due to differences in crystallographic anisotropy in different layers. Only in the torque-based measurement technique, the easy direction of aniso-tropic discs is misaligned on purpose in order to reduce the torque ripple (Figure 2.5) (Zhu et al., 1997a).

Several researchers used supplementary laminations (one or more). The additional laminations behave as 'shielding' and can reduce local normal component of the field (Figure 3.47) (Zhu et al., 1997b; Makaveev et al., 2000a; Jesenik et al., 2002; Mori et al., 2015; Wanjiku et al., 2015b). In such configurations, the 'shields' and the sample have the same dimensions, but this is not always the case (Leite et al., 2007).

Despite the claims of improved uniformity, it is difficult to generalise practical gains of such approach, as far as the measurement *accuracy* is considered. As can be seen from Figure 4.2b, the central part of the sample is magnetised to a lower value than its surface. If the shape was to be split into several layers, then the innermost layer will exhibit *lowest B*, as shown in Figure 4.6. Such results were calculated with FEM for laminations 60 mm long, 0.3 mm thick, separated by 1 mm. For example, if the B value in the outer laminations is 0.6 T, it is only 0.2 T in the middle one, so the difference is significant.

However, longitudinal uniformity within each lamination is still quite good. The central 30% of the sample is magnetised to the same B within $\pm 1\%$. The differences would be smaller for samples with larger diameter due to lower demagnetising effect. But in any open sample, such effect exists and it should be taken into account. It is clear that if the B values are different in neighbouring laminations then also the H values *must* be different too – including the tangential component. So the lack of nor-mal component of H does not mean that there is no gradient of H between the lami-nations (Zurek, 2015a). Therefore, an H-coil placed between such two laminations will detect some mean value, which might not be completely representative of the H inside the centrally placed sample. It would be better to employ the extrapolating technique (Figure 3.48) from more than one H-coil, as also suggested in Makaveev et al. (2000a). However, the design of such setup would need to solve the mechanical complexity, which itself could introduce a different source of errors.

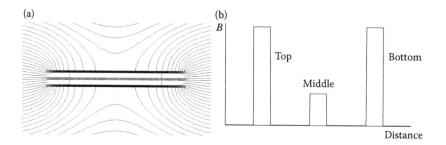

FIGURE 4.6   Symmetrical stack of three laminations with equal length will produce the lowest B in the middle: (a) plot of flux lines (not to scale) and (b) B values along vertical line through the centre of the setup.

Only ideally ellipsoidal bulk sample (i.e. not laminated) can produce completely uniform internal distribution of B. But a stack made from several laminations (as in Figure 4.5) is by definition separated by some small non-magnetic gaps between the layers (at the very least equal to the thickness of the non-magnetic coating). Worse still, B should be detected only in the central lamination. Search coils (Figure 3.1) required some space for the wire which increases air gap between the layers. This can be overcome to some extent by using deposition techniques (Figure 3.17) for search coils, but the problem remains for H sensors, which are even thicker and they also should be placed on the surface of the central lamination. A similar one applies if the rotational loss is to be measured by the thermometric technique (Figure 2.16) (Fiorillo et al., 1992a).

As a result, the gap between the laminations would have to be even larger than the thickness of the laminations. This leads to a configuration with large non-magnetic discontinuities and cannot result in uniform internal distribution of B.

Demagnetising field depends also on J (Equation 4.2) whose direction will depend on the distribution of crystallographic energy in anisotropic material (Fiorillo, 2004). As a consequence, if there are air gaps between the laminations, there is no longer a single ratio of diameters for all the layers which can result in uniform magnetisation at all excitation levels (Zurek, 2015a).

The question then arises how to define the correct cross section of such 'ellipsoid'. For the same dimensions the ellipse can be assumed differently, as shown in Figure 4.7. And as discussed above, the 'ideal' ratio can also depend on the excitation level, as well as the demagnetising effects of the yoke which surrounds the sample.

Quite good performance was demonstrated by using a 'sandwich' arrangement, with one sample and one 'shield' (similar to that shown in Figure 3.47) (Zhu et al., 1997b). It was shown that the normally expected variation of H above the sample surface is improved significantly without the need of extrapolation. However, it was also shown that the measured value of power loss is far less sensitive to the distance of the H-coils from the sample surface (Figure 3.48d). The H amplitude could vary by up to 400% with the H-coils positioned 2.5 mm above the sample, whereas the power loss varied only around 20% for the same displacement (Zhu et al., 1997b). The two-layer sample–shield 'sandwich' arrangement, if placed symmetrically, would produce the same B in both laminations, unlike a three-layer setup whose performance is illustrated in Figure 4.6b. On the other hand, for the two-layer case, each lamination would be excited *asymmetrically*.

Effectiveness of a given shield configuration cannot be assumed to be known for a given setup due to the many involved variables. In fact, it is possible to have

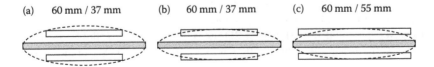

(a)   60 mm / 37 mm   (b)   60 mm / 37 mm   (c)   60 mm / 55 mm

FIGURE 4.7 Examples of three-layer approximations of elliptical-like cross sections (a and b) and depending on magnetising conditions version (c) can give optimum uniformity (drawings not to scale; there ratios give the sample diameter to shield diameter). (Adapted from Zurek S., *Przeglad Elektrotechniczny [Electrical Review]*, R. 91 [Nr 12/2015], 2015a, 43.)

such an unfortunate distance between the shield and the sample that the resulting non-uniformity is *worse* than for a setup without the shield, especially for higher excitations (Jesenik et al., 2002; Wanjiku et al., 2015b). It was also shown that even if the shields give improvement, the optimum performance depends on the spacing between the shields and the sample as well as the excitation level. For these reasons in the opinion of the author, such shields should be avoided.

4.1.3 BULK SAMPLE

There is no clear difference between a 'lamination' and a 'bulk' sample. The laminations are of great practical concern because all electrical steels and amorphous ribbons are manufactured as laminations with the thickness rarely exceeding 0.5 mm for most industrial applications.

However, a very thick lamination cut to smaller dimensions would become effectively a bulk sample, because the length of the sample would not be 'much greater' than the thickness anymore.

Bulk samples are typically used for materials which are not produced in laminations. This is the case for example for soft magnetic composites (SMCs) or powder cores.

A number of papers were published by Zhu and his colleagues, who studied properties under 3D magnetisation (Zhu et al., 2003; Zhong et al., 2005; Zhu et al., 2009; Li et al., 2011; Li et al., 2014a). The samples were made as cubes to fit in an appropriate yoke designed for the application of 3D magnetisation, as shown in Figure 2.39.

Bulk samples create specific requirements on the B and H sensors. The sensors are applied to each surface of the cube so that there could be up 12 sensors in total. Interestingly, the B-coils are not required to wound around the sample but instead can be placed on the appropriate faces so that they detect the normal component to the corresponding surface (Li et al., 2014a). Therefore, they use the same principle as shown in Figure 3.37.

4.1.4 LIQUID SAMPLE

There are few papers reporting studies on liquid samples under 2D magnetisation (Zeng et al., 2012; Petit et al., 2014; Guo et al., 2016).

Because of the lack of mechanical rigidity, such liquid samples have to be placed in suitable containers. For the same reason, the torque method is not quite suitable, so the fieldmetric method was used. Otherwise, all the discussion regarding the magnetising yokes and sensors is applicable.

However, practical importance of 2D measurements of ferrofluids or magnetorheological fluids is rather low, and they will not be discussed here in more detail.

4.1.5 SAMPLE CUTTING

The shape of the magnetising yoke dictates the shape of the sample. Therefore, the sample has to be either manufactured or machined in an appropriate shape in order

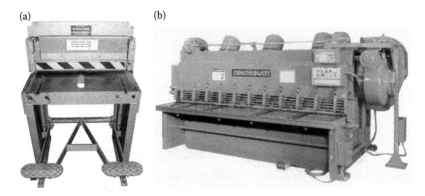

FIGURE 4.8   Guillotine (mechanical shear): (a) operated by pressing a pedal (equipment courtesy of Megger Instruments Ltd, the United Kingdom) and (b) motorised (photograph by Cincinnati Incorporated, Public Domain, Wikimedia Commons).

to fit the yoke. Manufacturing of electrical steels is carried out in large and very long sheets, which have to be cut to the required size.

There are many ways in which the samples can be cut, and this depends on the availability of a cutting method mostly from logistics viewpoint for a given laboratory.

Samples with straight edges (square, rectangular, hexagonal) can be cut inexpensively with a guillotine, which could be either manual or motorised (Figure 4.8). Significant mechanical stress is introduced around the cut edges (Loisos, 2002).

When using a guillotine, it is very important to appropriately control the dimensions and angles at which the sample is cut from a larger sheet. For instance, if the sample should be a square, then not only all the angles must be exactly 90° but also all the edges must have exactly the same length. As experienced by the author and other researchers, it might be difficult to achieve sufficient level of precision with manually operated guillotine (Pluta, 2013).

And if the sample is not cut with precise angles, then the magnetisation is usually more non-uniform, which can cause additional potentially uncontrolled and unknown measurement errors (Xu et al., 1997). Positioning errors of the sample (or if the sample is rectangular when it should be square) can lead to similar problems (Bottauscio et al., 2005).

Magnetising yokes based on stators of motors require circular samples (Zurek, 2005), which can be cut for instance with a lathe (Figure 4.9). A single sheet cannot be cut directly with a lathe due to problems with mounting or holding during lathe operation. Several samples can be cut simultaneously from laminations stacked up together. This usually requires some kind of sacrificial material which 'sandwiches' the stack of laminations between the chuck and the tailstock. The precision of cutting can be quite high, depending practically only on the type of lathe (0.01 mm is routinely achievable).

The whole process is based on mechanical machining of metal, so significant stresses are introduced around the cut edges.

There are several alternative cutting techniques which are equipped with computer numerical control (CNC). This eliminates errors caused by manual positioning

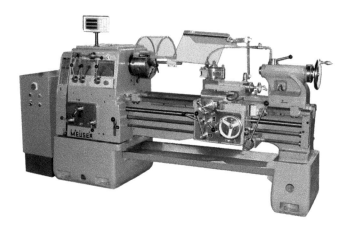

FIGURE 4.9   Typical lathe. (Photograph by Dipper, CC-BY-SA-3.0, Wikimedia Commons.)

of the sheet. The tolerances are dictated directly by the specification of the machine. As with any other technique, there are particular advantages and disadvantages, but also there is the question of availability of the machinery to a given laboratory.

Laser cutting (Figure 4.10a) is a popular technique used in the industry. The sheet to be cut is supported on a bed, and a head with the laser beam is moved in a controlled way above the surface and can cut an arbitrary shape at an arbitrary angle as driven by digital data. (For example, the samples in Figures 4.10d and 2.38a were cut with a CNC laser.) The type and power of the laser have to be chosen correctly to the type and thickness of the lamination to be cut (Loisos, 2002). If the parameters are not set correctly, then a very uneven edge can be produced similar to that shown in Figure 4.10d.

The material is removed as a consequence of generation of very high temperature, so thermal stresses are generated by this method.

Another technique is the electrical discharge machining (EDM) also known commonly as 'wire erosion' (Figure 4.10b). A thin continuous wire is passed in a very close vicinity of the cut material, in the presence of insulating liquid. Voltage is applied between the 'cutting' wire and the cut sample, which causes discharging and localised removal of the material. In modern machines, the positioning of the wire is controlled precisely and arbitrary shapes can be cut with ease, in sheets of any practical thickness (even tens of millimetres).

High temperature is not produced in the same sense as in the laser cutting method, so the wire erosion technique produces significantly less thermal or mechanical stress around the cut edges.

Waterjet cutting (Figure 4.10c) uses a precisely focused jet of water with abrasive particles, ejected at very high pressure and velocity. During cutting, the sample gets wet from the surrounding water mist, but if the sample is dried properly, there are no problems with rusting after the cutting (Pluta, 2013). The abrasion works very locally, and fairly small mechanical stress is created in the sample.

In industrial applications, plasma cutting is also used. However, this is usually applied to high-power cutting (thick materials) where high precision is not of prime importance or is too difficult to achieve. Very high temperatures are produced

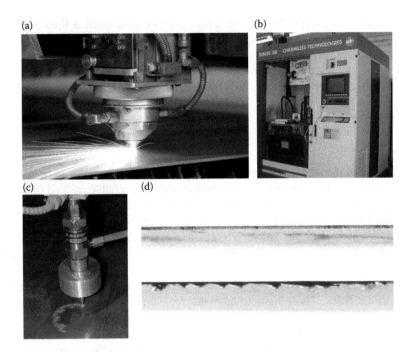

FIGURE 4.10 Cutting techniques: (a) laser cutting (photograph by Metaveld BV, CC-BY-SA-3.0, Wikimedia Commons), (b) wire erosion cutting machine (photograph by Glenn McKechnie, CC-BY-SA-2.0, Wikimedia Commons), (c) waterjet cutting (photograph by Waterjetter09, CC-BY-SA-3.0, Wikimedia Commons) and (d) lamination edge after correct and incorrect laser cutting (photograph by Joanna Kaczmarzyk, CC-BY-3.0, Encyclopedia Magnetica).

causing large thermal stresses around the cut edges, with possible blobs of metal which was molten away (similar to that shown in Figure 4.10d).

The samples could also be cut by milling, but again significant mechanical stresses would be introduced.

PRACTICAL COMMENT

Many of the techniques mentioned above introduce thermal or mechanical stress. This can significantly change magnetic properties of the sample, and thermal annealing might be required, as suggested by the manufacturer of the given sample material (e.g. annealing in hydrogen or nitrogen atmosphere).

Some of these methods allow creation of holes for B-coils, some do not. Interestingly, even if high precision of edge cutting can be achieved, it is not always possible to provide similar precision for the hole diameter. This is the case for waterjet cutting, which usually starts the cutting process at some distance, in a sacrificial part of the material, and only then moves towards the final destination for a high-precision cut. Thus, traditional hole drilling might be required anyway (described in the next section).

The great advantage of the numerically controlled methods is the ease of arbitrary shape, be it square, rectangular, hexagonal, circle or even a cross with slits (e.g. see Figure 2.60c).

However, cutting circular samples introduces an interesting non-trivial problem with positioning of the sample. This is discussed in detail Section 4.1.9 (see also Figure 4.14).

4.1.6   DRILLING

As discussed in the previous chapter, localised B-coils require holes to be 'created' in the sample (Figure 3.1b and c). The simplest and most reliable method of making such holes is drilling with small-diameter drill bits, as shown in Figure 3.13. Some drill bits have a thinner drilling tip and a thicker part to be held in a chuck.

Small-diameter drill bits are very fragile and break easily. They usually require high-speed rotation for correct operation, but in theory can be used even on a simple small drill press.

One of the most important and difficult to achieve parameters is the positioning of the drilled holes. These have to be made with high precision; otherwise large clockwise–anticlockwise power loss errors are introduced under rotational measurements (Zurek et al., 2006a,b). Therefore, in practice, such drilling should be performed only with a help of a precision positioning table, so the positioning of the holes can be controlled meticulously, or at least with the best practical precision.

The choice of diameter of B-coil holes is a compromise. On the one hand, smaller holes introduce smaller magnetic discontinuity and the uniformity of magnetisation is less affected. Therefore, the holes should be as small as possible. But there is less room for the B-coil wire to be wound so that only a single or at most a few turns can be made.

On the other hand, it is usually easier and less expensive to make larger holes. For instance, the B-coils shown in Figure 4.15 require rectangular holes, which can be cut by waterjet or laser, rather than by drilling (another possibility is milling). But it is much easier to make many turns of B-coil, so the signal-to-noise ratio of the induced signal is easier to measure accurately.

However, such large holes (several millimetres) undoubtedly create much greater non-uniformity in magnetisation than 0.3 mm holes (see also Figure 3.18 and Zurek, in press).

PRACTICAL COMMENT

Drilling produces sharp edges (burrs) around the hole, both on the top and bottom surfaces. As experienced by the author, such burrs can *easily* damage the thin enamel on copper wire of B-coils, or even thicker insulation of the wire. An easy practical way to remove the burrs is to use a slightly thicker drill bit and manually (i.e. holding the drill bit in fingers), delicately and slowly rotate and 'drill' into the hole until the burrs are removed, as shown in Figure 4.11. With conventional drilling, the exit burrs are usually more pronounced.

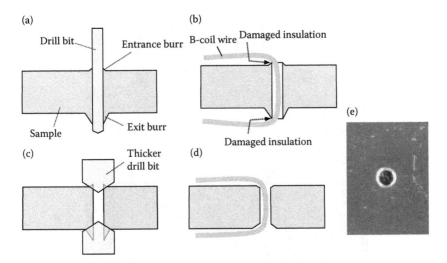

FIGURE 4.11 Drilling of holes: (a) drill bit produces sharp edges, (b) sharp edges can damage insulation of B-coil wire, (c) thicker drill bit removes sharp edges, (d) insulation is not damaged without sharp edges and (e) microscope view of a 0.5 mm drilled hole with manually de-burred edges with 1 mm drill.

4.1.7 ANNEALING

As mentioned above, mechanical cutting or shearing can introduce internal stresses, which can change magnetic properties of the sample. This is less important for larger samples because only the edges will be affected and the main volume of the sample will not. But for smaller dimensions, it can be quite significant, because in the worst case the influenced area around the cut edges can exceed 10 mm (Moses et al., 2000). The effect of mechanical stress can be quite drastic on some materials, such as high-permeability Ni–Fe alloys.

For many materials, the internal stresses can be relieved, and the pre-cutting properties can be restored by performing thermal annealing. The conditions of thermal treatment are usually specified by the manufacturers of a given material. For example, grain-oriented electrical steels could be required to be annealed at 800°C for 2 h in dry nitrogen, or mixture of nitrogen and hydrogen (ATI, 2014; ThyssenKrupp, 2015).

Because of such requirements, the annealing must be carried out in furnace with appropriate gas and temperature control – an example of such laboratory equipment is shown in Figure 4.12.

Annealing procedure is not possible for all materials. For instance, in order to achieve lower loss and higher permeability, some grades of grain-oriented electrical steels undergo the so-called domain-refining process, in which the domain spacing is optimised with *intentionally* added surface stresses by laser scribing or ball rolling (Soinski, 2001; Hilzinger et al., 2013).

Therefore, an attempt to remove cutting stresses will also remove the intentional stresses, and the global magnetic properties might be adversely affected by a larger

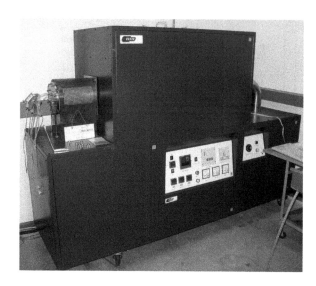

FIGURE 4.12   Laboratory furnace with controlled temperature and gas atmosphere. Equipment courtesy of Wolfson Centre for Magnetics, Cardiff University, Cardiff the United Kingdom.

amount than the degradation around the edges. For such materials, it is highly recommended to use low-stress cutting techniques such as waterjet. A similar one applies to amorphous and nanocrystalline laminations.

4.1.8   POLISHING

Magneto-optical Kerr effect (MOKE) microscopy requires very high level of surface smoothness, or the so-called 'mirror finish'. Polishing of large samples is difficult in practice (Williams, 2006), and even if it could be carried out with sufficient precision, then simultaneous observation of a large number of domains is impractical.

There are commercially available laboratory polishing machines (Figure 4.13) which provide all the components required for good-quality polishing. A system of this kind requires controlled motorised turntable, sample holder, and a dosing unit for automatic application of polishing suspensions and lubricants (Struers, 2008). Depending on the manufacturer and type of equipment used, the procedure can be referred to as *polishing*, *lapping* or *linnishing*.

Polishing procedure requires several stages, with decreasing size of grit in the suspensions. This process can be automated to some degree – as can be seen in Figure 4.13, there are several plastic bottles, each with different grade of grit, which can be automatically activated in a pre-programmed sequence.

The polishing process requires patience and cleanliness, because even a single larger particle can create a deeper scratch, which is difficult to eliminate without restarting the whole procedure. For instance, it can be seen in Figure 1.9 that there is straight scratch running at an angle from left to right, almost through the centre of the image. The scratch is clearly visible, but for this particular study, it did not seem to impact the distribution of domain walls, as none seem to be pinned to the line

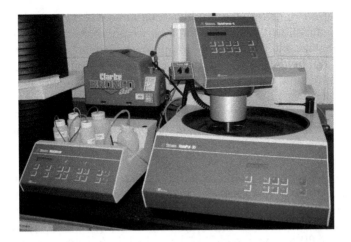

FIGURE 4.13   Laboratory equipment for polishing small samples. Equipment courtesy of Wolfson Centre for Magnetics, Cardiff University, Cardiff, the United Kingdom.

of scratch. Of course, such scratches should be avoided. This particular image was included here for better illustration of this problem.

4.1.9   SAMPLE AND SENSOR MOUNTING

The type of sample mounting depends strongly on the type of the magnetising yoke and the measurement technique. There are almost as many examples as there are prototypes of magnetising apparatus used by the researchers worldwide.

   In some cases, the sample is not mounted in any special way. For instance, as shown in Figure 2.24, the sample itself is also the magnetising yoke, so there is no need for a sample 'holder' as such, although some support of the weight might be required. A similar one applies to toroidal samples (Figure 3.43). In another example, the sample rests on a flat table and is kept down by the weight of the yoke (Figures 3.22 and 3.29).

   However, in many systems, some form of a sample holder is unavoidable. It defines correct positioning of the sample with respect to the magnetising yoke, ensures symmetrical air gap and holds the sample in place against vibrations, torque, etc. Several examples are shown earlier: Figures 2.3, 2.13 and 2.25.

PRACTICAL COMMENT

In author's experience, the torque developed on the sample is capable of rotating the sample if no clamping is provided. This is especially true for anisotropic materials such as grain-oriented electrical steels, due to significant magnetocrystalline energy. As can be seen from Figure 2.4, there can be a volumetric torque density exceeding 20 kN m/m^3, which for a 0.27-mm-thick sample, 80 mm diameter can produce a torque value of 0.03 N m. Assuming that this torque is acting at the arm length of a half of the sample radius (see also Figure 2.2), the resulting total force will be at the level of 1.4 N. The gravitational force from the total weight of the sample (0.01 kg) would be only around 0.1 N, so the rotation would definitely occur if the sample was not held in its place. Forces for less anisotropic materials would be smaller, but could still be comparable to the friction force of a freely resting sample.

Even more important problem is the attraction/repulsion between the sample and the yoke. If the sample is held exactly at the geometrical centre, all the radial forces would cancel out. In reality, some misalignment is always present.

A worst-case scenario would be if the sample is magnetised to 2.0 T, one edge of the sample is very close to the yoke and the other edge is much farther away from the other pole. A 0.5-mm-thick, 100-mm square sample has the edge surface area of 50 mm², and the force acting on this edge would be at the order of 80 N (comparable to 8 kg), which is definitely not only capable of moving the sample, but even inflicting serious damage to the operator's fingers.

If the shift happens between CW and ACW rotation, it is no longer possible to average the CW and ACW readings, because the measurement is unreliable (e.g. both values could be negative).

On the other hand, the sample cannot be clamped too hard, because this could introduce unwanted mechanical stress which could increase the measured power loss. In general, it is not possible to define the right balance for all the materials, and a good compromise should be found experimentally, especially for stress-sensitive materials such as Ni–Fe alloys as well as amorphous and high-permeability electrical steels.

A sample holder for circular samples is kind of a special case, as compared to other shapes. There are two main problems. Firstly, a circle does not have any natural straight edge or a corner which could stop the sample from rotating in its holder. Such rotation could be caused by the rotational torque or vibration from cyclic magnetisation.

The second problem and more important difficulty is the lack of clear definition of the reference direction of the sample. In order to keep the sample completely rotationally symmetrical, it should be machined as an ideal circle. But it is very difficult to align a circle *precisely* along a given reference direction (usually the rolling direction). This is not the case with other samples such as square or hexagonal, which have straight edges to be relied on. If a sample is machined with a clear mark (e.g. a little 'notch' cut into the edge), then this will produce local non-uniformity in the magnetic field, which might have unknown consequences on the accuracy of power loss measurement. A concept of such 'notches' is shown in Figure 4.14.

Of course, the direction can be marked non-invasively on the surface, for example with a permanent pen marker. But then the whole process of alignment relies on visual alignment 'by eye', which cannot be relied on for absolute preci-

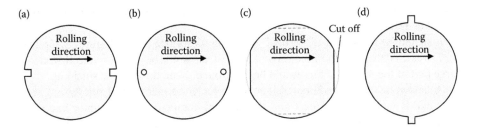

FIGURE 4.14 Proposed methods of mechanical reference for circular samples: (a) notches, (b) drilled holes, (c) straight edges and (d) alternative design with 'outward' notches; (drawings not to scale).

sion. For smaller samples, it is difficult to see alignment better than 1° with naked eye, let alone better than 0.1°, as required for minimisation of CW–ACW differences (Zurek et al., 2006a). The mutual alignment of B-coils can be made precise, because the holes can be machined at the same time as the sample is cut (e.g. by waterjet or laser, at least in theory). But 'external' sensors such as H-coils cannot be as easily referenced to the same fiducial point or direction of the sample, which will almost always lead to significant CW–ACW differences in the measured power loss.

It should be noted that a multi-lamination sample (Figure 4.5) can potentially suffer for accumulation of such angular positioning errors. This is less of a problem for non-oriented electrical steels or other weakly anisotropic materials. But for highly anisotropic samples (e.g. grain-oriented electrical steels), the angular misalignment between layers will cause different demagnetising fields between the layers so that the 'ellipsoidal approximation' will be adversely affected. Any such problems will be exacerbated by air gap between the laminations, as discussed above.

The positioning of a circular sample in the holder can be solved by compromising its shape slightly, for instance by adding two or more miniature notches (e.g. 0.5 mm) during the sample cutting, so that precise angular positioning can be established (Figure 4.14a and b). More experimental studies are needed on the effect of such notches on the overall magnetic performance of the sample.

The notches, holes or trimmed edges could be positioned at the rolling direction, because larger reluctance of the circuit can be tolerated in terms of power required for magnetisation of the sample. The size of notches and holes could be as small as 0.5 mm because they would be used only for location of the sample. Circular samples with notches were used by some researchers already (Miyamoto et al., 2011).

Figure 4.14c shows a circular sample modified by two straight cuts. Cutting off just 0.5 mm for a 150-mm-diameter sample would produce around 17 mm of a straight edge, which should be sufficient for locating the sample. The sample could be cut from four sides, thus producing the shape resembling an octagon. Loosing just 0.5 mm from each side should not significantly influence magnetising conditions, thus ensuring good uniformity, and it would still allow using the compact and efficient stator-like yoke. Such arbitrary shapes can be easily made, for instance with waterjet cutting, so that the angular alignment with the original rolling direction would be retained.

It should be noted that if 'vertical' shielding (described below, Figures 4.26c and 4.27) was used then effects of such notches/holes/cuts would be greatly reduced, because large part of the magnetic flux would be fed around them, through the shielding.

A hexagonal sample offers straight edges for good mechanical referencing, but the yoke is much less compact and thus cannot compete with the stator-like yoke for the level of magnetisation achievable with the same rating of power amplifiers. Perhaps the sample could be made as octagonal (or even with more edges) but the effects of interaction with the number of slots in the stator-like yoke would need to be investigated. In any case, expected uniformity of magnetisation for a polygonal sample with more edges (Figure 4.14c) should be better than for a sample with fewer edges (e.g. square or hexagon).

Sample holders should be carefully machined, with similar tolerances as the precision of the sample. A good sample holder should be properly designed so that it accommodates a given sample shape in a given yoke shape. An example of a typical sample holder is shown in Figure 4.15. Such assembly is placed inside of the magnetising yoke. The sample can be kept down with a weight (similar to that done

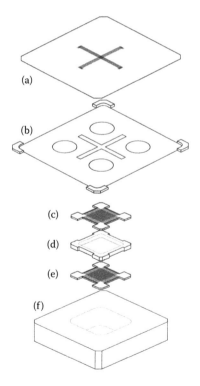

FIGURE 4.15 An example of integrated sample and sensor holder: (a) square sample with *B*-coils, (b) sample holder, (c) first set of *H*-coils, (d) *H*-coil spacer, (e) second set of *H*-coils and (f) *H*-coil holder and support for the sample holder. (Courtesy of Piotr Klimczyk, Brockhaus Measurements, Germany.)

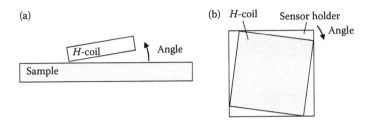

FIGURE 4.16 Angular positioning errors for horizontal (a) and vertical (b) alignment.

in Epstein frame (Figure 2.20b) or by other mechanical means (e.g. spring loaded surface).

As can be seen in Figure 4.15, the sample holder can have integrated space for holding the sensors as well. It was proved that differences in angular positioning of the sensors in the fieldmetric method produce large differences between rotational power loss measured in CW and ACW direction. It should be noted that even very small angular errors (less than 0.1°) can still produce large CW–ACW differences (Zurek et al., 2006a,b). Therefore, precision of the sensor holding also matters greatly.

The sensors are usually much smaller than the sample, so even if they are made with the same absolute mechanical precision, the influence on the absolute angular positioning will be greater than just for the sample. Some examples of possible angular positioning errors are shown in Figure 4.16.

For instance, let us assume that for the sample holder the machining precision is 0.05 mm. For the worst case, for a 100 mm square sample, the maximum angle will be atan (0.05/100) = 0.03°, which is a relatively small angle. But for a 20 mm sensor, it will be around 5 times greater or atan (0.05/20) = 0.14°. Such values can result in significant errors – as shown in Zurek et al. (2006a), an angle of 0.09° can easily produce *negative* loss (see also Figure 3.49c).

PRACTICAL COMMENT

Under normal conditions, the value averaged from CW-ACW is not affected in a significant way. But this applies *only* under the assumptions that the sample does not move between CW and ACW measurements. If the sample holder does not prevent movements of the sample, then in the worst case the sample could move by 2 · 0.03° = 0.06° and thus significantly influence the results. It is needless to say that the averaging would not eliminate such errors.

It should be noted that the holders should be made from non-magnetic non-metallic materials, for example plastics such as acrylic glass (Plexiglas, Perspex), or composites (Tufnol), in order to ensure that eddy currents are not induced and the distribution of magnetic field is not distorted. Additionally, any metallic support could come in electric contact with the sample, which would also drive currents external to the sample.

Plastics (and other non-metal engineering materials) are usually significantly less rigid than metals, and attaining high-mechanical precision of machined parts

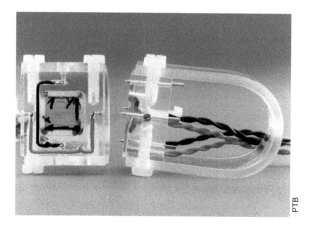

FIGURE 4.17   Integrated sensor holder, with double H-coil and two pairs of needle probes; there are four pairs of wires for connection to the data acquisition system. (Courtesy of Johannes Sievert, National Metrology Institute [PTB], Germany.)

is difficult. Firstly, during machining, the material flexes under the pressure of the cutting tool, and although the tool is guided with the required precision, the final tolerance is inferior. Absolute precision better than 0.05 mm is usually difficult to achieve in practice, even though 0.01 mm can be routinely achieved for metals. This is even more important for example when machining coil formers for H-coils. The thickness of the material is usually quite small (1 mm or less), and distortions due to bending can be exacerbated.

A second problem is that during frequent use the softer material will wear off quicker. Very frequent contacts with for example sharp edges of many samples can damage the precisely machined surfaces responsible for angular positioning (usually more important) as well as absolute (less important if other components of the system are symmetrical).

Some sensors such as B-coils are mounted directly on the sample and thus do not require any special mounting. However, provision must be made for the thickness of the wire with which the coil is made. This can be seen for instance in Figure 4.15 in which the part 'b' has cross-like cut-out to accommodate the B-coils.

Other sensors might require quite complex mechanical support. An example is shown in Figure 4.17. The integrated sensor head allows simultaneous measurement of four signals: B_x and B_y with two pairs of needle probes (also for B_x and B_y), and H_x and H_y with a double H-coil (Sievert, 1990).

The sensors also have to be aligned with the surface of the sample, and some researchers used for instance springing force from a soft sponge to ensure controlled pressure between the H-coils and the sample (Enokizono et al., 1990c).

4.1.10   Coil Winding (for Sensors)

All the various types of sensors were described in detail in Chapter 3. This section focuses only on practical considerations for winding the induction sensors, namely B-coils, H-coils and RCPs.

FIGURE 4.18 Winding of double H-coils (see also Figure 3.45), Richard Wakely (operating the machine) and the author looking to the camera. The photograph was taken during author's PhD studies in 2001, at the workshop of Wolfson Centre for Magnetics, Cardiff University, Cardiff, the United Kingdom.

Usually, local B-coils have few turns (Figure 4.15), or even just a single turn (Figure 3.1). The wire has to be threaded through the holes, one turn at a time. This is done manually and requires a certain amount of patience and care.

The situation is different for H-coils and RCPs. These have many turns, usually more than 100 (see Figures 3.45 and 4.17), and uniform winding is of great importance. Such winding can be done with sufficient precision only with the use of a machine. This is especially important for miniature H-coils which could have 350 turns wound on 4 mm width with 0.014 mm enamelled wire (Figure 6.21) (Aihara et al., 2011).

However, most researchers use larger H-coils, with around 20 mm length (or longer), wound with 0.05 mm wire or thicker. Such H-coils can be wound on fairly simple winding machines (Figure 4.18), equipped with control of rotational speed, pitch (linear shifting during winding) and tension on the wire.

PRACTICAL COMMENT

Wire tension control is especially important because the H-coil former is flat, and as it rotates, there are significant variations of the tension and the wire can snap easily. The winding speed must be thus *much* slower than for winding other types of coils.

An H-coil should be thin, and the thinnest possible configuration will be if exactly one layer of wire is wound, without any wires crossing each other. This can be achieved by careful control of the pitch of the winding, which should be set to a value slightly larger than the thickness of the wire. With ordinary winding machines, the pulley which feeds the wire does not offer particularly good control over positioning of the wire, which makes the winding more difficult. It is very important to mount the coil former in a symmetrical way, both axially (so the axis of rotation is not skewed) and radially (so the axis of rotation is exactly at the centre of the former).

After winding, the fine wire of the *H*-coil is exposed and can be easily damaged mechanically. Thicker terminating wires should be attached (e.g. by soldering, see also Figure 4.17), and the coils can be protected by a thin layer of low-viscosity varnish so that it penetrates the volume between the wires. The varnish must not be chemically aggressive in the long term, so it does not damage the wire insulation with time.

A single-layer winding requires for one end of the wire to return to the start of the coil (see Figure 3.50). Such return wire could be guided at the side of the coil (easiest option), through the centre of the former (very difficult to achieve mechanically) or diagonally.

Soldering of fine enamelled wires also requires some effort. The enamel cannot be removed mechanically, as it is the case for thicker wires. Most wires are made with 'self-fluxing' enamel, which melts when heated to sufficiently high temperature (e.g. 350°C). However, the thermally damaged enamel exposes copper even in places which are not covered by the solder. The copper oxidises (as well as reacts chemically with solder), and with time, the wire can break, which renders such coil useless. Other methods such as conductive paint can be employed for this purpose.

For soldering, it might be beneficial to carry out 'skeining' that is making a bundle of thin wire first, before soldering to a thicker wire. It is always a good idea to practice such soldering first, before applying it to the actual *H*-coil wires.

Winding coils for RCPs has its own difficulties. The winding uniformity is paramount for a correct operation of an RCP, as discussed in Section 3.2.2.2. However, as shown in Figure 3.53, the coil is not wound on a straight former, which makes the winding practically impossible in the final form of the sensor. For this reason, versions such as that shown in Figure 3.53c or Figure 3.53d were used by some researchers. The partial straight fragments can be wound with conventional methods.

But if RCP is made from a single piece (Figure 3.53a or Figure 3.53b), then winding can be done if the former is made from a flexible rod. The wire is then wound on a former which is initially straightened out, and only later the whole coil is bent into shape and fixed in position by the sensor holder. An example of such sensor is shown in Figure 6.23.

For all the sensors, the terminating wires should be as thin as practical and twisted very carefully (Figure 4.17). Such approach will help in minimising the pickup of unwanted signals, as illustrated in Figure 4.19 (Zurek et al., 2008a).

4.2 MAGNETISING YOKE

Magnetising yokes are probably the parts with greatest diversity as used in various measurement systems. The role is practically the same for all measurement techniques, but even within the same method, the yokes can be very different, with various shapes and configurations.

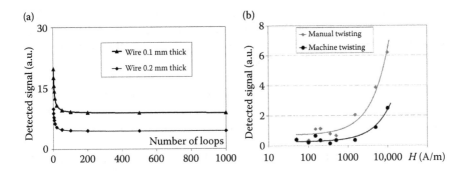

FIGURE 4.19 Unwanted signal in a pair of twisted wires exposed to external *H*: (a) effect of wire thickness and tightness of twisting for fixed *H* and (b) effect of twisting uniformity and various *H* levels. (Reprinted from Zurek S. et al., *Sensors and Actuators A*, 142, 2008a, 569. Copyright 2008, with permission from Elsevier.)

A magnetising yoke is used for the same reason as a magnetic core in any other electromagnetic device. Magnetic reluctance of the yoke is significantly lower than that of air and therefore a sample surrounded by the yoke can be magnetised to the same level with considerably less apparent power. However, if the power is of no special concern, then magnetic excitation can be applied just by coils suspended in air, without any yoke (see also Figure 2.8) (Nakata et al., 1992).

The apparent power (which is a product of RMS voltage and RMS current) required to excite a given system is proportional to the equivalent reluctance, because the active power loss in the sample under test is usually negligibly small by comparison. Relative permeability of air is unity, whereas relative permeability of yoke can reach a value of thousands. Under most conditions, the power dissipated in the yoke is also rather small. However, resistive loss in the yoke windings might not be negligible at high currents.

From such viewpoint, all the air gaps (or more precisely non-magnetic discontinuities) would be required to be eliminated or at least minimised to reduce power requirements. However, in general case, this is not possible. For instance, in the torque method, the yoke and the sample *must be* able to move relative to each other, which implicitly requires some spacing between them (Figures 2.3, 2.6, 2.10 and 2.13). For such measurements, the yoke is often a magnet or electromagnet with single pair of magnetic poles.

Spacing between the yoke and the sample can also be needed in the thermometric method. Ideally, the sample should be energised under adiabatic conditions, which means that it should be *inside* the vacuum chamber with the magnetising yoke placed *outside* (Figure 2.15).

The problem of magnetisation uniformity is widely recognised. In general case, it is not feasible in practice to obtain uniform distribution of *B* within the whole of the sample volume. This is possible theoretically only in exactly ellipsoidal samples (Figure 4.2a), but in practice such shapes cannot be made from electrical steels or other laminations. For this reason, *B*-coils should *not* be wound on the outside of the whole sample, because the detected *B* will be strongly affected by the non-uniformities. Such large sensors are sometimes referred to as 'pocket' *B*-coils (Figure 6.23b), but they should be avoided for the reasons listed above.

However, satisfactory uniformity can be usually attained at the centre of the sample (Figure 3.10), where *B* and *H* sensors would be typically placed. Therefore, the role of the magnetising yoke is not only to reduce the required power, but also to facilitate as good magnetisation uniformity as possible for a given sample type.

A great diversity of magnetising yokes can be found just within the fieldmetric technique. Even though the sample does not have to physically move, it was experimentally shown that greater magnetisation uniformity of the sample can be achieved if there is at least a small air gap between the sample and the yoke (Brix et al., 1984; Alinejad-Beromi, 1992; Xu et al., 1997).

Many types of magnetising yokes used for fieldmetric technique follow such rules. Some examples of the yokes used by researchers just for the fieldmetric method are shown in Figure 4.20.

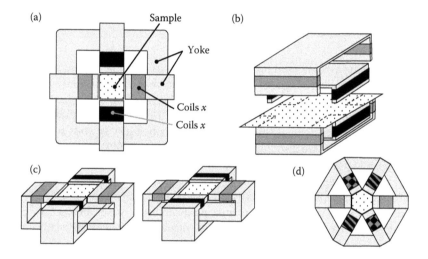

FIGURE 4.20 Some examples of various magnetising yokes (drawings not to scale, adapted from the cited publications): (a) with horizontal yokes (Adapted from Brix W. et al., *IEEE Transactions on Magnetics*, 20 [5], 1984, 1708; Enokizono M. et al., *IEEE Transactions on Magnetics*, 26 [5], 1990a, 2562; Stranges N. et al., *IEEE Transactions on Magnetics*, 36 [5], 2000, 3457; Zurek S., *Two-Dimensional Magnetisation Problems in Electrical Steels*, PhD thesis, Wolfson Centre for Magnetics Technology, Cardiff University, Cardiff, the United Kingdom, 2005a), (b) with double vertical C-yokes (Adapted from Brix W. et al., *IEEE Transactions on Magnetics*, 18 [6], 1982, 1469; Nencib N. et al., *Journal of Magnetism and Magnetic Materials*, 160 [1996], 1996, 174; Fonteyn K.A., *Energy-based Magneto-Mechanical Model for Electrical Steel Sheet*, PhD thesis, Aalto University, Espoo, Finland, 2010), (c) with double C-yokes (Adapted from Boon C.R. et al., *Proceedings of the IEE*, 112 [11], 1965, 2147; Gumaidh A.M. et al., Measurement and analysis of rotational power loss in soft magnetic materials, *Proceedings of 1st 2DM Workshop*, Braunschweig, 1993a, p. 173; Bajorek R. et al., A compact single sheet tester for investigation of both alternating and rotational magnetisation, *Proceedings of 6th 1&2 Workshop*, Bad Gastein, Austria, 2000b, p. 80), (d) three-phase horizontal yoke for hexagonal sample (Adapted from Hasenzagl A. et al., *Journal of Magnetism and Magnetic Materials*, 160 [1], 1996, 180; Ivanyi A. et al., *IEEE Transactions on Magnetics*, 34 [5], 1998, 3004; Parviainen A. et al., 2D measurement of core loss using a 3-phase excitation apparatus, *Proceedings of 6th 2DM Workshop*, Bad Gastein, Austria, 2000, p. 117),   *(Continued)*

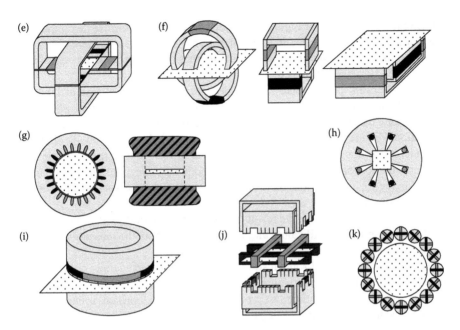

FIGURE 4.20 (Continued) Some examples of various magnetising yokes (drawings not to scale, adapted from the cited publications): (e) with double C-yokes and auxiliary horizontal yokes (Adapted from Kedous-Lebouc A. et al., On the magnetization process in electrical steel in unidirectional and rotational field, *Proceedings of 1st 2DM Workshop*, PTB, Braunschweig, Germany, 1992, p. 36; Nakata T. et al., *IEEE Transactions on Magnetics*, 29 [6], 1993, 3544; Okazaki Y. et al., 2-dimensional magnetic measurement for rectangular single strip of non oriented electrical steel, *Proceedings of 6th 1&2DM Workshop*, Bad Gastein, Austria, 2000, p. 76.), (f) with single C-yokes (Adapted from Gumaidh A.M. et al., Measurement and analysis of rotational power loss in soft magnetic materials, *Proceedings of 1st 2DM Workshop*, Braunschweig, 1993a, p. 173; Wan Mahadi W.N.L., *Design and Development of a Novel Two-Dimensional Magnetic Measurement System for Electrical Steels*, PhD thesis, Wolfson Centre for Magnetics Technology, Cardiff University, Cardiff, the United Kingdom, 1996; Engdahl S.G. et al., Measurements and modelling of 2-D magnetization and magnetoelasticity in silicon iron, *Proceedings of 5th 2DM Workshop*, Grenoble, France, 1997, p. 1), (g) yoke similar to motor stator (Adapted from Fiorillo F. et al., The measurement of rotational losses at I.E.N.: Use of the thermometric method, *Proceedings of 1st 1&2DM Workshop*, Braunschweig, Germany, 1992a, p. 162; Gorican V. et al., 2-D measurements of magnetic properties using a round RSST, *Proceedings of 6th 1&2DM Workshop*, Bad Gastein, Austria, 2000a, p. 66; Zurek S., *Two-Dimensional Magnetisation Problems in Electrical Steels*, PhD thesis, Wolfson Centre for Magnetics Technology, Cardiff University, Cardiff, the United Kingdom, 2005a), (h) round yoke with square sample (Adapted from Sugimoto S. et al., *Przeglad Elektrotechniczny [Electrical Review]*, R. 81 [NR 5/2005], 2005, 27), (i) double round vertical yoke without air gaps (Adapted from Ichijo N. et al., A new 2-D magnetic measurement method with vertical yokes, *Proceedings of 7th 1&2DM Workshop*, Bad Gastein, Austria, 2000, p. 197), (j) yoke with coils surrounding the sample (Adapted from Nakano M. et al., *IEEE Transactions on Magnetics*, 35 [5], 1999, 3965; Mori K. et al., *IEEE Transactions on Magnetics*, 41 [10], 2005, 3310; Miyagi D. et al., *Przeglad Elektrotechniczny [Electrical Review]*, R. 85 [Nr 1/2009], 2009, 47) and (k) coils arranged as a Halbach cylinder (Adapted from Alatawneh N. et al., *IEEE Transactions on Magnetics*, 48 [4], 2012a, 1445; Wanjiku J. et al., *International Journal of Applied Electromagnetics and Mechanics*, 8, 2015a, 255).

It should be stressed that Figure 4.20 shows just a few *generalised* examples and many more variants can be found in the literature. Practically, every laboratory uses a specific yoke so that even if the yoke from another laboratory is based on a similar concept, the exact realisation is different in several details.

For this reason, precise like-for-like comparison between different laboratories is still difficult and a path towards standardisation of rotational power loss measurement has not been defined yet (at the time of writing this book). The international intercomparison of measurements carried out by round-robin testing undertaken couple of decades ago showed fairly good agreement for measurements on non-oriented electrical steels. There were much larger discrepancies (tens of percent) for grain-oriented electrical steels between various laboratories (Sievert et al., 1996).

However, since the 1990s, the resolution and precision of digital voltage acquisition devices improved significantly and we know more about the measurement errors in rotational measurements.

Perhaps, a similar round-robin testing should be attempted again in order to investigate the likely improvement in agreement between various laboratories worldwide.

Some conclusions about the most promising path for an 'optimal' rotational apparatus are given at the end of this chapter.

4.2.1 WITHOUT YOKE

In some cases, magnetising yokes are not used. As mentioned above, a magnetising coil or set of coils can surround the sample with various degrees of proximity. For instance, the sample could be placed in a solenoid or in a Helmholtz coil (Brailsford, 1948; Grossinger et al., 2000; Kollar et al., 2000). However, due to the magnetic 'openness' of such circuit and the correlated demagnetising factor, high B values are difficult to obtain, and some authors moved away from magnetising the sample directly to an approach with a yoke (Geirinhas Ramos et al., 1995).

A second possibility is when the sample itself completes the magnetic circuit. An example of such configuration is shown in Figure 2.24. The sensors are placed only in one area where the rotational magnetisation is set up, but the rest of the sample is wrapped around so that the magnetic flux always travels inside of magnetic material.

However, it was shown experimentally (Brix et al., 1984) that the B distribution was significantly non-uniform in a cross-shaped sample (Figure 4.21). The effective area of the sample changes at the central area so that the flux has room to 'spread out'. Therefore, the magnitude of B at the centre of the sample is lower than in other locations. As a consequence, the outer part of such sample saturates before high B could be achieved in the central area, and this is ultimately a greater practical problem than the uniformity as such.

For instance, the cross-shaped sample described in Moses et al. (1973) could be magnetised only up to 0.875 T, which was '*the highest value of flux density which could be used in order to maintain sinusoidal flux in both arms*'. The sample-and-yoke approach began to be more popular in further research in order to achieve higher B values.

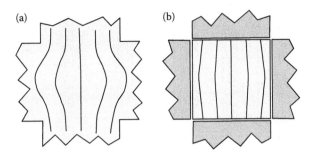

FIGURE 4.21   Flux distribution in cross-shaped sample (a) and a square sample in a yoke with some air gap (b). (Adapted from Brix W. et al., *IEEE Transactions on Magnetics*, 20 [5], 1984, 1708.)

4.2.2   FIELD UNIFORMITY AND EXCITATION LEVEL

A significant amount of effort was directed towards finding the optimum sample–yoke configuration with as good field uniformity as possible, but at the same time with improved level of attainable controlled magnitude of B so that the reduction of rotational power loss at high excitation could be studied.

In the 1990s, computer technology improved significantly and became more widely available at affordable cost. Finite-element modelling (FEM) was possible on personal computers, and several papers investigating the problem were published, mostly for two-phase and three-phase yokes with salient poles.

In one of such studies, it was shown that a circular sample in a two-phase yoke (as shown in Figure 3.10a) resulted with much worse non-uniformity than for a similar yoke with a square sample (Nencib et al., 1995). Similar simulations were carried out a decade later confirming the findings (Bottauscio et al., 2005).

A mostly arbitrary target of 1% deviation for uniformity was often assumed (Ichijo et al., 2000; Mori et al., 2005). 2D FEM simulations showed that some sample–yoke configurations provide intrinsically better uniformity than others, as shown in Figure 4.22. The best configurations were at the time concluded to be square (Figure 4.22c) and hexagonal samples (Figure 4.22f) with straight-edge poles, and circular sample with a stator-like round yoke (Figure 4.22d).

However, the validity of such 2D FEM studies was limited because all the geometrical aspects cannot be represented in two dimensions. The whole structure is effectively simulated as being 'infinitely deep' (Meeker, 2015), so the sample has the same thickness as the yoke, which is of course not true for a practical setup. As a result, there can be significant differences between the outcomes of 2D and 3D simulations (Ivanyi et al., 1998).

3D simulations are much more computationally expensive, and only recently, more detailed studies of uniformity were published (Jesenik et al., 2003b; Alatawneh et al., 2012b; Wanjiku et al., 2014; Borg Bartolo et al., 2015). (Some examples of 3D FEM simulations are shown in Chapter 7.) Nevertheless, qualitatively, the 3D FEM studies show that the *circular* sample–yoke configuration gives the best uniformity,

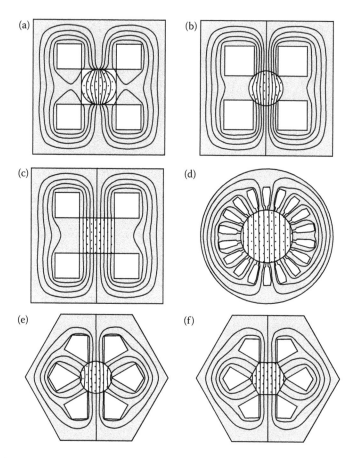

FIGURE 4.22 Various sample–yoke configurations as modelled in 2D FEM. (a) square yoke with flat poles and circular sample, (b) square yoke with shaped poles and circular sample, (c) square yoke with square sample, (d) stator-like yoke with round sample, (e) hexagonal yoke with round sample, and (f) hexagonal yoke with hexagonal sample. (Adapted from Bottauscio O. et al., *Przeglad Elektrotechniczny [Electrical Review]*, R. 81 [Nr 5/2005], 2005, 8.)

as compared to a square sample. No direct comparison of 3D FEM to hexagonal sample was shown within a single publication to date.

The problem of 'uniformity' can be discussed at various levels. As a wider picture, there is the uniformity of magnetisation of the whole sample. It is obvious that if the whole sample is magnetised uniformly, then also some local area is subjected to more uniform excitation (especially for isotropic materials).

However, non-uniformity outside of the measured area will not be detected by the sensors. Therefore, the uniformity is critical only within the measurement area exposed to the relevant sensors.

The lowest level is represented by the very local non-uniformities caused for instance by the presence of the *B*-coil holes (or other sensors). The holes introduce non-magnetic discontinuities, which inevitably affect distribution of magnetic field, for example as shown in Figure 3.18 (and also Figure 7.30).

Nonetheless, the global non-uniformity is driven by the shape of the sample–yoke system, and considerable research was carried out investigating the problem. Some aspects of various yoke configurations are described in the following sections.

4.2.3 Yoke for Cross-Shaped Sample

Cross-shaped samples are used with two-channel yokes (Nakata et al., 1992; Hubert et al., 2005). Sometimes, other aspects might be more important, for example when measurements of magnetic properties are carried out under applied mechanical stress. The shape of the magnetising yoke must allow not only magnetic excitation but also controlled application of the mechanical stress, also at an arbitrary direction within the plane of the lamination. Examples of cross-shaped specimens for such measurements are shown in Figure 2.60 (Basak et al., 1978; Hubert et al., 2005; Kai et al., 2011a), but see also Figure 4.20b (Fonteyn et al., 2009). A cross-shaped sample was also used with a planar yoke shown in Figure 4.20a (Ishihara et al., 1993).

When coupled magneto-mechanical measurements are required, then the choice is limited. But if only magnetic measurements are to be carried out, then such shapes should be avoided, because even a square sample usually gives better uniformity, as evident from Figure 4.21.

4.2.4 Yoke for Square Sample

Two-phase magnetising yokes were used for the first time over half a century ago (Boon et al., 1965). The sample is a square, and usually there is a small gap between the sample edges and the magnetic poles of the yoke. The magnetic circuit of the yoke can be wrapped around the sample in a several ways; some of them are shown in Figure 4.20.

Square sample can be inexpensively cut from larger laminations, for example by a guillotine (Figure 4.8). With an appropriate design of the yoke, the spacing between the magnetic poles can be adjusted so that samples of various dimensions can be accommodated or the sample–yoke air gaps can be set to various values (Figure 6.3). Non-uniformity of magnetisation in such a setup increases towards the corners of the sample as verified experimentally and illustrated for instance in Figure 3.10b.

Many researchers, including the author, struggled to achieve high level of *controlled* rotational B, typically not more than 1.7 T even with a powerfull magnetisation system (e.g. up to 1 kVA per channel) (Zurek, 2000; Gorican et al., 2003; Zurek, 2005; Mori et al., 2015). This is especially true for highly anisotropic materials such as grain-oriented electrical steels.

Such difficulty is caused by the pole-to-pole flux leakage through air so that the sample is partially bypassed (Krah et al., 2002). Additionally, for materials such as grain-oriented electrical steels, the hardest direction lies close to the diagonal of the sample (e.g. see Figure 1.22). Therefore, two channels (x and y) have to be energised partially but simultaneously in order to set up high value of B in that direction.

For these reasons, the magnetisation in the corners of the sample is very non-uniform, and in some cases, small parts of the corners are removed (Figure 4.15).

Therefore, the question of sufficient uniformity of magnetisation in such systems remains open. But a bigger problem seems to be the fact that high values of controlled B cannot be achieved similarly to other approaches (e.g. round yoke). Therefore, the phenomenon of the power loss reduction towards saturation under rotational magnetisation cannot be studied with sufficient detail.

However, yokes with square samples remain a popular choice in many magnetising apparatus, and they continue to be used by laboratory researchers worldwide as well as in commercial measurement systems. The cost of the samples remains an important parameter, especially for the commercial systems. Some examples are also shown in Chapter 6.

4.2.5   YOKE FOR RECTANGULAR SAMPLE

Sample cutting introduces additional costs and complications. More importantly, it makes more difficult comparing the results between various measurement systems (Sievert et al., 1995, 1996) or other standardised magnetic measurement techniques, such as Epstein frame (IEC, 2008). This is because large samples offer better 'averaging' of properties over larger volume of the material under test. Small samples are more prone to be effected by local properties, for example large grains in grain-oriented electrical steels.

Some researchers used rectangular samples for rotational power loss measurement (Sasaki et al., 1985; Ishihara, 1992; Okazaki et al., 2000). With such system, the measurements could be performed directly on Epstein frame strips (Figure 2.20), without any additional cutting. A simplified example of such approach is shown in Figure 3.11. For even greater ease of use, the B-coils are made as diagonal so that there is no need to drill holes in the lamination under test. In the described apparatus, one axis was magnetised by a coil surrounding the sample directly (solenoid). In the other direction, there was a magnetising yoke (similar to that shown in Figure 4.20c). However, the largest reported circular B for grain-oriented steel was only 1.2 T (Sasaki et al., 1985).

4.2.6   YOKE FOR OVERHANG SAMPLE

Many types of yokes shown in Figure 4.20 require precise shape and dimensions of the samples to fit between the magnetic poles of the given yoke. But precisely cut samples increase the cost of using such systems. The sensors–holder–sample–yoke system (Figure 4.15) requires careful assembly, which increases the overall setup time for each sample.

One way of overcoming this difficulty is to design the yoke in such a way that it can accept larger samples, extending beyond the size of the yoke, so that the edges 'overhang'. Examples of such yokes are shown in Figure 4.20b, f and i.

Similar approach has to be taken if the yoke is designed to be moved around the sample, which is placed on a flat surface. This allows investigation of local properties on one large sample, rather than investigating local properties of several small samples.

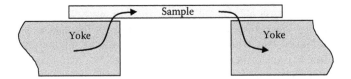

FIGURE 4.23 Asymmetric magnetisation in off-centre sample creates planar eddy currents. (Adapted from Krell C. et al., Rotational single sheet testing on sample of arbitrary size and shape, *Proceedings of 6th 1&2DM Workshop*, Bad Gastein, Austria, 2000, p. 96.)

An example of such small yoke is shown in Figure 3.29 and also described in several papers (Enokizono et al., 1997; Siebert, 2000; Hall, 2011a). The small yokes are practically the only way to apply magnetisation at arbitrary direction to a larger surface, for instance for non-destructive testing or surface scanning (Chady et al., 2009; Stupakov, 2013; Piotrowski et al., 2015). Some examples are shown in Chapter 7.

It is also possible to use a larger sample with a 'planar' yoke (Figure 4.20a and d). An example of such approach is shown in Figure 4.23 (Krell et al., 2000). Unfortunately, due to the large asymmetry of such system, there is a large normal component of *B* which leads to significant planar eddy currents. With such setup, '*systematic errors are to be expected however may be acceptable in practice*' (Krell et al., 2000). Such errors would be especially significant for the needle probe technique, as discussed in detail in Section 3.1.2.2. Therefore, it is best to avoid such configuration.

4.2.7 YOKE FOR HEXAGONAL SAMPLE

As discussed above, the yoke with a square sample does not allow achieving high *B* values in all directions. For highly anisotropic materials such as grain-oriented electrical steels, there is a great difference between 'easy', 'hard' and 'hardest' direction of magnetisation, as illustrated in Figures 1.22 and 1.24. The hardest direction located at around 55° with respect to the rolling direction requires the highest values of *H* in order to achieve the same level of *B*. However, typically, the samples are cut so that their edges are aligned with the rolling direction so that for a two-phase yoke (e.g. Figure 4.20a) the hardest direction would lie diagonally between the magnetic poles of the yoke.

As a remedy to this problem, a three-phase yoke was proposed (Figure 4.20d, also shown in Figure 6.13). The sample could be cut with similar orientation, but there were magnetic poles positioned at every 60° so that the hardest direction could be excited with greater power. Indeed, higher controlled *B* values can be achieved with such three-phase yoke, than with a two-phase yoke of similar size. The resulting uniformity of magnetisation is also claimed to be better than for square samples (Bottauscio et al., 2005).

A three-phase yoke requires controlled three-phase excitation, so the feedback algorithm is somewhat more difficult due to the fact that a two-phase information obtained from B_x and B_y coils must be translated into three-phase driving signals. From such viewpoint, two-phase systems are easier to operate.

At low excitation level and for materials with low anisotropy, the patterns can be obtained just by controlling amplitudes and phases of the magnetising currents, which can be achieved even without a dedicated feedback circuit or algorithm (e.g. by using three variable transformers) (Hasenzagl et al., 1995). Of course, controlled circular patterns (or more complex) cannot be obtained with such simplistic approach.

4.2.8   YOKE FOR ROUND SAMPLE

A round or circular sample can be accommodated in a yoke with salient poles, as shown in Figure 4.22. However, the published FEM simulations show that the uniformity of magnetisation in such configuration is rather poor (Nencib et al., 1995; Xu et al., 1997; Bottauscio et al., 2005). One of the few examples is shown in Appino et al. (1997), but no quantitative information on the uniformity of magnetisation was provided.

However, a better configuration is when a circular sample is magnetised by a circular yoke, without salient poles. It was shown in several publications by both simulations and experiments that such configuration yields superior uniformity as compared to other types of yokes (Gorican et al., 2003; Jesenik et al., 2003a; Bottauscio et al., 2005; Alatawneh et al., 2012b; Borg Bartolo et al., 2015). Additionally, such arrangement also allows achieving much higher excitation, and there are published examples of controlled circular B up to 2.0 T even for grain-electrical steels (Gorican et al., 2002; Zurek et al., 2006c; Mori et al., 2015).

A circular yoke can be made by utilising a stator of an induction motor; examples are shown in Figures 2.15, 3.9a and 4.20g (also in Chapter 6). The winding can be made as a more complex 'sinusoidal' with the phase windings overlapping (Gorican et al., 2000a), or as a simpler with windings not overlapping each other in the core slots (Zurek, 2005). The former one is likely to produce better uniformity, especially for samples with low anisotropy (Gorican et al., 2000a; Wanjiku et al., 2015a).

In either case, there are the so-called *space harmonics* in the magnetomotive force, whose frequency is directly related to the number of slots for the windings (Gorican et al., 2000a; Anuszczyk et al., 2009). Usually, there will be as many fluctuations per cycle as there are stator slots (and multiplicity of this number), because the flux must rotate through 360° encountering each slot subsequently. The variation in magnetic reluctance at each slot causes the space harmonics. This is an intrinsic phenomenon for all stators with slots, and it is also found in motors (Cochran, 1989) and generators (Radley et al., 1981).

PRACTICAL COMMENT

Without feedback, such space harmonics or fluctuations can be at the level of around 1% (Gorican et al., 2000a) and they would be definitely present if the magnetising currents were just sinusoidal. However, some of these distortions can be suppressed by the feedback mechanism.

For instance, the yoke shown in Figure 4.20g has 24 slots, so there would be a corresponding 24th harmonic (and multiplicity of it) in the detected signals of B and H. If the feedback is capable of controlling frequencies including 24 harmonics (and beyond), then the variations will be reduced by introducing compensating fluctuations in the magnetising currents. The currents will be appropriately distorted so that the space harmonics in the sample flux will be suppressed accordingly.

At 50 Hz, the 24th harmonic corresponds to 1.2 kHz. Therefore, all signal-processing and power-delivering circuits must be capable of processing at least such frequency, without significant attenuation or phase shift, and with the required performance for a stable feedback and measurement.

It should be noted that this would be an *even* harmonic and those are usually absent if the excitation is symmetrical.

A stator-like yoke could also be used with three-phase windings. An example of such system was shown for instance in Parviainen et al. (2000). However, the excitation was produced by a PWM inverter driving three-phase induction motor coupled with three-phase generator. There was no feedback, so the wave shapes were not controlled. A similar circular three-phase system but with feedback control was shown in de la Barriere et al. (2015).

A different approach was proposed in Ichijo et al. (2000). The yoke is illustrated in Figure 4.20i. The yoke consists of two identical halves: the upper and the lower. The sample is not required to fit inside, but it should be larger than the yoke. Therefore, the sample preparation is not as demanding, especially if needle probes are used for B measurement, rather than B-coils.

The half-yokes are placed directly onto the sample, without additional air gap. The reported uniformity was very good and confirmed with FEM simulations (Ichijo et al., 2000). However, the authors reported measurement only up to 0.75–1.0 T, and it is unclear if the setup allowed achieving higher controlled B. Nevertheless, the inner diameter of the yoke had 150 mm (hence the 'active' area of the sample), which is quite large as compared to other systems.

The shape and construction of such yoke could be especially applicable to industrial tests where the simplicity of the sample preparation procedure is a one of the key factors. However, the positioning of the B and H sensors with respect to the yoke would be probably problematic in practice.

An interesting approach with a circular yoke was presented for instance in Alatawneh et al. (2012c), and its concept is shown in Figure 4.20k. The circular sample is surrounded by a number of small cylindrical cores, each of which has a two-phase winding on it, and the two-phase windings are aligned at different angles, according to their position around the sample. The principle is similar to a Halbach cylinder, which produces a uniform field inside itself. All the coils of one phase are connected in series, and the same applies for the other phase. Application of a two-phase drive to such system produces rotating magnetic field, similarly to a round stator.

However, the presented Halbach-like yoke appears to suffer from greater non-uniformity of sample magnetisation and lower attainable B due to its elevated reluctance (each small cylinder is surrounded by even larger air gap), as compared to other round yokes with a more magnetically 'closed' circuit (Wanjiku et al., 2014).

4.2.9 SHAPE OF MAGNETIC POLES

The 2D drawings in Figure 4.22 do not show all the 3D aspects of the sample–yoke configuration. In planar yokes, the designer can choose various shapes of the yoke, in the proximity of the sample edge. This can also influence the uniformity of sample magnetisation and measurement errors, and some researchers studied the problem (Enokizono et al., 1990c); some results for a planar yoke (e.g. Figure 4.20a) are shown in Figure 4.24. The conclusion from such study was that the chamfered pole pieces gave better uniformity and lower *H* errors, and once again it is found that larger air gap is more beneficial as far as uniformity of magnetisation is concerned.

The study shown in Figure 4.24 assumed ideal symmetry between the sample positioning and the poles. However, in practice, this is not achievable, and some asymmetry, usually unknown, is always present. This will contribute to some additional errors, especially if needle probes are used, due to the planar eddy currents (Pfutzner et al., 2007). The effect is smaller, but similar to that shown in Figure 4.23.

PRACTICAL COMMENT

Uniformity of *H* can be somewhat improved by changing the shape of the pole pieces, but none of the configurations presented in Figure 4.24 *eliminates* the non-uniformity. Therefore, for attaining measurement with better accuracy, some other measures would have to be taken.

This could be an extrapolation of the *H* towards the sample surface (Figure 4.15, see also Chapter 3) or magnetic shielding described in the following sections. If the shielding is used, it might be the case that a flat-pole shape could be more beneficial for the shield performance, as discussed in Section 4.2.11.

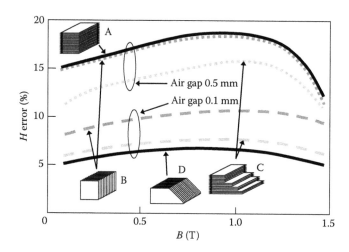

FIGURE 4.24 Effect of different types of pole shapes on the error of *H* measurement. (Adapted from Enokizono M. et al., *Anales de Fisica, Series B*, 86, 1990c, 320.)

It was also shown that for round 'deep' yokes (where 'depth' means the thickness of the lamination stack for the yoke shown in Figure 4.20g) the uniformity improves (Wanjiku et al., 2015b). Because such yoke is made similar or exactly as a stator of a motor, the 'pole faces' are flat by definition. Therefore, depending on the type of yoke, flat 'pole faces' can result in better uniformity than chamfered ones.

4.2.10 HORIZONTAL AND VERTICAL YOKE ARRANGEMENTS

Some yokes shown in Figure 4.20 are horizontal (planar) and some have vertical parts, which enclose the sample partially or completely. The yoke construction can be an important parameter if free access to the sample is required. This is the case for example with domain observation techniques, which require access to the surface of the sample (Figure 2.50).

It should be noted that the round yoke (Figure 4.20g) is quite robust and allows for such access, without introducing any asymmetry to the magnetic excitation. On the other hand, the C-shaped yokes (Figure 4.20c) give very good access to the top of the sample (McQuade et al., 1995), but there is a certain magnetic asymmetry, which is difficult to assess for a general case. This can have unquantified influence on the measurement errors and therefore it should be avoided. Similar reservations should apply to yoke types shown in Figure 4.20e and f, which all have certain degree of asymmetry, either for each double-C yoke in their own axis, or for the fact that there is only one 'C' yoke in each of the orthogonal axes.

However, if mechanical measurements need to be carried out at the same time, then neither the horizontal nor the round yokes are very suitable, with perhaps the exception of that shown in Figure 4.20i. There must be an allowance for transferring the mechanical forces to the sample, and this is easier to achieve with vertical yokes and cross-shaped samples, or similar (see also Figure 2.60).

PRACTICAL COMMENT

Figure 4.25 shows an example of FEM modelling of C-like yokes (compare with Figure 4.20f), in which the sample is positioned either symmetrically or asymmetrically in the yoke. The symmetrical yoke always produces symmetrical H around the sample surface (position zero means centre of the sample, which was set to 0.3 mm thickness). In all the cases, there is an H gradient, which has to be taken into account for precise measurement. However, at least the gradient is symmetrical and measuring the H above or below the sample would produce similar results.

This is not the case for asymmetrical yoke, which produces asymmetrical H gradient. Worse still, the gradient can even *reverse* between various excitation levels (Figure 4.25d). Therefore, measurement of H on either side of the sample surface can produce different results.

Qualitatively, similar behaviour will occur for any asymmetrical yoke (e.g. Figure 4.20c), but the exact asymmetry level will depend on the distribution

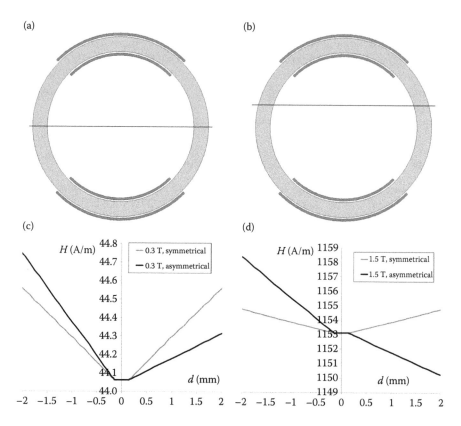

FIGURE 4.25 Symmetrical (a) and asymmetrical yoke (b) can cause changing H asymmetry at various excitation levels (c and d), as plotted through the central cross section of the sample.

of reluctances around the yoke and the sample. For quantification of such results, either the H gradient must be measured in the actual setup or at least a 3D FEM analysis would have to be carried out.

4.2.11   VERTICAL SHIELDING

One of the main tasks of the yoke is to provide as good excitation uniformity as possible. As discussed in the previous sections, the outcome can be improved by using multi-lamination samples, or samples with horizontal shielding. However, such shielding is not necessarily universally applicable, and under certain conditions, it can introduce *greater non-uniformity* than if the sample was left unshielded (Jesenik et al., 2002; Wanjiku et al., 2015b; Zurek, 2015a).

However, the shielding can be applied in a different way. The author carried out 2D and 3D FEM simulations of a shield (Zurek, 2017f), which also has vertically oriented parts. The concept is illustrated in Figure 4.26.

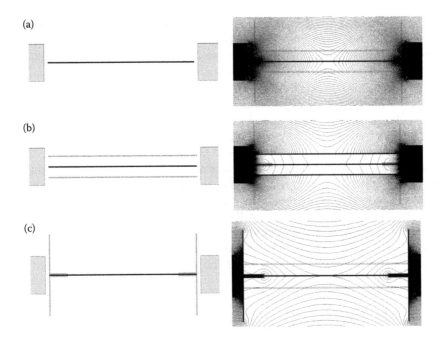

FIGURE 4.26 Shielding in horizontal yokes, the cross-sectional view and results of 2D FEM simulations: (a) just sample – no shielding, (b) two horizontal shields on each side of the sample and (c) 'vertical' shields.

The 'vertical' shield comprises a number of smaller parts (Figure 4.27), separated from each other and from the poles of the yoke by some small distance (e.g. 1 mm). Their horizontal parts are positioned close to the surface of the sample (e.g. 0.1 mm). Such arrangement results in reduced reluctance between the sample and the shield which allows for easier excitation, but larger reluctance between the shield and the yoke which improves uniformity of magnetisation.

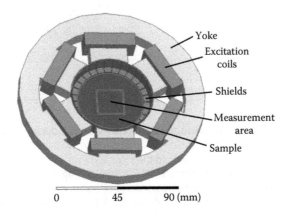

FIGURE 4.27 Vertical shielding used with three-phase horizontal yoke.

The vertical parts of the shield can be taller than the thickness of the poles of the yoke, which are *not* chamfered. The balance of all reluctances is such that the magnetic field lines prefer to enter the shield first, and then are transferred more effectively directly into the sample, rather than through the surrounding air (Figure 4.26c).

The shield must consist of a number of smaller parts, in a similar way as the slots are not closed up in the stator core of a motor. Otherwise, if there was a continuous ring, the flux could be guided around the cylinder, rather than into the sample. This also helps in maintaining a more controlled balance of reluctance when the field is rotating. Each small part could be made as a rectangle cut from a single lamination of electrical steel, and then bent at a right angle. The FEM simulations show that the pieces can be as thin as the sheet of the sample. If such pieces are added on the top and bottom of the sample, then they will not saturate before the sample, because they will always have the effective thickness twice as large as the sample.

The shield appears to be very effective at reducing the normal component of H (Figure 4.28a). The overall reluctance of the sample–shield–yoke is reduced, and

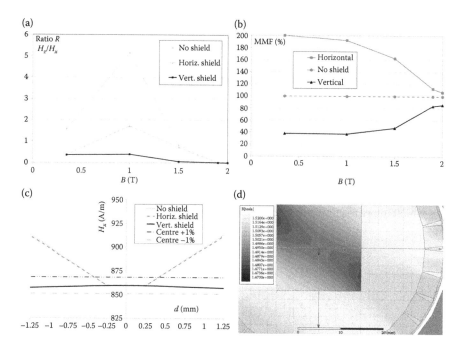

FIGURE 4.28 FEM-simulated performance of the 'vertical' shield: (a) reduction of the normal component of the magnetic field (H_z) in relation to the tangential component (H_x), (b) required *MMF* for exciting the sample to the same level for various shield configurations (100% is referenced to the value required for the corresponding B value without the shield), (c) reduction of the gradient of tangential component (H_x) above the surface of the sample at 1.5 T and (d) typical results of 3D FEM simulation at 1.5 T (the square in the centre represents area of 20 · 20 mm, with the rest of the circular sample hidden; the scale was set to show that the uniformity with the area of interest is better than ±1%).

significantly less power is required to drive the sample to similar level of B, for the same size of the sample and the same yoke.

It should be noted that the addition of horizontal shields (or multi-layer sample) produces a setup which always requires *higher* magnetomotive force for the same B as compared to the unshielded sample (Figure 4.28b). This is because some of the flux is shunted away from the sample, whereas in the 'vertical' shield, more flux is injected into the sample and thus *less* power is required.

The consequence of using 'vertical' shielding is that the gradient of the tangential H above the sample is significantly reduced (Figure 4.28c). For an unshielded sample, the gradient can be so high that a correct measurement of tangential H might be affected to a large degree, and worse still is very dependent on the yoke configuration, pole chamfering, etc. (see Figure 4.24). In order to overcome this problem, an extrapolation technique would have to be used (Figure 3.48).

The simulations show that 'vertical' shield greatly reduces the H gradient so that an H-coil as thick as 1 mm would be affected by less than 1% error *without extrapolation* (Figure 4.28c), therefore reducing or even eliminating the need for extrapolation to the surface. If this is confirmed experimentally, it would offer significant improvement of measurement accuracy, as compared to other configurations – compare for instance with Figure 4.24.

The performance of such shield can be further improved if its shape conforms better to the magnetic poles of the yoke – detailed description is presented in Zurek (2017f).

4.2.12 Core Materials and Construction

Magnetising yokes can be constructed from a number of suitable core materials. These are typically grain-oriented or non-oriented electrical steels, depending on the yoke configuration.

For instance, as shown in Figure 4.24, the poles of a planar yoke can be laminated in a number of ways. For such yoke, each of its part can be made from grain-oriented electrical steels, because the whole yoke is practically exposed only to an alternating, rather than rotational, magnetisation. An example of such yoke is shown in Figure 6.3, in which all the parts have laminations positioned vertically (as in parts B and D in Figure 4.24).

The pole pieces are made as stacks of straight laminations, and the four 'knee' parts are made as a wound core resembling a square shape, which is later cut into four pieces, as shown in Figure 4.29. All the parts of the yoke shown in Figure 6.3 were made from 0.27-mm-thick grain-oriented electrical steels.

Similar technique can be used for many yokes from Figure 4.20 – the parts can be made either as a stack of straight laminations or as wound cores, which are later cut in an appropriate way. This applies also to the system from Figure 4.20k in which each small cylinder is built from a stack of 56 laminations of M-15 gauge 29 (0.356-mm-thick) non-oriented electrical steels (Alatawneh et al., 2012a).

The laminations within each stack must be held together by some means, and this can be achieved by usual techniques of core impregnation used in industrial production of wound cores. The individual pieces of the yoke must also be held together,

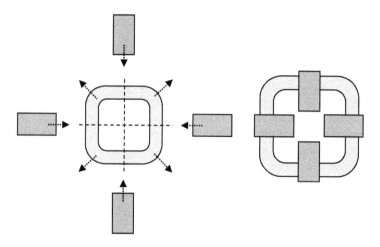

FIGURE 4.29   Construction of a horizontal or planar yoke for a square sample.

for instance by placing them on a suitable support and holding down by the top plate. Some examples are shown in Chapter 6.

A round yoke can be made inexpensively from a stator of an electric motor. For instance, the yoke shown in Figures 3.9 and 6.2 was made from a single-phase induction motor, whose bore diameter was 84 mm. Such inner diameter was suitable for using an 80 mm circular sample, which left 2 mm of air gap between the edge of the sample and the stator. The 2 mm gap represented additional magnetic reluctance, but it was helpful in improving the uniformity and minimising the space harmonics.

Stators of small motors are made from laminations of non-oriented electrical steels. The lamination stack has to be taken out from its typical aluminium or cast steel housing. The required depth of the yoke (the height of the lamination stack) was significantly smaller than the height of the motor stack, so the excess laminations had to be removed. The outside was welded in several places in order to ensure mechanical rigidity of the yoke. The vertical welds are visible in Figure 6.2b. Because these welds are on the outside of the core, they do not represent short circuits enclosing any significant magnetic flux; therefore, the potential eddy currents are negligible as proved in practical operation (Zurek, 2005). A stator of a motor (200 mm inner diameter) was also used for the magnetising setup shown in Figure 6.6. Similar solution was also employed in the yokes shown in Figure 6.9. Stators made from laminations cut especially to the given shape are shown in Figure 6.7.

The magnetic core of any yoke must be designed so that it leaves sufficient space for the magnetising windings and electrical insulation. The volume occupied by the windings can be comparatively quite large, as is dictated by the significant reluctance of all the active air gaps and the sample.

More information about the calculation of required space for the windings is given in the following sections.

4.3 EXCITATION

Excitation must be applied to the sample before a measurement can take place. Permanent magnets can be used for this purpose, but the most versatile method is a winding (or set of windings) driven by a suitably controlled voltage or current source(s).

4.3.1 PERMANENT MAGNET

In some cases, the magnetisation can be applied by means of permanent magnets (see Figure 2.3). The main advantage of permanent magnets is that they do not require any active power, there are no problems with overheating of the 'power source' and the attainable magnetic field strengths are sufficient even for very high-excitation levels, approaching saturation of the sample.

However, there are a number of drawbacks of using magnets, which is the primary reason why they are rather rarely used for rotational studies.

In order to perform rotational magnetisation, there must be a relative rotation of the magnetic field and the sample under test. Therefore, either the yoke with the magnets or the sample has to physically rotate, which significantly increases mechanical complexity of the apparatus.

Also, it is not possible to easily adjust the 'strength' of a given magnet which requires additional features such as regulation of air gap between the sample and the yoke, or within the yoke itself. Active feedback control of resulting B or H is not quite possible either, or at least there are no publications showing such systems.

PRACTICAL COMMENT

Last but not least, there could be quite prosaic problems with inserting the sample into the magnetising yoke. An accidental touching of the sample to the magnet would strongly magnetise the sample, and the setup would not allow for easy demagnetisation. The system would have to be designed such that the magnets could be short-circuited magnetically so that the sample could be put safely in its position before the magnetising field was applied.

4.3.2 MAGNETISING WINDINGS

Most of the difficulties of using permanent magnets described above can be avoided by using electromagnetic windings or coils. The currents in the coils can be driven in an almost arbitrary way (frequency, waveshape, etc.), and the amplitudes can be fairly easily controlled so that the excitation level and demagnetisation can be applied with great flexibility.

The biggest challenge with using magnetising coils is the measurements performed at very high-flux density, close to saturation. Such measurements require very high amplitudes of magnetising currents, which are problematic for at least two reasons: voltage amplitude required to drive the winding and the power loss in the windings.

4.3.2.1 Voltage Amplitude

Any coil of wire has some inductance associated with it. The inductance is increased by the presence of the magnetic core or yoke. There must be adequate voltage applied to such coil in order to drive the current sufficient for obtaining the required B or H for a given measurement. This is especially a problem for measurements at higher frequencies, because the impedance of the winding increases with frequency. Some researchers resort to using especially built 'high-frequency' systems with reduced number of turns so that the required drive voltage remains within a manageable range for a given power amplifier (Mthombeni et al., 2007; de la Barriere et al., 2015).

PRACTICAL COMMENT

For example, let us assume that the magnetising yoke is round, similar to that shown in Figure 4.20g. The inner diameter of the yoke is $l = 84$ mm and there are 24 slots in the core. The 'depth' of the yoke (thickness of the lamination stack) is $d = 40$ mm.

At the time instant when only one phase is powered, the magnetic circuit can be coarsely approximated with a flat-face coil-and-yoke system, as shown in Figure 4.30. For the worst case, it can be assumed that the inner diameter is equal to the distance between the pole faces. The worst-case cross-sectional area would be the length of the air gap times the depth of the yoke, thus $A = l \cdot d = 84$ mm $\cdot$ 40 mm $= 3360$ mm^2.

The reluctance of the yoke is negligible as compared with the reluctance of the air gap; therefore, the inductance of such winding will depend directly on the length l of the 'air gap'.

Let us also assume that the required peak $H = 23$ kA/m in order to achieve peak $B = 1.9$ T (see also Figure 2.27d). Hence, the required magnetomotive force (Equation 1.4) is $N \cdot I = H \cdot l = 23$ kA/m $\cdot$ 0.084 m $= 1932$ ampere-turns. The winding is located in 6 slots (Figure 4.30), so each part of the coil will have to generate 322 ampere-turns.

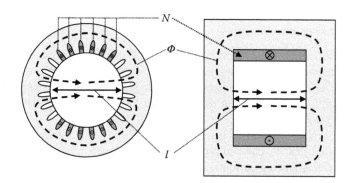

FIGURE 4.30 Round yoke can be coarsely approximated with a flat-pole system.

The size of each slot is such that it would be possible to use 30 turns of 0.56 mm enamelled wire (180 turns in total). Hence, the required magnetising peak current would be 10.7 A, and for mostly sinusoidal current around 7.6 A RMS.

Neglecting the presence of the sample and the reluctance of the yoke, inductance of such winding can be very coarsely estimated from the following equation:

$$L \approx \frac{N^2 \cdot \mu_r \cdot \mu_0 \cdot A}{l} \approx N^2 \cdot \mu_0 \cdot d \quad \text{(H)} \tag{4.3}$$

where N is the number of turns (unitless), μ_r is the relative permeability (unitless), μ_0 is the absolute permeability of free space (H/m), A is the cross-sectional area of the volume between the magnetic poles (m²), l is the distance between the poles (m) as shown in Figure 4.30 and d is the thickness of the yoke (m).

Assuming that the influence of the sample is negligible ($\mu_r = 1$) and substituting the values listed above, the calculated inductance is $L = 1.63$ mH.

However, the coarse calculations presented above do not take into account the so-called flux fringing, which would be present also outside of the main volume of interest. Such leaking flux increases the inductance of the main winding and thus higher required voltage to achieve the same magnetising current. One method of coarse estimation of the amount of fringing flux is to add the length of the gap (distance between the poles) to each dimension of the cross-sectional area (Hurley et al., 2013) so that the flux fringing factor (FFF) can be calculated:

$$FFF \approx \frac{(width + gap) \cdot (thickness + gap)}{width \cdot thickness} \quad \text{(unitless)} \tag{4.4}$$

Equation 4.4 is empirical and very imprecise, but we can use it here for order-of-magnitude estimations. With the values listed above the calculated $FFF \approx 6.2$, so the corrected inductance value would be $L \cdot FFF \approx 10.0$ mH.

Ignoring the resistance of the coil, which can only increase the impedance, and at 50 Hz the impedance of such winding would be $X_L = 2 \cdot \pi \cdot f \cdot L = 3.14 \; \Omega$, so a current of 7.6 A RMS would require voltage of around 24 V.

However, if the same yoke is to be used at 500 Hz to produce the same H of 23 kA/m, then the required voltage would also scale with the frequency, thus rising to 240 V. This example illustrates why sometimes specially constructed yokes have to be used in order to perform high-frequency measurements.

Voltages higher than 24 V can be very dangerous and should be avoided for safety reasons. More details on safety are given in Section 4.4.1.9.

For a yoke with given magnetic core, the required voltage can be lowered by using different winding but *only* at the expense of the increase in magnetising current, because the value of ampere-turns must be preserved in order to reach the required magnetisation level. This results directly from the amount of energy which needs to be stored in magnetic field within the volume inside of the yoke. Isolation transformers can be used for impedance matching, if necessary (see Section 4.3.4.1).

However, for the hypothetical values of 1.9 T and 23 kA/m, the corresponding relative permeability is around $\mu_r = 66$, which is still significantly higher than air, and the influence of the sample cannot be completely neglected. Nevertheless, the example discussed above is sufficient for a rough estimation of the minimum apparent power required to attain the target magnetic excitation. The power amplifiers must be thus chosen accordingly, with a practical sufficient overhead.

For a more accurate estimation, it is advised to use at least 2D FEM, which for the case without the sample will be significantly better than the approximated analytical calculations. There is a freely available user-friendly FEM package capable of such calculations called 'FEMM' (Meeker, 2015). Of course, a full 3D FEM simulation should result in even more precise values, and it is recommended before committing to the manufacturing of the yoke.

An optimisation procedure was discussed for instance in de la Barriere et al. (2015). A stator-like yoke for 80 mm sample magnetised at 1.5 T at 1 kHz required 127 V and 10 A, or 1.3 kVA per channel, which is a considerable amount of apparent power, with potentially dangerous level of voltage.

4.3.2.2   Power Loss

The wire of the magnetising winding has some resistance, and current flowing in such coil will create heat due to resistive loss. This loss is proportional to the square of the RMS current and becomes a problem at very high-excitation levels. The windings can get very hot, which can even cause their failure due to overheating. They can also elevate the surrounding temperature, which can directly influence the performance of the sample, whose resistivity (hence also the rotational loss to be measured) can change with temperature. An increase in temperature of a magnetising winding is visible for instance in Figure 3.61b.

This problem can be reduced by appropriate design of the magnetising coils so that their resistance is low enough not to cause dangerous power dissipation.

The temperature can also be lowered by forced cooling. (There is a fan visible at the bottom of the assembly in Figure 6.7.)

For the same reason, the magnetising coils need to be placed outside of the vacuum chamber so that the sample can be placed under conditions better approximating the adiabatic state. Otherwise, the heat could be passed from the windings (see also Figure 3.61b) to the sample for instance through radiation and thus falsify the results of the thermal method.

The resistance of the winding can be lowered simply by keeping the same number of turns but employing thicker wire. Many yoke designs from Figure 4.20 can accommodate increased volume for the magnetising coils.

However, the stator-like yokes (Figure 4.20g–i) are the least flexible. The coil space is pre-defined by the slots cut in the stack of laminations. This can be even more problematic if the yoke is made by reusing a stator of an existing motor – because it is not really possible to change the design.

The loss dissipation in the coils is an important practical factor and should be investigated at least through preliminary calculations before making the windings so that the safe operating range of the windings is identified.

PRACTICAL COMMENT

As with the inductance, the minimum resistance of the windings can be roughly estimated. It is possible to calculate the shortest length of a single turn in the given winding space, by using the dimensions of the yoke.

For instance, let us assume that the same round magnetising yoke is used as in the example above. The inner diameter of the yoke is 84 mm; hence, its inner perimeter is 264 mm. The winding is made from two phases so that the shortest turn which can be wound will have to straddle at least 6 slots (see Figures 4.30 and 4.33), which is a quarter of the inner perimeter or at least 66 mm. Therefore, as a minimum, the shortest length of a single turn is proportional to that distance and the thickness of the yoke (66 mm + 40 mm) · 2 = 212 mm. There would be 180 turns, so the minimum wire length for a single winding would be 38 m.

At room temperature, 0.56 mm enamel copper wire has nominal resistance of 0.0694 Ω/m (Pro-Power, 2015), which gives a minimum winding resistance of 2.64 Ω. Therefore, a current of 7.6 A RMS driven through such resistance would produce a significant real power loss of 152 W, which can be applied to the windings only for a very short time – otherwise the windings would overheat.

Volume of the wire can be calculated as its cross-sectional area 0.246 mm^2 multiplied by the length of 38 m, which gives 9348 mm^3. Specific density

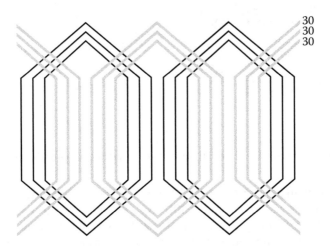

30
30
30

FIGURE 4.31 Two-phase 'simple' winding for a 24 slot stator; each slot contains 30 turns of the same phase; this winding was used for the research presented in Zurek (2005).

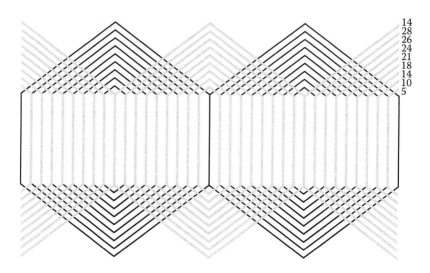

FIGURE 4.32 Two-phase 'sinusoidal' winding for a 36 slot stator; most slots contain some turns from both phases. (Adapted from Gorican V. et al., 2-D measurements of magnetic properties using a round RSST, *Proceedings of 6th 1&2DM Workshop*, Bad Gastein, Austria, 2000a, p. 66.)

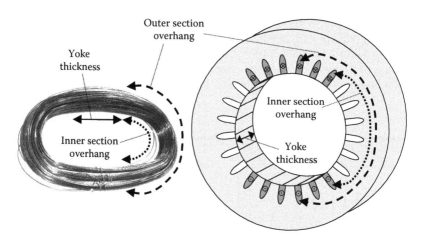

FIGURE 4.33 Three-section coil (left) ready to be inserted into stator-like yoke (right).

of copper at room temperature is 8960 kg/m³, so the minimum weight is 0.084 kg. Specific heat of copper is 390 J/(kg · K). Therefore, it would take around 1320 J to warm up the winding by 40 K above the ambient (a typical practical 'safe' temperature increase for copper windings). Assuming the worst case of no exchange of heat with the surrounding, by applying 152 W it would take less than 10 s to reach such a temperature increase. Actual performance close to such values was experienced by the author for the yoke shown in Figure 6.2, so the methodology described above gives reasonably good approximation at an order-of-magnitude level.

With such yoke, the high-excitation measurements had to be limited to short intervals at the highest amplitudes of current. Forced cooling could be applied, but it was most useful for faster cooling down *after* the power is switched off, simply because it was not powerful enough.

The overheating would be a much smaller problem for lower excitation. For example, at 1.6 T, the required H would be around 9 kA/m (Figure 2.27d). Assuming approximately constant yoke reluctance, the required current would reduce to 3 A (as compared to the 1.9-T case estimated above) and the power loss in the windings to 23 W. This is almost an order of magnitude less so it is much easier to handle thermally.

Restive losses can be reduced by using thicker wire – but *only* if the design of the yoke allows accommodation of proportionally thicker coils.

4.3.2.3 Coil Winding

If a magnetising coil does not have too many turns, it can be wound manually, by directly guiding the wire onto a given bobbin. However, it is difficult to keep appropriate tension on the wire and its positioning. As a result, the coil might be wound too loosely, and fewer turns can fit into the available space. But in many cases, it is not possible winding the coil in any other way, for instance because of the shape of the yoke, for which it is difficult to use bobbin (e.g. Figure 4.20f).

If a bobbin can be used, then it is usually possible to fit it onto a linear winding machine (Figure 4.18) and many turns can be wound with precision and good packing into the available volume.

Probably, the most difficult cases are the windings for round stator-like yokes (Figure 4.20g and i). There are commercial machines capable of making such windings automatically, but their use cannot be normally justified for winding a single prototype. Such windings can be made with a use of a special tool described below.

The stator-like yokes can be wound with a 'simple' winding, as shown in Figure 4.31 (Zurek, 2005). In a 'simple' two-phase winding, each slot has the same number of turns, and the winding sections for phase 'X' occupy exactly half of the number of slots, with the other half of the slots occupied by the 'Y' winding (Figure 4.31).

The winding can also be made as so-called 'sinusoidal'. The number of turns in a given phase is varied in each slot, and most slots contain sections from both phases (Figure 4.32).

Such distribution appears to give better excitation uniformity, especially for samples of non-oriented electrical steels (Gorican et al., 2000a; Wanjiku et al., 2014). The distribution of the number of turns in each section can differ between yokes and thus can be optimised (Gorican et al., 2000a; Kuczmann, 2008). Many other variations are possible, including some intermediate forms (Lancarotte et al., 2004), and also three-phase versions are equally feasible (Radley et al., 1981; Fiorillo et al., 1992a; de la Barriere et al., 2015).

PRACTICAL COMMENT

A special collapsible oval bobbin comprising several sections makes it easier to wind multi-section windings. Each section should be appropriately larger than the previous, as shown in Figure 4.33.

The dimensions should be calculated such that the length of the longer axis is appropriately longer than the thickness of the yoke. After insertion the straight part of the sections should extend somewhat above the surface of the yoke.

The length of the 'overhang' of the windings should take into account the fact that there are slots for the other part of the winding, so they should also be longer by a sufficient amount.

A better view of the final shape of such winding is shown in Figure 6.2. It should be noted that the slots must be insulated with foil or paper insulation before the winding is inserted. Also foil/paper insulators are inserted between the two phases. This is important if voltages exceeding 50 V will be applied to the yoke so that there is no voltage breakdown between the two (or three) phases.

4.3.3 Power Amplifier

Magnetising systems with active waveshape control require the use of power amplifiers. The voltage signals are generated by analogue or digital feedback circuits (described in the following sections), and they must be amplified in order to deliver the necessary apparent power to the magnetising coils.

4.3.3.1 Linear Power Amplifier

The most popular type of a power amplifier has a linear transfer characteristics within a given frequency spectrum. This means that the voltage input signal with an arbitrary shape is reproduced at the power output with very good amplitude linearity and fidelity of the spectrum.

A very wide choice of power amplifiers is available from the audio market. Such amplifiers offer very good linearity and fidelity of the amplified signals. For this reason, the nominal load impedance is usually at the order of 4 Ω (and generally between 2 and 8 Ω), which is a typical impedance for loudspeakers driven directly by such amplifiers. The choice of nominal power is very wide, from tens of VA to even tens of kVA rating, requiring three-phase power supply and build from several modules (Brockhaus Measurements, 2016; Crown, 2016).

For instance, a typical power amplifier Macro-Tech 3600VZ shown in Figure 4.34 has a nominal frequency range from 20 Hz to 20 kHz, total harmonic distortion (THD) smaller than 0.1% and nominal load impedance 2–8 Ω, up to 40 A or 100 V RMS, depending on load conditions. The weight of such device is substantial, around 24 kg (Crown, 2000).

Most commercially available audio amplifiers are built to have two independent signal amplification channels (Figure 4.34). Such option lends itself very well for using a two-phase magnetising yoke. A three-phase yoke (Figure 4.20d) requires

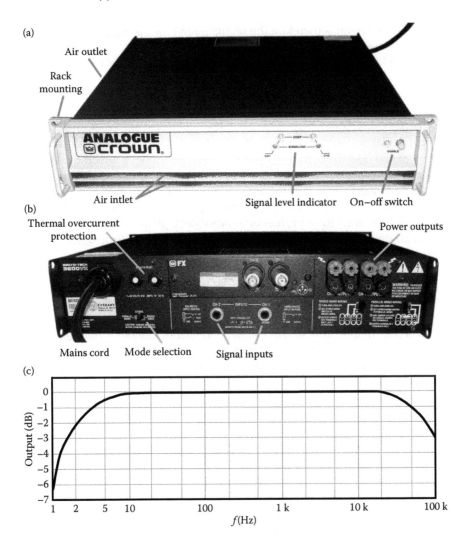

FIGURE 4.34 Power amplifier Macro-Tech 3600VZ (equipment courtesy of Megger Instruments Ltd, the United Kingdom); cooling air is taken through the vents at the front of the amplifier and expelled by the slots on the sides: (a) front, (b) back and (c) typical frequency response. (Adapted from Crown, *Macro-Tech MA-3600VZ, Power amplifier, Operation manual*, 2007.)

three independent channels, and therefore at least two power amplifiers, which immediately doubles the cost, if the same power-per-channel is to be provided.

PRACTICAL COMMENT

In most countries in Europe, the power supply is 230 V. In the United Kingdom, the mains sockets are rated at a maximum of 13 A. This means that the maximum power which can be taken from one single-phase socket is around 3 kVA, which is therefore the ultimate limit for the input power of a single-phase supply power amplifier.

It should be noted that the efficiency of a typical AB class amplifier is around 65% (Maxim Integrated, 2016), so the power available at the output will be only 1.9 kVA. This power has to be split into two channels, which makes it below 1 kVA per channel. This is common limit for many amplifiers.

The calculations shown above for the apparent power required for yoke at 500 Hz would be 240 V and 7.6 A, or 1824 VA per channel, which is obviously beyond what a single-phase amplifier could supply, even with a voltage step-up or impedance-matching transformer.

Any higher power would require a three-phase power supply, as it is the case for instance for a professional laboratory power amplifier PA-100, which can deliver up to 4 kVA or nominally 37 A, 100 V (Brockhaus Measurements, 2016).

It is very important to study the technical specification of a given power amplifier in details. For instance, the amplifier Macro-Tech 3600VZ has a built-in automatic 'mute' for low-frequency signals. Theoretically, the frequency spectrum can extend to 1 Hz and below, but the circuitry might be automatically 'muting' any signals below 1 Hz. Such 'muting' is different from just attenuation through the normal frequency spectrum in the sense that the very low frequencies can be actually actively suppressed rather than passively attenuated (Crown, 2000).

It should also be considered whether an amplifier with or without gain control is more suitable for a given application. Most power amplifiers, especially audio-related, offer gain control (in a similar sense as 'volume control' on a stereo set, see e.g. Figure 6.9). The advantage of such solution is that the output power can be limited for smaller, and hence more delicate, systems.

However, the disadvantage is that the operator must remember to dial the appropriate level of gain, which is usually controlled by a mechanical knob ('volume control'). If high gain is set by mistake, then damage to a given magnetising winding could occur if too high current is delivered unintentionally. A second potential problem is that the varying gain can upset the feedback circuit, which is then required to be stable over much greater dynamic range, which is not a trivial problem for an experimental setup.

In the author's experience and personal opinion, it is generally less troublesome to use an amplifier with a *fixed gain*, for example as shown in Figure 4.34, because then any guessing is eliminated.

Another practical problem with power amplifiers is their transfer function at low frequencies (Figure 4.34c), which is especially important for AC-coupled audio power amplifiers. The frequency response droops towards low frequencies, which means that the efficiency of signal amplification also reduces significantly. An attempt to drive high current for example at 1 Hz will dissipate significant power inside the amplifier, and in the extreme case, it can damage it in a catastrophic way. This problem is exacerbated by the fact that the impedance of the magnetising winding reduces significantly with frequency, so that the total impedance of the coil could reduce below the lowest allowed load impedance (e.g. below the typical nominal 4 Ω). The amplifier will then effectively drive an electrical short circuit with even more internal losses.

If the amplifier is not appropriately thermally protected, this will lead to its destruction (beyond repair), as it is the unfortunate *multiple* experience of the author. Modern amplifiers might have a special system of 'smart' thermal protection which lowers the output power without critically overheating the power stage of the amplifier and without sudden disconnection of the output current. This can be seen as the LEDs in Figure 4.34a marked as 'ODEP' (*Output Device Emulator Protection*), whose brightness indicates the level of 'thermal margin' during operation of the amplifier.

It should also be noted that a circuit breaker of any kind should be avoided, or it should be used only as a last resort. A sudden disconnection of high current from an inductive load can generate voltages dangerous or even *lethal* to the operator and to all the auxiliary equipment, due to the so-called 'fly-back voltage'. During sudden disconnection of inductive current, a voltage impulse reaching *thousands of volts can be easily generated*, which would almost certainly seriously damage any measuring equipment attached to it. So for instance, an input of a data acquisition device measuring voltage across a shunt resistor could be catastrophically damaged, together with the computer to which it is connected (see Figure 3.6c). This was also an unfortunate experience of the author, when after such sudden disconnection of high current one of the voltage inputs of the data acquisition device connected to the PCI bus was completely damaged. The computer survived.

In any case, it is a good idea to install some sort of overvoltage protection of all measurement channels, provided that leakage currents would not introduce unwanted measurement errors.

More discussion on safety implications is given in Section 4.4.1.9.

4.3.3.2 Digitally Controlled Power Amplifier

Audio power amplifiers are designed to amplify analogue signals. However, there are also *digitally controlled power amplifiers* also referred to as *programmable power supplies*. An example of such device is shown in Figure 3.4 (item 4).

Such amplifier can be thought of as a combination of a digital-to-analogue signal converter and a power amplifier, and it is usually possible to use it as an ordinary power amplifier, with an analogue input as well (Kepco, 2015).

The digital signal to be 'amplified' is communicated through a digital protocol such as Ethernet, GPIB, serial port or USB with newer designs. The input resolution is usually from 10 to 16 bits. The output is DC coupled, with the maximum parameters up to (depending on model) 100 V, 20 A, 400 W (Kepco, 2015). There are also versions (e.g. Kepco KLN series) which can deliver up to 3000 W (Kepco, 2016).

Kepco digital amplifiers were used for instance in experiments described in Grossinger et al. (2000) and Lancarotte et al. (2004).

4.3.3.3 Voltage-Mode and Current-Mode Amplifiers

A power amplifier can be designed and constructed to operate as a *voltage-mode amplifier* or *current-mode amplifier*. The difference between the two modes is the way the internal feedback operates.

A *voltage-mode amplifier* will strive to generate at its output voltage waveform, which is an exact copy of the voltage waveform supplied to its input. For audio amplifiers, which typically operate in the voltage mode, the fidelity can be at the level of 0.1% THD (see also Equation 4.13 in section 4.6).

Therefore, even if the circuit under test is highly non-linear, the amplifier will generate the requested voltage at the output, regardless of the impedance of the load – as long as the peak voltage, peak current or the signal bandwidth do not exceed the rating of the amplifier. There is no limitation on maximum load impedance, so voltage-mode power amplifier will happily operate with its output not connected to any load or open circuit, but not into a short circuit (Crown, 2000).

For a *current-mode amplifier*, the input signal is also supplied as a voltage signal. However, the amplifier is designed to generate an exact replica of its input voltage but in the shape of an output current. The fidelity can be at the level of 1% THD (Amp-Line, 2016).

The current-mode amplifier will operate happily into a short circuit (zero-ohm load). However, it will be unable to operate into open circuit or high impedance, because it will attempt to drive the requested current by increasing the output voltage to infinity. Since the output voltage level is limited (the so-called *compliance voltage*), it is not possible to control the current if the compliance voltage was reached.

With a similar design and rating, both types of amplifiers can deliver the same level of power. Impedance matching (described in the following sections) can be used equally well in both cases. The peak current and peak voltage limitations will be identical if both amplifiers were to drive load of identical impedance.

The main difference is the fact that the internal signal feedback in current amplifiers is derived by different means, usually resulting in lower sensitivity for the current-mode amplifier; hence, THD specification or noise level is worse. Therefore, current-mode amplifiers might produce more noisy output waveforms.

Voltage-mode amplifiers appear to be more commonly available commercially, probably due to the large market for high-power audio applications.

4.3.4 Elimination of DC Bias

Low frequencies (below 20 Hz) usually pose challenging magnetising conditions for power amplifiers. This is especially true for AC coupled power amplifiers, because the transfer characteristic is reduced at lowest frequencies (Figure 4.34c). Impedance of the magnetising yoke also reduces significantly and can easily become smaller than the nominal load value for the power amplifier.

Therefore, for measurement at very low frequencies (e.g. 1 Hz and below), DC-coupled power amplifiers must be used. These are readily available in the market, but a DC-coupled device can introduce an unwanted DC offset in the magnetising windings. If no DC offset is allowed, as it is the case more often than not,

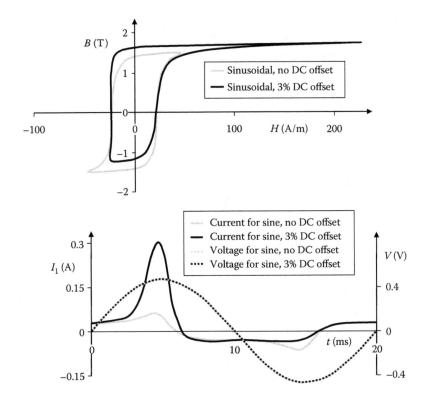

FIGURE 4.35 Typical effect of 3% DC offset in the input voltage. (Adapted from Moses A.J. et al., Effect of DC voltage on AC magnetisation of transformer core steel, *Presented at 13th ISEM*, East Lansing, Michigan, USA, September 2007.)

then some means must be employed for reducing or eliminating it. Otherwise, there will be even harmonics introduced into the waveforms and asymmetry of a *B–H* loop might ensue, even though the output voltage waveform will be controlled to be sinusoidal (Figure 4.35) (Moses et al., 2007). This problem is more pronounced in magnetising system with low reluctance (no air gap), but in the experience of the author even in gapped systems (like most rotational yokes) this can be troublesome enough to require some method of DC offset reduction.

PRACTICAL COMMENT

The problem of DC offset is exacerbated because for DC the impedance of the magnetising winding is just its resistance, which is significantly lower than the impedance at normal measurement frequency. For instance, as estimated in Section 4.3.2.1, the impedance of the winding at 50 Hz might be 3.14 Ω. In order to produce rotational circular magnetisation of 1 T, the required magnetic field strength could be around 30 A/m for easy direction of grain-oriented steel (Figure 2.27b). The required amplitude of AC current for 180 turns and 84 mm magnetic path length would be just 14 mA.

DC offset at the output of a power amplifier Macro-Tech MA3600VZ could be up to 10 mV (Crown, 2000). Such voltage would drive only the DC resistance of the magnetising winding, which in this example would be at the level of 2.6 Ω, so the resulting DC current would be 3.8 mA. It is a relatively small value, but in this case it would reach 27% of the AC amplitude. Such ratio is likely to introduce visible asymmetry in the relevant waveforms, so such DC offset has to be eliminated for correct measurements.

The problem is usually worse for amplifiers rated for higher power, because larger residual DC offset can be present, the value of which might be drifting with temperature (Analogue Associates, 2001). This is because they are designed to handle higher load currents so that the control of small signals might not be as good.

4.3.4.1 Isolating Transformer

A proven method of DC offset elimination is by using *isolating transformer* or *separating transformer* (Figure 4.36) (Lyke, 1982; Qu, 1984; Ishihara et al., 2000; Kondo et al., 2003; Zurek, 2005; Patel, 2009; Owzareck, 2015). The rating of the transformer should be sufficient to allow the highest expected excitation level or the power rating of power amplifier or the power source.

Off-the-shelf isolating transformers with 1:1 ratio are available commercially at a reasonable price. Typical nominal voltage is around 240 V, sometimes with step-up (415 V) and step-down windings (110 V or 55 V). Most of these would be rated to operate at 50/60 Hz.

There are several practical advantages of using isolating transformers. Some of these features are discussed below in this chapter, but to just name a few: galvanic

FIGURE 4.36 Isolating transformer, rated at 500 VA, 50 Hz, weight about 5 kg (see also Figure 4.39). Equipment courtesy of Megger Instruments Ltd, the United Kingdom.

isolation of circuits, elimination of DC bias in the magnetising current, impedance matching and low-pass filtering behaviour.

They also introduce some unwanted aspects: increased non-linearity of the system, phase delay which reduces stability of digital feedback, limited operating range at lower frequencies.

Nevertheless, in the opinion of the author, isolating transformers are extremely useful in magnetising apparatus, and their benefits strongly outweigh the negatives.

PRACTICAL COMMENT

Magnetic cores of transformers are designed to operate at maximum possible B, but without saturation at the nominal voltage and frequency. With decreasing frequency, the impedance of the primary winding decreases and saturation of the transformer core will occur at proportionally lower voltage. Therefore, maximum power rating also decreases proportionally with frequency, because safe current level is dictated by the cross-sectional area of the wire in the windings.

Hence, if the measurements are to be carried out at the frequency reduced from 50 to 1 Hz, then the hypothetical nominal voltage would also reduce by a factor of 50, or from 240 to 4.8 V. An attempt to drive higher voltage into such transformer will saturate it, hence reducing the power available at its output and adversely impacting on the overall non-linearity of the system so that the digital feedback might become unstable.

Fortunately, the apparent and active power required to drive the magnetising yoke also reduces with frequency, in the first approximation by the same factor – which is a direct consequence of Equation 3.3. Measurements even at 0.5 Hz were performed with this method, although with some difficulties for shape control (see Figure 1.17) (Zurek, 2005).

On the other hand, elevated frequencies can also be used through such a transformer. Eddy current loss in the laminations is proportional to the square of frequency, but also to the square of B, as evident from Equation 1.36. But it follows from Equation 3.3 that if voltage amplitude is to be kept constant (i.e. not exceeding the nominal 240 V) then doubling the frequency will mean halving B in the magnetic core. Thus, the eddy current loss will not increase extensively even at elevated frequencies. Hysteresis loss increases linearly with frequency (Figure 1.28), but it is also proportional to $B^{1.6}$, which means that this component of loss will also decrease appropriately, if the voltage amplitude is kept constant, because $(B/C_x)^{1.6}$ reduces faster than $f \cdot C_x$. A similar one applies to the excess loss (Figure 1.30).

The total loss of such transformer will probably increase somewhat with frequency, because equations such as Equation 1.36 are applicable only if the magnetisation through lamination is uniform (negligible skin effect), which will definitely not be the case at kHz frequencies for a 50 Hz transformer. More importantly, the losses in the sample and the rest of the system increase with frequency, so the excitation must be increased to compensate for it.

However, in practice, such transformer can be reliably employed for measurements at elevated frequencies so that a transformer rated at 50 Hz can be used even up to 25 kHz and beyond (Zurek et al., 2005b).

Barkhausen noise measurements could be potentially influenced by the Barkhausen noise produced not by the sample but by the isolating transformer, due to the presence of ferromagnetic core with substantial volume and potentially similar magnetic properties. Therefore, some researchers resorted to the use of an air-cored transformer (i.e. transformer without magnetic core) (Patel, 2009).

Such transformer could be constructed for instance by winding two coils on a long tube. The magnetic coupling is much poorer, and significantly higher apparent power is required to drive the whole setup. Poorer coupling means that the efficiency is also reduced and high-power loss can be expected, which requires forced cooling of the transformer windings for high excitations. For more efficient magnetic coupling, the primary and secondary layers should be interleaved as much as possible (Patel, 2009).

PRACTICAL COMMENT

Such an air-cored transformer creates an additional problem to be solved. Namely, the magnetic field is not confined inside the transformer, due to the complete lack of magnetic core. A winding on a tube is effectively a solenoid and therefore generates magnetic field which radiates over potentially large distances (Figure 1.3).

This radiated field can easily influence sensitive magnetising apparatus if placed in the close vicinity. For this reason, some form of magnetic shielding would have to be placed around the solenoid, for instance in a form of a large drum made from a high-permeability material (Patel, 2009).

However, direct comparison to a performance with ordinary transformer showed negligible differences in the influence on the Barkhausen noise measured on the sample (Patel, 2009). Therefore, it is questionable if transformers with magnetic cores have to be specifically avoided even for such sensitive measurements.

4.3.4.2   Series Capacitor

Barkhausen noise measurements could be potentially influenced by the Barkhausen noise produced by the isolating transformer, due to the presence of ferromagnetic core with substantial volume and potentially similar magnetic properties to those in the investigated sample.

Therefore, some researchers investigated the possibility of using AC coupling capacitor for such purpose (Figure 4.37) instead of the isolating transformer (Patel, 2009).

PRACTICAL COMMENT

There are two major disadvantages of using a coupling capacitor. Firstly, the capacitance will always form a resonating circuit with the inductance of the magnetising yoke. In the worst case, the resonance might fall very close to

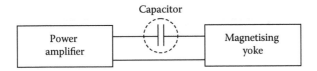

FIGURE 4.37 DC blocking with AC coupling capacitor.

the measurement frequency or its harmonics, which could pose problems in applying the controlled excitation.

The second problem is that significant magnetising current has to be coupled which requires high value of capacitance rated for high AC currents and relatively high voltages (tens of volts). Such approach is therefore costly and potentially dangerous due to the amount of energy which can be stored in the capacitance.

As with the air-cored transformer, direct comparison to a performance with ordinary transformer showed negligible differences in the influence on the Barkhausen noise measured on the sample (Patel, 2009).

4.3.4.3 Active DC Compensation

A typical magnetising regime is to generate sinusoidal B in the sample. In the non-linear region of the sample properties, the magnetising current will be heavily distorted, but it should be completely symmetrical if DC offset is not present.

The situation changes if DC offset is present. However, even though the currents can be severely distorted, the flux density (and thus the induced voltage) can be still controlled to be perfectly sinusoidal (Figure 4.35).

This is a direct consequence of Faraday's law or Equations 3.1 and 3.2. The induced voltage is proportional to the *derivative* of the magnetic flux. Therefore, it is not possible to sense the DC component of B in the output voltage, because the derivative of DC is zero by definition.

Some information about the level of DC offset is transferred through the signal chain, because the output voltage will contain even harmonics due to asymmetry. However, the feedback algorithm will suppress such harmonics so that the output signal will be purely sinusoidal (within the precision of the feedback). But the DC offset and the even harmonics will still be present in the magnetising current and will not be eliminated. As a result, the real B–H loop can be offset from the origo as shown in Figure 4.35, but it is not possible to know how large this offset is.

It is therefore not possible to control the DC offset directly through the same feedback loop which is used for the waveform shape control, and another feedback loop must be used for DC control based on the asymmetry of the magnetising current (Figure 4.38).

It should be noted that such asymmetry cannot be controlled by using the signal from H-coil or Rogowski–Chattock potentiometer (RCP) because they also rely on signals induced in the coils, so Equations 3.1 and 3.2 still apply. Thus, the actual magnetising current should be monitored (e.g. through a shunt resistor, Figure 3.43), even if it is not used for measuring the magnetic field strength.

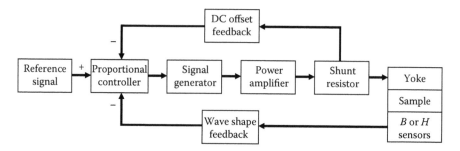

FIGURE 4.38   Block diagram of the two feedback loops.

PRACTICAL COMMENT

There are two feedback loops in Figure 4.38, and each of them must have their own value of gain. The DC offset might not be actually a constant value and might change with the level of load connected to amplifier (level of excitation). Therefore, during continuous generation of the magnetising current, the DC offset feedback must be active all the time and must counteract any arising DC component of the current. However, an adjustment to the already flowing slightly asymmetrical current will immediately influence all the other waveforms, so both feedback loops must work simultaneously to produce the required level of waveshape control as well as elimination of the DC offset. The two feedback loops might compete with each other and a kind of 'oscillation' might result. Therefore, careful operation is required with this approach.

The discussion presented above was aimed at elimination of any DC offsets in the magnetising current. However, some measurements are carried out under magnetising conditions with *intentional* DC offset introduced in the magnetisation (see also Figure 4.35) (Matsuo et al., 2000; Enokizono et al., 2003; Yoshida et al., 2006; Moses et al., 2007; Zhao et al., 2010; Yanase et al., 2011; Hasan et al., 2012; Ishikawa et al., 2012).

Therefore, by definition, the magnetising setup must operate in a DC-coupled mode and the offset must be controlled, by some means, for instance by the double-feedback method as shown in Figure 4.38 (Enokizono et al., 2000) or by applying DC magnetisation through separate windings (Ishikawa et al., 2012). The latter method requires slightly more complicated impedance matching, because the AC signal will couple into the DC coil, but it would have to be blocked from the DC supply.

4.3.5   IMPEDANCE MATCHING

As mentioned above, the nominal load impedance for power amplifiers can be at the order of 4 Ω, and this applies for the whole frequency range of the amplifier (Crown, 2007). But the impedance of the magnetising yoke with the sample can vary significantly, in the first approximation linearly with frequency. Therefore, from 2 Hz to 20 kHz (Figure 4.34c), the load impedance could change through 4 orders of magnitude.

However, the required value of magnetomotive force (ampere-turns) must be delivered at any frequency for the controlled magnetisation to take place. If the reference is taken at 2 Hz, then the voltage will have to increase with frequency. Thus, the requirement for apparent power will grow at least proportionally to the magnetising frequency.

Any power amplifier has a limited peak voltage it can deliver. Therefore, at sufficiently high frequency, the maximum voltage can be reached and therefore no higher excitation can be applied. It should be noted here that power amplifier might be perfectly capable of delivering the amount of apparent *power* required, even at a much higher frequency. But the restriction will come from the limited *peak voltage* it can supply.

On the other hand, for very low frequencies, the impedance of the magnetising yoke decreases, and it might reduce below the nominal value for the given amplifier. From the amplifier viewpoint, the load will become 'short circuit', which makes the operation of the amplifier very inefficient. An attempt to drive large amount of power under such conditions can overheat the internal circuits of the amplifier and even cause an irreparable damage.

Therefore, in order to provide optimum operating condition for the power amplifier, the load should be ideally always equal to its nominal load (e.g. 4 Ω), or greater. This can be achieved by *impedance matching* for instance with the help of the above-mentioned isolation transformer, which has several windings allowing various turn ratios (N_1/N_2). The load impedance is 'reflected' from the secondary to the primary winding with the square of turn ratio $(N_1/N_2)^2$.

If the load represents low impedance, then the isolating transformer can be set to perform a step-down function so that the load impedance is reflected to the primary winding and hence the amplifier as a much higher value.

On the other hand, if the load impedance is very high, it can be 'scaled down' by setting the transformer to a step-up operation, thus improving the amount of power it can be delivered.

For example, let us assume that given transformer windings can be changed between 50 turns and 250 turns, independently for primary and secondary. If the ratio is set as 250:250 then 1:1 coupling is provided and the amplifier is presented practically with the actual impedance of the yoke.

For a setting of 250:50, the ratio is 5:1 and the impedance of the yoke will be increased by a factor of 25. For a setting of 50:250, the ratio is 1:5 and the scaling will decrease the impedance by a factor of 25. Therefore, with just three turn ratios, the impedance range can be matched by a factor of up to 625, which is almost 3 orders of magnitude.

An example of such application for a single-phase supply is shown in Figure 4.39. This auxiliary apparatus contains the isolating transformer with windings configurable as 50 or 250 turns, two switchable shunt resistors (1 and 47 Ω), polarity reversing switch and a digital circuit for automatic detection of the shunt connection. The analogue output provides voltage drop across the active shunt resistor. The power connections are for primary and secondary of the isolating transformer.

For rotational applications, there would need to be as many such apparatus as there are magnetising channels.

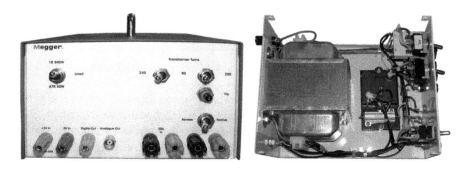

FIGURE 4.39 Auxiliary apparatus containing isolation transformer, two shunt resistors (1 and 47 Ω) with digital detection of their selection, impedance-matching selectors and normal/reverse switch. Equipment courtesy of Megger Instruments Ltd, the United Kingdom.

PRACTICAL COMMENT

The concept of impedance matching can also be stretched to 'power matching'. Under very demanding magnetisation conditions, the power amplifier might be unable to deliver the required excitation, for instance at very high level of flux density in the sample (close to saturation).

Even when the impedance of the magnetising yoke is optimised and the impedance matching is perfect between the yoke and the amplifier, the required level of apparent power might be simply greater than the rating of the amplifier.

For instance, a rotational magnetising yoke optimised for its impedance might require 1.3 kVA per channel (de la Barriere et al., 2015), which is quite a significant amount.

Many modern power amplifiers allow additional configurations of the power stages, with three commonly used modes: stereo, parallel mono and bridge mono (Figure 4.40).

'Stereo' is the default operating mode. Each channel runs independently and amplifies its own input signal. The available voltage and current is 100% of the rated value from each channel.

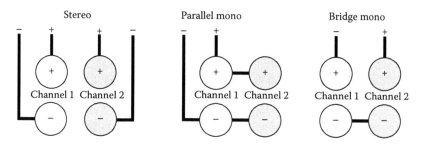

FIGURE 4.40 Example of configurations for two-channel power amplifier Macro-Tech 3600VZ.

'Parallel mono' connects both channels in parallel. Usually, analogue input of channel 1 is the signal to be amplified, with the second input disconnected internally. Because of the parallel connection, the available output voltage is 100% and current is 200%. Therefore, just one powerful channel is created, and another amplifier would be necessary if a two-phase yoke was to be driven.

'Bridge mono' connects the two output channels in series. Again, analogue input of channel 1 is amplified, with the second input being disconnected. The phase of channel 1 is reversed so that the voltages add up. As a result, the available voltage is 200% and the current is 100%.

Apart from connecting the power outputs accordingly, an additional switch must also be selected to choose the right mode (Crown, 2007).

These modes of operation *might not be available* in all audio power amplifiers. It is extremely important that the power amplifiers are connected *only* in the way allowed by the manufacturer. For instance, if the power amplifier had a variable gain, independent for each channel, then an attempt to connect in parallel its two outputs would most likely damage the amplifier, as the two channels would 'fight' each other, probably with all the available power.

For two identical power amplifiers (or two identical power channels), the 'parallel mono' connection can also be achieved by using the isolating transformer (Figure 4.41). Each channel can be connected to a separate primary of the isolating transformer so that they remain galvanically separated. The power is taken from a single secondary winding. The inputs of both amplifiers are joined and thus driven by exactly the same input signal. Such configuration

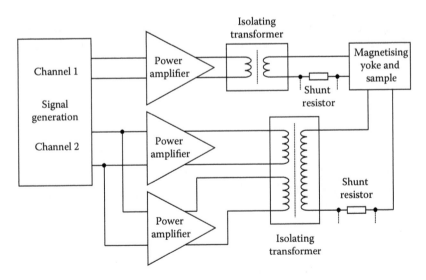

FIGURE 4.41 Means of using two identical power amplifiers to drive one winding by coupling with a three-winding transformer. (Adapted from Zurek S., *Two-Dimensional Magnetisation Problems in Electrical Steels*, PhD thesis, Wolfson Centre for Magnetics Technology, Cardiff University, Cardiff, the United Kingdom, 2005a.)

was used to obtain the data in Figure 2.28 where more power was required for the channel driving the hard magnetisation direction (Zurek, 2005).

Of course, such connection can be used only if *identical* power amplifiers are used, and they are set to the same operating conditions (e.g. the same gain).

Last but not least, the isolating transformer also provides galvanic separation of the ground of power amplifier from the rest of the circuit. This allows for each part (sample, sensors) to behave as truly floating voltage sources. This improves flexibility of connections and makes the measurements not only more accurate (Figure 4.71) but also safer (see Figure 3.43 and Section 4.4.1.9 for more detail).

4.3.6   DEMAGNETISATION

Reproducibility of magnetic measurements is affected by a correct demagnetisation procedure, and for precise measurement, this is paramount (Fiorillo, 2004). Demagnetisation is recommended by several magnetic standards; for instance, IEC (2008) states that '*The test specimen shall then be demagnetized in a decreasing alternating magnetic field of an initial level higher than used in previous measurements*'. Another standard gives the following recommendation (ASTM, 2011): '*The [demagnetising] frequency used should be the same as the test frequency*'.

The amplitude of the alternating (or rotating) excitation should be reduced gradually, over many cycles to ensure that the remaining flux density (remanence B_r) is reduced to a value as close as possible to zero. If reduction of the amplitude is too fast, then some remanence will be present at the end of the procedure, but the actual state cannot be guaranteed.

An example is shown in Figure 4.42. The results were recorded for a wound toroidal Mumetal core (high-permeability material) magnetised at 50 Hz up to 0.5 T

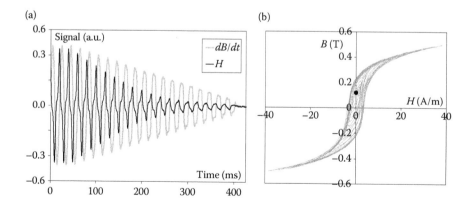

FIGURE 4.42   Typical waveforms during demagnetisation (a) and the corresponding $B-H$ trajectory (b); the waveforms were recorded for a Mumetal wound core, magnetised at 50 Hz and demagnetised within 20 cycles, resulting with incomplete demagnetisation.

(controlled sinusoidal B). The demagnetisation was applied by reducing the amplitude of the magnetising current within 20 cycles. Each cycle has slightly reduced amplitude, and after the last cycle, the magnetising current is switched off.

This particular demagnetisation was done deliberately in order to show the outcome of such less-than-ideal procedure. The waveforms shown in Figure 4.42a were recorded with an oscilloscope, with the 'H' waveform measured as the voltage on the shunt resistor in series with the primary winding. The dB/dt waveform was measured as the voltage induced in the secondary winding. As can be seen, the dB/dt is initially sinusoidal (controlled sinusoidal B) and the 'H' waveform is quite distorted.

The B–H trajectory in Figure 4.42b was calculated from the waveforms shown in Figure 4.42a. As can be seen at the end of the procedure, there is still a significant value of B_r (the black circle) in Figure 4.42b. Therefore, such demagnetisation was not very successful, but it was recorded here on purpose, for better illustration of the issue. Many more cycles would have to be used to ensure that the B_r value is much closer to the origo.

There is no prescribed procedure which can be applied to all materials and can always guarantee perfect demagnetisation. The number of cycles should be chosen experimentally for each setup (type of material, yoke configuration, etc.) – unfortunately mostly by trial-and-error.

The demagnetisation can be carried out by applying DC current with reversing polarity and reducing amplitude, and this is often done in industry for large cores (like transformers) for which nominal frequency procedure would be too difficult due to the required reactive power (Baran et al., 1996; Zurek, 2015c). But if well controlled, then the switched-DC method can be very successful even for precise measurements (Fiorillo, 2004). Therefore, it is not necessary to provide waveshape control during demagnetisation, as long as the gradual decrease in alternating amplitudes is performed with sufficiently small step. Obviously, it is not necessary to record the waveforms during demagnetisation (Lundgren, 1999).

PRACTICAL COMMENT

It is a good idea to design the magnetising system in such a way that the demagnetisation is applied at the end of every experiment. This ensures that the sample is returned to its original state and experimentation consumes less time in the future. Moreover, requirements of both IEC (2008) and ASTM (2011) are met, because the same frequency will be applied as well as the applied excitation will be at least equal to the last measured point (which will be actually within the same experiment if the demagnetisation is applied with magnetising current generated continuously after the last measurement was recorded).

In the experience of the author, this works very well in practice, especially if the excitation is controlled by the computer. The concept is illustrated in Figure 4.43. The measurement was carried out on a toroidal sample of Mumetal, whose power loss was measured up to 0.5 T (above the 'knee' point for this material), at 50 Hz and controlled sinusoidal B. The measurement part finishes as soon as the required level of control is achieved, for instance for

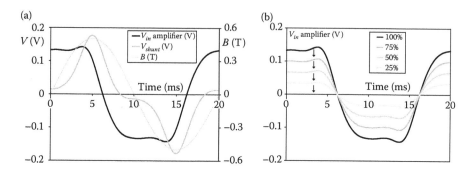

FIGURE 4.43 Typical waveforms recorded during an experiment and demagnetisation: (a) input signal to the power amplifier, voltage across the shunt resistor and sinusoidally controlled flux density B during measurement just before demagnetisation, and (b) input signal to the power amplifier reduced gradually during demagnetisation.

B_{peak}, and distortion errors are reduced below certain level. The data is saved, but the magnetising current is generated in a continuous way, because the appropriate voltage input signal is provided to the amplifier (Figure 4.43a).

The software then switches from the measurement mode into demagnetisation mode, without interrupting the alternating current. There is a variable with the number of cycles for demagnetisation (e.g. 20 steps were used in Figure 4.42).

The amplitude of the amplifier input signal is reduced in a linear way so that at each iteration the remaining waveform is reduced by the given step, by subtracting an exact copy of the shape, but with 1/N amplitude, hence 1/20 (5%) in this case. Therefore, the signal is changed from 100% at the beginning of demagnetisation to 5% at the last step, to 0% at the end. This is also illustrated in Figure 4.43b for just four steps (for better clarity).

It should be noted that the *shape* of the amplifier input signal is *not* changed in order to ensure smooth transition between any two subsequent levels of excitation. The advantage of such approach is that the speed of reduction can be set to a very low number and thus a very good demagnetisation can be obtained.

Nevertheless, there are two drawbacks. Firstly, just before the demagnetisation stage, the initial shape is such that the flux density is sinusoidal. But at the following steps, this is no longer the case, because shape of the magnetising current changes due to the non-linearity of the sample. This can be noticed in Figure 4.42a, especially for the lowest amplitudes. However, as mentioned above, the shape control is not a requirement during demagnetisation.

The second drawback is the fact that the last step does not approach zero asymptotically, but the last step is a change from some amplitude (proportional to 1/N) to zero. However, the size of this final step is dictated by the total number of cycles, so it can be reduced to a very small number. It can be deduced from Figure 4.42b that asymptotic decrease after the final step would not result in any better demagnetisation, because the previous iterations already introduced an asymmetrical trajectory, which would never end at origo, regardless

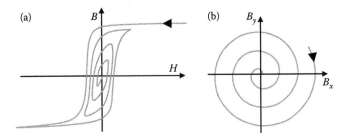

FIGURE 4.44 Simplified illustration of demagnetisation: (a) in a uni-directional system is a spiral on the B–H plane; (b) in a 2D system is a spiral on the B_x–B_y plane.

how the amplitude is reduced after the 1/20 stage. Therefore, the reduction through the previous cycles is more important, and more cycles should be chosen for the demagnetisation procedure.

Another interesting observation can be made from Figure 4.42a. Namely, the amplifier input signal is reduced *linearly*, but the peak current does *not* reduce in a linear way. This is a consequence of the fact that voltage-mode amplifier was used. The *current* produced by the amplifier is related to the load impedance, and this changes owing to varying permeability during demagnetisation. The *voltage* produced by the amplifier is a replica of its input voltage.

In general, the higher the amplitude to which sample was magnetised towards the saturation, the more steps are required for correct demagnetisation. For grain-oriented electrical steels magnetised up to 1.7 T in a closed magnetic circuit (no air gaps), around 200 cycles are usually sufficient, but it would be prudent to use many more steps (even more than 1000) for higher excitations.

The waveforms in Figure 4.43 were recorded for uni-directional magnetisation, for clarity of presentation. The procedure produces a spiral trajectory in the B–H plane, as shown schematically in Figure 4.44a.

The same principle applies for demagnetisation under rotational excitation. The output waveforms are gradually reduced from their maximum values to zero. If the starting point was circular B, then the demagnetisation procedure will create a distorted inward spiral on the B_x–B_y plane (Figure 4.44b), as well as differently shaped spiral on H_x–H_y plane. Obviously, the direction of rotation during rotational demagnetisation will dictate the direction of the spiral (clockwise or anticlockwise).

4.4   SIGNAL CONDITIONING

The topic of signal conditioning is quite extensive and very much dependent on the type of specific magnetisation conditions, magnetic sensors, frequencies, etc. Even the term 'signal conditioning' is not very well defined. It is usually used with connection to the signals generated by the sensors and transducers, but this is by no means exclusive.

Only a few examples will be described below to highlight the importance of the likely problems. Some typical difficulties and solutions were discussed in Chapter 3 where each type of sensor was described in more detail.

The most important aspect in conditioning of low-level signals is suppression of any unwanted signals coupled by inductive and capacitive means from the surrounding components. Inductive coupling can be reduced for instance by using appropriate twisting of the connecting wires (Figure 4.19).

In general, it is always a good idea that useful signals should be as large as possible so that the signal-to-noise ratio is as high as it can be for a given setup. This is important and applicable for very low signals, for which noise level might be comparable or even greater than the amplitude of the 'useful' signal (see Figure 3.56).

4.4.1 OUTPUT SIGNAL CONDITIONING

There are several 'output signals' in a magnetising apparatus. An 'output signal' is created at any point at which there is information passing from one module to the next. Even information passed entirely in a digital domain is also a kind of output signal, and as discussed below, it might require special signal conditioning such as digital filtering for various reasons.

In this section, the 'outputs' are all those information and signals which contribute to generation of suitable excitation conditions for the sample. However, the output signals of the sensors become the input signals to the data acquisition module. These are described in a separate section below.

4.4.1.1 Analogue Output Glitches

Some signal generation devices (especially the older types or lower-cost) or rather their digital-to-analogue converters might generate unwanted glitches in the output voltage signal (Maxim Integrated, 2001; NI, 2016a). Newer devices might have de-glitching circuits built-in, but they might need to be activated by the software (NI, 2007c). In any case, the technical specifications of such devices should be *always* studied for information about such behaviour.

Such glitches are a *normal* behaviour for these devices and occur when the most significant bit of the code is switched, which can be in the middle of the range or near the zero crossing. The amplitude is quite significant and it might be for instance up to 100 mV. They can be suppressed by applying the sophisticated external techniques with sample-and-hold technology (Maxim Integrated, 2001), but they might not be suppressed at all, depending on the hardware.

The rise time of the glitch will be comparable to the sampling frequency. Therefore, the cut-off frequency of the anti-glitch filter should be set as low as possible so that it is capable of absorbing the glitch energy, but not too low in order not to introduce additional phase shift for the useful part of the frequency spectrum.

PRACTICAL COMMENT

A simple RC circuit can be used as a low-pass filter to significantly reduce the size of the glitch (Baker, 2011). An example is shown in Figure 4.45a.

(a) (b)

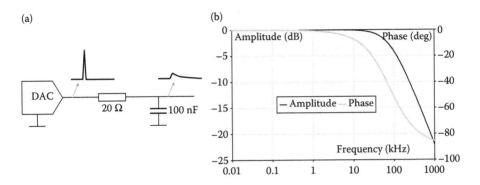

FIGURE 4.45 Simple RC circuit acting as a low-pass filter for deglitching: (a) Circuit. (Adapted from Baker B., *Use an RC filter to deglitch a DAC*, 2011, accessed online 2016-02-07, http://edn.com/electronics-blogs/bakers-best/4368256/use-an-rc-filter-to-deglitch-a-dac) (b) Amplitude and phase response of a basic low-pass RC filter.

By applying a resistor of 20 Ω and a capacitor of 100 nF to the output of a digital-to-analogue converter, the 100 mV glitch lasting for 1 μs can be reduced in this case from 100 to 20 mV (of course depending on the harmonic content of the glitch). The time constant for such RC circuit is 2 μs. Signals up to 10 kHz would be attenuated by a negligible amount, and also the phase would not be affected by more than 10° (Figure 4.45b).

However, at 100 kHz, the attenuation is −5 dB (around 50%) and the phase shift is 50°, which would almost certainly cause problems with the stability of the feedback loop for such harmonics, if they are present in the signals.

4.4.1.2 Resolution

Unfiltered output voltage generated by a digital-to-analogue converter is not smooth but consists of a series of distinctive steps dictated by the digital resolution of the device (see also Figure 4.48). The voltage resolution or the smallest step V_{res} can be calculated from the following equation (Mancini, 2002; Pelgrom, 2012):

$$V_{res} = \frac{V_{max} - V_{min}}{2^n - 1} \approx \frac{V_{max} - V_{min}}{2^n} \quad (V) \qquad (4.5)$$

where V_{max} is the maximum voltage of the range (V), V_{min} is the minimum voltage of the range (V) and n is the number of bits (unitless).

Analogue-to-digital data acquisition devices usually have a few ranges, with the typical lowest as ±200 mV; the digital-to-analogue generators often have just one output range, for instance from −10 to +10 V. Hence, for example, the output resolution of a 16 bit device can be calculated as $(+10 \text{ V}) − (−10 \text{ V})/(2^{16} − 1) = 20 \text{ V}/65{,}535 = 305.2 \text{ μV}$. And it is the smallest change which can be generated by such output.

The limited resolution can be a problem for generation of very low signals. As shown in Figure 4.46, an attempt to generate 1 mV of output signal required by the

TABLE 4.1

Voltage Resolution versus Bit Resolution for Popular Types of Configurations

Bits	Discrete Levels	Resolution of ±10 V Range	Resolution of ±200 mV Range
8	256	78 mV	1.6 mV
10	1024	20 mV	0.39 mV
12	4096	4.9 mV	97 μV
16	65,536	0.31 mV	6.1 μV
18	262,144	76 μV	1.5 μV
24	16,777,216	1.2 μV	24 nV

power amplifier will produce only three steps for each polarity due to the 0.31 mV resolution (Table 4.1). Each step is amplified and delivered to the magnetising yoke, but with a limited bandwidth. Some low-pass filtering takes place, but the distortions in the excitation (Gozdur, 2004) and hence B waveform are still clearly visible and they match the staircase waveform very well. However, the spikes in the induced voltage ('V_2' in Figure 4.46) completely dominate the shape of the voltage waveform, which ideally should be a pure sine.

PRACTICAL COMMENT

The situation could be improved in several ways, and five examples will be given below. They can be used separately or combined. The choice should be made carefully, because each of them has some advantages and drawbacks.

1. The gain of the power amplifier could be reduced as this would require higher input voltage in order to produce the same output current. Hence, the waveform would be composed from more steps and

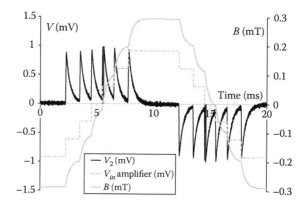

FIGURE 4.46   Limited resolution of 16 bit output can produce visible steps and spikes in the processed signals. (From NI, *National Instruments, NI 6115/6120 Specifications*, 2008b.)

the relative distortions would be reduced. However, for amplifiers with fixed gain, this option obviously cannot be used.

2. For small excitations, the value of the shunt resistor could be significantly increased for instance from 1 to 50 Ω, because it does not directly influence the accuracy of current measurement (Zurek, 2015b). If the resistance is significantly higher than the impedance of the yoke, then significantly larger voltage will have to be produced by the amplifier, hence also by the signal generator. However, it should be noted that the yoke represents some inductance L and the shunt resistor some resistance R. The time constant responsible for the settling time in an L–R circuit is proportional to a ratio of L/R. The inductance of the yoke does not change but resistance increases, so the settling time gets *shorter*. This makes the low-pass filtering behaviour less effective, and even though there will be more steps, the changes in the current will occur faster, likely producing *sharper* spikes, but more of them and with smaller amplitude. In practice, this method works quite well.

3. The ratio of the impedance-matching transformer can be changed so that higher primary voltage will be required to produce the same secondary voltage.

4. The output of the signal generator can be connected to a simple resistive voltage divider (Figure 4.47). In this way, the full range can be scaled down to an almost arbitrarily small number. For instance, a series connection of an $R_1 = 999$ kΩ and $R_2 = 1$ kΩ resistors theoretically allows scaling ± 10 V into ± 10 mV, retaining the full 16 bit resolution, but improving it from 305 μV to 305 nV per step. Of course, the total available range is also reduced by a factor of 1000, but this can be a very powerful method of improving the resolution for generation of very small signals, for instance when studying the Rayleigh region, for example below 10 mT (Zurek, 2007; Zurek et al., 2008b,c).

   The total resistance of the divider must have sufficiently high base resistance in order not to overload the DAC output. High resistance introduces additional thermally caused voltage noise (Equation 4.6) whose amplitude might exceed the scaled-down resolution (Tumanski, 2006).

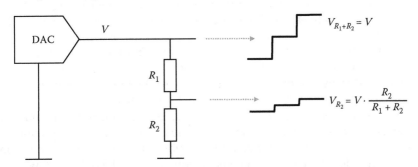

FIGURE 4.47 Improving resolution by scaling the signal with a simple voltage divider.

$$V_{th} = \sqrt{4 \cdot k \cdot T \cdot R \cdot \Delta f} \qquad (V) \qquad\qquad (4.6)$$

where k is the Boltzmann constant ($k \approx 1.38 \cdot 10^{-23}$ J/K), T is the absolute temperature (K), R is the resistance (Ω) and Δf is the frequency bandwidth (Hz).

According to Equation 4.6, a 1 kΩ resistor at 20°C for a 50 Hz bandwidth would produce random noise of around 28 nV, which is roughly an order of magnitude smaller than the 305 nV of the scaled resolution, which should be sufficient in practice.

5. The voltage steps in the output signal can also be reduced by an appropriate low-pass filter. The filter shown in Figure 4.45 has a time constant of 2 μs which corresponds roughly to a cut-off frequency of around 80 kHz. The value of $R \cdot C$ could be increased so that the cut-off frequency is set much closer to 50 Hz, so the steps are smoothed out to a much larger degree. However, the additional phase shift from such filter might make the waveshape control impossible, due to the additional phase delays.

However, even if the bit-by-bit distortions are fairly small, they can still create problems. For instance, if the waveshape feedback uses the *dB/dt* information, the small distortions might introduce jitter of signal triggering and thus introduce instability of the feedback.

But more importantly, the instabilities due to the staircase-like waveforms can influence the measured value of loss (Matsubara et al., 1996). Both the *dB/dt* (induced in *B*-coils) and *dH/dt* signals (in *H*-coils) will detect such transients as shown in Figure 4.46. The transients will appear in both types of sensors (*B* and *H*), but they might appear with a different harmonic content due to the effective bandwidth of the given signal-processing channel and the associated parasitic parameters such as self-capacitance, local load impedance, etc. These effects can introduce effective phase shifts between the *B* and *H* signals, and these phase shifts might differ with changing level of excitation or frequency.

As a result, the calculation of power loss might be affected by a large error; up to 30% was shown in Matsubara et al. (1996) (and also experienced by the author of this book). A remedy for the problem could be filtering out the steps by a low-pass filter. But the effect occurs mostly because the steps in the *generated* voltage are *synchronised* with the *acquisition* of the input signals. It is therefore possible to use different number of points for generation and acquisition in order to 'desynchronise' the sampling frequencies for the input and output. This gives sufficiently good results, so the low-pass filtering does not have to be used (Matsubara et al., 1996; Zurek, 2005).

However, the penalty for the 'desynchronisation' method is that the number of input and number of output points per cycle are different. Direct point-by-point calculation of the waveforms is not possible, and one waveform must be re-scaled in order to allow the implementation of a digital feedback. A waveform can be re-calculated to a different number of points for example by using

Fourier transform, and then inverse Fourier transform with a different number of points. Another approach is to use methodology based on the precise post-triggering method described in Section 4.4.2.6, where the whole waveform can be re-calculated to a different number of points.

There are several other factors which can affect the actual resolution of a digital device, and many of them are applicable to both generation and acquisition (Pelgrom, 2012). Out of these, probably ENOB (*effective number of bits*) is the most important in practice. Other factors might be more relevant for input, but not necessarily for output, due to accuracy implications discussed below.

Effective resolution reduces if one cycle of signal is generated with fewer points. The waveform in Figure 4.48a is a single cycle of sine, generated with 100 points with a hypothetical 4 bit resolution or 16 discrete levels of voltage (the amplitude is

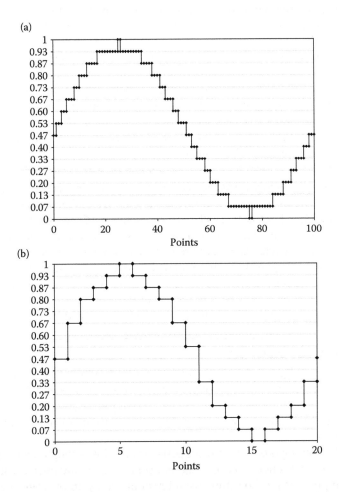

FIGURE 4.48 Generating signal with fewer points reduces effective resolution (ENOB); an example showed for signal generated with 4 bit resolutions and (a) 100 points and (b) 20 points.

normalised between 0 and 1). As can be seen, each increment of resolution is exercised through the waveform – so full 4 bit resolution is available.

However, if the same sinusoidal waveform is generated with the same resolution of the converter (4 bits, or 16 levels) but with just 20 points per cycle, then it can be seen that some levels are skipped during generation. The sampling frequency is not fast enough and the time interval between each point is so long that some levels are omitted altogether. Therefore, the effective resolution of waveform from Figure 4.48b is lower than that from Figure 4.48a even though both are generated by the same device. In the extreme case of 2 points per cycle (Nyquist limit), any resolution could be reduced to just 1 bit – because only the minimum value (0) and the maximum value (1) could be captured and represented.

PRACTICAL COMMENT

The ENOB value becomes more important for measurements at higher frequencies. The maximum sampling frequency for a given device has a maximum value. The measurement frequency must be sufficiently lower than the maximum sampling of the device in order to provide sufficiently stable generation, without jitter and significant degradation of signal quality. Using the example of Figure 4.48b, a data generation device with maximum sampling frequency of 50 kS/s (50,000 samples per second) could generate a maximum measurement frequency of 50 kS/s/20 points = 2.5 kHz. Waveshape control would be severely limited, because theoretically only 10 harmonics can be controlled, but in practice definitely fewer than that.

Interestingly, the digital voltage resolution is *not* critical in practice for accuracy of measurement as far as power loss is concerned, at least in uni-directional measurements (Zurek et al., 2007a). It is true that with reduced resolution the repeatability worsens significantly. However, the average power loss changes very little and remains practically within the confidence interval for a full resolution measurement (Figure 4.49a).

The waveform in Figure 4.49b shows secondary voltage of Epstein frame, which was driven with a resolution *artificially* lowered to 8 bits, but with the input channel working at full 16 bit resolution. The steps in the generated waveform are clearly visible (Figure 4.49b), yet the average power loss value is affected by less than 0.5% (Figure 4.49a).

PRACTICAL COMMENT

As evident from Figure 4.49, the output resolution might not be as critical a feature as other parameters. For data acquisition device selected for a given measurement system, lower resolution can be chosen, but for instance a higher-sampling frequency, which would be more useful for high-frequency measurements.

The data from Figure 4.49 also suggests another interesting conclusion. It would be more beneficial to use a higher-*absolute accuracy* device with 12 bit input, than perhaps for the same cost a lower accuracy 16 bit device (Zurek et al., 2007a).

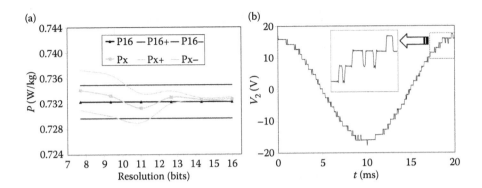

FIGURE 4.49 Effect of output resolution on average power loss measured in Epstein frame at 50 Hz for grain-oriented electrical steel (a) and large jumps in the induced voltage caused by coarse resolution of the output (b). (Reproduced with permission from Zurek S. et al., *Przeglad Elektrotechniczny [Electrical Review]*, R. 83 [NR 4/2007], 2007a, 50. Copyright © 2007 *Przeglad Elektrotechniczny [Electrical Review]*.)

4.4.1.3 Gradual Adjustments

In order to execute correct magnetic excitation, the magnetising current must be generated in a continuous way so that the alternating current waveforms are produced without stopping.

Digital-to-analogue signal generation devices achieve the continuous generation by using a data buffer. Such buffer can be loaded with a single cycle of waveform, which is then generated repeatedly (Figure 4.50a) so that continuous alternating waveform is produced at the output.

However, the shape and amplitude of the output signal must be adjusted to a new level, without breaking the continuity of the generation. When a new data is loaded to the buffer at the point of changeover, a glitch will be generated (Figure 4.50b).

PRACTICAL COMMENT

When the digital feedback is active, the data is recalculated for each iteration, so the buffer is updated for each iteration and the glitch is produced at *every* changeover, albeit with a very small jump.

Luckily, a steady state of amplitude and shape of the waveforms need to be achieved around the measured point, so the changes are indeed very small to the point of being negligible. But large glitches should be avoided throughout the measurement procedure anyway, both for the increase or the decrease in the amplitudes. Therefore, the gain of the feedback or the step of demagnetisation cannot be large because otherwise large jumps could be produced.

In theory, it should be possible to avoid glitches if the changeover happened exactly at zero crossing of signal. However, in practice, this is normally impossible to be controlled, because the buffer loading is carried out by the software (relatively very slow) and the continuous generation is handled by the low-level hardware (very fast).

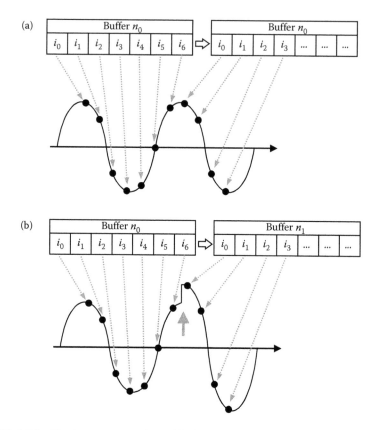

FIGURE 4.50   Continuous generation with the same data buffer (a) and with the buffer updated 'on-the-fly' (b).

Additional difficulty also comes from the fact that normally the waveshape, amplitude or phase of the generated signal is irrelevant from the measurement viewpoint. Only the controlled waveform is critical, for example dB/dt (i.e. voltage). Therefore, the shape and phase of the output waveform will be continually adjusted by the feedback in order to accomplish the required shape of the controlled waveform. Thus, the zero crossing of the adjusted output signal might no longer coincide with the first point in the data buffer and hence the changeover at zero cannot be executed.

The same applies for rotational magnetisation. With at least two channels used for generation, the data buffers for both channels will be changed over at the same time. But the actual waveforms are likely to be such that if channel x crosses zero then channel y is around its maximum.

It should be noted that in practice the size of this glitch is *not* related to the resolution of the device, but to the amplitude of the digital waveform submitted to the data buffer. Therefore, in a practical approach, it is best to ensure that the changes are small and gradual. For this reason, a feedback algorithm which converges *too fast* might be actually detrimental to the reproducibility of measurements.

It should also be stressed that the fact that the glitch is not visible in the acquired waveforms does not mean that it is not present. Such situation can arise if the acquisition is initiated after the update has already taken place (refer also to Figure 4.82). In the 'continuous generation' interval, no glitch will be measured, because the glitch had occurred before the acquisition would have been started. For this reason, it might be necessary to use independent equipment such as an oscilloscope to verify the behaviour during the buffer updating (Zurek, 2017a).

4.4.1.4   DC Offset

Digital-to-analogue signal generators used for rotational measurements would be normally DC coupled. Therefore, there is a possibility of a permanent unwanted DC offset generated together with the useful signal. This offset might reach several millivolts as specified by the manufacturer (NI, 2008b), and its presence should be taken into account to ensure that no adverse effects are introduced into the magnetising system.

This is especially important if the power amplifier is also DC coupled. In such case, the DC offset in the generated signal will be amplified and delivered to the magnetising yoke. This is why DC offset elimination, for instance by means of transformer (as described above), is beneficial.

4.4.1.5   Digital Filtering

Low-pass filtering of the output signal might be beneficial for operation and stability of the negative feedback (described in the following sections). Waveshape control is most important for the lowest harmonics, and above certain level, the harmonics can be neglected as far as the feedback is concerned. In practice, it will never be possible to control all the harmonics over infinitely wide spectrum, and there is no practical need to do so. Therefore, the output waveform could be digitally filtered (i.e. before it is converted from digital numbers to voltage), for instance with the help of Fourier transform (Figure 4.51).

The original waveform is decomposed into harmonics by means of Fourier transform (Figure 4.51a). All the harmonics above certain number are set to zero

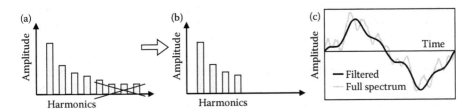

FIGURE 4.51 Schematic illustration of digital filtering by means of Fourier transform: (a) spectrum before filtering, (b) spectrum after filtering and (c) waveforms.

(Figure 4.51b). The waveform is then composed back through the inverse Fourier transform.

Therefore, the output waveform (Figure 4.51c) can be considered to be passed through an 'ideal' low-pass filter, because all the harmonics below cut-off frequency are retained with 100% amplitude, and all the harmonics above are completely eliminated.

Of course, such filtering should be performed only for the digital feedback data responsible for the waveshape control. Any signal use in the measurement channel should *not* be filtered in this way, as this would lead to measurement errors.

4.4.1.6   Output Accuracy

The absolute *accuracy* of the output voltage of a signal generator is practically *irrelevant*. This is because the signals must be processed by a number of components (Figure 4.1) including the non-linearity of the magnetising yoke and/or the sample.

If such parameters like the above mentioned output DC offset or voltage resolution and sampling frequency are satisfactory, then the accuracy is unimportant. For instance, connecting any load to the signal output will always affect the absolute value of the generated voltage. But as long as the load is within safe limit for the output, its value is of little importance. After all, devices such as the voltage divider from Figure 4.47 by definition change the output voltage by a large factor, and the system can still operate normally, without any other changes, because of the action of the feedback. Hence, no calibration would be even required for such a voltage divider.

Additional low-pass filtering can be really problematic only if introduces phase shifts large enough to impact the stability of the feedback loop.

In any case, the amplitude, phase and shape of the output signal will be iteratively corrected by the negative feedback so that all the non-linearities and inaccuracies of the output are taken out of the equation – within workable range of the feedback algorithm and signal levels.

This is of course not the case for the *input* channels, where the absolute accuracy is of prime importance.

4.4.1.7   Mains Hum and Other Distortions

Power amplifiers are not ideal devices and they can be affected by the mains frequency they are supplied with. Sometimes, such information can be found in the technical specification of the device (Crown, 2007), but this is not always the case.

A power amplifier has an internal DC power supply which is derived from mains (e.g. 50 Hz or 60 Hz), which is a sinusoidal supply. The filtering between the sinusoidal input and the internal DC rail is not infinitely good, and some modulation by mains can be present. This modulation might eventually find its way into the output current of the power amplifier and become a nuisance during measurements. The modulation can also be caused by less than ideal effectiveness of input filtering of the given power amplifier (also discussed in the next section).

Such modulation at mains frequency is called 'hum'. The name comes from audio applications in which its presence would generate audible noise in loudspeakers, which would make humming noise at the mains frequency (50/60 Hz) or its double (100/120 Hz).

PRACTICAL COMMENT

An example of such phenomenon is shown in Figure 4.52. As soon as a power amplifier was connected to a rotational yoke, there was an output current of 4 mA peak (grey waveform in Figure 4.52) flowing in the circuit, even if the input signal was zero. This unwanted current was caused by the less-than-ideal internal filtering in the power amplifier.

This caused a problem for measurements at low excitations, at which the 4 mA peak showed a significant distortion, 'travelling' around the rotational pattern. Under normal conditions, it was not possible to suppress such distortion by the feedback loop, because such distortion did not represent itself as steady state due to a frequency mismatch (described in the next section).

After consultation with the manufacturer, the power amplifier was sent back for modification of the internal filtering so that the amplifier was improved, with the hum being suppressed roughly by an order of magnitude (black waveform in Figure 4.52).

It should be noted that the specification of the values given in the data sheet of a power amplifier for signal distortion might not give enough information about such effects. For instance, in the specification (Crown, 2007), the THD was given as 0.1% at 1 kHz and at rated power, for each channel. Rated power in this case (230 V, 50 Hz supply) was 1520 W into 2 Ω load, which means an RMS current of 28 A. In the worst scenario, if all the distortion came from hum, then its amplitude could be as high as 0.1% of 28 A, which is 28 mA, so the 4 mA peak was well within the original specification. Yet it was significantly too big to be useful for precise magnetic measurements.

Data sheets of audio amplifier would also specify such distortion-relevant values such as intermodulation (IM or IMD) and crosstalk. Intermodulation is the mutual modulation of frequencies within the same channel and should also have a very

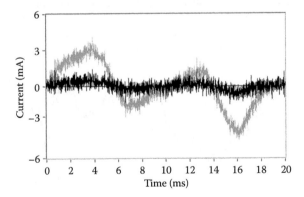

FIGURE 4.52 An example of 50 Hz hum in power amplifier connected to a rotational yoke. (Adapted from Zurek S., *Two-Dimensional Magnetisation Problems in Electrical Steels*, PhD thesis, Wolfson Centre for Magnetics Technology, Cardiff University, Cardiff, the United Kingdom, 2005a.)

small value, at the level of 0.1%. In practice, this is an irrelevant number because any stationary small-amplitude distortion in the magnetising current will be tackled by the feedback loop, provided that the signal processing has enough resolution.

On the other hand, crosstalk quantifies how much signal in one channel affects the signal in the other channel. A typical value could be for instance −98 dB, which is quite a low value. This means that if one channel is driven with full amplitude, for example 100 V, then the other channel (whose input is held at zero) could amplify some of this signal, but lower in amplitude by −98 dB, which is roughly 5 orders of magnitude, so for the 100 V example, it would be only around 1 mV. Because such distortion would represent a steady state, it should be normally dealt with by the feedback loop, if it was within the required amplitude resolution.

4.4.1.8 Mains Frequency Mismatch

The mains frequency is driven by the national power network and can vary by a noticeable amount. As shown in Figure 4.53, a variation by ±0.2 Hz is quite common within an hour, and the specification for the United Kingdom is ±1% or ±0.5 Hz (National Grid, 2016). This will be very similar in most countries.

Digital signal processing is used in modern measuring equipment. The output signal is often synthesised by a digital-to-analogue converter, which is normally driven by a precise and very stable internal clock. It is not uncommon to see base clock accuracy at the level of ±0.01% (NI, 2008b). Not only is this clock an order of magnitude more precise than the mains frequency, but it is also much more stable so that its internal frequency does not change with time.

Therefore, if a waveform at 50 Hz frequency is set to be generated from such digital device, in a general case it will be a frequency slightly *different* from the 50 Hz of mains. This mismatch of frequencies creates a problem for measurements at small amplitudes and mains frequency. (Of course, the same applies to 60 Hz installations.)

As discussed in the previous section, the mains frequency might not be filtered out completely by the power amplifier (Figure 4.52) or other circuitry or even picked up from unshielded cables. In such case, the magnetising current might be dis-

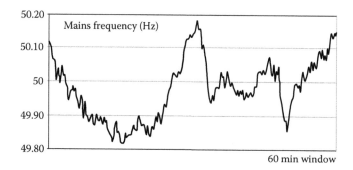

FIGURE 4.53 An example of 50 Hz mains frequency variation in the United Kingdom. (Adapted from National Grid, *Frequency response services*, 2016, accessed online 2016-02-10, http://www2.nationalgrid.com/uk/services/balancing-services/frequency-response)

First iteration   Second iteration   Third iteration

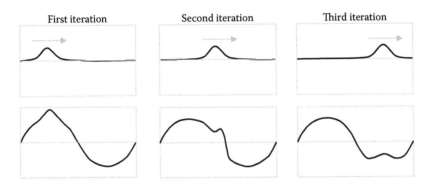

FIGURE 4.54 An example of mains hum 'floating' with respect to the measurement frequency; each window shows 20 ms frame, or one cycle of 50 Hz: top, just mains 'hum'; bottom, resultant waveform.

torted, modulated or otherwise affected by the mains frequency. By definition, such modulation will be locked in phase with the mains frequency, whatever it might be at the given instant of time (Figure 4.53).

For clarity of illustration, let us assume that the mains-related distortion appears as an asymmetrical waveform, as shown in Figure 4.54, top. For precise signal processing, the internal frequency must be driven by the internal clock, and usually internal triggering would also be synchronised with it, but *not* with the mains frequency.

However, because the mains frequency can be slightly different, the phase of the distortion will appear as 'floating' (moving) with respect to the acquired waveforms (Figure 4.54). The direction of floating will depend whether the mains frequency is lower than the measurement frequency (distortion floats to the left) or vice versa (distortion floats to the right).

Generation of the output waveforms is continuous, and every 20 ms, the data buffer will be regenerated in the same way (see also Figure 4.50), due to the hardware action of digital-to-analogue converter. But the acquisition can happen at much slower pace, for instance one iteration per second.

Therefore, at each acquired iteration, the mains distortion will appear at a slightly different position, and as a consequence, the acquired waveforms will be distorted at a different position of the cycles (Figure 4.54, bottom).

Hence, such distortion represents a kind of a slow transient, which will affect the performance of a non-real-time digital feedback.

The problem arises because the digital feedback can act only on the information available to it at the given instant of time, and the information would have been obtained from the *previous* iteration. The difference between the real and the target waveform will be calculated and applied to the output, for the next iteration.

But in the next iteration the distortion is located at a *different* position in the cycle. The new position of the distortion is not taken into account (because it had not existed in the previous iteration) and therefore the correction is applied in the *wrong* place. As a result, the controlled waveform might be distorted by even a *greater* amount (Figure 4.55) as far as THD value is concerned.

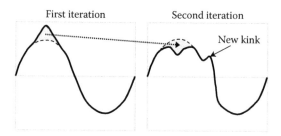

FIGURE 4.55   Operation of feedback on a transient distortion can lead to more distortions because the correction will be applied in the wrong place in the subsequent iteration.

PRACTICAL COMMENT

The floating distortion can be a real nuisance in low-amplitude measurements, and it is surprisingly difficult to be eliminated by purely digital means. Several observations can be made.

Firstly, if the feedback is unable to react quickly enough, then the waveform (e.g. B) remains distorted, which can adversely affect the accuracy of the measured properties. For instance, relative amplitude permeability is based on B_{peak} and H_{peak} values. It is evident from Figure 4.54 that the peak values will change depending on the actual phase of the distortion.

Secondly, such distortion cannot be averaged out easily. In order for averaging to suppress the effect, the distortion would have to pass equally through every location through the cycle, throughout the whole averaging procedure. But this cannot be controlled because the frequency variation of mains appears to be fairly random (Figure 4.53) and thus it is not possible to know beforehand how many cycles should be used for averaging for the effect to be smoothed out. Some improvement is obtained by averaging, but for significant gains, the values would have to be calculated from hundreds if not thousands of operations, which is not practical due to the very long time required for such averaging. Counter-intuitively, the effect is more difficult to work with if the frequency mismatch is small, but not very small. This is because it takes more iterations for the distortions to pass through all the positions, and thus it would require many more iterations from which the data should be averaged. If the frequency mismatch is larger, then it is easier to average it out.

Thirdly, fortunately, the feedback has more chance to act on the distortion if the frequency mismatch is very small or ideally zero. Under such conditions, the distortion appears more like a steady state and the floating is so slow that active feedback can suppress it to a large degree, or at least to an acceptable level.

A consequence of the above point is that making the iterations faster achieves the same effect because there is less amount of floating from one iteration to the next. The extreme case is completely real-time feedback control, which is achievable by analogue circuits (described in the following sections).

But sometimes with digital feedback, there is very little it can be done to speed up the iterations. The data acquisition time is dictated by the type of device used for processing the multiple signals simultaneously with large amount of data; it is difficult to achieve the processing time faster than around 10 iterations per second (at mains frequency).

It should be noted that the distortion shown in Figures 4.54 and 4.55 was chosen to be asymmetrical only for the sake of clarity of illustration. In reality, the modulation could have mostly a symmetrical shape and then the only effect would be the cyclical change of amplitude of the whole waveform, as the two frequencies 'beat' (mains and measurement). However, this 'beating' is as troublesome as an actual distortion, because it is much more difficult to control the peak of the waveform to the exact required target value. At every iteration, the feedback algorithm is fed with slightly incorrect information and keeps 'hunting' for the perfect control, which cannot be achieved due to the slow transient.

If the problem is really difficult to be overcome in a given setup, then as a last resort the values can be measured at two slightly different frequencies, for instance 47 and 53 Hz, for each of which the averaging will work flawlessly. The final value of 50 Hz could be used as the average of the two values. With more measured frequency points, suitable curve fitting can be employed so that the 50 Hz data is interpolated.

The frequency mismatch problem is significant in practice, because in most cases it makes it more difficult to measure at the most important frequency for electrical steels – the mains frequency. It would be easy to measure the samples at 60 Hz, for instance in Europe, and 50 Hz in the United States, but this is hardly a practical solution. Perhaps Japan is ideal with the mixture of 50 and 60 Hz in the national grid.

4.4.1.9   Overcurrent Protection and Safety Implications

The subject of overcurrent protection was mentioned above, but it is repeated here because of its importance and the safety implications.

Some means of overcurrent protection is necessary in every high-power system. Even for medium-size rotational yokes, the total magnetising power can reach thousands of volt-amperes (kVA) and such power can be not just *dangerous* but outright *lethal*. Apart from design precautions and considerations, every experiment should be preceded with a formal risk assessment due to the level of involved voltage, current and power. There is a risk of electrocution as well as fire.

Magnetising conditions can vary vastly in an experimental circuit. Worse still, the behaviour of the feedback circuit or algorithm cannot guarantee absolute stability, and under extreme conditions, the action might turn into effectively a positive feedback, which will demand full power from the power amplifier, usually in an uncontrolled way if it happens.

> Implementing the overcurrent protection between the power amplifier and the load with a simple circuit breaker or a fuse should be avoided.

A sudden disconnection of high current into an inductive load can generate voltages lethal to the operator and to all the auxiliary equipment, due to the so-called 'flyback voltage'. During sudden disconnection of inductive current, a voltage impulse reaching *thousands of volts* can be easily generated, which would almost certainly destroy any measuring equipment attached to it.

And if an input of a data acquisition device measuring voltage across a shunt resistor is catastrophically damaged, then likely also will be the computer to which it is connected (see Figure 3.6).

PRACTICAL COMMENT

Some 'protection' mechanism can be built into the software. There are several conditions which can be detected before dangerous levels of power are generated. The following points should be noted, mostly applicable to a digital feedback. This is a non-exhaustive list:

- High value of the feedback gain should be avoided, as it is the usual culprit for feedback instability.
- The peak value of the digital output signal can be checked against a pre-defined safety level. If sudden increase in the amplitude is detected in calculations, then the software can pause or shutdown before the output is actually generated.
- When instability occurs and the action of the feedback becomes positive, the controlled waveform becomes very quickly distorted. In many cases, it is therefore enough to monitor the THD of the controlled waveform and shut down the generation if certain value (e.g. THD > 50% of *dB/dt*) is exceeded.
- The value of the shunt resistor is usually known as well as its maximum allowed dissipation power. The software can monitor the power developed on the shunt resistor(s) and shut down the generation if dangerous levels are detected.
- The input voltages can be monitored for clipping (Figure 4.56). If levels close to being dangerous are measured, then the generation can be shut down automatically by the software.
- It is also a good idea to configure one button on the keyboard (e.g. the *Escape* button) and another mechanical safety switch (e.g. latching switch), which will unconditionally shut down the power generation (e.g. the electricity supply to the power amplifier).
- The shutting down should not be a simple 'switch off' or 'zeroing' of the *output* current because this would be comparable to a circuit breaker action. Ideally, the amplitude should be reduced very quickly, but *gradually* from the peak to zero (e.g. as shown in Figure 4.42). This will minimise the potential for generation of excessive 'flyback' voltages.
- The hardware could be additionally protected by using back-to-back (bipolar) high-power Zener diodes, which would clamp the input voltage to a safe value (e.g. 24 V). Another way of protecting against overvoltages is the so-called *crowbar circuit*, based on a triac or thyristor

(Wikipedia, 2016). However, in any case, the leakage current through such circuits should be very carefully controlled in order not to influence the operation and accuracy of the sensors.

4.4.2 INPUT SIGNAL CONDITIONING

Some aspects of analogue and digital processing of input signals were discussed in Chapter 3. However, before the appropriate processing can be applied, the signals must be presented to the relevant inputs in a correct form. The following sections discuss some of the most important aspects of the signal conditioning, which can be pertinent to the overall measurement accuracy.

4.4.2.1 Resolution and Input Ranges

Voltage resolution of analogue-to-digital conversion (ADC) follows similar rules as for the digital-to-analogue devices described in Section 4.4.1.2, and Equation 4.5 and Table 4.1 apply in the same way. Factors such as ENOB are also relevant (Platil et al., 1999), and the reduction could be as much as from 14 to 11.5 bits (Platil et al., 2003) or even from 12 to 6.5 bits (Platil et al., 2000) at high-sampling speeds.

The main difference is perhaps that there are usually several input ranges selectable by the software separately for each input channel with voltage levels: 0.2, 1, 5 and 10 V (NI, 2013a) or 0.2, 0.5, 1, 2, 5, 10, 20 and 42 V (NI, 2008b).

In a given device, the number of bits for each range is the same; therefore, lower input range allows measuring the voltage with proportionally finer resolution. Hence, in order to utilise the full resolution, it is recommended that the lowest possible range is used for a given measurement. The difference between a range of 0.2 and 10 V is a factor of 50, and between 0.2 and 42 V a factor of 210, which is certainly worth implementing. Usually, the ranges can be set dynamically during the measurement, for instance between two iterations.

Some researchers employed oscilloscopes (Bertotti et al., 1993; Lundgren, 1999; Gorican et al., 2000b; Barbisio et al., 2003; Enokizono et al., 2003; Kondo et al., 2003) or power analysers (Fiorillo et al., 1990; Kai, 2011b) as means of digitising or processing the signals, for both uni-directional and 2D measurements. Some implications of using digital oscilloscopes were described in Chapter 3.

PRACTICAL COMMENT

Most data acquisition devices (Figure 3.6) are incapable of changing the input range automatically. It is up to the user to write the software in such a way as to detect and switch to the required range.

High resolution is usually desirable, but there are practical limitations. Devices offering very high-sampling frequency usually have lower resolution. For oscilloscope, this could be as little as 8 bits, so checking technical specification is advised.

There are also devices with very low-sampling frequency, for example 14 S/s, used for thermocouple measurements (NI, 2014b). Such devices are equipped with internal analogue filters, suppressing 50 and 60 Hz noise, and the measurement is aimed at detecting very slowly varying or DC signals. An attempt to read 50 Hz signal at 14 S/s will mean for instance only around one sample per three cycles, so acquired waveform would consist almost entirely of aliased data (see also Figure 3.5), provided that it is not completely blocked by the internal filter.

Therefore, the user still must be aware of any such properties for the device in question, and the implications of the associated signal processing.

Even if the devices can operate at very high-sampling frequency with high resolution, there is a problem of amount of data to be processed. For instance, 24 bit resolution requires at least a 4 byte variable to be stored for each value. If the data was sampled at 1 MS/s, then acquiring just 1 s of data (e.g. because the magnetisation is carried out at 1 Hz) of continuously acquired signal would produce 4 MB of data. So if a measurement was to be carried out for four signals (B_x, B_y, H_x, H_y) for 60 s, this would require memory of almost 1 GB, for measurement of just a single run in such experiment.

Such data is not a problem for hard drives, but at the time of acquisition, it has to be stored usually in the internal memory of data acquisition devices, which have much smaller buffers. This is therefore another parameter which needs to be taken into account when designing an experimental setup.

4.4.2.2   Input Ranges and Overvoltage Protection

The maximum signal level (peak voltage) must not reach the maximum value of the range; otherwise, the measured signal will be 'clipped' digitally and hence erroneous (Figure 4.56).

Such clipping should be avoided and fortunately it is relatively easily detectable in software when it occurs. For instance, it is enough to compare the highest and lowest converted points (max. and min.) to a digital numbers of the range limits, and if any point exceeds, for example 95% of the range, then the next available higher range should be selected.

This is especially important for applications such as magnetic measurements due to the character of the excitation. In many experiments, the level of excitation is continuously increased, for example when measuring the power loss under sinusoidal magnetisation from 0.1 to 1.8 T, so a factor of around 20 just for B, and much more

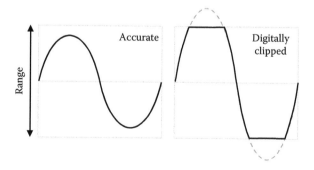

FIGURE 4.56   Only signal from within the range can be measured accurately; otherwise, it is clipped.

for H at higher excitations. Therefore, with each new measured point, the output signals increase significantly, and as a result, several ranges have to be used in order to complete the whole experiment within one procedure.

The ranging of the inputs also has some safety implications. For very high excitations or high frequencies, the voltages induced in the system or supplied by the power amplifier can easily exceed values safe to the user or to the equipment. Most precise data acquisition devices are not designed to handle excessive voltages. The overvoltage protection circuits offer only limited protection, and they would be unable to handle high energy. Typically, the overvoltage protection is only marginally higher than the highest input range (NI, 2013a) and sometimes just equal so that no overvoltage is allowed at all (NI, 2008b).

4.4.2.3   Aliasing and DC Offset

Aliasing and input DC offsets were partially explained in Section 3.1.1.2. The aliasing phenomenon is illustrated in Figure 3.5. The effect can be reduced by using appropriate low-pass filtering so that any harmonics above sampling frequency are suppressed by a sufficient amount.

Some data acquisition devices have anti-aliasing filters built-in permanently; in others, it can be switched on by the software (but not by default), and there can be even more than one filter to choose from (NI, 2008b). It is advised that such filters are used because if there are harmonics susceptible to aliasing then they cannot be measured correctly anyway. In many cases, they can be neglected from the view point of power loss measurement around mains frequencies. In other words, the harmonics might be present as such but their amplitude would be below level discernible from the viewpoint of reproducibility of the power loss measurement.

As discussed in Section 4.3.4, DC offsets in the output signals can be physically eliminated by using AC coupling by means of transformer (Figure 4.36) or capacitor (Figure 4.37). But this is not the case for DC offsets in the *input* signals, for instance those generated by the sensors.

Any component inserted between the sensor and the voltage-measuring device might significantly influence the measurement accuracy. Moreover, even if it were possible to insert an ideal DC-blocking component, still not all the DC offsets could be eliminated in this way. This is because the internal circuits of the data acquisition device might introduce additional *internal* offset and there is no physical way of stopping it from the outside of the device.

The magnitude of such DC offset can be at the level of 0.2 mV for the lowest input range, and the value could be higher with the anti-aliasing filter enabled (NI, 2008b).

Such value could be comparable with or even greater than the total amplitude of the input signal, especially at low frequencies, and hence it cannot be ignored because it would severely impact the signal integration (Figure 3.8).

The only way to tackle this problem is to remove the input DC offset digitally – by numerical calculations. If the processed signal is expected signal to be *exactly* symmetrical, with *exactly* integer number of cycles, then the offset can be removed by subtracting an arithmetic average of all the points in the given waveform.

But this is not always possible and other numerical methods should be employed. These would very strongly depend on the type of measurement and shape of acquired waveforms.

The applied method can be checked by shorting the input to the voltmeter and using the DC removal technique as normal. In any case, the processed output should not show any ill effects, like for instance the 'ramping' due to integration (Figure 3.8d).

PRACTICAL COMMENT

However, under certain conditions, the apparent DC offset is not just the artefact of the signal processing. The Earth's magnetic field has amplitude of around 40 A/m, which can produce a noticeable effect similar to an 'offset' in the H values, especially for very soft magnetic materials measured in magnetically closed yokes (Gorican et al., 2000c). Such effect is not just a problem with signal processing, but it affects the sample under test in a real magnetic way. Therefore, such problem cannot be solved just by some signal processing or filtering.

The effect is usually very small ($\pm 0.4\%$ as reported for the power loss), but it is large enough to be easily correlated with the angular positioning of the yoke (Gorican et al., 2000c). The recommendation is to change the physical position of the yoke, for instance, so that the long edge of the sample is aligned with the east–west direction. For very precise measurements, a magnetic shielding chamber might be required. Luckily, the effect is far less pronounced for systems with significant air gap in the magnetic circuit, because of the influence of the demagnetising coefficient discussed at the beginning of this chapter.

4.4.2.4 Input Signal Amplification

Input signals from sensors commonly used in rotational magnetisation have small amplitudes around mains frequencies. This is true for B-coils, needle probes, H-coils, RCPs, etc.

For instance, by rearranging Equation 3.3, it is possible to calculate that a B-coil made as a single turn with 20 mm width, on a 0.5-mm-thick sample, at 50 Hz and sinusoidal $B_p = 1$ T will produce a voltage of around 3 mV peak. With 6 µV resolution (± 0.2 V range, 16 bit), there would be over 500 steps to represent such voltage which should be still sufficient for relatively good fidelity of the signal (neglecting noise).

But if the measurement is to be made on 0.27-mm-thick sample at 1 Hz, then the expected voltage would be only 34 µV, or just six steps of resolution. As can be seen from Figure 4.46, so few steps do not translate into very 'smooth' representation.

Some researchers used signal amplification (*pre-amplification*) in order to boost the signal amplitude expected from the sensor (Sievert, 1984; Salz, 1994; Bajorek et al., 1999; Gorican et al., 2000a; Shimamura et al., 2000; Kai, 2011a).

Such pre-amplification can be achieved by using precise instrumentation amplifiers (similar to operational amplifiers) which have low-noise and good gain and phase stability. In theory, amplification by a factor up to 1000 is possible and various gain settings can be applied (Burr-Brown, 1993). An example of signal amplifiers can be seen in Figures 6.4 and 6.5 (small white boxes next to the magnetising yokes).

PRACTICAL COMMENT

However, any such amplification inevitably introduces additional errors. The gain error is not necessarily constant or linear; changes of 10 ppm/°C or more can occur. Additional noise and DC offsets are also introduced. Some of these effects can be calibrated out, while some cannot.

One performance feature which is difficult to control is the phase error. Using a typical device (Burr-Brown, 1993) as an example,[*] the frequency bandwidth values for the gains 1, 10, 100 and 1000 are 1 MHz, 80 kHz, 10 kHz and 1 kHz, respectively. It should be noted here that the term 'frequency bandwidth' refers to the specification in which the signal amplitude can reduce by −3 dB, which represents an amplitude error of 29% and phase shift of 45°.

For a gain of 10, the upper frequency at the −3 dB point is 80 kHz. In the first approximation, the amplitude and phase would follow similar trends as shown in Figure 4.45 for which the −3 dB falls at around 80 kHz, but the corresponding phase error would be around −45°.

However, even at much lower frequency, the phase error remains significant, as far as rotational measurements are concerned. For the same 80 kHz bandwidth, the phase error at 50 Hz would be around 0.036°, and as can be seen from Figure 3.49, a phase shift of just 0.09° can cause the detection of *negative* loss under rotational magnetisation.

For a gain of 100, these values also change by an order of magnitude. The −3 dB point is at 10 kHz, and at 50 Hz, the phase error becomes 0.29°, which is a value definitely capable of influencing the calculations so that large apparent negative power loss is measured (Figure 3.49b).

For a gain of 1000, the phase error at 50 Hz would exceed 2.8°. With such phase errors, the measured values of rotational power loss could reach several hundreds of W/kg even though the average value would be less than 10 W/kg (see also Figure 7.13b) (Zurek, 2005).

The main problem lies in the fact that there are at least four signals to be measured: B_x, B_y, H_x and H_y. If the phase errors *differ* between these four input channels, then the measurements will be affected by an unknown factor. As shown in the example above, the phase shift at various gain settings can differ indeed by an order of magnitude, and this should be certainly avoided.

At the very least, if amplification is used, all the channels should be set to the *same gain* so that similar phase errors are introduced in *all* input channels. But this is still not a remedy, because the actual phase errors can differ

---

[*] It should be noted that this signal amplifier is used here only as an example, not for critique of the component performance – quite the opposite. The author used PGA204 in the past (Zurek, 2000), and at that time, it offered as accurate and reliable performance as expected and guaranteed in the data sheet by the manufacturer (Burr-Brown, 1993).

between signal amplifiers due to the manufacturing tolerances – the data sheet gives only *typical* values (Burr-Brown, 1993).

For this reason, the author did not use any amplification during his studies of rotational power loss measurements (Zurek, 2005). Very small signals were measured by the data acquisition device *directly* from the sensors, even though in some cases the amplitude of the expected signal was *below the resolution* of the data acquisition device. This is possible and was carried out routinely by using dithering and signal averaging, described in the next section.

At low excitations, the signals from *H*-coils reduces even more than that of a *B*-coil. For instance, the *H*-coils used by the author (see Figure 3.45d) were wound on a substrate with dimensions 0.25 · 20 · 20 mm, with 230 turns. At 1 T and 1 Hz (see also Figure 4.59a), the measured *H* for the 'easy' direction of grain-oriented steel could be as low as 50 A/m (Zurek et al., 2003; Zurek, 2005). In air, this corresponds to 62 µT, and by using the *H*-coil constant of 2364 mm^2 (Zurek, 2005) and Equation 3.3, the peak output voltage was therefore only 0.9 µV. It should be stressed that such voltage is around *7 times lower* than the resolution of the 16 bit data acquisition device used for the measurements (see also Table 4.1), but it still could be detected with reasonable signal-to-noise ratio (Zurek et al., 2003).

4.4.2.5   Waveform Averaging

Averaging is a very powerful and widely used means of improving the signal-to-noise ratio in many measurements, including common multimeters. The measured value is re-acquired many times, all the instantaneous readings are averaged, and only the final average value is displayed. Some multimeters allow 'fast' and 'slow' modes (see also Figure 3.74b), so the user can choose how many samples should be used for averaging.

The averaging works so well, because paradoxically the signals are not clean but noisy. The effect of *dithering* comes into play, and if enough 'randomness' is included, then the averaged signal comes out with better resolution than it would be possible for 'clean' signals (Aumala, 1999; Ando et al., 2001; Hanks, 2016).

Let us consider signals as shown in Figure 4.57. The expected ideal signal amplitude (Figure 4.57a) is only twice the available step of resolution. Under 'clean' acquisition, the real signal would not be represented with high fidelity but it would look like the one shown in Figure 4.57b. Moreover, an infinite amount of averaging would not improve the situation in any way, because the clean quantised signal would be always the same, and an infinite amount of averaging would always produce the same staircase-like values.

However, the real signal is usually of such a small amplitude that a significant amount of random noise is also present. In this case, the noise amplitude is equal to the ideal expected amplitude (Figure 4.57c) so that the maximum measurable value (the highest peak) can be double of the ideal waveform. When such signal is quantised with the same resolution, the same voltage steps are discerned as before with the same resolution, but of course with all the noise present (Figure 4.57d).

Yet, this time there is a difference if averaging is applied. Each spike of noise is completely random (e.g. white noise), so each acquisition produces different values, which can be averaged on a point-by-point basis, from many waveforms.

Let us assume that 50 waveforms were used for averaging. The new averaged noisy waveform, acquired with a 3 bit resolution but averaged from 50 waveforms, becomes as shown in Figure 4.57e. This looks remarkably similar to the noisy waveform acquired with a 6 bit resolution but with appropriately reduced noise level

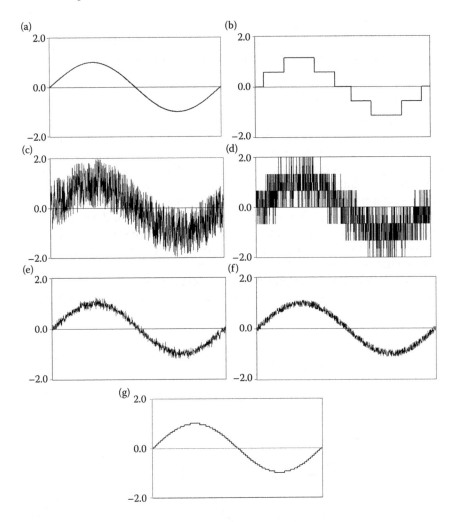

FIGURE 4.57 Simulation of improving resolution by means of dithering and averaging: (a) ideal sinusoidal signal, (b) ideal signal quantised with 3 bit resolution, (c) noisy 'real' signal, (d) noisy signal quantised with 3 bit resolution, (e) noisy signal quantised with 3 bit resolution and averaged from 50 waveforms, (f) noisy signal quantised with 6 bit resolution but with 7 × smaller noise and (g) ideal signal quantised with 6 bit resolution. (Adapted from Zurek S., *Two-Dimensional Magnetisation Problems in Electrical Steels*, PhD thesis, Wolfson Centre for Magnetics Technology, Cardiff University, Cardiff, the United Kingdom, 2005a.)

(Figure 4.57f). If the noise was not present, then the ideal 6 bit resolution would present a waveform as shown in Figure 4.57g.

It is therefore clear by comparing Figure 4.57d and e that under certain conditions dithering and averaging can improve the resolution of the given device *beyond* its nominal resolution.

The theoretical improvement in the noise reduction is $\sqrt{n}$, where n is the number of 'readings' used of the averaging. If $n = 50$, then the resolution step improves by $\sqrt{n} = 7.07$, which would correspond to around $\log_2\left(\sqrt{n}\right) = 2.8$ bits of additional resolution. Therefore, the new effective resolution would be at the level of $3 + 2.8 = 5.8$ bits, which is very close to the 6 bits shown in Figure 4.57g. In the same way, a 16 bit resolution can be improved to 19 bit resolution, and beyond.

PRACTICAL COMMENT

For measurements of very low signals, the number of readings can be set to 1000 or even 10,000 (Zurek et al., 2008b,c), thus giving the improvement by a factor of 31.6 (5.0 bits) or 100 (6.6 bits), respectively. As can be seen in Figure 4.58, the averaging can increase the final voltage resolution well below the nominal resolution. The arrows in Figure 4.58b indicate the voltage levels of the original resolution, and the averaged signal is much 'smoother'.

Therefore, a device with a nominal resolution of 16 bits and input range of ±0.2 V theoretically can have its resolution boosted from 6.1 µV (Table 4.1 in Section 4.4.1.2) to 191 nV (1000 average = effective 21 bit resolution) or even to 63 nV (10,000 average = 22.6 bit resolution).

In such a case, the 0.9 µV input voltage estimated in the previous section for an *H*-coil sensor can be easily detected without any amplification, even

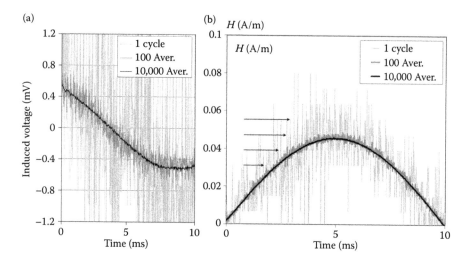

FIGURE 4.58   Improvement of resolution vs. averaging for (a) *B* waveform and (b) *H* waveform. (Reproduced with permission from Zurek S. et al., *Journal of Electrical Engineering*, 59 [7/s], 2008c, 7. Copyright 2008, *Journal of Electrical Engineering*, Elektrotechnicky Casopis.)

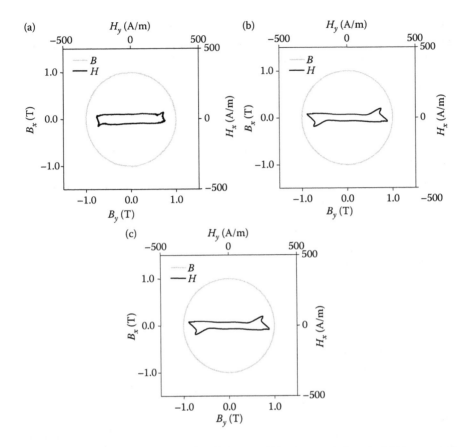

FIGURE 4.59 Measurements without signal amplification carried out at 1 T for a sample of grain-oriented electrical steel at (a) 1 Hz, (b) 50 Hz and (c) 1 kHz; the *B*-coils were made as 20-mm-wide single-turn coils; the *H*-coils were wound on a former 0.25 · 20 · 20 mm. (Adapted from Zurek S. et al., Digital feedback controlled RSST system, *Proceedings of Soft Magnetic Materials Conference*, Dusseldorf, Germany, 2003, p. 365.)

though it was *7 times lower* than the nominal resolution. Yet, the shape of the *B* loci (Figure 4.59a) is controlled to be circular (for a highly anisotropic material) and the complex shape of *H* loci is represented reasonably well.

For the dithering to work, the amount of noise must be such that its amplitude is larger than at least the smallest step of the given resolution. If not enough noise is present, then artificial 'noise' can be added, for instance by means of summing the measured voltage and some high-frequency signal with an instrumentation amplifier.

It should also be noted that for the inductive sensors (*B*-coil, needle probes, *H*-coils, RCPs), there is an additional effect which improves the apparent performance of the sensors. This is explained in Chapter 3 and illustrated in Figure 3.56.

Namely, the voltage produced by an inductive sensor must be integrated by analogue or digital means in order to produce signal proportional to the detected quantity, such as B or H. In a general sense, the integration behaves like a low-pass filtering and suppresses the noise component to a large degree. Therefore, if digital processing is applied, the first step of noise reduction comes from averaging and the second step from calculating an integral of the averaged signal (Equations 3.9 and 3.10, Figures 3.7 and 3.8).

PRACTICAL COMMENT

An example of effectiveness of integration as a low-pass filter is shown in Figure 4.46 where series of sharp voltage spikes are transformed into monotonic slopes of the B waveform. Another example is shown in Figure 3.56b in which a very noisy signal is smoothed out significantly just by the action of integration. In both examples, the integrals were performed digitally in the software.

As a result, even for a very noisy signal, the waveform for the primary quantity such as B or H can be recovered without the need for, and the problems associated with signal amplification.

Several papers related to rotational measurements or magnetic measurements refer to averaging (Gorican et al., 2000b; Ishihara et al., 2000; Zurek et al., 2008c).

However, the averaging process must be carried out under very well-controlled conditions in order to provide sufficiently accurate performance. Two of these conditions are absolutely critical for accurate processing: precise triggering and stability of the processed signals.

As illustrated in Figure 4.60, an average of otherwise identical waveforms, but whose phase is not equal, leads to misrepresentation of both: the peak values and the overall shape of the waveform. For this reason, if averaging is used, then the triggering must be very precise, and fuller discussion is given in the next section.

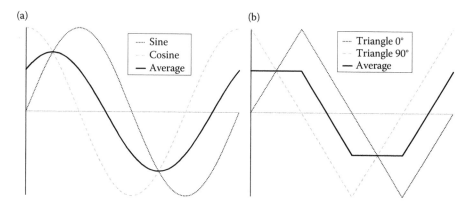

FIGURE 4.60   Average of two imprecisely triggered waveforms can reduce the peak value of sinusoidal waveform (a) and can misrepresent the shape in the case of non-sinusoidal waveforms (b); large phase shift is shown for better clarity.

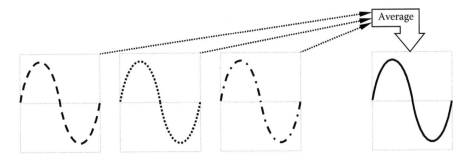

FIGURE 4.61   Averaging from single cycles.

Stability of the signals (both amplitude and phase) has a similar effect. If the shape of the signal changes from one iteration to the next, then by definition the average shape will be different from all the constitutive waveforms.

Therefore, the input waveforms must remain stable. If transient effects can be neglected, then the action of the digital feedback can be 'frozen' when gathering all the waveform 'readings' for calculation of the average. Alternatively, the feedback can be kept active, because around the measurement point all the important parameters (like B_{peak}, FF or THD) should be controlled to the required level and only very small adjustments would be necessary just to keep the excitation at the controlled level.

Waveform averaging function can be performed in one of the two practical ways: repeated single cycle acquisition or continuous acquisition.

In the first case, the software is designed so that only one cycle of the waveform is acquired at a time (Figure 4.61). Each cycle is remembered, and after the required number of acquired cycles, they are all used for calculation of the averaged waveform. The average is calculated on a point-by-point basis. Several waveforms are acquired; the average value of all first points constitutes the first point of the averaged waveform, the average value of all second points is the second point of the averaged waveform, and so on.

If the process is executed with sufficient precision, the averaged waveform can be calculated even from 10,000 cycles, as shown in Figure 4.58.

The other method of averaging relies on a continuous acquisition. One long waveform is acquired, with the window length designed to fit the required number of cycles (Figure 4.62). Once the continuous acquisition is complete, the software must then identify the length of each cycle, separate the data into single cycles, and use the new single cycles for calculation of the averaged waveform.

PRACTICAL COMMENT

From the author's experience, the second method (Figure 4.62) appears to give less reliable results, and the first method (Figure 4.61) produces more stable basis for averaging. This is probably because each new cycle can be triggered in the same way, and the 'zero crossing' point can be defined more reliably for each cycle.

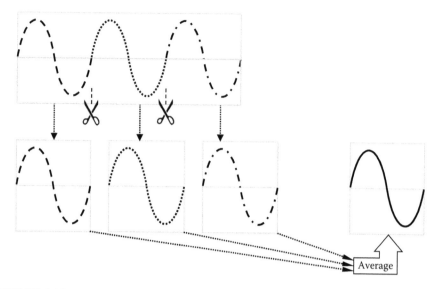

FIGURE 4.62 Averaging from continuous acquisition.

On the other hand, the continuous acquisition relies on the absolute stability of generated frequency, and also on the fact that the input and output frequencies are absolutely synchronised. This is not always the case. For instance, the case of 'desynchronisation' between the input and output digital waveforms was mentioned above (Matsubara et al., 1996; Zurek, 2005). In such a case, the output and input frequencies might differ by a very small amount (e.g. less than 0.01%), but they will *differ*. Therefore, after 1000 cycles, there could be a significant phase mismatch, because $1000 \cdot 0.01\% = 10\%$.

Another problem with the continuous acquisition is that the internal memory of the data acquisition device must be capable of storing the whole length of the acquired window. The available computer memory and available disk space are not the problem generally, but the internal memory of a data acquisition card is usually more limited, for instance to 16 MS (mega-samples) (NI, 2008b). This might or might not be sufficient for acquisition of all input channels simultaneously for prolonged periods, certainly not for 10,000 iterations.

When working with waveform averaging, it is important to note the time required for a given measurement. For instance, 100 'readings' at 50 Hz will take *at least* 2 s because 100 cycles must be acquired, each lasting 20 ms. In reality, this will take much longer, because a single iteration can last 100 ms, so one averaging from 100 readings could take around 10 s.

However, the signal reduces with frequency so more readings are required, but the length of the cycle increases as well. Therefore, if 10,000 readings were required at 1 Hz, this would mean acquisition time of at least 10,000 s or around 3 h to complete a single measurement point. The author attempted such measurement in the past, and indeed it took several hours for each operating point.

For this reason, it is advised to have much faster acquisition times just for the feedback to control the waveform and get it stable within close vicinity of the target point, and to turn on the long averaging only when the steady state was reached so that a precise measurement can be executed.

4.4.2.6   Triggering and Post-Triggering

In many magnetic measurements, it is enough to acquire a single cycle of all wave-forms (or a single cycle averaged from many cycles). In order to acquire a single cycle, the triggering can be applied in hardware or software, as allowed by the data acquisition device (NI, 2002).

However, for very low signals, built-in hardware or low-level software automatic triggering might be impossible, because of the minimum signal level which must be presented for the automatic features to work. In such case, the signals must be pro-cessed by the general software, without external triggering.

This can be achieved by allowing the general software to acquire the input data as soon as possible with the next iteration. It does not matter at which phase point the acquisition is started, as long as at least one window with two cycles is acquired (Figure 4.63).

Then the triggering can be carried out 'manually', for instance by scanning through the data for the first rising-edge crossing the trigger value v_L. For instance, it would be sufficient to find a point for which two neighbouring values v_i and v_{i+1} in the waveform array meet simultaneously the two criteria (see also Figure 4.66), for example,

$$\begin{cases} v_i \leq v_L \\ v_{i+1} > v_L \end{cases} \quad \text{(V)} \tag{4.7}$$

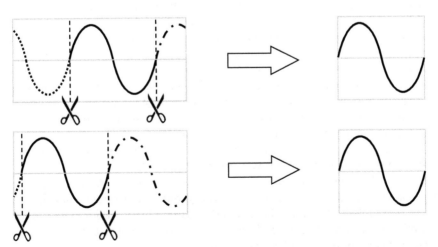

FIGURE 4.63   'Manual' triggering from at least two cycles, from a signal with arbitrary phase.

This method can be applied as soon as the feedback is at a steady state around the measurement point. Searching for a triggering point is easier in a 'smooth' waveform, because the points 'before' and 'after' trigger can be defined in a much easier way. For this reason, it might be beneficial to apply the triggering to the B waveform which has additional filtering due to the integration of the signal.

Additionally, a low-pass filter based on Fourier transform can be added, as shown for instance in Figure 4.51. It should be stressed that just for the triggering procedure the signal can be processed in an arbitrary way so that a stable triggering point can be detected. But a separate thread should be used in which the waveforms are *not filtered* so that the magnetic measurement calculations can be carried out without any loss of information (apart from isolating a single cycle).

Once the triggering point is found, then all the remaining input channels can be triggered using the same information – because all inputs will be acquired with the same sampling clock. The precise post-triggering can also be applied accordingly, because the last pre-trigger point will be known.

By using this method, the basic single cycle can be isolated in most cases, regardless of the phase of the raw acquired data. Each isolated cycle can be then post-triggered with a greater precision with the method described below.

PRACTICAL COMMENT

For averaging with the two-cycle technique from many 'readings', the acquisition time would be doubled, because half of the data would be discarded (Figure 4.63).

An alternative practical approach is also possible. A single cycle is acquired, with an arbitrary phase shift (Figure 4.64). The waveform is scanned for the triggering point, at which the data waveform is separated in two parts, and then re-assembled in a 'corrected' order so that the newly produced data is a single cycle, starting on the trigger crossing as intended. (This can be achieved for example with a pre-programmed software library called 'rotate array', as it is available in LabVIEW.)

This method can be applied only if the feedback reached a complete steady state and no changes to amplitudes are being applied. Otherwise, the 're-assembled' data might not constitute a completely 'smooth' waveform and the measurement accuracy can be adversely affected. However, if the first point of the waveform does not match exactly the last point of the waveform, then such measurement is incorrect anyway, because it signifies some kind of transient taking place, rather than a stationary AC excitation.

The method can be applied even for asymmetrical waveforms, as long as exactly one cycle of continuous waveform is acquired.

It should be stressed here that the acquisition must be timed *precisely* so that *exactly* one full cycle of waveform is acquired and that the feedback action is disabled. Only then the 'cut and re-assemble' procedure can produce a continuous waveform practically synonymous with the isolating a single cycle from two or more cycles (Figure 4.63).

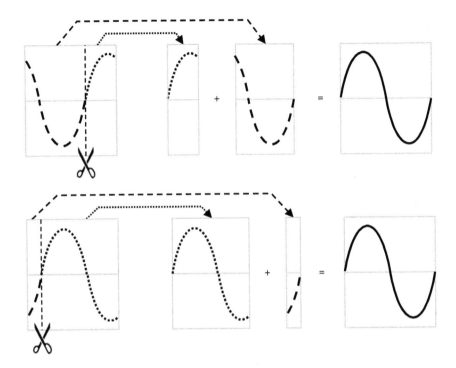

FIGURE 4.64 'Manual' triggering of a single cycle from a single cycle of arbitrary phase.

It is worth noting that in all the cases (also for the post-triggering method discussed below) only one signal needs to be triggered precisely from its own information, and the triggering position can be passed simultaneously to all other channels. Therefore, such analysis can be carried out for just one channel.

Precise triggering is required for at least two important reasons. Firstly, the averaging described in the previous section cannot be carried out with sufficient precision, so immediately the measurement repeatability would be reduced.

Secondly, the feedback algorithm will have a harder task of controlling the waveforms if at each iteration the triggering is not precise, due to apparent minute phase-shift errors. The triggering as described above *might not* guarantee sufficient precision of triggering.

The signal of interest should ideally cross the trigger level *precisely* at the moment of triggering (Figure 4.65a). Under normal conditions, such event is infinitely unlikely to occur. In a general case, the acquired points will straddle both sides of the trigger level (Figure 4.65b).

Therefore, the acquired waveform with imperfect triggering will appear as to have some phase shift with respect to the ideally triggered waveform (Figure 4.65c), because for further processing the first acquired point of the waveform is always going to be the first point in the data array (see also Figure 3.7).

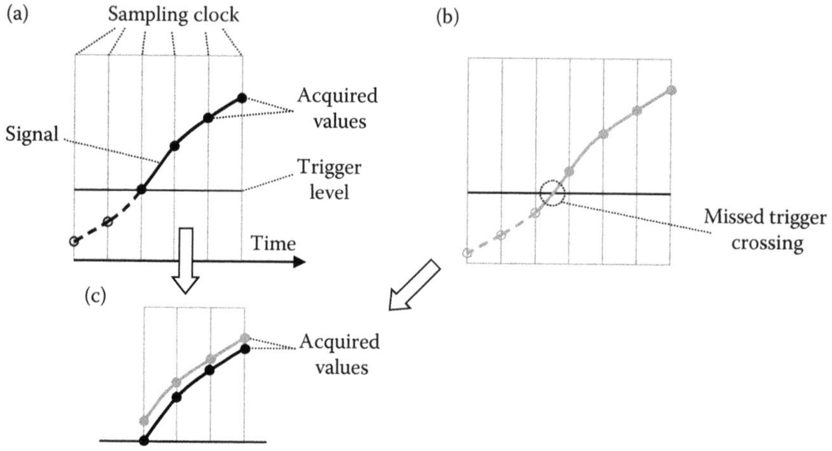

FIGURE 4.65   Triggering: (a) ideal, (b) real and (c) comparison of ideal and real triggered signals.

However, the digital feedback algorithm will strive to control the first point to be exactly at the zero crossing, but the first point will be *always* shifted by some amount and will continue to jitter between the iterations even if the shape of the waveform is controlled perfectly (THD $= 0\%$, amplitude error $= 0\%$).

PRACTICAL COMMENT

For most devices, the non-ideal triggering will happen even if reliable hardware triggering is used on a high-amplitude signal. This is because the acquisition parameters of the device are set before the waveform is captured. The values are sampled only at the instances dictated by the clock, and this procedure is started *before* the triggering information is presented to the device.

If the trigger crossing happens between the sampling instances, then the best the sampling can do is to acquire the next point immediately after the trigger (Figure 4.65b). Sometimes, hardware (analogue) triggering is not even possible due to the involved signal levels. For instance, in the device NI 6120, the trigger resolution is 5 mV so the signal must change by more than this value for the analogue hardware triggering to be activated (NI, 2002).

Interestingly, the non-ideally triggered waveform will appear to *always* lead in phase with respect to the hypothetical ideally triggered waveform (Figure 4.65c). The maximum possible phase shift is related to the number of samples per cycle, and for example with 360 points per cycle, it could be anywhere from 0° to 1°, because the whole cycle would be divided into 360 sampling instants.

The apparent phase shift between ideal and non-ideal triggering (Figure 4.65c) can be corrected mathematically, by applying numerical precise post-triggering (Zurek et al., 2005c). For clarity of illustrations, let us assume that the signal is monotonic

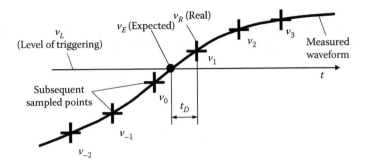

FIGURE 4.66 Illustration of imprecise triggering and definition of terms. (Reproduced with permission from Zurek S. et al., *Przeglad Elektrotechniczny [Electrical Review]*, R. 81 [No. 5], 2005d, 78. Copyright © 2005 *Przeglad Elektrotechniczny [Electrical Review]*.)

and it passes the triggering level at some point as shown in Figure 4.66. The data is acquired after as well as before the trigger point (*pre-trigger data*) (NI, 2002).

Under ideal conditions, the first value after the ideal trigger level (v_1 in Figure 4.66) would be synonymous with the trigger level v_L (so the expected value v_E would be equal to v_L) because it would be acquired exactly at the trigger level; hence, it would be that $v_1 = v_R = v_E = v_L$.

However, in a general case, the value v_1 will be acquired some time after crossing the trigger level, for instance after time t_D. Hence, $v_1 = v_R \neq v_E = v_L$.

In the first approximation, the acquired waveform can be assumed to change linearly from one point to the next (and in any case not much more information about the shape between the points is available anyway – see also Figure 3.7). Therefore, a straight-line approximating function can be stretched between any two neighbouring points, also crossing though the trigger level, as shown in Figure 4.67.

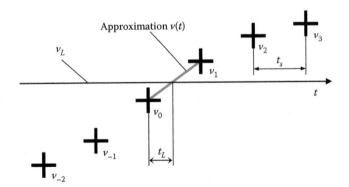

FIGURE 4.67 Illustration of straight-line approximating function between two neighbouring points. (Reproduced with permission from Zurek S. et al., *Przeglad Elektrotechniczny [Electrical Review]*, R. 81 [No. 5], 2005d, 78. Copyright © 2005 *Przeglad Elektrotechniczny [Electrical Review]*.)

The following three values are known: v_L, v_0 and v_1. For the two neighbouring points as shown in Figure 4.67, the time can be assumed as $t = 0$ at the instant of v_0. Hence, the straight-line approximating function $v(t)$ can be defined as

$$v(t) = \frac{v_1 - v_0}{t_S} \cdot t + v_0 \quad \text{(V)} \tag{4.8}$$

Therefore, it results from Equation 4.8 and Figure 4.67 that

$$v_L = \frac{v_1 - v_0}{t_S} \cdot t_L + v_0 \quad \text{(V)} \tag{4.9}$$

which allows calculation of the triggering correction time t_L as

$$t_L = \frac{v_L - v_0}{v_1 - v_0} \cdot t_S \quad \text{(s)} \tag{4.10}$$

The sampling time interval t_S is constant as dictated by the sampling frequency, which is known by definition. Therefore, the value t_L from Equation 4.10 can be used for correction of *all* the pairs of points in the waveform. For the pair v_0 and v_1, we can calculate the corrected value v_1', which of course must be equal to the trigger value v_L:

$$v_1' = v(t_L) = \frac{v_1 - v_0}{t_S} \cdot \frac{v_L - v_0}{v_1 - v_0} \cdot t_S + v_0 = v_L \quad \text{(V)} \tag{4.11}$$

However, as mentioned above, the same procedure can be carried out for any pair of neighbouring points throughout the whole waveform. Equation 4.11 may be used in the same way, but with an approximating function stretched between the two points of interest v_{i-1} and v_i. The general equation can be written as follows:

$$v_i' = (v_i - v_{i-1}) \cdot C_x + v_{i-1} \quad \text{(V)} \tag{4.12}$$

where $C_x = (v_L - v_0)/(v_1 - v_0)$ is the scaling coefficient (unitless) calculated from the first pair of points straddling the trigger level.

As seen in Figure 4.68, the approximated waveform retains the approximated shape of measured waveform, but the triggering time delay is greatly minimised so that $v_1' \approx v_E = v_L$.

Such precise post-triggering procedure was described above just for one channel, but with simultaneous sampling, the time instances will be the same for all the involved channels. Therefore, once the triggering correction time is known (Equation 4.10), if required it can be applied in exactly the same way for all the input channels by the virtue of Equation 4.12.

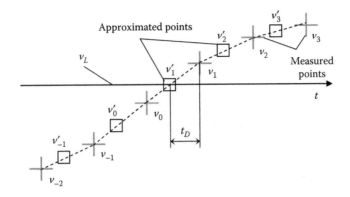

FIGURE 4.68 Approximated waveform calculated from the measured waveform. (Reproduced with permission from Zurek S. et al., *Przeglad Elektrotechniczny [Electrical Review]*, R. 81 [No. 5], 2005d, 78. Copyright © 2005 *Przeglad Elektrotechniczny [Electrical Review]*.)

PRACTICAL COMMENT

The precise post-triggering method requires the knowledge of at least one pre-trigger point (e.g. v_0 in Figure 4.67). This can be achieved in two ways. With 'manual' triggering, the waveform can be measured in such a way that the last before-trigger point is also acquired.

The post-triggering procedure becomes even more required at higher frequencies, at which the number of points per cycle decreases due to limited sampling frequency of the data acquisition device. For instance, in a rather extreme case, if just 36 samples per cycle were to be used, then the imprecise triggering would generate random phase jitter reaching up to 10°.

Further improvement could be achieved similarly to the Simpson's rule for calculating the integrals by using a second-degree polynomial instead of a straight line for the approximating function in Figure 4.67, as discussed in Chapter 3 (see also Figure 8.3).

The post-triggering procedure can also be reliably applied for correcting the phase errors resulting from non-simultaneous sampling, which can take place in simpler and less expensive, multiplexed data acquisition devices. The multiplexing time delay is usually known because it is related to the number of active input channels. And if the multiplexing delay time is known, then it can be substituted instead of the value t_L in Equation 4.9 and the calculations can be carried out in an appropriate way. It should be remembered that the delay time will be *different* for each multiplexed channel, with each subsequent channel having greater delay (Zurek et al., 2005c).

This method can significantly reduce the multiplexer phase errors, which can be significant even for non-rotational measurements. An example is shown in Figure 4.69 for a single-strip tester. With two input channels sampled at 500 samples per cycle, the estimated multiplexer phase error becomes $360°/500/2 = 0.36°$.

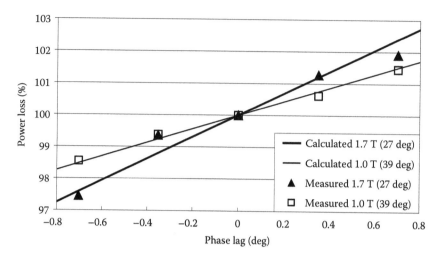

FIGURE 4.69 Effect of multiplexer delay on power loss measurement for uni-directional measurements in a single-strip tester (see also Figure 3.16). (Reproduced with permission from Zurek S. et al., *Przeglad Elektrotechniczny [Electrical Review]*, R. 81 [No. 5], 2005d, 78. Copyright © 2005 *Przeglad Elektrotechniczny [Electrical Review]*.)

It should be stressed that under rotational magnetisation the multiplexer phase error normally will *not* average out from clockwise and anticlockwise measurements. This is because the averaging relies on the reversal of all phases of the signals between CW and ACW measurements. But for normal settings, the acquisition sequence will be the same and for instance the B_y waveform will be acquired with the same phase shift with respect to the B_x waveform. As a result, the errors will be additive and will not average out – this is illustrated in Figure 3.16. But using the precise post-triggering method is capable of reducing such errors by a great amount (Figure 3.16) (Zurek et al., 2005c).

The triggering based on level-crossing can be used reliably only away from the peak of the waveform so that a well-defined 'edge' of the signal is available. This cannot be achieved with as good repeatability close to the peak of a sinusoidal waveform.

Therefore, in a two-phase rotational system, the phase with 'sine' would be normally used for such purpose, because otherwise 'cosine' would have to be triggered around its peak. The designer of the system must then make a decision as to which signal will be responsible for triggering. For example, the author used the signal from a B_x coil (Zurek, 2005).

However, such approach does not offer a universally usable solution. If the properties are to be measured as uni-directional at different inclination angles (see Figure 2.37), then the phases of x and y channels will be the same (0°) but their amplitudes will be adjusted accordingly. For example,

a. Magnetisation at 0° (along the x axis) requires $B_x = 100\%$ and $B_y = 0\%$.

b. Magnetisation at 45° requires $B_x = 71\%$ and $B_y = 71\%$.
c. Magnetisation at 90° (along the y axis) requires $B_x = 0\%$ and $B_y = 100\%$.

In the last case, the input signal for the B_x has zero amplitude and therefore it will not be possible to execute triggering on that signal, either in hardware or in software. If nothing is changed in the software, then measurements at 90° cannot be performed with such system. For this reason, the curves in Figure 2.37 do not have values at 90° because they could not be measured with this particular configuration.

This can be fairly easily measured by swapping the x and y signals, but it requires significant changes in configuration and might have impact on the measurement accuracy or reproducibility of the results. It should be noted that with such measurements there is not a single signal from sensors (B_x, B_y, H_x, H_y) or from power amplifiers (V_x, V_y, I_x, I_y) which could be used for this purpose because either x or y channel will have zero amplitude in at least one configuration. To make the triggering defined for all cases, an auxiliary triggering signal would have to be generated and acquired.

4.4.2.7 Input Accuracy

Measurement accuracy is directly affected by the accuracy of the input channels of a data acquisition device. However, the absolute accuracy can be difficult to define for the measurement of complex waveforms (as discussed in Chapter 5).

The accuracy statement for a data acquisition device can be very complex and can include multiple terms related to gain, range, offset, noise, thermal drifts, non-linearities, calibration drifts, quantisation errors, filtering, aliasing, common-mode rejection ratio, crosstalk, signal settling time, source impedance and so on (NI, 2002).

All these factors cannot be easily investigated by the user. The best estimation comes probably from the overall accuracy of a given device as specified by the manufacturer, which can come in absolute (e.g. volts) or in relative units (percents).

It should be noted that the absolute accuracy is often specified only for given measurement conditions, and there could be different measurement accuracy for each input range. The absolute accuracy could also be specified for single-shot measurements as well as averaged from a number of samples (e.g. 100) (NI, 2007b, 2013a). The implications of absolute accuracy and uncertainty are discussed in more detail in Chapter 5.

PRACTICAL COMMENT

It should be noted though that high resolution does not necessarily mean high accuracy. This concept is illustrated in Figure 4.70. High-resolution device can discern very small changes of the input signal, but still the reported values might not show good agreement with the *absolute* expected value. On the other hand, a high-accuracy device might have lower resolution, but the value reported by the measurement would be closer to the absolute expected value.

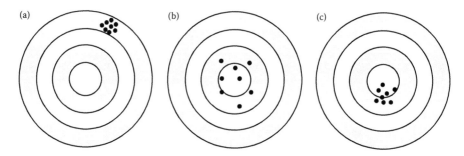

FIGURE 4.70   Illustration of (a) higher resolution or precision but lower-absolute accuracy, (b) higher-absolute accuracy and (c) high resolution and high-absolute accuracy. (Adapted from Auty F. et al., *Beginner's Guide to Measurement in Electronic and Electrical Engineering, Good Practice Guide No 132*, National Physical Laboratory, The Institution of Engineering and Technology, 2014.)

The drawings were made in such a way that the resolution of Figure 4.70a is 4 times better than Figure 4.70b, but still the accuracy of Figure 4.70b is twice as good as that of Figure 4.70a.

Therefore, it might be more important to choose a device which provides better absolute *accuracy* rather than just higher resolution (Zurek et al., 2007a). For ultimate precision, the devices should be calibrated regularly, for example once a year, or as per the manufacturer recommendation.

In some cases, the accuracy of the input cannot be taken for granted just from the manufacturer specification. For example, the input impedance is usually very high at the level of 1 MΩ. But some data acquisition devices employ a different means of signal processing (like internal resistive voltage dividers) in order to measure voltages greater than 10 V, and for these ranges, the input impedance is lower, for example 10 kΩ (NI, 2002).

If the signal to be measured can be driven with sufficiently low impedance, then the device will measure it correctly. But this cannot be always assumed. If a voltage drop is to be measured across a 50 Ω shunt resistor (Figure 4.39), then the 50 Ω value will be changed by 10 kΩ connected in parallel with it. The change will be 0.5%, which is a non-negligible number in the overall uncertainty budget (Zurek, 2015b).

4.4.2.8   Input Modes

At the heart of each digital signal acquisition device, there is an analogue-to-digital converter, which has a fixed range, referred to the internal ground (NI, 2009).

But the input signal is connected to this A–D converter through intermediate circuitry responsible for instance for impedance matching and amplification. Usually, there are several possibilities of signal connection depending whether the source of the measured signal is 'floating' or 'referenced' to some ground level.

A 'floating' signal source has no galvanic connection with the rest of the circuit. An example is a *B*-coil or *H*-coil. The voltage is generated in the coil by means of induction (see Equations 3.1 and 3.2), and it will be generated even if the coil is not connected to any other circuit, and not referenced to any ground.

A 'referenced' signal source is such that it is electrically connected to some other circuit with a ground present. An example is voltage across a resistor R_2 in Figure 4.47 (see also Figure 3.43). The voltage has to be measured with respect to the local ground, which might also be connected to the neutral or earth of the mains network.

There are various ways of measuring the voltage depending on the type of the signal course (floating or referenced), its internal impedance, expected signal level and noise level. Some examples are shown in Figure 4.71.

Differential connection with floating signal source (Figure 4.71a) may be used if the following conditions are present (NI, 2009):

- The signal source is floating.
- Differential input is available (AI+ and AI−).
- Low-expected signal (less than 1 V).
- Signal leads are subjected to noisy environment.
- Long leads placed in noisy environments.

Differential signal processing has better common-mode noise rejection and generally reduces noise pickup. The signal is allowed to float within the common-mode limits of internal amplifiers.

However, if the signal source is truly floating (like a *B*-coil or *H*-coil), then it might float to a value outside of the allowable range for the internal signal amplifiers. In order to prevent this from happening, there must be a connection provided to the analogue ground (AI GND). This connection (*bias resistor*) is to provide a path for a 'bias current' required for correct operation of the internal signal amplification.

For DC-coupled signals and sensors with low-internal impedance (e.g. <100 Ω), the connections can be such that R_2 in Figure 4.71a is open (not present) and R_1 represents a low value (short circuit). Such connection would be typically possible for a *B*-coil, whose resistance would be usually well below 100 Ω.

For sensors with higher internal impedance (e.g. *H*-coil which with fine wire can have internal resistance up to several kiloohms), it would be recommended to use such a value for which R_1 in Figure 4.71a would be around 100 times greater than the source impedance (and R_2 = open circuit).

Such a single-resistor connection makes the configuration unbalanced and thus less immune to noise pickup. For a fully balanced configuration, it would be better to use two resistors of the same value. Therefore, if the internal source impedance is 2 kΩ, then the two bias resistors in Figure 4.71a would be $R_1 = R_2 = 100$ kΩ (much larger than the source resistance, but equal to each other).

However, it must be remembered that loading a source of 2 kΩ with two resistors 100 kΩ each effectively connected in series would load the sensor with 200 kΩ and produce an amplitude error of around 1% (NI, 2009). Therefore, if such connection is used, then the calibration of the sensor should be carried out with the whole configuration in place so that any gain error would be calibrated out. If the sensor itself is AC-coupled (e.g. through a capacitor in series), then even higher values of the bias resistances could be required (NI, 2009).

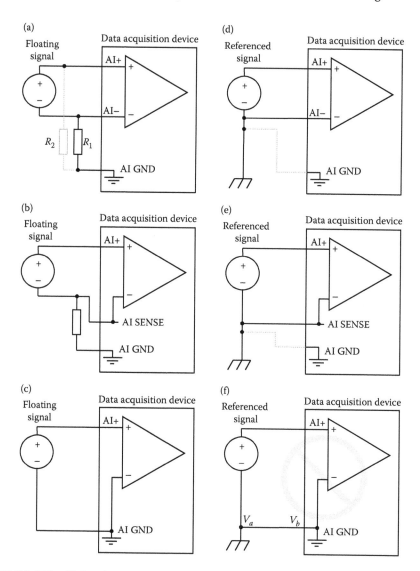

FIGURE 4.71   Various input modes for data acquisition devices: (a) differential for floating signal, (b) non-referenced single-ended for floating signal, (c) referenced single-ended for floating signal, (d) differential for ground-referenced signal, (e) non-referenced single-ended for ground-referenced signal and (f) referenced single-ended for round-referenced signal (to be avoided) diagrams. (Adapted from NI, *National Instruments, NI USB-621x User Manual*, 2009.)

Non-referenced single-ended connection with floating signal source is shown in Figure 4.71b. The requirement for the bias resistor is similar as described above. The 'negative' end of the signal source should be connected to the reference point AI SENSE or AISN (which is an appropriate separate input connection on the data acquisition device). The single-ended connection is less immune to noise and is recommended only for larger signals, for example exceeding 1 V (NI, 2009).

Referenced single-ended connection with floating signal source is shown in Figure 4.71c. This version can be used if the input signal comes from a floating source, but it *can share* a common reference point such as AI GND with other signals. Again, this configuration is recommended only for larger signals (>1 V).

Differential connection with ground-referenced signal source is shown in Figure 4.71d. The signal source can be connected directly to the differential inputs AI+ and AI−, and no additional bias resistors are necessary. The differential connection provides better rejection of noise (including common-mode noise) and is thus suitable for low signals (<1 V). An example could be a *B*-coil whose signal is amplified by a pre-amplifier connected to an isolated laboratory DC power supply. For such a configuration, the output impedance is low, signal relatively high, and although the signal source is floating, it can be connected to a common reference point without adverse effects on the accuracy.

Non-referenced single-ended connection with ground-referenced signal source is shown in Figure 4.71e. In all single-ended modes, more electrostatic and magnetic noise can couple in the signal circuit than in the differential configurations. Therefore, this mode is recommended only for large signals (>1 V). It could be used for instance for shunt resistors which might be connected directly to ground-referenced power amplifiers (Figure 3.43).

The last option shown in Figure 4.71f is *referenced single-ended connection with ground-referenced signal source* which does not offer good noise rejection and therefore should be *avoided*. Both the signal source and the data acquisition device are connected to ground or earth but at different locations. There can be a potential difference between the two points, for instance due to some current flowing through the ground loop. As a result, the voltage difference between V_a and V_b cannot be rejected and it will affect the measured signal. The error could be as much as 100 mV (NI, 2009).

As described above, there can be numerous schemes for signal connection in order to achieve accuracy of measurement as specified by the data acquisition device manufacturer, or even just to avoid completely erroneous results. Therefore, the technical specification of each device should be studied very carefully before the connections of a given configuration are made.

PRACTICAL COMMENT

The description of various input configurations was given only as an example. Manufacturers will use different terminology, which might or might not mean the same configuration in a given context. Even the same manufacturer might refer to seemingly the same or very similar configurations in different ways. For example, for data acquisition device NI USB-6210 (NI, 2009), the terminology is *differential input connection for ground-referenced signal source* whereas for NI 6115 (NI, 2008b) the description appears as *pseudo-differential input connection for ground-referenced signal source*. The actual difference is in the internal electronics of the given device.

It is therefore very important to study carefully the manufacturers' recommendations in order to ensure that no ill effects are created by incorrect wiring.

This is especially relevant for multi-terminal sensing techniques, for example as shown in Figure 3.67.

However, as transpires from the description given above, for a two-terminal floating sensor the differential connection is almost always superior and should be implemented if possible for a given configuration (Figure 4.71a). Symmetrical bias resistors should be used for best results.

4.4.3   Air Flux Compensation

Measurement of power loss can be based on the detection of B–H characteristics, as well as J–H characteristics. The relationship between B and J is given in Equation 1.9.

As discussed in Chapter 1, the volume of a ferromagnetic material is filled with atoms responding to the magnetic field and thus giving rise to the ferromagnetic behaviour (Figure 1.6), with very high relative permeability $\mu_r \gg 1$ (Figure 1.7). The part of this volume occupied by atoms is responsible for this behaviour and gives rise to the J component in Equation 1.9.

However, anything else inside of a material which does not contribute to the internal ferromagnetic interactions is effectively non-magnetic. For instance, all other electrons contribute to some amount of diamagnetism with $\mu_r \approx 1$ (Bozorth, 2003), and the empty space between the atoms is by definition non-magnetic with $\mu_r = 1$.

Therefore, all the non-ferromagnetic volume remained within material gives rise to the $\mu_0 \cdot H$ component in Equation 1.9. As a result, B will continue to increase with the applied excitation even if all the ferromagnetic contributions are fully saturated so that J reaches the highest possible value for such material, namely J_{sat}. This is illustrated in Figure 1.12 (upper right corner).

Therefore, the total response is obtained when these two components are combined so that $B = \mu_0 \cdot H + J$. Magnetic losses (e.g. hysteresis) are directly related to the J component. The $\mu_0 \cdot H$ component usually contributes to only small increase in B, the magnitude of which is responsible for instance for eddy current loss. Therefore, eddy current loss will increase with increasing H, even though full saturation (J_{sat}) of material was reached because B will keep increasing.

If the volume of the sample under test is well defined and the cross-sectional area A_1 of the sample under test is negligibly close to the area A_{coil} of the sensing winding so that $A_1 \approx A_{coil}$ (Figure 4.72a), then the measurement can be based on B–H characteristics, as it is typically carried out for toroidal samples (Figure 2.19) (IEC, 2003).

PRACTICAL COMMENT

It can be deduced from Figure 2.27d that for the easy direction of a typical conventional grain-oriented steel at $B = 1.9$ T there is around $H = 5$ kA/m. Therefore, it can be estimated for such values that $\mu_0 \cdot H = 6.3$ mT, so the difference between B and J at 1.9 T is only around 0.3%, and for this

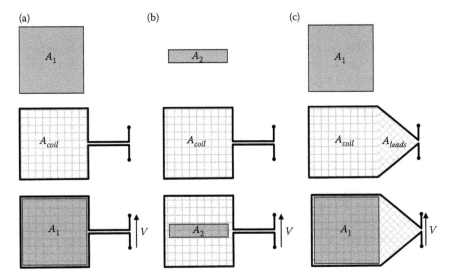

FIGURE 4.72   Schematic illustration of air flux (see also Figure 3.1): (a) negligible air flux, (b) significant air flux due to the coil area and (c) significant air flux due to leads.

reason, it is often assumed that $B \approx J$ so that air flux compensation is not required.

Of course, even such a small error should be taken into account in the full uncertainty budget, but in most practical research experiments, such error is almost negligible, especially for lower amplitudes of excitation.

The effect is greater for lower permeability because of increased H values. For the hard direction and the same conditions (Figure 2.27d), the respective values would be $H = 23$ kA/m and $\mu_0 \cdot H = 29$ mT, so the difference would be 1.5%.

However, it should be stressed that in certain cases even for toroidal samples such assumptions must not be made universally. For instance, as shown in Figure 2.61, the cross-sectional area of the sample is significantly smaller than the area of the coil, due to additional mechanical components over which the coil is wrapped. Under such conditions, it might be required that the air flux compensation is implemented (see description below). A similar one applies for any other measurements where the area of the coil is greater than the area of the toroidal sample (Morishita et al., 2011).

The assumption $B \approx J$ cannot be used when the area of the coil is significantly greater than the area of the sample so that $A_{coil} \gg A_2$ (Figure 4.72b). This is the case for instance for Epstein frame (see Figure 2.20d) because the area of the coil is fixed, and it is designed to accept samples with few laminations (IEC, 2008). Hence, the contribution of the $\mu_0 \cdot H$ component definitely cannot be neglected simply because of the differences in cross-sectional areas and must be compensated out for correct measurement.

PRACTICAL COMMENT

In Figure 4.72, if $A_{coil} \gg A_2$, then the estimation of errors the effective areas must be taken into account. The same excitation will be used as in the example given above: 1.9 T and 5 kA/m.

If the sample is a stack of four laminations, each 0.3 mm thick, then the area is $A_2 = 4 \cdot 0.3$ mm $\cdot$ 30 mm $= 36$ mm². Such sample will produce output signal whose amplitude is proportional to the component $A_2 \cdot J = 68.4$ T $\cdot$ mm².

The minimum cross-sectional area of an Epstein frame coil is $A_{coil} = 30$ mm $\cdot$ 30 mm $= 900$ mm². But the contribution to the output signal will be scaled accordingly through the active area of the coil, thus $A_{coil} \cdot \mu_0 \cdot H = 5.7$ T $\cdot$ mm². Therefore, this time the error is at the level of $5.7/68.4 = 8.3\%$ and definitely cannot be neglected even for the easy direction.

For the hard direction (23 kA/m), it would be as much as 38%. This is why the air flux compensation *must be applied* in the Epstein frame and single-sheet tester (Figure 2.22), as well as similar methods.

The idea of the compensation method is to produce signal which is proportional only to the dJ/dt component (rather than to dB/dt). This can be achieved if all the following conditions are met:

a. The cross-sectional area of the sensing coil is fixed (does not change within the same experiment or between the experiments).
b. The sensing coil is available without the sample inside.
c. There is a direct proportionality between the primary current I_1 and the produced magnetic field strength H.

If *all* these conditions are met, then the measurement system can be modified from non-compensated B-detecting (Figure 4.73a) to air flux compensated J-detecting (Figure 4.73b).

The $\mu_0 \cdot H$ component causes some magnetic coupling between the primary and the secondary winding. The air is magnetised (also other gas, or vacuum, as well as other non-magnetic components such as a coil former) and thus produces some signal in the sensing coil – even when the sample is absent. When the sample is present, the produced voltage V_2 is proportional to B or $V_2 = f(J + \mu_0 \cdot H)$ as shown in Figure 4.73a.

However, it would be possible to use an identical 'dummy system' supplied by the same primary current, but whose sensing coil is connected in anti-phase with the first-sensing winding (Figure 4.73b). The voltage induced in the 'dummy system' will produce a voltage proportional only to the $\mu_0 \cdot H$ component, because there will be no sample in it, so that $V_0 = f(\mu_0 \cdot H)$.

Because the two voltages are connected in series opposition, they will subtract so that the total output voltage will be produced as $V_3 = V_2 - V_0$ and therefore the output voltage will be proportional only to the dJ/dt so that $V_3 = f(J)$, with the $\mu_0 \cdot H$ component eliminated.

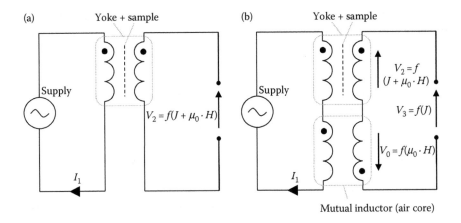

FIGURE 4.73 Uncompensated (a) and air flux compensated system (b).

PRACTICAL COMMENT

The same method is used in the Epstein frame and single-sheet tester. The recommended level of compensation is to be better than 0.1% of the voltage under normal operating conditions (IEC, 2008, 2010). Namely, if the uncompensated voltage without the sample is 100% at a given excitation level, then the 'dummy system' should be adjusted so that the remaining voltage is only 0.1% or less, from the original 100%. Perfect compensation is not feasible in practice (Tang et al., 2013), mostly due to problems with phase shifts between the signals.

In practice, the 'dummy system' does not have to be identical to the main magnetising system. The only requirement is that it produces the same voltage as the main system without the sample. For Epstein frame, this is achieved by using a round non-magnetic bobbin with two windings wound directly onto it – referred to as *mutual inductor* (Beckley, 2000a; Tumanski, 2011). The number of turns of the windings is adjusted by a trial-and-error method such as to produce the 0.1% quality of compensation. Because of the 'air core', the operation is very linear and reliable. An example of such mutual inductor is shown in Figure 2.23, next to a single-strip tester.

The mutual inductor produces magnetic field around itself and should be positioned in such a way as not to influence the main system, because the magnetic field spreads quite far around such coil (see also Figure 1.3). For Epstein frame, the mutual inductor is placed inside of the square, with axis perpendicular to the plane of the square (IEC, 2008). For other yokes, the mutual inductor should be positioned sufficiently far away from the main setup.

However, for the method to work, all three conditions must be met, as listed above. The area of the coil must be fixed, because the induced voltage due to $A_{coil} \cdot \mu_0 \cdot H$ must remain the same as for the mutual inductor. The sensing coil must be available without the sample inside, because this is required for tuning the mutual inductor in the first place. Therefore, it cannot be used for a coil wound directly onto the sample, or through the holes in the sample (Figures

3.1 and 3.9) because the area enclosed by the wires might not be known with sufficient precision to apply the compensation (Tang et al., 2013). And for the coil wound through drilled holes, it is not possible to produce *identical* coil without the sample present (Figure 3.12).

And the final condition makes this method applicable practically only for the systems with a closed magnetic circuit. When there is a considerable gap between the magnetising yoke and the sample, the primary current is not directly proportional to the H in the area of interest in the sample. This is typical for rotational systems, and this is why such compensation is not used and measurements of B are performed, rather than J. Otherwise, H could be measured directly from the magnetising currents, but in most systems, sensors such as H-coil or RCPs are used.

The compensation of air flux can also be applied numerically. As evident from Figure 4.73, the component which needs to be compensated out is directly proportional to $\mu_0 \cdot H$ and the scaling coefficient is proportional to the area of the coil. During magnetic measurement, the value of H is measured anyway. Therefore, it is possible to define the scaling coefficient, which will be fixed for a given coil or configuration. The condition is that the cross-sectional area of the coil must be known as this dictates the amount of voltage induced due to the $\mu_0 \cdot H$ component.

PRACTICAL COMMENT

The compensating voltage $V_0 = f(\mu_0 \cdot H)$ can be computed numerically and subtracted from the uncompensated voltage $V_2 = f(J + \mu_0 \cdot H)$. This works well in practice for magnetically closed circuits (Morishita et al., 2011) but can also be applied to circuits with an air gap, as long as H is measured with sufficiently high precision. For instance, extrapolation of H values towards the sample surface or vertical shielding might be required as discussed above and in Chapter 3.

The problem with the air flux can also arise from the leads or connections, as shown in Figure 4.72c. The sensing coil might be positioned tightly around the sample, but the leads could still enclose substantial area. This can be especially important for thin laminations (relatively small cross-sectional area) and sensors such as the needle probes. The connecting wires must be above the surface of the sample (Figure 3.20), and even if they are positioned as close as possible, there is always some area for the air flux to couple into the loop, for instance even due to the thickness of the insulation on the wire (Figure 4.17).

PRACTICAL COMMENT

For a fully assembled sensing head (like the one shown in Figure 4.17), it could be possible to derive the performance of the sensor without the sample. This could be achieved for instance by placing the sensor on a non-magnetic material and apply the same H as for a normal sample under test. For such a configuration, it could be possible to create a mutual inductor compensation, but again only for a configuration for which H is directly proportional to the magnetising current I_1, so only for magnetically closed circuits. Such a circuit is shown for example in Figure 3.22. However, no air flux compensation was applied to that circuit (Konadu, 2006). Numerical compensation could be an alternative in such case.

4.4.4 Measurements at Various Temperatures

Experiments at temperatures significantly different from room temperature have their own quite specific requirements. There are several examples of such measurements in the literature.

4.4.4.1 Cryogenic Temperatures

Magnetic properties at low temperatures are sometimes investigated (Soinski, 1984b; Miyagi et al., 2010; Miyamoto et al., 2011).

A setup for measurement of rotational and vector hysteretic properties was described in Miyamoto et al. (2011). A stator-like round magnetising yoke was used with an inner diameter of 77 mm and a rated power of 200 W. The whole yoke with the sample was installed inside of a cryostat capable of cooling down to liquid nitrogen temperature of 77 K.

Because of the thermal coefficient of expansion of the sample under test, the cross-sectional area of the B-coils was corrected accordingly. The correction could reach up to 7% (Miyagi et al., 2010). There was no correction for H-coils, as from previous experiments, it was not deemed to be necessary – only around 1% was expected (Miyagi et al., 2010).

The findings were that at higher excitations (>1 T) the permeability at 77 K was higher than at room temperature, but at the same time the power loss was also greater by up to around 15%. A possible explanation of the higher loss is the decreased resistivity of the sample so that the eddy current component is elevated accordingly. Similar behaviour was observed for uni-directional as well as rotational magnetisation (Miyamoto et al., 2011).

An interesting detail is that although the sample was circular (76 mm diameter) it had as many notches cut around the perimeter as there were slots (24) in the stator/yoke (compare also with Figure 4.14).

From a general viewpoint, the requirement for a rotational system operating at low temperature would be similar as other systems operating in such conditions, for example single-sheet tester (Miyagi et al., 2010). For obvious reasons, the tester must be designed not only to survive the temperature, but even direct submerging in liquid nitrogen. A miniaturisation of the system might be advantageous due to the volume of cooling liquid required (Miyagi et al., 2010).

4.4.4.2 Elevated Temperatures

Similarly as for the low temperatures mentioned above, there are few papers for elevated temperatures, for which uni-directional magnetisation experiments are published (Minov et al., 2011; Morishita et al., 2011).

The temperature range used for such measurements dictates the technical solutions which must be applied to the magnetising setup. For instance, typical enamelled copper wire with nominal thermal class as 155°C has the so-called 'cut through' temperature at 230°C (Elektrisola, 2016). At such temperature, the enamelled insulation gets damaged, so the wire cannot be used anymore, either for the magnetising coil or even for the sensing winding. Similar limitation applies to other typically encountered insulations such as PVC or even Teflon-based, which can have higher temperature rating.

Therefore, if the measurement temperatures are to exceed these values, then the wires must be additionally insulated or come with the appropriately thermally rated insulation.

For instance, the experiments described in Morishita et al. (2011) were performed up to 700°C, which is quite demanding for electrical insulation. In order to withstand such temperatures, the wire was furnished with a special insulation rated up to 1000°C.

The windings (magnetising and sensing) were not wound directly onto the sample, but on a special ceramic frame, also capable of withstanding up to 1000°C. This particular measurement was performed for toroidal samples, so the supporting frame had to be designed in such a way as to accept the toroidal sample, yet fully enclose it so that the windings were not touching the sample directly (Figure 4.74).

The reported results show that the maximum permeability significantly increases (up to a factor of 3) for elevated temperatures. At the same time, power loss reduces, mostly because of the reduced conductivity of the material under test. The tests were carried out on non-oriented electrical steel, 6.5% Si electrical steel, and generic construction steel (Morishita et al., 2011).

However, it is known that saturation magnetisation decreases at elevated temperatures (Soinski, 2001; Bozorth, 2003), so the usable part of the B–H magnetisation curve reduces for practical applications.

No results for rotational power loss at elevated temperatures were available at the time of writing this book.

4.5   COMPUTER AND MEASUREMENT SOFTWARE

The type of computer and measurements software used for digitally controlled measurements is practically irrelevant, as long as appropriate functionality if ensured by the given system.

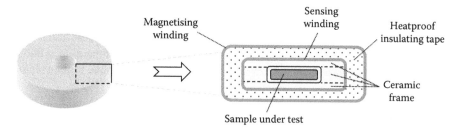

FIGURE 4.74   Special frame for measurement at elevated temperatures. (Adapted from Morishita M. et al., *Przeglad Elektrotechniczny (Electrical Review)*, R. 87 (N. 9b), 2011, 106.)

Many rotational magnetisation publications do not describe the type of computer or simply refer to a 'personal computer' as being part of the measurement system.

Over the years, the personal computer technology developed tremendously, from hardly existent or usable before 1980s, to multi-processor machines being commonly available at low cost for every measurement system.

The software underwent similar transition. Initially, data acquisition functionality had to be executed by low-level programming (like C+), which was more time consuming to develop. Memory size and speed of processing could be a problem to carry out fully automated rotational measurement, especially with slower computers in the past (Alnigrad, 1999).

In the twenty-first century, much better and more automated data acquisition devices became available. High-level programming software such as LabVIEW was developed sufficiently so that very complex programming and signal processing with lot of computer memory could be implemented and commissioned relatively quickly. Examples of a front panel of measurement systems developed in LabVIEW are shown in Figures 6.18 and 6.28 (Zurek, 2005; Piotrowski et al., 2015). LabVIEW is a product of National Instruments, which also manufacture data acquisition devices, so that a fully self-compatible hardware–software system can be completed.

Other researchers used for instance dSPACE (Wanjiku et al., 2014) or MATLAB® which also provides a data acquisition toolbox (Kai et al., 2015; MathWorks, 2016).

However, as mentioned above, it is irrelevant which type of software or hardware is used as long as the appropriate measurement procedure can be implemented. A very simplified algorithm for measuring the power loss is shown in Figure 4.75.

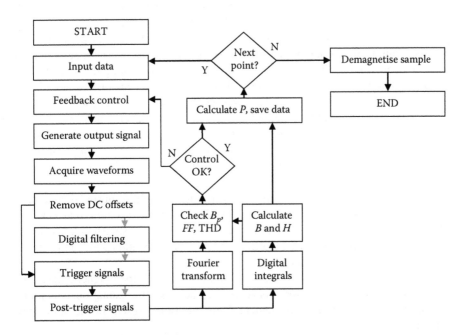

FIGURE 4.75 Example of simplified power loss measurement algorithm implemented in software; similar algorithm was followed for research presented in Zurek (2005).

This diagram does not include functionality of averaging, interaction with the operator, protection features, and many others. It is shown here just to indicate that even a fairly simple concept of power loss measurement might require a multi-step algorithm, because some data is used just for triggering, whereas other data for accurate measurements.

A very important part of the measuring software for rotational magnetisation is the feedback control of shape of the waveforms, for instance in order to achieve controlled circular magnetisation. Of course, the problem does not depend on the type of software, but there are several difficulties which must be tackled before stable and reliable operation is achieved. The waveshape control is described in detail in the following sections.

4.6 WAVESHAPE CONTROL

The problem of controlling the shape of exciting waveforms arises because magnetic properties of various samples can be directly compared only if the measurements are carried out under identical experimental conditions. Otherwise, the measured power loss and permeability values will differ and thus it would be impossible to definitely state if one material is better (e.g. less lossy) than a different material excited under different conditions.

As illustrated in Figure 4.1 a typical measurement system might comprise a number of hardware components. These components might differ between laboratories, and it is not possible to ensure that all of them will perform in *exactly* the same way. Magnetic properties of materials are known to strongly depend on many factors such as excitation, temperature, processing, mechanical stresses, etc.

Therefore, some means of standardisation of such magnetic measurements was required. In uni-directional alternating magnetisation, it was internationally agreed to use sinusoidal flux density as the reference criterion for creating the reproducible excitation conditions. This is the case as defined in the international standards for toroidal samples, Epstein frame and single-sheet tester (IEC, 2003, 2008, 2010).

The sinusoidal B condition stems from the fact that conventional power transformers in the electricity grid are fed with a sinusoidal voltage waveform at 50 Hz or 60 Hz. Therefore, sinusoidal B produced in the sample under test was a means of ensuring that the operating conditions similar to those expected in real transformers could be achieved.

The voltage waveform in the mains network is not perfectly sinusoidal, and there is a noticeable amount of odd harmonics, but these typically account to around 3% (Flanagan, 1993). Therefore, the flux in the core is even more sinusoidal, because it is proportional to an integral of the primary voltage (see Equation 3.5), so the higher harmonics are adequately reduced.

Hence, with the B waveform controlled to be sinusoidal, the magnetic measurements can be made much more *reproducible*, because the excitation conditions will be made the same and independent from variations present in the laboratory equipment.

A magnetic sample under test exhibits non-linear behaviour, especially above the 'knee' of the B–H curve. For lower excitations (e.g. below 1.2 T for electrical steels at 50 Hz), the system behaves in a quasi-linear way. But the higher the excitation, the more difficult it is to maintain the sinusoidal excitation, with some difficulties present at 1.7 T and extremely difficult above 1.9 T.

Sinusoidal magnetisation conditions were problematic before the dawn of commonly available transistor-based electronics because feedback circuits could not be implemented easily and low-output-impedance power supplies were used instead for this purpose (Drake, 1982; Beckley, 2000a).

Nowadays, such non-linearities are 'linearised' by using feedback, and this is recommended in the relevant standards for uni-directional magnetisation (IEC, 2003, 2008, 2010). The criterion of sufficiently sinusoidal shape is that the value of form factor of the *output voltage* waveform (*dB/dt*) is within $FF = 1.111 \pm 1\%$, which is the ratio of RMS to AVG value of a pure sine ($\pi/\sqrt{8} \approx 1.1107$). This criterion was devised for easy execution under fully analogue conditions, for instance in a system such as Epstein frame.

In order to verify if the value of FF is kept within the required range, it is enough to measure the secondary voltage with an RMS voltmeter and an AVG type voltmeter, with an accuracy of each better than 0.2%. When both voltages are measured, they can be manually calculated to check if the $FF = 1.111 \pm 1\%$ is met. However, the standard still recommends that the shape of the waveform is to be observed on an oscilloscope to ensure that it is sinusoidal (IEC, 2008). This is because even though the condition of $FF = 1.111 \pm 1\%$ is fulfilled the secondary voltage can be visibly distorted (Figure 4.76).

Because of the lack of any better set of criteria, similar reference approach was used for rotational magnetisation.

In a fieldmetric method, the data can be acquired as two orthogonal waveforms B_x and B_y. For a purely circular B, these two components would be sinusoidal and cosinusoidal, respectively (see also Figure 6.28). Hence, similar criterion of $FF = 1.111 \pm 1\%$ can be applied for each dB_x/dt and dB_y/dt simultaneously.

It is generally held that digital technology is capable of providing more stable feedback than purely analogue electronics, especially for magnetisation of highly anisotropic materials (Beckley, 2000a; Tumanski, 2011).

The following sections discuss in details several technical solutions applied by various researchers and laboratories.

PRACTICAL COMMENT

If the distortion comes just from the saturation of the sample for uni-directional magnetisation, then even small distortion can be detected by using the *FF* value. It is shown below (Figure 4.79) that a small local departure of the waveform from ideal can easily produce *FF* error exceeding 1%.

However, the *FF* criterion is insufficiently sensitive to certain types of distortion. For example, if the 100% of fundamental harmonic has 20% of 20th harmonic added to it, then the *FF* value will still differ by just 0.95% (Figure 4.76a). Clearly, the waveform is severely distorted and the *FF* criterion is incapable of detecting it. This is why the IEC Epstein frame standard includes a comment about visual inspection of the waveform.

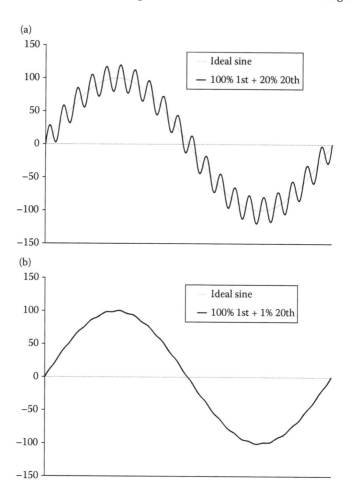

FIGURE 4.76 Comparison of distortions: (a) THD $= 20\%$ and *FF* error $= 0.99\%$; (b) THD $= 1.0\%$ and *FF* error $= 0.005\%$.

With digital technology, it is easy to apply another criterion called *total harmonic distortion* (THD). THD quantifies the ratio of amplitudes of all higher harmonics to the fundamental (Tumanski, 2006) and for ideally sinusoidal signal is 0% (zero distortion).

$$\text{THD} = \frac{\sqrt{h_2^2 + h_3^2 + \cdots + h_n^2}}{h_1} \cdot 100\% \qquad (\%) \qquad (4.13)$$

where h_1 is the amplitude of the fundamental frequency (V) and h_2, h_3, ..., h_n are the amplitudes of subsequent higher harmonics (V).

As shown in Figure 4.76b, a waveform controlled to THD $\leq 1\%$ is guaranteed to resemble very well the ideal sinusoidal waveform.

In general, the signal distortions under rotational and 2D magnetisation have much more complex character than those under uni-directional excitation. For this reason, the use of THD method is far more robust, as evident from Figure 4.76 and experienced by the author (Zurek, 2005, 2017a).

4.6.1 No Waveshape Control

There are several publications in which the results were acquired without waveshape control. In some cases, such approach is justified, as discussed below.

4.6.1.1 Single Phase

A single-phase supply or rather DC excitation can be used for torquemetric measurements (Figures 2.3, 2.6, 2.10 and 2.13). The gap between the magnetic poles of the electromagnet (or permanent magnet) is large, and the presence of the sample does not change the resulting H significantly. Therefore, no control is needed during the rotation of the sample within the field as long as the value is kept constant by the power supply (Zhu et al., 1997a; Anuszczyk et al., 2009).

This is especially applicable if the excitation is so high that it saturates the sample (Figure 2.10). The sample is saturated at all times during rotation, so the control of 'waveshape' is not really needed.

4.6.1.2 Two and Three Phases

If all parts of the magnetising apparatus are mechanically stationary, then the rotation of magnetic field requires two or more phases. Probably, the first rotating field applied to rotational magnetisation was produced by a three-phase system, as shown in Figure 2.8 (Brailsford, 1948).

Only later, the two-phase systems were used, in which the phases were set to sine–cosine and their amplitudes were varied. Both circular (Boon et al., 1965; Moses et al., 1973) and elliptical rotating fields (Young et al., 1960; Strattan et al., 1962; Young et al., 1966) could be produced this way. Anisotropic materials cannot be measured with larger amplitude because the 'circularity' of the produced rotation cannot be guaranteed, especially at higher excitations (Moses et al., 1973).

Demagnetising coefficient of the sample can be used as an advantage in a system with an air gap between the sample and the magnetising yoke. A large demagnetising coefficient of the sample will not significantly influence the rotating magnetic field at different directions despite large anisotropy of the material (Brix et al., 1982). Such effect was used for instance in a three-phase system with a large gap (Figure 2.15). Circular and elliptical fields can be set up just by controlling the amplitudes of all three phases, without an actual waveshape control applied to the magnetising currents (Fiorillo et al., 1992a). The amplitude control was also used by other researchers (Hasenzagl et al., 1995), with frequency control executed by a motor–generator setup (Parviainen et al., 2000).

Also, by definition, waveshape control is not used when studying the magnetisation patterns occurring in real magnetic cores (Moses et al., 1972; Radley et al., 1981; Moghadam et al., 1989; Moses, 1992; Krismanic et al., 2000; Kitz et al., 2003).

4.6.1.3 Other Non-Controlled

Waveshape control is not always practical or even not required. For instance, impulse (Takada et al., 1996) or industrial PWM excitation (Moses et al., 2003; Leicht et al., 2008) is not controlled, at least not in the same sense as sinusoidal control required for Epstein frame or similar methods. However, PWM excitation can be reproduced in a fully controlled way, if necessary (Zurek et al., 2004, 2007b; Zurek, 2017a).

As mentioned above, some waveshape 'control' could be achieved by using power supplies with sufficiently low-output impedance. In a magnetically closed circuit, if the resistance of the primary path and other impedances (like mutual inductor, Figure 4.73) is negligibly small compared with the reactance of the primary winding, then the applied excitation will be shaped accordingly to the voltage supplied from the power source. For instance, if this voltage is kept sinusoidal, then the resulting flux will also be driven to be sinusoidal without any actual 'active' feedback control (Drake, 1982; Matsubara et al., 1995a; Beckley, 2000a; Owzareck, 2015).

If the applied voltage could be kept constant, then the resulting flux density will continue to increase linearly as long as the supply voltage is held constant and any other impedance is negligible. Very high values of 'controlled' excitation can be applied in this way, reaching 100 kA/m for toroidal samples (Owzareck, 2015).

With such method, the supplied voltage is very well defined, because the internal impedance of the source is very low, so its behaviour approximates an *ideal voltage source*. The magnetising current flowing in the primary winding results from the equivalent impedance of the load. As the sample begins to saturate, the impedance of the primary winding reduces so the current increases accordingly, because the voltage does not depend on the current drawn from the source but is controlled independently. Similar effect is illustrated in Figure 4.35 in which the primary current increases rapidly when the magnetisation is pushed beyond the knee of the curve – yet the voltage is held sinusoidal.

When approaching the saturation of the sample, the reactance of the primary winding reduces enough so that other impedances become non-negligible and non-linear errors begin to creep in (Owzareck, 2015). Therefore, for this method, it is better to construct the primary winding in such a way as to create the highest possible reactance so that other errors are reduced by comparison. The idea is that at all times the internal impedance of the voltage source is much greater than the impedance of the load.

For instance, the IEC standard for single-sheet tester recommends parallel connection of the multiple sections of the magnetising winding (IEC, 2010), and this *reduces* its reactance and effectiveness of the low-impedance method. If the sections of the windings are connected in series, then the distortions are noticeably reduced with the low-impedance method (Matsubara et al., 1995a). Of course, the penalty is requirement for higher voltage required to achieve the same level of magnetisation.

The low-impedance source method was also demonstrated to be used together with an analogue feedback circuit (described below) (Sankaran et al., 1983).

However, the low-impedance source method cannot be used for controlling the magnetic circuits with an air gap. The magnetic coupling is affected for instance by a demagnetising factor of the sample, and the relationship between the primary voltage (driving the magnetising current) and the secondary voltage (output voltage of a B-sensing coil) might not be linear. Therefore, even if the supply voltage is sinusoidal, the output voltage might not be, especially for highly anisotropic materials.

Another example when waveshape control is not strictly required is measurements under very low-magnetising frequency for uni-directional excitation. A single cycle might take up to 300 s or around 3 mHz (Anderson, 2008). Under such quasi-static conditions, the shape of magnetisation is not really relevant because only hysteresis loss component is present. Therefore, the excitation can be applied effectively as triangular H, with a constant dH/dt (Anderson, 2008). It should be noted that even at such a slow magnetisation the waveshape still *can be* controlled if necessary (Anderson, 2011), for example by means of digital feedback, discussed in the following sections.

4.6.2 ANALOGUE FEEDBACK

Analogue feedback could be implemented more easily once inexpensive transistor-based electronics became commercially available (Beckley, 2000a). Negative feedback could be implemented with operational amplifier circuits (so-called *opamp*), together with amplification, impedance matching and filtering (Hollitscher, 1969; Zhu, 1994; Loisos, 2002; White et al., 2012).

The basic concept of analogue feedback is illustrated in Figure 4.77. The operating point is controlled by the reference input signal (reference voltage V_{ref}) which is supplied to the summing point with a positive sign (Figure 4.77a). The signal passes through the controlled object and produces an output signal (output voltage V_{out}). The output signal is connected back to the summing point, but with a negative sign.

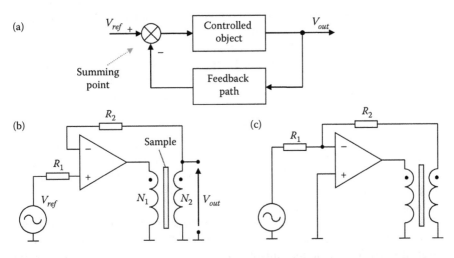

FIGURE 4.77 Concept of analogue electronic feedback: (a) basic block diagram of a closed-loop system with negative feedback, (b) non-inverting implementation and (c) inverting implementation. (Adapted from Zurek S., *Przeglad Elektrotechniczny [Electrical Review]*, R. 93, NR 7/2017, 2017a, p. 16.)

The summing point acts as an error detector. The reference and the output signals are fed back (hence the name *feedback*) to the summing point; they are compared and an error signal is created to correct for any deviation of the output from the desired state given by the reference. As a result, the output can be controlled even if the object is non-linear and external disturbances such as noise or environmental changes are present (Bakshi et al., 2010).

The action of the summing point can be conveniently realised by employing an operational amplifier (Cox, 2002) and can be used for controlling the waveshape of a magnetising system with a non-linear sample (Kubach, 1966; Mazzetti et al., 1966; Baldwin, 1970; Blundell et al., 1980; Lyke, 1982; Beckley, 2000a; Tumanski, 2011; Zurek, 2017a).

As illustrated in Figure 4.77b, the sinusoidal reference voltage V_{ref} can be connected to the non-inverting input (positive input of the summing point) and the output voltage V_{out} is connected to the inverting input (negative input of the summing point). The signal generated by the amplifier drives the magnetising current into the primary winding N_1. The primary current will be automatically adjusted in such a way as to compensate for any non-linearities so that the output signal induced across the secondary winding N_2 will closely follow the reference signal in shape and amplitude.

Functionality of a negative feedback loop can also be achieved by using the inverting implementation (Astrom et al., 2010) (Figure 4.77c). For all practical purposes, the inverting circuit performs identically, but the controlled output waveform has reversed polarity (Loisos, 2002; Tumanski, 2011). The resistors R_1 and R_2 are in the circuit in order to control the gain and match input impedances for correct operation of the operational amplifier.

The main advantage of the analogue feedback is that it executes the control truly in *real time*. The shape of the waveform is controlled instantaneously 'on the fly' so that the output follows the reference waveform from one *instant* of time to the next – hence the circuit operates as the so-called 'voltage follower' (Sankaran et al., 1983).

The real-time action of the analogue feedback is more desirable in some applications than that of the digital feedback (described below). However, the real-time control executed by very wide frequency bandwidth of the circuit can also be a drawback. For instance, it was observed that large-amplitude Barkhausen jumps (Figure 2.63) can be suppressed by analogue feedback circuit (Baldwin, 1970; Fiorillo, 2004).

The wide frequency bandwidth of the feedback loop also makes it susceptible to instabilities, especially for highly non-linear materials (Beckley, 2000a; Fiorillo, 2004). Such self-induced oscillations can arise very quickly and generate uncontrolled waveform into the power amplifier, which will attempt to amplify them – at the full available power in the worst case.

There are further difficulties in using the analogue circuits. The 'summing point' device must be a high-quality amplifier, low-noise, low DC offset, etc. A common universal operational amplifier might not be sufficiently good, and a much more sophisticated differential amplifier might be required (Zhu, 1994). Because of the DC coupling of the signal path, the DC offset might become a problem and AC coupling might be required through a high-pass filter (e.g. allowing for frequencies higher than 0.3 Hz). At the same time, there is little need for controlling the

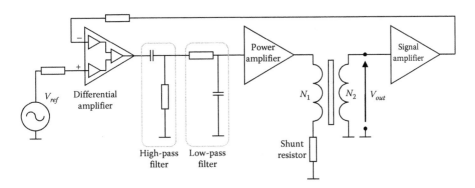

FIGURE 4.78 More detailed circuit diagram for analogue feedback. (Adapted from Zhu J.G., *Numerical Modelling of Magnetic Materials for Computer Aided Design of Electromagnetic Devices*, PhD thesis, University of Technology, Sydney, Australia, 1994; Loisos G., *Novel Flux Density Measurement Methods of Examining the Influence of Cutting on Magnetic Properties of Electrical Steels*, PhD thesis, Wolfson Centre for Magnetics, Cardiff University, Cardiff, the United Kingdom, 2002.)

full-frequency bandwidth and very high frequencies can be attenuated by means of a low-pass filter, as indicated in Figure 4.78.

For a small *B*-coil, the output signal could be very small and extremely noisy (see also Figure 4.58) and thus difficult to process accurately, especially at lower frequencies. Such signal might be required to be amplified before it is fed to the feedback loop so that it can be compared more effectively with the reference waveform (Loisos, 2002).

Analogue feedback can also control heavily distorted waveforms, not just sinusoidal (Hollitscher, 1969). This is because the feedback operates in real time and thus it will simply follow the shape of the references signal, whatever it might be.

The simplest analogue feedback circuit can be built on just a single operational amplifier (Zurek, 2017a), for example OP177 (Analog Devices, 2012). For a single-channel system (uni-directional magnetisation of a toroidal sample), the circuit can be as simple as shown in Figure 4.77b (the circuit does not show only the bi-polar DC supplies and the de-coupling capacitors on the supply rails as commonly used for operational amplifiers). The resistors R_1 and R_2 were 4.7 kΩ. The signal V_{ref} was connected to a sinusoidal signal generator whose frequency was set to 50 Hz and amplitude was slowly varied. An example of waveform acquired with such analogue feedback is shown in Figure 4.79a. The target waveform is achieved but only up to 1.55 T with an *FF* error of <1%. An attempt to obtain higher excitation ended up with either waveforms distorted with *FF* > 1% or oscillations.

With this particular operational amplifier, it was not possible to maintain sinusoidally controlled waveform for higher *B*. It should be noted that the power of the amplifier was sufficient because the peak current was delivered as required by the feedback. The distortion took place *after* the peak current, on the fastest slope. The reason was that although the full bandwidth of the operational amplifier was specified up to 0.4 MHz, the limitation of available voltage swing starts developing as low as 3 kHz, as shown in Figure 4.79b (Analog Devices, 2012).

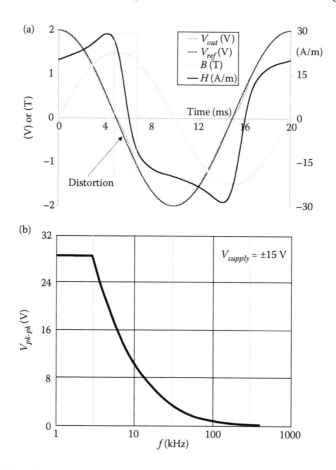

FIGURE 4.79 Signals recorded for a toroidal sample of grain-oriented electrical steels: (a) Magnetised at 1.55 T and 50 Hz, form factor of 'Vout' diverged by 0.99%, THD = 2.7%. (Reproduced with permission from Zurek S., *Przeglad Elektrotechniczny [Electrical Review]*, R. 93, NR 7/2017, 2017a, p. 16. Copyright © 2017 *Przeglad Elektrotechniczny [Electrical Review]*.) (b) Maximum output voltage swing vs. frequency for OP177. (Adapted from Analog Devices, *Ultraprecision Operational Amplifier OP177, Data sheet, Rev. G*, 2012.)

Analogue feedback can perform sufficiently well only under specific operating conditions. Control theory requires that for a system with a negative feedback loop to be stable it is required to have sufficient combination of both gain and phase margins, for example 6 dB gain margin with 30° phase margin (Baran et al., 1996; Bakshi et al., 2010). (This entire topic is too wide to be discussed here in full detail, and the reader is advised to refer to the plentiful literature on control theory.)

For example, it is obvious that if the system had gain of unity (0 dB) and the phase shift was 180° then the summation point in Figure 4.77a would add both signals with a positive sign, rather than having a negative feedback loop. Therefore, a condition of positive feedback would arise and the system would be *inherently unstable*. And because of the real-time operation, such instability could *demand instantaneously full power* from the power amplifier. This is potentially dangerous condition for both equipment and the user as discussed in Section 4.4.1.9.

From this viewpoint, analogue electronic feedback is thought to be inferior to digital feedback (Bertotti et al., 1993; Beckley, 2000a). There are indeed difficulties when applying the analogue feedback to rotational measurements, especially of highly anisotropic samples and/or low frequencies. Successful control would require even wider active frequency bandwidth, which is difficult to attain in such circuits. The effects are exacerbated by the variable amplitude characteristics (gain margins) of the whole setup, as discussed below for digital feedback.

In order to improve the performance, some researchers came up with a hybrid approach, combining analogue with digital feedback (Ludke et al., 2001), also for simultaneous multi-channel control (White et al., 2012). Also, it was shown that controlled magnetisation at very high-flux density can be obtained with just analogue circuitry but carefully designed (Qu, 1984).

However, most recent rotational and 2D research tends to employ digital feedback, as evident from the several last International 1&2-Dimensional Magnetic Measurement and Testing Workshops.

4.6.3 Digital Feedback

The concept behind digital feedback closely follows that of the analogue feedback described above. The controlled signal is compared with a reference waveform, and the exciting signal is modified accordingly in order to reduce discrepancy between the controlled and the reference signal (Zurek, 2017a).

The operation of multi-channel feedback is based on exactly the same principle as a single-channel feedback. Therefore, for simplicity, detailed description below is given for a single-channel version, with the simplest implementation based on a proportional controller. Multi-channel implications as well as more complex implementation are discussed later.

4.6.3.1 Iterative Principle

In digital domain, it is rather difficult to obtain real-time control at mains frequencies because of the very fast processing required. This is because the turnaround time for acquire–calculate–generate would need to be executed on-the-fly between two neighbouring sampling points. With 1000 points per cycle at 50 Hz, this would call for a processing speed of just 20 μs.

The method has been demonstrated in practice (De Wulf et al., 1998; Grote et al., 1998), but the published results were only up to 1.5 T and 100 Hz. Also, there is a danger of instability because the signal processing happens during real-time, so a small error (e.g. due to unforeseen glitch in the signal) could make it unstable – in the same way as for the analogue circuit.

Most commonly used data acquisition devices and their firmware do not support such fast turnaround times for acquire–generate processing. Fortunately, there is no strict requirement for real-time processing, because for magnetic measurements, full waveshape control can be achieved in an iterative way, as extensively referred to in the literature (Birkelbach et al., 1986; Bertotti et al., 1993; Matsubara et al., 1995b;

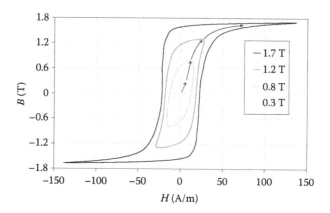

FIGURE 4.80 Intermediate values must be passed through before the maximum target value is reached during iterative control. (Reproduced with permission from Zurek S., *Przeglad Elektrotechniczny [Electrical Review]*, R. 93, NR 7/2017, 2017a, p. 16. Copyright © 2017 *Przeglad Elektrotechniczny [Electrical Review]*.)

Spornic, et al., 1998; Makaveev et al., 2000b; Mehnen et al., 2000; Gorican et al., 2002; Usak, 2002; Barbisio et al., 2003; Kondo et al., 2003; Zemanek, 2003; Zurek et al., 2005a; Zhang et al., 2007; Baranowski et al., 2009; Anderson, 2011; White et al., 2012; Stupakov et al., 2013; Zurek, 2017a).

Because of the iterative operation, the process of control is somewhat similar to demagnetisation, but applied in a reverse order – the controlled waveform starts from zero and is gradually increased to the target waveform, through the intermediate values, as illustrated in Figure 4.80 (compare with Figure 4.44).

4.6.3.2 Description of Simple Digital Feedback

The iterative principle of digital feedback is illustrated in Figure 4.81. At the beginning, (iteration 0) the generation has not started yet, so the generated signal S_{gen} is zero, and hence there is no magnetising current yet.

The target waveform S_{target} is set to the desired amplitude, but the acquired signal to be controlled $S_{control}$ is still zero. The difference signal S_{diff} between these two waveforms (for iteration i) can be computed as

$$S_{diff(i)} = S_{target} - S_{control(i)} \quad \text{(arbitrary units, e.g. V)} \tag{4.14}$$

where the notation with bold font means that the variables are digital waveforms (arrays, collections of values in the computer memory), and the subtraction is performed on a point-by-point basis so that the first point of $S_{control}$ is subtracted from the first point of S_{target}, the second from the second and so on.

The difference waveform S_{diff} is multiplied by some correction gain G_{corr} (single value, i.e. not waveform):

$$S_{corr(i)} = G_{corr} \cdot S_{diff(i)} \quad \text{(arbitrary units, e.g. V)} \tag{4.15}$$

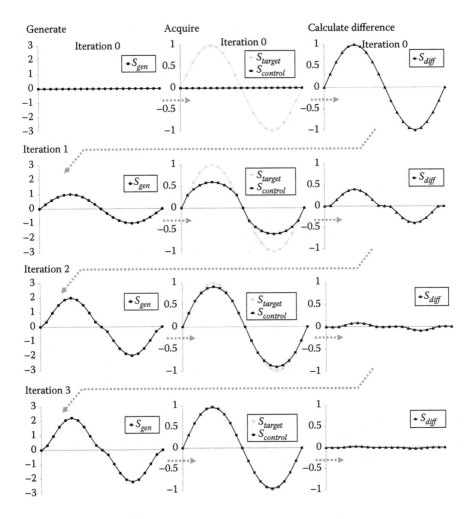

FIGURE 4.81   Simplified illustration of iterative feedback control.

and the result is added to the data from the previous iteration in order to calculate the signal to be generated:

$$S_{gen(i)} = S_{gen(i-1)} + S_{corr(i)} \quad \text{(arbitrary units, e.g. V)} \qquad (4.16)$$

Therefore, in the first iteration, $S_{gen(1)}$ with some amplitude is generated (iteration 1, Figure 4.81).

The digital signal S_{gen} is generated as a physical voltage V_{gen}, converted by the power amplifier into some excitation signal, the sample under test is magnetised to some level and some output signal is produced from the sensors. However, because the magnetising system and/or sample can be non-linear, the to-be-controlled signal coming back from the sensor might be distorted and have smaller amplitude as compared to the target waveform (iteration 1, Figure 4.81).

Nevertheless, the difference between the target and the controlled waveform is reduced, so the amplitude of $S_{diff(1)}$ (next iteration) is smaller than $S_{diff(0)}$ (previous iteration).

In each subsequent iteration, the process repeats, until the desired level of control is achieved. In each step, the information obtained from the difference waveform is used to modify the generated signal. With each step, the difference is reduced and the controlled waveform approaches the target waveform. If the system is non-linear, then the generated signal will be distorted accordingly so that the controlled waveform will achieve the shape and amplitude of the target waveform (iteration 3 in Figure 4.81).

As the acquisition and computation takes place, the algorithm dwells at a given current amplitude for at least one cycle (typically more than one) so that the current continues to be generated with a constant amplitude (Figure 4.82).

For an ideally linear system, the target waveform could be achieved theoretically in just one iteration if the gain of the digital feedback is equal to unity. However, this is not the case in practice because real systems are never completely linear.

The type of operation illustrated in Figure 4.81 is synonymous with a proportional controller. The consequence of a non-linear system and the feedback gain $G_{DF} < 1$ is that the amplitude of the difference reduces exponentially. Theoretically, it would take infinitely long time to suppress the difference (control error) to 0%. However, in reality, the error does not have to be zero, but just smaller than some pre-determined value, for example 0.1%. Therefore, the convergence time is finite and acceptable in practice (Zurek et al., 2005a; Zurek, 2017a), especially that additional small corrections in the software can help in reaching the goal quicker (e.g. aiming for slightly higher amplitude, etc.)

4.6.3.3 Importance of Phase Relationship

A simplified block diagram of a single-channel digital feedback is shown in Figure 4.83. The mathematical principle is contained in Equations 4.14 to 4.16, which

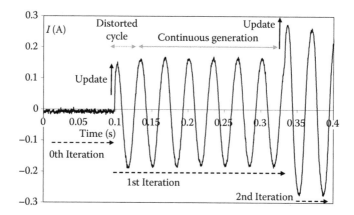

FIGURE 4.82 Example of magnetising the current measured through initial iterations. (Reproduced with permission from Zurek S., *Przeglad Elektrotechniczny [Electrical Review]*, R. 93, NR 7/2017, 2017a, p. 16. Copyright © 2017 *Przeglad Elektrotechniczny [Electrical Review]*.)

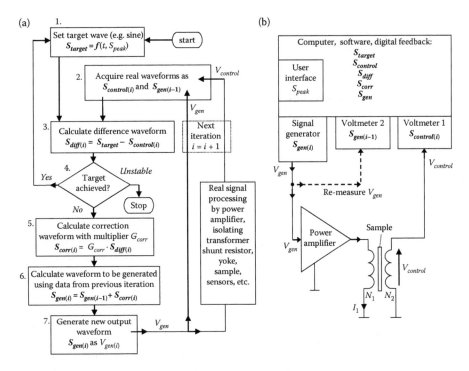

FIGURE 4.83 Simplified block diagram (*i*, iteration; *S*, digital signals processed in the software): (a) block diagram of signal processing and (b) block diagram of hardware implementation. (Adapted from Zurek S., *Przeglad Elektrotechniczny [Electrical Review]*, R. 93, NR 7/2017, 2017a, p. 16.)

are also shown in Figure 4.83 for better illustration. However, these two diagrams do not include the information critical for correct operation of digital feedback: the phase relationship between the processed signals.

For a given target point, the reference or target waveform S_{target} is kept constant in amplitude and phase as shown in Figure 4.81. Hence, in order to simplify the processing the controlled signal $S_{control}$ can also be triggered in such a way as to match the phase, as also shown in Figure 4.81. This guarantees that these two signals will have the same phase and the difference waveform can be calculated by a direct subtraction, as per Equation 4.14 or step 3 in Figure 4.83.

However, in a general case, there will be always some phase shift between the controlled signal $S_{control}$ and the signal S_{gen} generated as V_{gen} for the amplifier. Such phase shift is clearly visible in Figure 4.84, because at the zero crossings of $V_{control}$ the V_{gen} occurs at different instants of time.

The significance of this fact is that it *might not be possible* to infer the phase of the *physical* signal V_{gen} from the corresponding *digital* signal S_{gen} (NI, 2014d). If this is not possible for a given hardware, then such information must be obtained by some other means, for instance by re-measuring the generated physical signal V_{gen} as shown in Figure 4.83b. This can be achieved simply by splitting the signal and

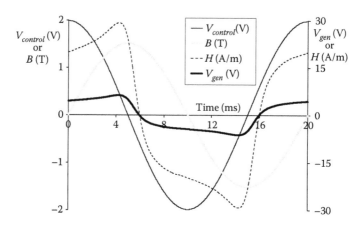

FIGURE 4.84 Typical phase relationship between various signals controlled by a digital feedback (compare with the signals shown in Figure 4.79a controlled by an analogue circuit, for the same sample and magnetising conditions); only $V_{control}$ signal is shown for clarity (the waveshape discrepancy was very small, with THD < 1%).

connecting one path to the input of the power amplifier, and looping back the other path to one of the analogue inputs.

With such approach, the data used for calculation of the new to-be-generated signal ($S_{gen(i)}$) should be based on the $S_{gen(i-1)}$ from the previous iteration, which is the re-measured voltage V_{gen}, rather than on purely digital data held in some S_{gen} buffer (which will have a different phase). Therefore, in each iteration, the S_{gen} signal must be re-measured from V_{gen}, in order to keep track of the phase differences after each new iteration.

In this way, the acquisition is carried out by triggering the controlled signal; therefore, the phase relationship between all the signals is always captured correctly.

PRACTICAL COMMENT

The action of re-measuring V_{gen} is affected by the processing errors of both generation and acquisition. The physical generated signal can differ from the ideal intended value because of the accuracy of the signal generator. And the acquisition is also less-than-ideal and further error is compounded. As a result, the processing of the digital signal $S_{gen(i)}$ generated as the physical signal V_{gen}, which is reacquired as $S_{gen(i-1)}$ for the next iteration, introduces some discrepancies which would be absent if just purely digital data was employed.

These small deviations are completely insignificant from the view point of stability of a digital feedback, measurement accuracy or waveshape control. After all, any non-linearity of the system (e.g. isolating transformer or non-linearity of the sample) is eliminated by the very action of the feedback.

However, the convergence can be adversely affected as far as the amplitude control is concerned. For instance, the feedback could converge to within 0.5% of the peak target value, but will be incapable of getting any closer, even after infinitely long time. This would happen because in each iteration the 0.5% difference would be 'lost' in the system so that it could be never compensated properly.

Of course, this can be eliminated by introducing an appropriate amplitude correction which takes into account the additional output–input discrepancies either globally or per each range. For instance, the amplitude of the generated signal can be artificially increased by some small amount, for example by a factor of 1.003 (or +0.3%). This will ensure that the digital feedback will attempt to generate a value slightly *higher* than the target, but it will pass *through* the target (see also Figure 4.80). But when all the coefficients are controlled, the data can be recorded and the measurement performed as soon as all the convergence criteria are met (e.g. *FF*, THD and B_{peak}), before any higher signals are generated. This is possible because at each iteration the feedback dwells for a few cycles, as illustrated in Figure 4.82.

Such small correction can also be added dynamically in a similar sense as the integral component 'I' in a proportional–integral–derivative (PID) controller. The software can very slowly increase/decrease the correction factor as required until all the convergence criteria are met.

The signal generator and voltmeters in Figure 4.83b can be a part of the same multi-functional data acquisition device, for example equipped with several analogue outputs and analogue inputs.

4.6.3.4 Gain of Simple Digital Feedback

It was stated above that the total feedback gain should be below unity so that $G_{DF} < 1$; otherwise, overshooting may occur, which will generate oscillations of amplitude. For correct measurements, the amplitude should be increased monotonically as recommended in the international standards (IEC, 2003, 2008); otherwise, in the vicinity of the target point, the sample cannot be treated as properly demagnetised.

For example, for uni-directional measurement, the value of permeability at 1 T for electrical steels is usually higher than at 1.7 T. Therefore, the feedback gain G_{DF} must be either dynamically adjusted during magnetisation or initially set sufficiently lower than unity so that during passing through the lower point with higher permeability (Figure 4.80) it always remains $G_{DF} < 1$. Otherwise, instability of the fundamental or some higher harmonics could occur.

It should be stressed that the value of the correction gain G_{corr} in Equation 4.15 is *not* synonymous with the total feedback gain G_{DF} so in a general case $G_{corr} \neq G_{DF}$.

As shown in Figure 4.83, there are several signal levels, and hardware as well as software components processing signals with various amplitudes. Each processing can contribute to the overall system gain G_{sys}, but some values might be impossible to define with sufficient accuracy. For instance, if the power amplifier has manually adjusted gain, then it will not be possible to know its *exact* value for an arbitrary position of the gain knob.

However, the system gain can be estimated through a fairly simple procedure (Zurek, 2017a). By the way of example, the assumption used in the description below is that the controlled signal is the secondary voltage from a *B*-coil.

With the digital feedback disabled (e.g. with $G_{corr} = 0$), some sinusoidal signal should be generated, with sufficiently small amplitude to avoid any saturation effects, so that all the signals will have shapes close to sinusoidal, including the generated physical signal V_{gen} and the to-be-controlled signal $V_{control}$ with their digital representations S_{gen} and $S_{control}$, respectively (Figure 4.83). Peak values of both these signals can be extracted, and the approximated value of the system gain can be calculated as (RMS values could also be used)

$$G_{sys} \approx \frac{S_{control,peak}}{S_{gen,peak}} \quad \text{(unitless)} \tag{4.17}$$

Therefore, the overall gain of digital feedback can be expressed as

$$G_{DF} \approx G_{corr} \cdot G_{sys} \quad \text{(unitless)} \tag{4.18}$$

and by rearranging Equation 4.18, the value of G_{corr} can be calculated as

$$G_{corr} \approx \frac{G_{DF}}{G_{sys}} \quad \text{(unitless)} \tag{4.19}$$

For oscillation-free operation, it is required to have $G_{DF} \leq 1$; hence,

$$G_{corr} \leq \frac{1}{G_{sys}} \quad \text{(unitless)} \tag{4.20}$$

The G_{corr} value estimated in this way can be fixed for all amplitudes, from small (quasi-linear operation) to high (strong non-linearity, Figure 4.85), for sinusoidal and highly distorted waveforms (like PWM), provided that there is a sufficient phase margin in the system (discussed below).

PRACTICAL COMMENT

It should be stressed that the value of G_{corr} and hence the value of G_{sys} should be re-evaluated for *each* configuration of the system (Zurek, 2017a), for example if *any* of the following changes (non-exhaustive list):

- Gain of power amplifier
- Ratio of isolating transformer
- Value of shunt resistor
- Type of magnetising yoke
- Number of turns of primary winding
- Number of turns of secondary winding (or B-coil)
- Effective cross-sectional area of the sample (or B-coil)

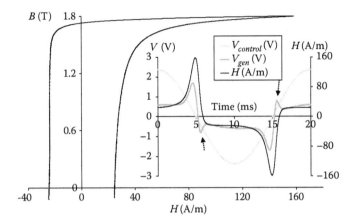

FIGURE 4.85 Results measured with digital feedback under controlled sinusoidal B at 1.8 T (see also Figure 4.86) performed with isolating transformer; V_{gen}, output voltage of signal generator; V_{out}, output voltage of B-coil (equivalent to $V_{control}$ hence with sinusoidal shape); $G_{DF} = 1$, $G_{corr} = 0.3$, $G_{sys} = 3.3$.

- Type of magnetic sample (change of permeability)
- Gain of signal pre-amplifier (Figure 4.78)
- Attenuation of voltage divider (Figure 4.47)
- Changing between controlling dB/dt (voltage), B or H
- Operating frequency (Figure 4.88)
- Changed connections (polarity of the wires could get reversed)

4.6.3.5 Convergence of Simple Feedback

An example of measurement controlled by digital feedback is shown in Figure 4.85. A toroidal sample of conventional grain-oriented electrical steel was magnetised up to 1.8 T at 50 Hz under controlled sinusoidal B (hence also sinusoidal voltage $V_{control}$).

It should be noted that an isolating transformer introduced a significant non-linearity between the shape of V_{gen} and the shape of the magnetising current (which would be the same as the shape of H in Figure 4.85, but scaled by the value of the shunt resistor and magnetic path length). The dotted arrows show that significant instantaneous voltage peaks with *reverse polarity* (as compared to the corresponding instantaneous values of H) had to be generated by digital feedback in order to reduce the magnetising current quickly enough to keep B sinusoidal.

The shapes of V_{gen} signal are compared in Figure 4.86 for the same measurement (1.8 T, 50 Hz) on the same sample, but performed with and without the isolating transformer. Both the shape of the V_{gen} waveform and its amplitude are completely different, because the system gain and the system non-linearity changed between the two configurations. Therefore, the feedback gain value had to be adjusted accordingly, in this case from $G_{corr} = 0.3$ to $G_{corr} = 0.036$, in order to ensure that in both cases $G_{DF} \leq 1$.

With the isolating transformer, the convergence from zero to 1.8 T, with an FF error $<0.2\%$, THD $< 1\%$ and B_{peak} error $<0.1\%$, took 231 iterations lasting 18 s

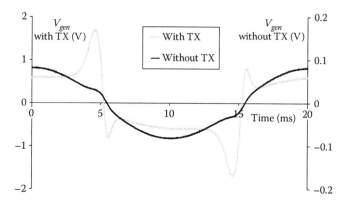

FIGURE 4.86 V_{gen} signals for configuration with ($G_{DF} = 1$, $G_{corr} = 0.3$, $G_{sys} = 3.3$) and without isolating transformer ($G_{DF} = 1$, $G_{corr} = 0.036$, $G_{sys} = 28$) – note the change of scales on the vertical axes; other signals were identical to those in Figure 4.85 so they are not shown for clarity (TX denotes transformer). (Reproduced with permission from Zurek S., *Przeglad Elektrotechniczny [Electrical Review]*, R. 93, NR 7/2017, 2017a, p. 16. Copyright © 2017 *Przeglad Elektrotechniczny [Electrical Review]*.)

(Figure 4.87a). On the other hand, without the isolating transformer, the same control (1.8 T, 50 Hz) converged in just 22 iterations or 2 s (Figure 4.87b).

For lower excitation, the convergence times are significantly shorter. On the other hand, certain hardware configurations do not permit achieving higher controlled excitation. For instance, the turn ratio of the isolating transformer used for measurements shown in Figure 4.85 called for the peak of V_{gen} voltage of around 1.7 V. The maximum output of this particular signal generator was 10 V (NI, 2008b), so due to high non-linearity of the sample, not much higher controlled excitation could be applied, because 1.9 T would require V_{gen} in excess of 10 V, which could not be generated with this setup. The isolating transformer was set to turn ratio 5:1 (see also Figure 4.39).

PRACTICAL COMMENT

As evident from Figure 4.86, because of the additional voltage drop due to the isolating transformer (as well as the shunt resistor), the large current peak (large H peak) required generation of appropriately high amplitude of V_{gen}. However, the B_{peak} error was reduced to below 10% within the first 10 iterations (Figure 4.87). Therefore, with each following iteration, the correction was <10% (as per Equation 4.14), and this was partly the reason why so many iterations were required. The convergence was completed with very good control with $FF < 0.25$ and THD < 0.9% (Figure 4.87a) – it just took longer because no feedback acceleration techniques were used.

When the isolating transformer was removed, the additional impedance in the magnetising path was reduced significantly, so the required voltage to drive the same current was also reduced (around a factor of 10, as evident from the vertical axis scales in Figure 4.86). The gain of the power amplifier

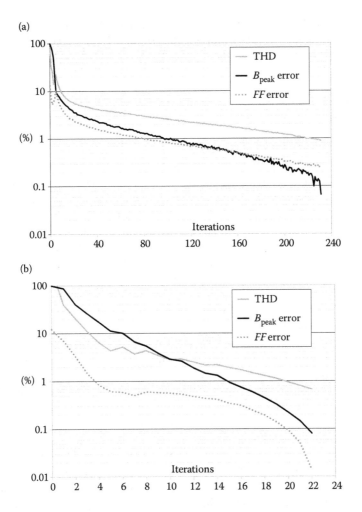

(a)

(b)

FIGURE 4.87 Typical convergence curves for a toroidal sample of conventional grain-oriented steels (see also Figure 4.86) from zero to 1.8 T at 50 Hz: (a) with isolating transformer and (b) without isolating transformer.

was fixed, but with lower required drive voltage, proportionally lower V_{gen} was also required. With the isolating transformer, the peak of V_{gen} was around 1.7 V (Figure 4.86), whereas without the transformer it was only around 0.09 V (around 19 times smaller).

Therefore, even if the same correction of <10% in each iteration is used, then the peak voltage will be reached in significantly fewer iterations. Indeed, the number of iterations required for the convergence reduced from 231 to just 22 (around 10 times faster). The wave shape parameters were even better with $FF < 0.02$ and THD < 0.7%.

It should be noted here that the speed of convergence was a direct effect of the impedance in the magnetising path and *not* the value of gain used in digital feedback. In both cases, the value was set so that $G_{DF} \approx 1$, which was achieved by setting $G_{corr} = 0.3$ in the first case, and $G_{corr} = 0.036$ in the latter. Therefore, in the second case, the G_{corr} value was around *10 times lower*, yet the feedback converged *10 times quicker*. This shows that the speed of convergence is dictated mostly by other factors, not the gain of the digital feedback, which must be set at such value that always $G_{DF} \leq 1$ (Zurek, 2017a).

4.6.3.6   Amplitude and Phase Margin

Full analytical amplitude and phase margin analysis is difficult and time consuming for a system containing multiple components with non-linear amplitude and phase characteristics (Bakshi et al., 2010). Very good working knowledge of control theory would be required to carry out such analysis in order to fully derive all the involved functions and stability criteria. The stability criteria would change for *each* hardware configuration (e.g. gain of power amplifier, number of turns of B-coil, value of shunt resistor, etc.). Such comprehensive analysis is beyond the scope of this book. This is the case especially that mathematical equations for some of the required components are not possible to derive in a universal way, like the cross-coupling between two or more channels for rotational magnetisation, which is highly non-linear and different for each type of material of the sample.

However, the problem can be tackled on a semi-empirical basis, with some extreme cases serving as examples. As evident from Figure 4.77a, the to-be-controlled signal is subtracted from the reference signal (negative feedback). Therefore, if the system introduced additional phase delay of 180°, then subtraction would change into addition and positive feedback would take place, with a *guaranteed* instability. This is an extreme case, but in practice, a phase shift greater than 45° becomes increasingly difficult to control, if possible at all.

At a specific frequency of operation, the basic gain of digital feedback can always be adjusted by changing the value of G_{corr} (Equation 4.15 and Figure 4.83). However, in general, the gain value for other harmonics is not the same and this can cause problems. Typical phase and gain characteristics for a single-channel magnetising system are shown in Figure 4.88. It can be seen that the black curves (no isolating transformer) result in the smallest phase shift and the most constant gain for all harmonics. For such configuration, the digital feedback is most stable and adjustment of just G_{corr} is sufficient to avoid instability in most cases, even when controlling waveforms with large amount of harmonics present (e.g. when emulating PWM-type waveforms).

On the other hand, the setting which resulted with the widest changes of both phase and gain changes was for isolating the transformer set to 250:250 ratio. The gain for higher harmonics dropped significantly as compared to 50 Hz and at the same time the delay increased. At 2 kHz, the gain was around 10 times lower (than at 50 Hz) and the phase delay exceeded 45°. As a result, the convergence is very slow for such configuration, and at higher excitation, the condition of controlled excitation might be impossible to achieve. The value of phase margin required for stability often quoted in control theory is 45° (Baran et al., 1996; Bakshi et al., 2010); therefore, with smaller phase margin, there can be problems.

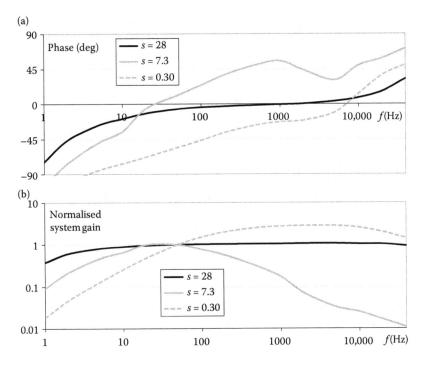

FIGURE 4.88 Typical phase (a) and gain (b) characteristics for various configuration of a magnetising system: $s = G_{sys} = 28$ ($G_{corr} = 0.035$) without isolating transformer, $s = G_{sys} = 7.3$ ($G_{corr} = 0.137$) with isolating transformer 250:250, and $s = G_{sys} = 0.30$ ($G_{corr} = 3.3$) with transformer 250:50; gain curves normalised to 50 Hz. (Reproduced with permission from Zurek S., *Przeglad Elektrotechniczny [Electrical Review]*, R. 93, NR 7/2017, 2017a, p. 16. Copyright © 2017 *Przeglad Elektrotechniczny [Electrical Review]*.)

PRACTICAL COMMENT

The curves shown in Figure 4.88 are fairly easy to measure in practice. With the feedback disabled, sinusoidal V_{gen} is driven at one frequency, for example 50 Hz with an arbitrary chosen value of amplitude, for example 100 mV. The phase delay between V_{gen} and $V_{control}$ is measured, as well as their amplitude ratio (i.e. the system gain as per Equation 4.17). Then, a frequency is changed and the amplitude of V_{gen} is adjusted, so the amplitude of $V_{control}$ is the same as for 50 Hz, and the phase and gain are re-measured at the new frequency. The operation is repeated for the whole operating spectrum, as shown in Figure 4.88.

(Alternatively, the V_{gen} amplitude could be held constant for various frequencies. However, this would change the level of excitation in the sample, which is linked to changed permeability, which in turn impacts the detected system gain. But it should be still good enough for qualitative investigations.)

Such purely empirical analysis can be useful for finding out why particular configuration is more prone to instabilities. For instance, there is a dip in the phase curve for $s = 0.30$ in Figure 4.88a, between 1 and 10 kHz. It is likely that this is caused by some partial self-resonance of the isolating transformer, which can exacerbate any instability issues.

As evident from Figure 4.84, the phase delay between B-coil voltage and B calculated from it is exactly 90° (B lags V). This is true for any frequency and any shape of voltage because it arises from mathematical relationship (B is an integral of V). Therefore, if B was to be used as the basis for feedback, then the phase characteristics in Figure 4.88a would have an additional delay of 90° on the top of the presented curves for voltage. This creates a phase response, which in most part represents phase shift greater than 45°, and as such will be unstable when attempted to be controlled by a simple proportional feedback.

Theoretically, B could be used as a target waveform for feedback, but appropriate phase compensation modules in the feedback algorithm would have to be added in order to compensate for the additional phase shift (Zurek et al., 2005a).

4.6.3.7   More on Triggering and Signal Processing

Some aspects of signal triggering were already discussed in Section 4.4.2.6. However, there are features specific to digital feedback, or just practicality of using it with various magnetisation waveforms.

As discussed above for the digital feedback, the output voltage of B-coil can be used as the signal to be controlled. Therefore, it would be ideal to trigger on such signal, for example *precisely* at its zero crossing. However, in practice, this might be impossible to achieve, not only precisely, but even *coarsely*.

For instance, let us consider the voltage waveform 'V_2' shown in Figure 4.46. The waveform consists of a series of sharp 'spikes', and some other rather noisy portion remaining around zero. It is impossible to define *precisely* at which point the zero crossing of interest occurs, especially with low-level hardware or software.

For such waveform, repeatable triggering will not be possible not only for zero but also for any level. Indeed, there are fast rising and falling edges which would be normally very suitable, but the point to note is that there are *several* of such edges in each half-cycle. It is therefore not possible to ensure that the triggering will occur at the 'first' edge, because it is impossible to know when a given iteration will finish and when the circuit will be 'armed' for the next triggering instance.

Similar one applies to multi-pulse waveforms such as PWM (Figure 4.89) or very noisy waveforms (low-frequency, small B-coil with single turn) as shown in Figure 4.58.

However, as evident from Figure 4.46, the 'B' waveform corresponding to the 'V_2' waveform is far less 'noisy', with only one rising-edge zero crossing, which would be quite suitable for triggering. If the triggering is executed by the high-level programming, then it is irrelevant if the voltage or B waveform is triggered, because the calculations will be performed in a similar way (see Figures 4.61 to 4.64), but the change in phase of the signal must be taken into account accordingly.

B waveform is calculated from voltage by means of numerical integration which inherently suppresses any high-frequency events such as spikes or noise (Figure 3.56). In most cases of a 'simple' waveform, it would be therefore sufficient just to use the

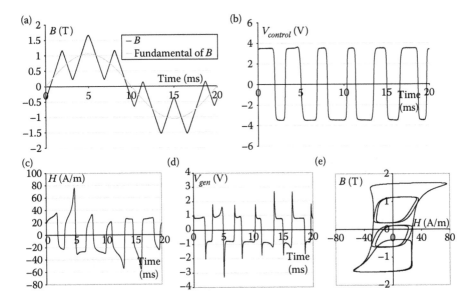

FIGURE 4.89 Example of waveforms with emulation of a two-level PWM excitation with 6 pulses and modulation index of 0.4: (a) B and its fundamental, (b) $V_{control}$ (proportional to dB/dt), (c) H, (d) V_{gen} and (e) asymmetrical B–H loop with several minor loops.

B waveform as the basis for triggering. For instance, the 'B' waveform in Figure 4.46 would be perfectly suitable for finding a well-defined and single-valued zero crossing.

However, under certain conditions, even using the 'smoothed' B waveform may be insufficient. An example is shown in Figure 4.89 for a two-level PWM excitation. The voltage waveform has several pulses, each of which crosses zero (Figure 4.89b), so triggering cannot be based on it. But similar one takes place for the B waveform because *several* slopes cross zero and it is not possible to guarantee that triggering on such waveform will produce the same phase for each iteration of the digital feedback.

The solution is to use additional filtering based on Fourier transform as shown in Figure 4.51. First, the voltage (Figure 4.89b) is converted to the B waveform (Figure 4.89a). Then a Fourier filter is applied to the B waveform, so as to extract only its *fundamental* harmonic. As a result, pure sine is extracted from the waveform with multiple zero crossing, and a solid base for active triggering is produced because there will be *precisely* one rising-edge zero crossing (Figure 4.89a), by definition of the Fourier transform.

However, one more operation might be required in order to ensure that the triggered signal and the reference waveform have *exactly* the same phase. Namely, a complex reference waveform might need to be filtered and actively triggered in the same way, and then the original reference waveform needs to be triggered passively with the timing information extracted from the filtered waveform. This needs to be carried out just once for each measurement point, because the reference waveform is constant by definition. A block diagram of such implementation is shown in Figure 4.91.

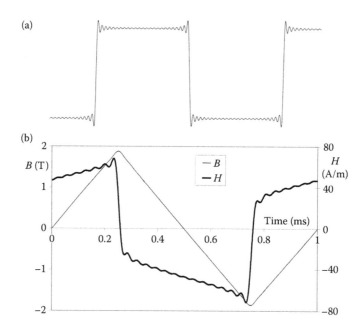

FIGURE 4.90   Gibbs phenomenon for 49 harmonics (a) and its influence on *B* and *H* waveforms (b). (Reproduced with permission from Zurek S. et al., *Przeglad Elektrotechniczny [Electrical Review]*, [NR 12/2005], 2005c, 5. Copyright © 2005 *Przeglad Elektrotechniczny [Electrical Review]*.)

PRACTICAL COMMENT

The filtering based on Fourier transform shown in Figure 4.51 works very well for mathematically 'smooth' waveforms. But if the filtered waveform has sharp corners such as rectangular, trapezoidal or triangular, then apparent distortions are produced around such corners due to the so-called Gibbs phenomenon (Jerri, 1998) which can produce 'ripples' in the magnetising waveforms, extending up to very high frequency, and in theory to infinity (Figure 4.90). For this reason, Fourier transform filtering is less useful for such waveforms and other techniques might be more effective in practice, such as low-pass Butterworth filter (NI, 2016b), or combination of both. Such filter was used to avoid the 'ripples' which would otherwise arise in the $V_{control}$ waveform in Figure 4.89b (so the corners became 'round').

Of course, all other signals will not be filtered in the same way, so the power loss can be calculated correctly, but they will be all triggered passively by using the timing information generated by the active trigger (Figure 4.91).

It should be noted that with such approach the actual phase of the voltage signal becomes irrelevant for the correct operation of digital feedback, as long as appropriate phase relationship is preserved between *all* the signals. As can be seen in Figure 4.84, the single cycle of the controlled voltage waveform does not start from zero crossing, but rather from its peak, because it is the B waveform which is used for triggering.

4.6.3.8 Acceleration of Convergence

Proportional digital feedback described above is the simplest possible implementation. There are many other ways in which digital feedback can be achieved, also including such techniques as evolutionary algorithm (Mehnen et al., 2000), contraction mapping principle (Ragusa et al., 2006) or artificial neural networks (Baranowski et al., 2009).

Another way of achieving faster convergence is to employ a PID controller (Baranowski et al., 2009). Several other techniques relying on the knowledge of circuit parameters were also presented (Matsubara et al., 1995b; De Wulf et al., 1998; Makaveev et al., 2000b).

However, it is not straightforward to find a global optimum for stability of a PID controller (De Wulf et al., 1998), and for some configuration might not be possible at all, due to inherent instability conditions (Bakshi et al., 2010). Thus, the problem of stability is pertinent to *any* type of negative feedback, including proportional. If the value of gain is set to too high, then oscillations *will* occur even in an *ideal linear system* with an iterative proportional controller. The problems are usually exacerbated when controlling the non-sinusoidal waveforms or highly non-linear systems.

Therefore, even for a proportional feedback, all the relevant variables must be set accordingly to avoid oscillations or instabilities. It was shown that the simple proportional digital feedback can be very robust (Zurek, 2017a), and in such case, it is sufficient to adjust only a single value (correction gain G_{corr} in Figure 4.83) in order to ensure stability, but provided that the phase margin of the system is adequate. But as discussed above, this value must be adjusted for *each* system configuration, and if incorrect value is set, then at best the convergence is very slow, and at worst the system will become unstable.

The same applies for more complex techniques. Algorithms employing several variables, such as PID coefficients or circuit parameters, will require adjustment of all these values simultaneously for each new system configuration. This procedure might not be too difficult to apply in practice, but the operator must remember to choose the right set of coefficients in order to avoid problems. Some researchers used an automatic procedure for detecting the system parameters (Bajorek et al., 2000b; Makaveev et al., 2001), which also use difference between predicted and measured H for acceleration of convergence. Acceleration can speed up convergence for example by a factor of 6 (Matsubara et al., 1995b).

PRACTICAL COMMENT

The proportional feedback described above uses only the difference proportional to the output voltage of the B-sensor so that the correction has some proportionality only to the B waveform (due to mathematical relationship between B and voltage). Towards the end of convergence, the difference between the target and controlled waveforms is very small and thus the correction is also very small, especially at high excitation like 1.9 T for electrical steels (Zurek, 2017a). One of the ways to accelerate convergence would be to scale the correction by the V_{gen} waveform, on a point-by-point basis. In a very rough approximation, there is some resemblance between the shape and amplitude of V_{gen} and the H waveforms (Figures 4.84 and 4.85).

However, such additional instantaneous scaling will modify the total feedback gain and care must be taken that the correction gain G_{corr} is set to an appropriately lower value in order to ensure that *at all times* the condition $G_{DF} < 1$ is met. This can be achieved for instance by having another branch in the block diagram in Figure 4.83a, which will deal with the acceleration by scaling through the V_{gen} waveform. Such extra feedback branch can have its own independent correction gain. Clearly, this leads to more potential for unstable operation.

In a wider sense, such approach can be optimised with separate threads, each working for the inductive or resistive part of the voltage drop across the primary winding circuit, as demonstrated in several publications (Matsubara et al., 1995b; Spornic, 1998; Barbisio et al., 2003).

However, the main goal of all feedback acceleration techniques is by definition to converge to a given target waveform with fewer iterations. Such techniques should be used carefully because of problems arising from a completely different viewpoint, due to the nature of the way the digital signals are converted to analogue voltages. This topic was discussed in some details in Section 4.4.1.3 (see Figure 4.50).

Namely, the update of the generated waveform is usually not smooth but sudden, and a glitch can be produced when a new buffer data is generated as illustrated schematically in Figure 4.50 and shown with real waveforms in Figure 4.82. The size of the glitch will be directly proportional to the size of the adjustment, and it will be *greater* if *fewer* iterations are required for convergence, as it will be the case for any acceleration technique (Zurek, 2017a). Needless to say that such glitches should be avoided or at least minimised for better reproducibility of magnetic measurements.

Therefore, it might be better to ensure that no acceleration technique is used and the gain of the proportional controller is *reduced* so that the upper size of the glitches is limited to some value, for example 1% between subsequent iterations. As evident from Figure 4.84, the zero crossings of $V_{control}$, B and V_{gen} generally do not coincide. Therefore, regardless whether the digital feedback is based on voltage or on flux density waveform, the current *will have* an update glitch. This can be avoided only if the update is applied at the zero crossing, but generally it is not the case (Zurek, 2017a). However, it might be possible when the controlled waveform is H, because different phase relationship will exist between the controlled and the generated waveforms.

A combination of solutions is also possible, whereby at lower excitation (quasilinear region) the gain is substantially lower, and the acceleration is used only in the final part of convergence, where the changes are much smaller and take many more iterations (see also Figure 4.87).

One of the cases when convergence acceleration is a highly desirable feature is when the power loss dissipation in the sample is so significant that it is capable of considerably increasing the temperature of the sample.

During long convergence of the feedback, the temperature of the sample could change by a large amount, and this could significantly impact reproducibility of measurements. Elevated temperature can increase the resistivity of metals, and the eddy current loss component could be detectably lower than at non-elevated temperature. Such problems can arise especially at higher excitations frequency.

A rotational system described in de la Berriere et al. (2015) can operate up to 5 kHz, and at such frequencies, the losses in the sample can reach significant values. The system is equipped with forced-air cooling (see also Figure 6.7). Measurements on samples submerged in cooling oil can be another way of reducing the influence of excessive heating.

PRACTICAL COMMENT

Another practical problem with the acceleration of convergence might be auto-ranging of input ranges. The acquisition device cannot know beforehand what level of voltage to expect and it can auto-range only *after* given voltage was applied to its input. If a correct range is not set, then the signal could be acquired as clipped (Figure 4.56), even though in reality it is not. If such signal is passed into the feedback node, then a wrong correction will be calculated and generated, thus potentially further increasing the size of glitches or even directly leading to oscillations or other dangerous instabilities.

For this reason, it is a good idea to apply at least some sort of filtering to the signals used in the digital feedback, as shown in Figure 4.91. This includes the filtering just before the output signal is generated. In this way, the frequency spectrum is limited to only the harmonics important from the viewpoint of control and a potential for instability is reduced.

But this example also illustrates why a convergence which is too fast might be problematic in practice, and if it is not absolutely necessary, it should be avoided.

4.6.3.9 Protection against Instability

The risk of feedback instability was already mentioned multiple times. It is very easy to set the gain to a wrong value or connect the signals with wrong polarity which will produce unstable operation or positive feedback demanding full power from the amplifier in just few iterations. Such conditions could be dangerous to user and/or equipment, and suitable protection should be built in the software in order to minimise the consequences.

PRACTICAL COMMENT

In the experience of the author, it is even possible to damage the power amplifier when worst-case instability occurs. The algorithm triggers and calculates differences automatically and also generates the output signal V_{gen}

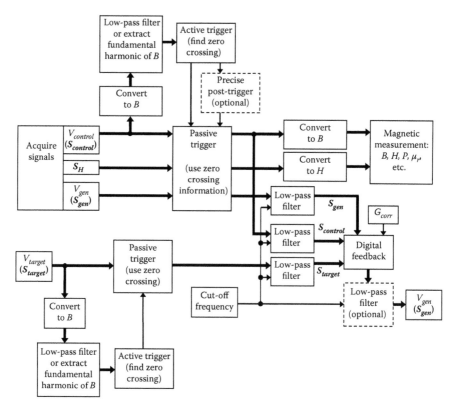

FIGURE 4.91   Block diagram for illustration of active and passive triggering; thick lines represent waveform data, and thin lines single value data.

automatically (Figure 4.75). An unforeseen condition can always arise, for instance a glitch induced in the measured voltage due to some high-power device switched on in the vicinity of the measurement setup. If the triggering is impacted so that it triggers the signals out of phase, then for a large gain positive feedback will be applied and the output voltage will reach 'full deflection' in just a few steps. Such instabilities are usually exacerbated by any feedback acceleration technique, which will produce *higher* signals *faster*.

As a result, the power amplifier will attempt to amplify such a high signal, which might mean full power delivered immediately to the magnetising yoke. This is potentially a dangerous condition and should be prevented.

For instance, as shown in Figure 4.83a, at each iteration, the stability of the feedback should be checked. If instability is detected, then the operation should be shut down automatically and immediately by the software, for example by setting the output signals to zero or initiating a demagnetising procedure.

The detection can be based for example on the value of THD. For a sinusoidal waveform, the value of THD is usually quite low, even when distorted with some high harmonics due to saturation. Therefore, a criterion of THD < 50% (or some other value) can be used, and if exceeded, the protection should shut down the operation (Zurek, 2017a).

Another useful condition is the amplitude of the output signal calculated for generation. If this value exceeds some amplitude, for example 90% of the full range, then also the feedback algorithm should be terminated.

Also, as discussed in Section 4.4.1.9, the voltages generated across the shunt resistor or from sensors such as the B-coil might exceed the safe value for overvoltage protection of data acquisition device. Hence, if the detected amplitude reaches the extremes of the highest input range, then the procedure should be stopped.

Yet another example would be power dissipated on a shunt resistor (Figure 3.42). The amount of dissipated heat can damage such resistor. However, the power rating of the resistor and its heat sink are usually known, so the software can calculate the power dissipated at each iteration and terminate the measurement if the rating is exceeded, even though no actual instability is detected.

The reader is encouraged to read again Section 4.4.1.9, because powerful amplifiers can generate *lethal* voltages and all steps should be taken to ensure that any risk is reduced to safe level. For instance, the power amplifiers used for rotational magnetisation described in de la Barriere et al. (2015) can deliver up to 150 V, 40 A and such power source would be outright *lethal* if touched directly, even under controlled conditions.

4.6.4 CONTROL OF *B* OR *dB/dt*

The choice between controlling *B* or *dB/dt* (voltage) waveform is a matter of convenience rather than any real difference. As evident from Figure 4.79a for an analogue feedback and Figure 4.84 for a digital feedback, the relationship between *B* and *dB/dt* is defined *completely* by the mathematical functions of integral or derivative (depending if converting *V* to *B* or vice versa).

Therefore, if one waveform is controlled, automatically the other is controlled to the shape resulting from mathematical calculations.

Sinusoidal waveforms are special in this regard, because sinusoidal *V* results with sinusoidal *B*, just with different amplitude and phase. However, because of the derivative character of *V*, any discrepancies between the target and the real waveform especially for higher harmonics are kind of 'amplified'. The result is akin to having the derivative 'D' component in a PID block. The feedback acts faster on such changes and thus faster convergence can be achieved, without actually implementing such functionality on purpose.

It is simply enough to use *V* instead of *B* to have two very important advantages: firstly, better response to higher harmonics, and secondly, better phase margin for such system (Figure 4.88a).

4.6.5 CONTROL OF *H*

The principle behind controlling the *H* waveform is the same as controlling *B*, with the same functionality (the same components in the block diagram in Figure 4.83). However, obviously, the target signal has different units and different amplitudes, and this must be taken into account in calculations.

In theory, the control can be based either on *H* or on *dH/dt* depending whether for a given system *H* is directly proportional to the magnetising current (shunt resistor or Hall-effect sensor) or the *dH/dt* information is available (e.g. *H*-coil). This depends on the hardware configuration of a given system, and appropriate choices have to be made and implemented accordingly in the software.

There are a few publications reporting results obtained under controlled rotational *H* (Gumaidh et al., 1993b; Hasenzagl et al., 1996; Zhu et al., 1997a; Zurek, 2005; Zurek et al., 2006c). For materials with low anisotropy, the difference between controlling *B* or *H* is very little, and hence the achievable level of controlled excitation is also very similar.

The situation is different for highly anisotropic materials, which represent more demanding conditions for controlled excitation.

Some examples of controlled *H* rotational measurements are shown in Chapter 7.

PRACTICAL COMMENT

In practice, controlled *H* conditions might be more difficult to achieve than controlled *B* magnetisation for circuits with air gaps, especially for materials with high anisotropy under rotational magnetisation (Zurek, 2005).

Stability of *H* control can also be affected by the phase margin, depending on the type of signal used as the controlled waveform and hardware components, as evident from Figure 4.88.

4.6.6 CONTROL OF NON-SINUSOIDAL WAVEFORMS

Controlling of non-sinusoidal signal is similar to controlling a sinusoidal waveform. It is just that the target waveform has different shape and the feedback operates in practically the same manner, even if the target waveform has a very complex shape.

For instance, a PWM waveform can be emulated by controlling the shape of the applied voltage, as shown in Figure 4.89. It should be noted that the voltage waveform (Figure 4.89b) was the *controlled* shape rather than generated as a series of voltage pulses. This is evident from the shape of the V_{gen} waveform, whose shape is far from rectangular pulses, because of compensation for the phase and amplitude characteristics of an isolating transformer (Figure 4.89d).

Non-sinusoidal waveforms are sometimes used because they can offer better 'diagnostic' tool for studying the material behaviour at specific controlled magnetising conditions. For example, uni-directional magnetisation can be carried out with minor loops (Son et al., 1996; Spornic, 1998; Barbisio et al., 2003; Kondo et al., 2003), with DC offset (Moses et al., 2007), asymmetrical first-order reversal curves

(Zirka et al., 2006) and emulation of real PWM conditions (Zurek et al., 2007b); under rotational and 2D conditions, there can be uni-directional at various inclination angles (Gorican et al., 2002; Zurek et al., 2009a), with DC offset (Enokizono et al., 2000), distorted elliptical (Sakata et al., 2003) and circular (Zhang et al., 2015), asymmetrical triangular and trapezoidal (Zurek et al., 2003, 2005a), etc.

Elliptical rotational excitation simplifies to controlling sinusoidal waveform in both channels, but with different amplitudes and phases (Young et al., 1960; Boon et al., 1965; Zhu et al., 1993; Hasenzagl et al., 1996; Zurek, 2000); therefore, it is similar to controlling sinusoidal and circular magnetisation. The same applies for 'circular' 3D excitation (Zhu et al., 2003).

The B waveform in Figure 4.89a was purposely set to be *asymmetrical* – the positive triangular peak falls at the positive peak of the sine, whereas the two negative peaks straddle the negative peak of the sine. As a result, the H waveform is also visibly asymmetrical, and obviously so is the B–H loop (Figure 4.89e). More examples of complex and asymmetrical controlled waveforms are shown in Chapter 7 (see Figure 7.8).

PRACTICAL COMMENT

Control of non-sinusoidal waveforms requires proportionally wider frequency spectrum to be controlled. Waveforms with mathematically 'sharp' corners have theoretically infinite frequency spectrum and of course this cannot be processed in reality. Some low-pass filtering might be required in order to limit the required bandwidth and to ensure that stability is retained. In general, it can be expected that highly distorted waveforms cannot be controlled to the same peak flux density as sinusoidal, due to the limited phase margin for higher harmonics. For example, for an audio-class amplifier, the upper frequency range will be around 20 kHz, and therefore the phase margin for higher frequencies cannot be maintained (Figure 4.88).

With a limited frequency bandwidth, the control might be insufficient so that higher harmonics might be visible in the given waveform (Figure 4.90).

Another problem is that if the controlled waveform has a triangular peak, then it will be impossible to sufficiently control its peak value because the real waveforms will be always smoothed out to some degree. This can be seen as 'rounding' of the peak of the triangular B waveform visible in Figure 4.90b. Therefore, a much larger peak error can be expected for such waveforms, and this error will be always negative; that is, the real 'rounded' peak will always be lower than the ideal 'sharp' peak.

One way to exert some influence on the outcome would be to filter the ideal waveform with similar bandwidth so that the same 'rounding' will be present in the target waveform as well.

4.6.7 MULTI-CHANNEL CONTROL

Description of digital feedback given so far focused primarily on a single-channel feedback, which is directly applicable to uni-axial magnetisation.

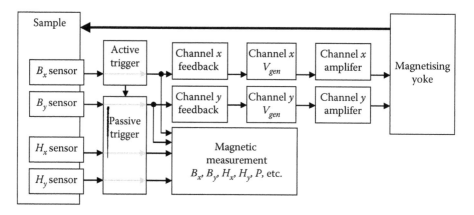

FIGURE 4.92 Simplified block diagram of a two-phase digital feedback (see also Figure 4.91).

However, in order to establish the rotating field, at least two phases are required, and some apparatus employ even three phases. Therefore, the digital feedback must be capable of providing controlled magnetising condition with such poly-phase magnetising yokes.

There is a certain degree of cross-coupling between the phases in such systems. Such coupling could take place directly in the sample as well as in the magnetising yoke, because the phase windings can overlap (Figures 4.31 and 4.32).

In the simplest approach, for a two-phase system, the digital feedback will be designed to have two *independent* feedback paths. Therefore, the block diagrams from Figure 4.83 would be simply doubled, but still with a single signal for triggering of all channels (Figure 4.92). And for a three-phase system, there would be one for each channel. There is no immediate need for the cross-channel coupling to be taken into account because it will be 'perceived' as an unwanted distortion which must be corrected by the feedback, independently but simultaneously for each channel, as shown schematically in Figure 4.92.

Therefore, all the corrections will be calculated independently for each channel, and with the iterative process, appropriate control will be established, even for very complex waveforms of B or H, highly anisotropic materials, as well as two- and three-phase systems. All the examples of rotational excitation measured by the author and shown in Chapter 7 were acquired *without* any cross-coupling taken into account in the feedback (Zurek, 2005; Zurek et al., 2005a).

Some papers describing the digital feedback refer to the method of including the cross-channel coupling into the algorithms (Spornic, 1998; Makaveev et al., 2000b). Such techniques can be thought of as a type of acceleration of convergence. They might be useful, but are not essential for operation and convergence, and might be actually detrimental to stability for highly anisotropic materials, for which the exact function of cross-coupling will be difficult to determine. In some cases, multi-channel control might not be achievable, but this is related more to the phase margin problem rather than to the cross-coupling as such.

PRACTICAL COMMENT

The simplified block diagram of a two-channel feedback in Figure 4.92 shows an important feature required for correct operation. Namely, all the signals must be actively triggered from a *single* 'master'' signal, which in Figure 4.92 is the signal for the B_x channel. All the other signals are triggered passively on the information provided by the active trigger.

It is not important which channel is used as the 'master' channel, but one channel must be designated as such. This also puts a requirement on maintaining a correct phase relationship of *all* the measured signals.

As discussed above, it is beneficial to have the main target waveform as starting from zero (Figure 4.84). Therefore, a circular clockwise rotation can be established by having the B_x target waveform set as sine (i.e. starting from zero), and the B_y target waveform as cosine. Then the H_x and H_y must have correctly connected phases so that the both B and H vectors are measured as rotating in the same direction. An easy practical way to ensure that all the signal have correct phases is to display the four waveforms (B_x, B_y, H_x, H_y) and compare if the phase of H_x matches that of B_x and H_y matches B_y (Figure 6.28a). A similar one applies to a three-phase system.

With such configuration, it is then sufficient for example to change the amplitude of the B_y channel from positive to negative in order to switch from a clockwise to an anticlockwise direction of rotation (Zurek, 2005).

Any digital feedback, be it single-channel or multi-channel, should be constructed and tested gradually so that each component is proven to work before deploying it in the whole system. Further blocks (like low-pass filtering) can be added after the main functions are implemented. Similarly, the measurement and data exporting are irrelevant from the viewpoint of the digital feedback and can be added only after the development is completed.

In a rotational system, it might be required to re-measure two signals (Figure 4.83b) in order to gather full information about phases of all waveforms. This would mean that the data acquisition device would have to have at least six input channels.

Three-phase magnetising system requires information for generation of appropriate signals for three power amplifiers. However, in the fieldmetric method, the input information is available only as two phases for the x and y directions.

Therefore, the information must be suitably transformed from a two-phase system of coordinates to a three-phase system. This can be achieved for instance by using the Clarke and inverse Clark transformations, which is extensively described in the literature, for example in Microsemi (2013).

4.7   WHITHER ROTATIONAL APPARATUS?

At the time of writing this book, there are no international or national standards defining the measurements of power loss under rotational magnetisation. In the past, some work through round-robin experiments was carried out by several laboratories, but discrepancies between the measurement results were too great (tens of percent) to be practically useful (Sievert et al., 1995, 1996).

As described above, there are several principles with which such measurements can be carried out, based on mechanical, temperature or electromagnetic effects. Because of the practical reasons, the mechanical and temperature methods are thought to be more difficult to be applied with sufficient universality, so the above-mentioned preliminary standardisation work focused on the fieldmetric method.

However, even within the fieldmetric technique, there is a plethora of ways in which the measurements can be carried out. Just the magnetising yoke can be made in a number of shapes (Figure 4.20), so the question remains – which type of apparatus is *the best*?

To answer this question, firstly we would need to define what is meant by 'the best'? This can be attempted by looking at other methods of magnetic measurements, for instance the Epstein frame (IEC, 2008), single-sheet tester (IEC, 2010) and the toroidal sample (IEC, 2003), but there are also the comparable ASTM standards.

It should be stressed that these standards focus primarily on the measurement of *power loss* and permeability, and not magnetostriction or other effects, which would be standardised by different documents. Power loss measurements are certainly easier to carry out with better reproducibility, and this should be probably the first direction in which rotational measurements could be standardised.

The main concern of such methods is *reproducibility* of results between various laboratories. This means that not only the signals have to be well defined and well controlled, but also the apparatus must provide the optimum excitation conditions to achieve repeatability for different materials and different samples.

For example, the toroidal sample standard (IEC, 2003) calls for the magnetising windings '*to be wound uniformly over the whole length of the test specimen*'. This ensures that the whole volume of the sample is magnetised as uniformly as *practically* possible. Similar hints about uniformity are given in the other magnetic standards.

This is probably the first requirement which should be taken into account. The magnetising yoke should be chosen of such type that ensures the best practically possible uniformity of magnetisation. From the work carried out recently by various researchers, it is very clear that yokes with square or rectangular samples are inferior (Gorican et al., 2003). Better uniformity is achieved for hexagonal yoke–sample (Figure 4.20d) and for yokes with circular samples (Figure 4.20g, i and k).

However, the magnetising yoke should also allow obtaining sufficiently high-controlled excitation without excessive power supply, and the published results show that the yokes in Figure 4.20i (circular placed on the sample) and Figure 4.20k (Halbach-like cylinders) were not capable of achieving the high-excitation levels.

This leaves two types of yokes: three-phase hexagonal (Figure 4.20d) and two-phase circular stator-like (Figure 4.20g). The round yoke can also be made with three-phase windings (Figures 6.6 and 6.7).

Three-phase power supply gives some advantage; because three power sources can be used to drive the excitation, very high excitation can be achieved with less difficulty. However, the complexity and cost of the system are increased. The latter one should not be disregarded. The Epstein frame system is a much less expensive solution, and some laboratories do not possess the single-sheet tester, just because it is much more expensive to purchase.

Two-phase system offers similar advantages, but it still can reach high-excitation levels. It was demonstrated that up to 2.0 T is routinely achievable for both non-oriented and grain-oriented electrical steels in different laboratories (Gorican et al., 2002; Zurek et al., 2006c), which should be sufficient for most *practical* measurements. A two-phase system requires only two power sources, and this can be achieved by using a *single* power amplifier with two independent channels (e.g. audio stereo). This minimises unnecessary cost, complexity and size of the whole setup. For example, it should be noted that the sheer size and weight of any of the planar yokes from Figure 4.20 are significant (see also examples in Chapter 6).

If the magnetising yoke is 'efficient' with converting the primary current into magnetisation of the sample, then a single two-channel power amplifier can be supplied with a single-phase power supply. High-rating power amplifiers requiring three-phase power supply are much more expensive, and the three-phase mains might not be available in all laboratories. Therefore, the ordinary single-phase mains supply is a good feature if it can be retained for such a system.

Also, the digital feedback algorithm is simplified. For rotational measurements, at least four signals need to be measured: B_x, B_y, H_x and H_y. Assuming that B_x and B_y are controlled to achieve circular flux density loci, then the digital feedback has two separate paths, each for controlling its own signal and driving its own power amplifier, as shown in Figure 4.92. Such approach would help in simplifying the software, improving the robustness and reducing the variability between different systems. Otherwise, the two-phase signal detection (x, y) must be translated into three-phase power generation (Hasenzagl et al., 1996; de la Barriere et al., 2015). It is perhaps not very difficult to do, but it does increase complexity, which is best avoided, for the ease of operation of such system.

On the basis of all the information as discussed in this book, the two-phase stator-like yoke appears to offer the *optimum compromise* between achievable uniformity of excitation, highest level of magnetisation and minimum complexity of the system. Therefore, it is the personal opinion of the author that such yoke should be the basis for standardisation of rotational power loss measurements, but of course only if further work confirms obtaining sufficiently good reproducibility of measurements.

There is still the unknown effect of the type of winding which should be used for such yoke: 'simple' (Figure 4.31) or 'sinusoidal' (Figure 4.32) or perhaps even some intermediate between the two. Both are capable of achieving very high-flux density (2.0 T), but excitation uniformity should be investigated. It was shown that the 'sinusoidal' winding offers very good uniformity for non-oriented electrical steels (Gorican et al., 2003). But similar investigation should be carried out *especially* for highly anisotropic samples such as high-permeability grain-oriented electrical steels. Owing to large deviations in directional reluctance of the sample, the magnetising current must also be significantly distorted in order to control circular loci of the B vector. It is therefore not immediately obvious that the sinusoidal winding will produce equally good results for highly anisotropic materials (Gorican et al., 2002), and simply more studies are necessary.

The size of such yoke will dictate the size of the sample. The typically reported size of circular samples was 60–80 mm, so this is definitely achievable in practice. However, it will be difficult to accommodate significantly larger samples to ensure averaging from a larger area, like it is the case for the standardised single-sheet tester, with *minimum* sample dimensions required as $300 \cdot 500$ mm (IEC, 2010). The reason is that such a large stator-like yoke with air gap would require massive amount of reactive power to drive, in order to achieve high-flux density in the sample. Values of 600 VA are required per channel of a two-phase yoke to magnetise the circular sample with 80 mm diameter. If scaled linearly, for a 5 times larger yoke (400 mm sample), the required power would be around 2.5 kVA per channel. Such powerful amplifiers are commercially available and indeed were used by some researchers (de la Barriere et al., 2015). However, at the time of writing, no publications were available which showed rotational studies on such large samples.

Horizontal magnetic shielding (Figure 4.26b) should *not* be used as it offers limited improvement for a *general* case. It is possible that a 'vertical' shielding (Figures 4.26c and 4.27) might be much more beneficial, both in terms of improved uniformity and the required excitation power. The latter point might be especially relevant for larger samples (diameter $>$ 150 mm).

Additionally, the 'vertical' shield could result in reduction of H gradient above the sample so that extrapolation would not be required. The same applies to the H component perpendicular to the surface of the sample. More experimental studies are needed to clarify any potential benefits or to prove its inferiority.

The next question is the type of sensors. It is evident that uniform magnetisation in rotational apparatus can be set up only in the middle of the sample. Therefore, B should be measured only for the local, uniformly magnetised area, rather than by winding a B-coil over the whole sample.

For local measurement, B-coil or needle probes can be used. Both have their merits and their disadvantages.

The biggest drawback of the needle probe method is that the voltage induced in the sensor is not *strictly* related to the distance between the needles (Equation 3.12), but even more importantly the induced voltage can be severely affected by asymmetrical eddy currents (Loisos et al., 2001; Pfutzner et al., 2004).

On a given setup, good repeatability can be achieved, and with a suitable 'sensor head', the needles can be applied with much more ease (Figure 6.22) than it is the case for a B-coil, whose wire has to be threaded through holes. However, the needle probe method would be probably insufficiently robust and immune to positioning errors, as far as *absolute accuracy* is concerned because symmetry of magnetisation cannot be absolutely guaranteed (and also because of Equation 3.12).

The B-coil approach also has significant disadvantages. It is much more destructive, because the sample would have to be drilled; hence, non-magnetic discontinuity is put in the vicinity of the measured area. The effect of drilled holes can introduce systematic errors for the detection of B, as illustrated in Figure 3.18.

However, if the drilled holes are small enough and positioned away from each other by a sufficient distance, they would still deliver the most reliable *practical* method. For instance, in order to keep the systematic error below 1%, the ratio of the hole diameter to the hole spacing would also have to be at the level of 1%, so for instance hole smaller than 0.22 mm for a 20-mm-wide B-coil (Zurek, in press). As experienced by the author, the smallest practical hole diameter which can be made with conventional drilling is around 0.3 mm (Figure 3.13). Therefore, for 1% error, the width of the B-coil would have to be at the order of 30 mm, but for 0.1%, it would have to be 300 mm. It is immediately obvious that the B-coil wire cannot have thick insulation, which in practice calls for enamelled copper wire. However, the increase size of the sample mentioned above would nicely accommodate the increased width of such B-coils.

As discussed above, such size would be probably the upper limit what is possible to achieve in practice without excessive cost implications. Even if the active sample area was 10 times smaller than for the standardised methods of SST or Epstein frame, this would still call for a sample diameter of 230 mm. (Further accuracy implications are discussed in more detail in Chapter 5.)

It should be noted that it was already demonstrated experimentally that 150 mm square samples could be magnetised rotationally to 1.7 T (Mori et al., 2015). Therefore, a more efficient round yoke should be capable of achieving the similar results for even larger samples.

This would allow larger holes to be drilled and much better averaging of properties. A square measurement area of $120 \cdot 120$ mm would be 36 times greater than $20 \cdot 20$ mm, so the improvements in averaging over many grains should be quite substantial.

The other problem with local B-coils is that the location of the drilled holes would have to be known as precisely as possible, because this directly impacts the systematic measurement uncertainties. This calls for drilling the holes with precision positioning capabilities. In order to achieve 0.1% for a 120-mm-wide B-coil, the positioning would have to be better than 0.06 mm. Typical mechanical tolerance which can be achieved without significant difficulty is 0.05 mm, and this would guarantee the 0.1% uncertainty for a 120 mm coil. This is another reason why samples should be as large as possible.

Such small hole diameter also puts limitation on the number of turns of B-coil. However, in the opinion of the author, with modern high-resolution data acquisition equipment, a single turn should be perfectly sufficient, because it was demonstrated that it produces usable signals even at low frequencies for a 20 mm B-coil. The additional very important benefit is that for single-turn B-coils the local thickness of the two crossing wires is as small as possible so that the tangential sensor of H can be positioned as close as practically possible to the surface of the sample. This impacts the measured values of both H and P, as evident from Figure 3.54.

The toroidal sample standard (IEC, 2003) warns against a problem with damaged insulation of the windings: 'Care shall be taken to ensure that the wire insulation is not damaged during the winding processes'. This point is even more valid for the B-coil which would require threading thin enamelled wire (hence with very thin electrical insulation) through a small hole with potentially very sharp edges. This problem would have to be solved somehow for each drilled hole. An example of a practical solution is shown in Figure 4.11.

For H-sensors, probably the best results can be obtained by using flat H-coils (Figures 3.45 and 6.22) or RCPs (Figure 6.22). It was demonstrated that the difference between them is quite small in practice, as far as rotational power loss is concerned (Xu, 1995). However, there are discrepancies especially for measurement of H, even for the same setup and like-for-like comparison (Xu, 1995).

In practice, it is probably easier to manufacture a good-quality flat H-coil than a good-quality RCP, because the latter relies heavily on good uniformity of the winding, and this is more difficult to achieve on a semi-circular or multi-part former (Figure 3.53). Mutual angular positioning between H_x and H_y sensors is also important, and smaller potential for discrepancies is for the double orthogonal H-coil, because both of the coils can be wound on the same precisely machined former.

Any angular errors between H_x and H_y coils should also be smaller for larger coils, which would be an additional advantage of a larger sample.

In the opinion of the author, signal amplifiers for B and H sensors should be avoided. They inevitably introduce additional phase errors, which impact the measured values, and are difficult to calibrate out. Waveform averaging, when executed with precise triggering, can significantly improve the signal-to-noise ratio so that even *very* low signals can be measured (Figures 4.58 and 4.59). Anyway, this would automatically improve proportionally for larger samples, because larger B-coils and H-coils could be used.

Suggested maximum ratio of the B-coil size to the diameter of a circular sample is around 75% (and a smaller ratio would be certainly better for uniformity). Therefore,

FIGURE 4.93 Suggested configuration of the optimum magnetising yoke and sample for measurements for rotational power loss (see also Figures 6.24 to 6.26): large sample diameter (>150 mm) with trimmed straight edges for angular and linear sample positioning, large sensing area (B-coils >100 mm), PCB-based 'ideal' H-coils with grooves for accommodating B-coil wires, with optional 'vertical' shielding (partially shown on the left-hand side of the image). (Courtesy of Andrew Gilham, Megger Instruments Ltd, the United Kingdom.)

TABLE 4.2
Optimum System for Rotational Power Loss Measurement

Component or Parameter	Recommended Values	Comments
Level of controlled excitation	Not greater than 1.9 T	This would be similar to currently standardised Epstein frame and SST
Magnetising yoke	Two-phase round stator-like	Reduced complexity as compared to three-phase yoke, good compactness requiring less power for excitation
Size of yoke/sample	Circular with at least 150 mm diameter	Recommended 150–200 mm, even better if 300 mm was possible, further work needed
Magnetic shielding	None (or 'vertical')	'Vertical' shielding could be used if experimentally confirmed
Sample positioning	Good linear and angular positioning	Some means of definitive orientation of a circular sample should be provided, for example straight trimming of the edges
Magnetising windings	'Simple' or 'sinusoidal'	'Sinusoidal' very good for NO steels, further study required for highly anisotropic materials
Air gap	At least 0.5 mm	Air gap improves uniformity of excitation
B sensors	Single-turn B-coils with ≤ 0.1 mm wire	Single turn reduced thickness at wire crossing and with modern data acquisition devices still provides sufficient signal amplitude
B-coil	<0.25% systematic error	Suggested ratio of hole diameter to coil width should be at most 0.3 mm/120 mm
H sensors	Flat H-coils, 0.5 mm thick	Easier to manufacture than RCPs Larger H-coils will provide sufficient signal, PCB technology allows production of 'ideal' H-coils with in-built structure for extrapolation
H sensors positioning	On top of B-coils	With groove for the B-coil wires
Size of B and H sensors	Max. 75% of sample diameter	The same for B and H sensors
Integration of signals	Digital	The same for B-coil and H-coil signals
Signal amplification	None	Signal amplification adds phase-shift errors
Power amplifiers	Voltage mode, supplied from a single-phase socket	With two independent channels, at least 1 kVA each (larger yokes might require higer power)
Digital feedback	Iterative	Without excessive acceleration of convergence
Signal acquisition	At least 12 bit	16 bit resolution (and higher) recommended
Sampling frequency	>50 kS/s	At least 1000 points per cycle at 50 Hz
Waveshape control based also on THD	$FF < 1\%$, THD $< 1\%$	THD $< 1\%$ might be too challenging in practice, but the THD criterion should be used
Power loss calculation	Average of CW and ACW	

for a sample diameter of 230 mm, the B-coils could be up to 120 mm wide, which would significantly boost the signal (as compared to the 20 mm version). It is probably the best to keep the same size of B-coils and H-coils.

PCB technology could be employed for making H-coils, which should generate enough signals for such large areas. Off-axis sensitivity could be significantly improved (Figure 3.50) and the construction could be such that it would accommodate the crossing wires of the B-coils (Figure 6.26). PCB technology would also allow the integration of needle probes to create 'sensing head' in a similar sense to that shown in Figure 6.22. Such integrated sensor could be put very close to the sample surface, and it could also facilitate *quadruple* H-coils for the purpose of extrapolation of H, if necessary. The concept of such PCB-based H-coils is shown in Figure 4.93 (see also Figures 6.24 through 6.26). Their performance was proved to be significantly more immune to off-axis H components, and the mechanical positioning would also be very well defined.

However, even for a 230 mm sample, the measurement area on single lamination would be at best only around 120 · 120 mm, which might not be enough for sufficient 'averaging' of magnetic properties, as it is the case for the standardised Epstein frame or single-sheet tester. Large misaligned grains are capable of redirecting the flux away from smaller B-coils so that the measured value can differ by up to 15%–30% (Figure 7.31). Therefore, the amount of scattering for such measurements depends on the crystallographic structure of the sample and for a general case would remain significant.

The matter of final *reproducibility* needs investigation through round-robin testing between several laboratories.

To summarise, from the currently available state-of-the-art, the suggested 'best' magnetising apparatus would be capable of measuring the power loss under rotational circular flux density with the components and parameters as defined in Table 4.2.

It should be noted that most of these concepts are supported by results already reported in the literature. Just five ideas would require further experimental validation: maximum practical sample size, type of yoke winding, circular sample positioning (notches/holes/edges), vertical shielding and PCB-based H-coils.

Only after combining all of the objectively best approaches, it might be possible to achieve the reproducibility and uncertainty (described in Chapter 5) at the level acceptable for international standardisation.

5 Measurement Uncertainty

The topic of *measurement uncertainly* is recognised for its difficulty. Nevertheless, almost each physical measurement is affected by uncertainties and hence each student, researcher and scientist should be familiar with the ramifications.

It is not possible reproduce here the breadth of the information required for complete[*] knowledge on the subject. Instead, this chapter focuses on a clear and practical description of the most fundamental concepts and their practical implementation. Several examples of step-by-step calculations are also included for the benefit of the reader.

5.1 INTRODUCTION

The subject of *measurement* is inseparable from *measurement uncertainty* (Taylor, 1997; JCGM, 2008). A measurement, however carefully carried out, cannot be infinitely accurate and some discrepancy between the measured value or *measurand* and the 'true' value will always be present. The 'true' value is not only unknown, but also might not even exist at all.

For instance, when measuring the length of an object with ever greater precision, eventually the size of atoms and their vibrations become the limiting factor, so the value cannot be technically measured with infinite precision to a fixed, 'true' number.

Therefore, if the 'true' value is never known, then the *measurement error* is also never known; hence the difference between the measured value and the 'true' value is uncertain (Figure 5.1). For this reason, when discussing the measurement accuracy, it is recommended to use the word *uncertainty* rather than 'error'. According to the recommendations outlined in JCGM (2008), the word 'error' in such meaning should be avoided.

An expression of the measured value, or *measurand* (JCGM, 2008), is almost meaningless without also giving some information about the estimated accuracy of measurement. The accuracy is estimated by the calculation of *uncertainty*, which should always be given together with the results of the measurement.

The correctly measured value is believed to be sufficiently 'close', so that the interval defined by the uncertainty will encompass the 'true' value with sufficient

---

[*] The methodology presented in this chapter is based mostly on *An Introduction to Error Analysis, The Study of Uncertainties in Physical Measurements* (Taylor, 1997), *Evaluation of Measurement Data – Guide to the Expression of Uncertainty in Measurement* (JCGM, 2008) as well as *The Expression of Uncertainty and Confidence in Measurement* (UKAS, 2012). Other publications can also be very helpful in explaining the difficult concepts (Cook, 2002). The author strongly recommends reading *at least* these fundamental works, which define in much greater detail the whole topic of measurement uncertainty. It should also be noted that there is a continuing progress in the field of methodology of uncertainty estimation and therefore the latest publications on the matter should be studied accordingly.

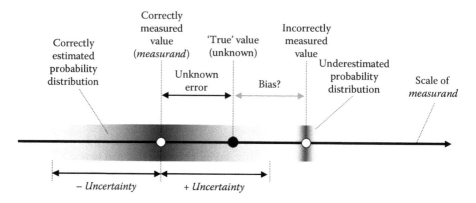

FIGURE 5.1 Illustration of measurement uncertainty with correct coarser-resolution measurement (correctly estimated uncertainty) and incorrect finer-resolution measurement (incorrectly estimated uncertainty).

probability. The 'true' value remains unknown, but at least it is possible to claim that it lies somewhere within the given interval with given probability. It should be stressed that the distance to the 'true' value remains unknown, so the 'error' cannot be quantified. Only a 'degree of belief' can be expressed through the value of the uncertainty, and there may be a small probability that the 'true' value lies just outside of the specified range. This is indicated in Figure 5.1 by the probability distribution extending beyond the interval of uncertainties.

As shown in Figure 5.1, an incorrect measurement might also be equally 'close' to the 'true' value (compare the lengths of 'unknown error' and 'bias'). However, the measurement could be rendered incorrect simply by underestimation of the uncertainty interval.

Using equipment with *finer resolution* could give higher perceived precision of measurement, but it does not guarantee higher *absolute accuracy* (Auty et al., 2014). Other factors can impact the absolute accuracy, for example, due to unknown bias resulting from using different equipment or limitations of the measurement method, or many other factors.

PRACTICAL COMMENT

The difference between resolution and accuracy was mentioned in Chapter 4. But the importance of correct estimation of the uncertainty interval is better illustrated in Figure 5.2. If the 'true' value was known, then it would be very easy to judge which measurement is more accurate, as it is evident from Figure 4.70. But if the 'true' value was known, then there would be no need to carry out the measurement in the first place!

The true value is *unknown* and therefore it is *not* possible to judge the accuracy of the measurement simply by the position or even the spread of the measured values. The reader is encouraged to compare Figure 5.2 with Figure 4.70. Careful estimation of uncertainty is required in order to quantify which of the two results has a narrower interval of uncertainty.

FIGURE 5.2 'True' value is never known, so it is not possible to tell which set of values is more accurate (compare with Figure 4.70 for better illustration of the implications): (a) very narrow scattering with clustering around some position, (b) very wide scattering – is this measurement really less 'accurate'? and (c) medium scattering – are these values more or less accurate than the two previous measurements?

The uncertainty must be re-estimated separately for each new measurement method and new equipment, and if relevant also even for each new sample or measurement. Therefore, only general methodology can be described, and the exact technique must be developed appropriately for each experiment.

Uncertainty informs about the degree of *probability* that the measured result falls between the claimed limits. It is expressed by a positive number, which in most cases extends from the central measured value in both directions (Figure 5.1). The probability distribution can extend beyond the uncertainty interval, as explained below.

PRACTICAL COMMENT

There is one exception in which the value can be measured absolutely *precisely*, namely, if the measured value can only take integer quantities, such as number of turns in a winding. If all the turns are exposed to visual inspection, then the 'measurement' can be carried out simply by visually counting the turns and thus the true value can be 'measured' without any uncertainty, and hence without any error.

However, if there are many turns, then it is still possible to miscount them and thus an incorrect value can be introduced by a human-made mistake. If such an error is probable to occur, then it should also be taken into account in the total uncertainty budget.

5.2 PROBABILITY DISTRIBUTIONS

Probability distribution can be expressed by different functions, depending on the type of measurand, physical processes involved or type of equipment. The three functions most frequently encountered in practice are: normal distribution, rectangular and triangular, as illustrated in Figure 5.3 (but there are more).

Normal distribution (also referred to as Gauss, Laplace–Gauss or bell-shaped) is usually applied when the scattering of values is random in frequency and amplitude. The curve is 'bell-shaped' (Figure 5.3a) and represents probability Q of occurrence of an independent variable v (horizontal axis in Figure 5.3a). The *average value*

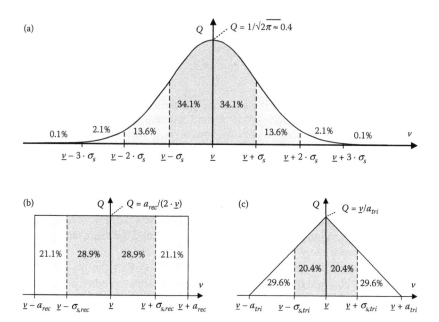

FIGURE 5.3 Probability distributions: (a) normal, (b) rectangular and (c) triangular.

$\underline{v}$ is located at the centre of the distribution, which is described by the following function:

$$Q(v, \sigma_s) = \frac{1}{\sigma_s \cdot \sqrt{2 \cdot \pi}} \cdot \exp\left[-\frac{1}{2} \cdot \left(\frac{v - \underline{v}}{\sigma_s}\right)^2\right] \quad \text{(unitless)} \quad (5.1)$$

The characteristic value of a normal distribution is a *standard deviation* σ_s. It is calculated by using a statistical approach described in detail in the next section.

An interval marked by $\pm 1 \cdot \sigma_s$ encompasses a probability of 68.27% (that is twice of 34.13%, Figure 5.3a). This means that 68.27% of randomly obtained values will lie within this interval and the remaining 31.73% will be outside of it.

An interval with two standard deviations ($\pm 2 \cdot \sigma_s$) represents a probability of 95.45% and for three standard deviations, it is 99.73%. The multiplicity of standard deviations is referred to as *coverage factor* k_Q. For instance, if the value is $k_Q = 2$, then the *level of confidence* is 95.45% and so on (JCGM, 2008). Non-integer values of k_Q can also be applied, as required.

The horizontal axis of the normal probability function extends to infinity. For example, the probability of finding a value greater than $4 \cdot \sigma_s$ is only around 0.003%, but nevertheless such values could still occur for the normal distribution.

PRACTICAL COMMENT

Let us assume that we need to measure an RMS value of a very noisy signal (see Figure 3.56). The instantaneous values vary randomly due to the presence of noise. Hence, there is a certain amount of variation of the RMS value from once cycle to the next. Most measured values will be around some central number, but smaller and greater values will also be present. Therefore, a graph similar to that shown in Figure 5.3a will be produced when enough samples are acquired and processed statistically.

The central value for normal distribution represents the average of infinitely many values. This would be synonymous with the 'true' value; therefore, the hypothetical error would be zero. For a mathematically ideal, continuous random variable, the central value of the distribution can be calculated as an average $\underline{v}$ of all the values.

Rectangular distribution (also called *uniform distribution*) is shown in Figure 5.3b and can be applied, for example, for equipment with a digital readout (uncertainty resulting from resolution of the last digit). The characteristic value of standard deviation for rectangular distribution $\sigma_{s,rec}$ is defined as (JCGM, 2008)

$$\sigma_{s,rec} = \frac{a_{rec}}{\sqrt{3}} \quad \text{(unit the same as } a_{rec}) \tag{5.2}$$

It should be noted that for the rectangular distribution, the width of the interval is *finite* and probability drops to zero for values smaller than $\underline{v} - a_{rec}$ and greater than $\underline{v} + a_{rec}$. In other words, 100% of values are expected to lie within such limits.

Rectangular distribution is sometimes referred to as 'distribution of minimum knowledge' because it can be applied in the absence of better information (Cook, 2002).

PRACTICAL COMMENT

For example, if a display of digital scales shows a value of 193 g, then it can be concluded that the value is somewhere between 192.5 g and 193.5 g because otherwise the values would be rounded down to 192 g or up to 194 g, respectively. It is not possible to know what is the actual value with any greater precision; therefore, all values between 192.5 g and 193.5 g are *equally probable* and in such a case, the limits of the distribution (see also Figure 5.3b) are characterised by $a_{rec} = (193 \text{ g} - 192 \text{ g})/2 = 0.5 \text{ g}$.

Triangular distribution (Figure 5.3c) also has *finite* limits defined by the value a_{tri}. The standard deviation can be calculated as (JCGM, 2008):

$$\sigma_{s,tri} = \frac{a_{tri}}{\sqrt{6}} \quad \text{(unit the same as } a_{tri}) \tag{5.3}$$

PRACTICAL COMMENT

Triangular distribution can be sometimes applied to a temperature-controlled environment (A2LA, 2014) but this is not always the case (JCGM, 2008).

For detailed analysis, many other types of distribution can be applied: triangular asymmetrical, trapezoidal, U-shaped, etc. However, the depth of this topic is beyond the scope if this book and only the simplest cases are discussed for illustration. It should be borne in mind though that, for example, the U-shaped distribution has more probability to find value *away* from the average and thus it is even 'worse' case than the rectangular distribution, with the so-called divisor having a smaller value (divisors are explained in the following sections).

The reader is advised to study other publications such as JCGM (2008).

5.3 TYPE A AND TYPE B UNCERTAINTIES

Measurement uncertainty arises from many factors. A quite comprehensive list (but of course non-exhaustive) is given in JCGM (2008) (Copyright 2008 Joint Committee for Guides in Metrology and Bureau International des Poids et Mesures. Reproduced with permission. Courtesy of Bureau International des Poids et Mesures):

a. *incomplete definition of the measurand*
b. *imperfect realization of the definition of the measurand*
c. *nonrepresentative sampling – the sample measured may not represent the defined measurand*
d. *inadequate knowledge of the effects of environmental conditions on the measurement or imperfect measurement of environmental conditions*
e. *personal bias in reading the analogue instruments*
f. *finite instrument resolution or discrimination threshold*
g. *inexact values of measurement standards and reference materials*
h. *inexact values of constants and other parameters obtained from external sources and used in the data-reduction algorithm*
i. *approximations and assumptions incorporated in the measurement method and procedure*
j. *variation in repeated observations of the measurand under apparently identical conditions*

Some or all of these factors can be at play simultaneously, and they should be evaluated in the best possible methodology. For infinitely large set of data produced by various methods and instrumentation, all uncertainties could be evaluated by statistical means (JCGM, 2008) but in practice this is not possible. Therefore, practical means must be used for the evaluation of all components of uncertainty.

In the past, the uncertainties were referred to as 'random' and 'systematic'. Nowadays, such terminology is perceived to be misleading and it is discouraged (JCGM, 2008). The estimation of uncertainty is performed by using Type A and Type B uncertainties.

Therefore, care should be taken when referring to older publications on the subject because they might not represent the current state of the knowledge or recommendations for processing the uncertainties.

Type A uncertainties are not 'random' but are merely estimated by statistical means (Cox et al., 2010). On the other hand, Type B uncertainties are not 'systematic', but are estimated by other means, which are not statistical. If there is a known 'systematic' difference (e.g. Bias? in Figure 5.1), it should be taken into account and corrected for – and thus eliminated from the uncertainty budget. However, such 'systematic' difference could be calculated, for instance, with linear regression, which has its own additional uncertainty, but some aspects of it might still require a statistical treatment. For these reasons, the name 'systematic' is discouraged (JCGM, 2008).

All the uncertainty components should be treated in the same way because no distinction can be made between uncertainties arising from various sources (whether they are perceived to be 'random' or not) – all the components are combined in the same way, and all will contribute to the overall interval of probability. A 'random' component in one context might become 'systematic' in a different mathematic model (JCGM, 2008).

Both Type A and Type B are treated equally and combined appropriately when calculating the overall uncertainty of measurement (JCGM, 2008).

5.3.1 Type A Uncertainty

For a number of *independent* observations that vary *randomly*, the probability distribution follows the normal distribution (Figure 5.3a).

Therefore, the best estimation of the measured quantity is the arithmetic average $\underline{v}$ of n observations v_i, namely, v_1, v_2, v_3, ..., v_n through all the items i, from 1 to n (JCGM, 2008):

$$\underline{v} = \frac{1}{n} \cdot \sum_{i=1}^{n} (v_i) \qquad \text{(unit the same as } v_i) \qquad (5.4)$$

There are several characteristic values that can be calculated by statistical means. A degree of scattering of the whole population of values can be estimated by calculating the *standard deviation* σ_s. Such a value can be defined for any shape of probability distribution (JCGM, 2008):

$$\sigma_s = \sqrt{\frac{1}{n-1} \cdot \sum_{i=1}^{n} (\underline{v} - v_i)^2} \qquad \text{(unit the same as } v_i) \qquad (5.5)$$

However, this parameter gives information only about the scattering of the whole set of data. A more interesting value is the *standard deviation of the mean* σ_{sdom} (Fornasini, 2008; JCGM, 2008):

$$\sigma_{sdom} = \frac{\sigma_s}{\sqrt{n}} \qquad \text{(unit the same as } v_i) \qquad (5.6)$$

$$u = \sigma_{sdom} \qquad \text{(unit the same as } v_i) \qquad (5.7)$$

The *standard deviation of the mean* is therefore synonymous with the Type A *uncertainty u* of the performed measurement.

PRACTICAL COMMENT

The standard deviation for other shapes of distribution is also calculated by using Equation 5.5, but for rectangular and triangular distributions, it can be shown that the calculations simplify greatly to simple formulas (5.2) and (5.3), respectively.

Standard deviation σ_s gives information about the scattering of the whole set of the values v_i. However, the calculated average value $\underline{v}$ will always lie close to the centre of a symmetrical distribution.

If another series of measurements was made, then the scattering of the data represented by σ_s would have similar width, with the random values falling close or far from the average. But the average value $\underline{v}$ will again fall very close to the centre of the whole distribution. Therefore, there is a much smaller chance that the position of the average value will differ significantly, than it is for the individual values. Hence, the *uncertainty u* (equal to the *standard deviation of the mean* σ_{sdom}) of the average value $\underline{v}$ is significantly smaller than the standard deviation of the whole set.

For an infinitely large set of random values, the value of $u = \sigma_{sdom}$ approaches zero because of the additional $\sqrt{n}$ in the denominator (Equation 5.6) – there is increased 'sureness' that the average represents the 'true' value even if the input data is very noisy (see also Figure 4.58). On the other hand, even for infinitely large set of data, σ_s retains a significant non-zero value, characteristic of the dispersion of the random data (e.g. proportional to the level of the 'noise').

Statistical calculations offer quite a well-defined tool for processing the values, but do not necessarily offer superior 'accuracy' of estimation because of the limited number of observations that can be made in practice. This leads to 'uncertainty of uncertainty' (JCGM, 2008), which can be surprising large, and, for example, for $n = 5$ observations, it is 36% (Table 5.1), but even for $n = 50$, it is still as much as 10%.

This illustrates that for practical values of n, the accuracy of standard deviation evaluation through statistical means (Type A) is not necessarily superior to those evaluated by other means (Type B, described below).

5.3.1.1 Student's Distribution

The normal distribution is described by Equation 5.1. However, such a function can be calculated *only* if the standard deviation and the average value are known precisely because both σ_s and $\underline{v}$ appear in the function.

This is not the case for practical measurements. It can be suspected that random variations follow the normal distribution function, but it is impossible to know the *precise* values of σ_s and $\underline{v}$. They are mathematically defined (Equations 5.4 and 5.5) and thus can be calculated even from a population of just $n = 2$, but they would not *precisely* represent the actual values required for Equation 5.1, as it is evident even from Table 5.1.

TABLE 5.1

Standard Deviation of the Experimental Standard Deviation of the Mean (σ_s of σ_{sdom})

Number of Observations, n	σ_s of σ_{sdom} (%)
2	76
3	52
4	42
5	36
10	24
20	16
30	13
50	10

Source: JCGM, Evaluation of measurement data, Guide to the expression of uncertainty in measurement, *Joint Committee for Guides in Metrology, JCGM 10: 2008*, 2008. Copyright 2008 Joint Committee for Guides in Metrology and Bureau International des Poids et Mesures. Reproduced with permission. Courtesy of Bureau International des Poids et Mesures.

Therefore, *Student's distribution* (also *t-distribution* or *Student's t-distribution*) is used when there is a limited number of readings in the Type A evaluation (JCGM, 2008).

For infinitely large set of data, Student's distribution approaches the normal distribution as shown in Figure 5.4. But for small population sizes, it is appropriately

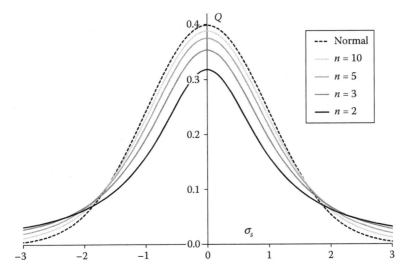

FIGURE 5.4 Comparison of Student's distribution with normal distribution; the horizontal axis is scaled in σ_s of the normal distribution (normal distribution is synonymous with $n = \infty$).

more probable that a value will fall outside of the standard deviation range as compared to the normal distribution.

Student's distribution is parametric and depends on the size of population (n – number of readings) or more precisely on the characteristic parameter called *degrees of freedom* d_f defined as

$$d_f = n - 1 \qquad \text{(unitless)} \tag{5.8}$$

The distribution is defined by the following function (Lukacs, 2014):

$$Q_t(v, d_f) = \frac{\Gamma\left(\dfrac{d_f + 1}{2}\right)}{\Gamma\left(\dfrac{d_f}{2}\right) \cdot \sqrt{d_f \cdot \pi}} \cdot \left(1 + \frac{t^2}{d_f}\right)^{\frac{-(d_f + 1)}{2}} \qquad \text{(unitless)} \tag{5.9}$$

where the Γ function is

$$\Gamma(n) = (n - 1)! \qquad \text{(unitless)} \tag{5.10}$$

PRACTICAL COMMENT

Factorial function is defined only for integer arguments, but the Γ function can be interpolated for continuously varying arguments, for instance, by using built-in spreadsheet functions (Zaiontz, 2016).

In Student's distribution, the minimum practical value is $d_f = 1$, so at least two readings must be taken ($n = 2$).

The calculation of the average value v, the standard deviation σ_s and the standard deviation of the mean σ_{sdom} are made by using the same equations as for the normal distribution, namely, Equations 5.4 through 5.6, respectively.

In Student's distribution, the central part takes lower values and the 'tails' are heavier (Figure 5.4). Hence, the interval required to cover 68.27% must be wider than for the normal distribution and a single standard deviation represents a different width (Fornasini, 2008).

In order to take this increased width into account, the correction is made by multiplying the calculated standard deviation by a special correction factor $k_{Q,t}$ usually taken from pre-calculated tables available in the literature. An example of such a table is given in JCGM (2008), and it is included here in its abridged form as Table 5.2.

Therefore, when using Student's distribution, the *standard uncertainty* should be calculated as

$$u = k_{Q,t} \cdot \sigma_{sdom} \qquad \text{(units the same as } \sigma_{sdom}) \tag{5.11}$$

where σ_{sdom} is a value calculated from Equation 5.6.

TABLE 5.2
Student's Distribution Correction Factor $k_{Q,t}$ for Various Degrees of Freedom and Confidence Levels

Number of Observations n	Degrees of Freedom d_f	Fraction of Q (%)		
		Confidence Level 68.27%, $k_{Q,t} = 1$ (Interval $\pm 1 \cdot \sigma_s$)	Confidence Level 95.00%, $k_{Q,t} = 1.96$ (Interval $= \pm 1.96 \cdot \sigma_s$)	Confidence Level 95.45%, $k_{Q,t} = 2$ (Interval $= \pm 2 \cdot \sigma_s$)
2	1	1.84	12.71	13.97
3	2	1.32	4.30	4.53
5	4	1.14	2.78	2.87
10	9	1.06	2.26	2.32
20	19	1.03	2.09	2.14
30	29	1.02	2.04	2.09
51	50	1.01	2.01	2.05
101	100	1.005	1.984	2.025
Infinity	Infinity	1.000	1.960	2.000

Source: JCGM, Evaluation of measurement data, Guide to the expression of uncertainty in measurement, *Joint Committee for Guides in Metrology, JCGM 10: 2008,* 2008. Copyright 2008 Joint Committee for Guides in Metrology and Bureau International des Poids et Mesures. Reproduced with permission. Courtesy of Bureau International des Poids et Mesures.

PRACTICAL COMMENT

The data in the '68.27%' column correspond to a single standard deviation of the normal distribution. Therefore, for infinitely large number of observations, $k_{Q,t} = 1$, which is synonymous with the normal distribution. Similarly, the '95.45%' column corresponds to two standard deviations, hence for an infinitely large set of data, the factor is $k_{Q,t} = 2$.

However, an interval frequently quoted in the literature is 95% (i.e. 95.00%) and thus to be precise the $k_{Q,t}$ factor should be somewhat smaller in order to account for the 0.45% difference, so that for infinity, it is 1.960 and not 2.000. This distinction is made, for instance, in A2LA (2014), where the value for a single standard deviation is rounded down to 68% (i.e. 68.00%) and thus the value for infinity is calculated as 0.9945 instead of 1.000 for 68.27%.

Interestingly, many books on statistics, but not related to measurement uncertainty, show Student's factor tables for several other values but not for single, double or triple standard deviations. The lack of values for a standard deviation (68.27%) is especially inconvenient, so the usefulness of such tables is limited for calculation of Type A *standard* uncertainty, which is a very important parameter.

As can be seen in Table 5.2 for larger values of the degrees of freedom, Student's distribution gets very close to the normal distribution. In practice, for $d_f \geq 30$, the normal distribution can be assumed because for $k_{Q,t} = 1$ and

$k_{Q,t} = 2$, the differences are only 2% and 4.5%, respectively. Such small values are usually negligible because the uncertainty should be reported to one or a maximum of two significant digits (Taylor, 1997). This is partly a consequence of Table 5.1.

For calculations with complex numbers, the coverage factors are *not* the same, but significantly larger (Hall, 2011b).

Student's distribution is also applied to the uncertainty combined from Type A and Type B assessments, as described below.

5.3.2   Type B Uncertainty

Type B uncertainty is not assessed by statistical means in the same sense as Type A. Other methods or mathematical tools must be employed. Such techniques cannot be generalised because they strongly depend on the type of the variable, physical quantity, instrumentation or even measurement method. Therefore, only some examples are discussed below, with the hope that they will sufficiently illustrate practical approaches required for Type B uncertainty estimation.

5.3.2.1   Imported Values

A very important uncertainty to recognise in the overall budget are the so-called *imported values* and the uncertainties associated with them (JCGM, 2008). This is usually the case for most commercially available equipment for which the manufacturers guarantee some level of accuracy. The measurement uncertainty of such devices can be expressed in a multitude of ways and the operator must be able to extract the relevant uncertainty information from technical specifications or data sheets.

This is usually quite a tedious task because the format of the data differs wildly between manufacturers, devices and even ranges on the same device.

In order to be able to use the accuracy statement for uncertainty calculations, it is usually required to extract information about the *standard uncertainty* (i.e. not maximum or typical) and such information is typically *not* given directly for commercial devices.

Therefore, careful manual calculations are required to extract such information from the available data so that values such as 'maximum uncertainty' or 'typical accuracy' can be converted to the required *standard uncertainty*. Three simple examples of such imported information are given below.

PRACTICAL COMMENT

For the sake of illustration, these three examples of imported uncertainty were worked out for a single value. In reality, this would be too tedious to apply in practice because the full recalculation would have to be repeated for each new measured value. For example, if power loss was to be measured at all the points from 0.1 to 1.8 T in steps of 0.1 T, then calculations from Table 5.3 would have to be repeated 18 times, which hardly lends itself to a practical application.

TABLE 5.3

Excerpt from Accuracy Specifications for Agilent 34410A

Accuracy Specifications ± (% of Reading + % of Range)[a]

Function	Range[b]	Frequency, Test Current or Burden Voltage	24 Hour[c] Tcal ±1°C	90 Day Tcal ±5°C	1 Year Tcal ±5°C	Temperature Coefficient/°C 0°C to (Tcal −5°C) (Tcal +5°C) to 55°C
True RMS AC voltage[d]	100.000 mV to 750.000 V	3–5 Hz	0.50 + 0.02	0.50 + 0.03	0.50 + 0.03	0.010 + 0.003
		5–10 Hz	0.10 + 0.02	0.10 + 0.03	0.10 + 0.03	0.008 + 0.003
		10 Hz–20 kHz	**0.02 + 0.02**	**0.05 + 0.03**	**0.06 + 0.03**	**0.005 + 0.003**
		20–50 kHz	0.05 + 0.04	0.09 + 0.05	0.10 + 0.05	0.010 + 0.005
		50–100 kHz	0.20 + 0.08	0.30 + 0.08	0.40 + 0.08	0.020 + 0.008
		100–300 kHz	1.00 + 0.50	1.20 + 0.50	1.20 + 0.50	0.120 + 0.020

Source: Figure 3.74b, Keysight Technologies, 3410A and 34411A Multimeters, Data sheet, 2015. © Keysight Technologies, Inc. Reproduced with permission. Courtesy of Keysight Technologies.

[a] Specifications are for 90-minute warm-up and 100 PLC.

[b] 20% overrange on all ranges, except DCV1000 V, ACV 750V, DCI and ACI 3 A ranges.

[c] Relative to calibration standards.

[d] Specifications are for sinewave input >0.3% of range and >1 mVrms. Add 30 μV error for frequencies below 1 kHz. 750 VAC range limited to 8×10^7 V-Hz. For each additional volt over 300 Vrms, add 0.7 mVrms of error.

For this reason, it is usually easier to split the imported uncertainties into relative and absolute parts and estimate them as separate values, but for the whole given input range (Carriage, 2017; Matthews, 2017).

Needless to say, it is absolutely paramount not to confuse the absolute and relative parts of uncertainty. In standard calibration laboratory procedures, these are typically assessed *separately* and might not be even combined in the final figure (Carriage, 2017; Matthews, 2017). As shown in Table 5.3, the manufacturer of the given device can give the specifiation as a set of absolute and relative values, without combining them into a single value. Such notation is used because it is possible to generalise the uncertainty for the whole range, whereas if converted to a single relative value (e.g. in percent), it could give far too pessimistic uncertainty limits.

However, for simplicity and clarity of the examples, they will be combined in the examples given below, so that the absolute uncertainty will be expressed with the corresponding value of a relative uncertainty. This will be made clear from the context.

Example 1 – Type B Uncertainty of a Voltmeter

The voltmeter Agilent 34410A shown in Figure 3.74b has the technical speci-fication for its AC voltage measurement as shown in Table 5.3 (Keysight Technologies, 2015).

The whole data sheet does not give any information about the confidence level. There is also no knowledge of the character of the uncertainties (type of distribution).

However, it is quite common for commercially available devices that the manufacturers give a specification that covers *all* the values, so that it is *guaranteed* that the measured value will always be within the limits stated by the specification. This usually implies a rectangular distribution (Figure 5.3b) and therefore Equation 5.2 with factor $\sqrt{3}$ should be used.

Equation 5.2 requires the value of a_{rec} and this must be calculated from the specification. This can be done only for the signals that are *within* the specification of the device. For example, there is no specification for signals above 300 kHz, and if such values were measured, then the uncertainly level will be simply unknown, and cannot be derived from the data in Table 5.3.

As a first thing, it should be noted that the specification applies *only* to measurements performed within 1 year from calibration. For this reason, annual recalibration of measurement instruments is required (or as specified by the manufacturer); otherwise, uncertainty cannot be estimated. Let us assume that the measurements are made 6 months after calibration so that the '*1 year*' values apply.

Also, 90-minute warm-up and the averaging parameter set as PLC $= 100$ (the comment under the table marked with superscript 'a') are required. The acronym PLC denotes 'power line cycles'; hence at 50 Hz, the value of PLC $= 100$ means averaging for $100 \cdot 20$ ms $= 2$ s. If such averaging is not used when the measurements are carried out, then the performance of the device is *not* specified and the uncertainty cannot be assessed on the basis of data shown in Table 5.3. If this is discovered *after* all the measurements are made, then they would have to be repeated with the correct setting. Alternatively, clarification could be sought from the manufacturer.

The next comment (superscript 'b') gives some additional relevant information. The ranges on this instrument (not shown in Table 5.3) are 100 mV, 1 V, 10 V, 100 V and 750 V. The relevant part is that there is an overrange of 20% on lower ranges.

This means that if the amplitude of the signal increases slowly, then a voltage of 109.73 mV will be measured on the 0–100 mV range, not on the 100 mV to 1 V range. This has an immediate impact on the uncertainty values because the heading of the table refers to '*% of the range*'. Hence, if an incorrect range is assumed for calculations, then the uncertainty could be overestimated (or underestimated) by a factor of 10.

Let us assume that the average amplitude of the measured signal is $\underline{v} = 109.73$ mV and the frequency is 50 Hz. Therefore, another set of limitations (superscript 'd') is that the voltage must be greater than 0.3% of the range (condition met for this range) and '*>1 mVrms*' (also condition met). But we need to add '*30 μV error for frequencies below 1 kHz*'.

The measurements were carried out at 23°C, which was within 5°C from the '*Tcal*' value, so that no further temperature coefficient is required.

All these criteria eliminated most of the data, so that only the values of '*0.06 ± 0.03*' are relevant, as underlined in Table 5.3. Therefore, assuming a

TABLE 5.4

Type B Uncertainty Estimation for 109.73 mV Value Measured at 100 mV Range

Deviation of the value	$0.06\% \cdot 109.73 \text{ mV} = 0.0658 \text{ mV}$
Deviation due to range	$0.03\% \cdot 100 \text{ mV} = 0.03 \text{ mV}$
30 μV error	0.03 mV
Rectangular distribution parameter a_{rec}	$0.0658 \text{ mV} + 0.03 \text{ mV} + 0.03 \text{ mV} = 0.1258 \text{ mV}$
Type B uncertainty $\sigma_{s,rec}$ (from Equation 5.2)	**0.07 mV**

TABLE 5.5

Type B Uncertainty Estimation for 109.73 mV Value Measured at 1 V Range

Deviation of the value	$0.06\% \cdot 109.73 \text{ mV} = 0.0658 \text{ mV}$
Deviation due to range	$0.03\% \cdot 1 \text{ V} = 0.30 \text{ mV}$
30 μV error	0.03 mV
Rectangular distribution parameter a_{rec}	$0.0658 \text{ mV} + 0.30 \text{ mV} + 0.03 \text{ mV} = 0.3958 \text{ mV}$
Type B uncertainty from Equation 5.2 $\sigma_{s,rec}$	**0.23 mV**

symmetrical rectangular distribution, the final relevant values are as shown in Table 5.4.

However, if the value was obtained by reducing the signal amplitude, then the measurement was made on the 100 mV–1 V range and the deviation due to the range would be different, as summarised in Table 5.5.

Therefore, measuring the very same value with the very same instrument but on a different input range can result in significantly different uncertainty (in this case, more than three times greater).

If a triangular distribution was used instead of a rectangular one, then the Type B uncertainties in the two cases shown above would be 0.05 mV instead of 0.07 mV, and 0.16 mV instead of 0.23 mV, respectively. However, it should be noted that as recommended (Taylor, 1997; JCGM, 2008), all the final values were rounded to one or two significant digits.

Example 2 – Type B Uncertainty due to a Digital Display

As discussed above for a number shown on a digital display, its value can lie anywhere between half of the previous digit to half of the next digit, and a rectangular distribution can be applied (see Figure 5.3b). If a value of a single digit of resolution d_r extends from $-a_{rec}$ to $+a_{rec}$ (Figure 5.3b), then according to Equation 5.2, the standard deviation is (JCGM, 2008)

$$\sigma_{s,rec} = a_{rec}/\sqrt{3} = 0.5 \cdot d_r/\sqrt{3} = d_r/\sqrt{12} \approx 0.29 \cdot d_r \qquad (5.12)$$

Therefore, the contribution to the uncertainty is 0.29 or 29% of the resolution, purely because of using digital display. An example of 1 g resolution was mentioned above – the uncertainty resulting just from a digital display would therefore be 0.29 g.

Let us also assume that five repeated observations were made, resulting in the values: 193, 194, 193, 193, 193. (And all other contributions were assumed to be negligibly small and the weighing was repeated, for instance, with disturbing the scales every time so that any friction errors are averaged out.)

The average value (Equation 5.4) is $\underline{v} = 193.20$ g. The standard deviation of the mean (Equation 5.6) is $\sigma_{sdom} = 0.20$ g. If the Student's distribution correction is applied (Equation 5.11) for 4 degrees of freedom then $u = 0.23$ g. Therefore, the Type B standard deviation due to digital resolution would have a larger contribution to the uncertainty (0.29 g) than the Type A estimated with a statistical method (0.23 g). Of course, if all five values were 193 g, then the Type A uncertainty would be 0.00 g, but the Type B uncertainty would remain at 0.29 g.

It should be emphasised that the Type B estimation was not made with statistical tools, but it was still rooted in the probability distributions. This illustrates that sometimes statistical basis can be used for Type A and sometimes for Type B uncertainties – and a certain amount of personal judgement based on knowledge must be used accordingly as to which approach should be applied.

Example 3 – Type B Uncertainty of a Digital Balance

Let us assume that an electronic digital balance (laboratory scale) model PFB 1200-2 (KERN, 2013) was used for weighing samples. The specification extracted from the operating manual is as shown in Table 5.6.

Once again, no information about standard deviation, coverage factor or confidence level is given in the technical specification.

The specification informs that the maximum measurable value is 1200 g, the warm-up time is 2 hours and there should be at least 3 s before oscillations are damped and the data can be read out from the display.

TABLE 5.6

Excerpt of Technical Specification of PFB 1200-2 from KERN, 2013

Readability (d)	0.01 g
Weighing range (max)	1200 g
Reproducibility	0.01 g
Linearity	±0.03 g
Warm-up time	2 hours
Stabilisation time (typical)	3 sec

Source: KERN, PFB, Operating Manual, Precision balance, Technical specification, 2013. Copyright KERN and Sohn Gmbh. Reproduced with permission. Courtesy of KERN and Sohn Gmbh.

TABLE 5.7

Imported Type B Uncertainty, with Contribution of Digital Resolution

Reproducibility Contribution	Linearity Contribution	Rectangular Distribution Parameter a_{rec}	Type B Uncertainty $\sigma_{s,rec}$ (Equation 5.2)
0.01 g	0.03 g	0.01 g + 0.03 g = 0.04 g	**0.023 g**

In Table 5.6, the resolution is referred to by the manufacturer as 'readability' with the units of 'd' (meaning 'digits').

Some uncertainty results from the finite resolution as discussed above, and because in this case the resolution is $d_r = 0.01$ g, the uncertainty due to digital display would be 0.0029 g. This value can be combined into the total imported uncertainty or it can be combined later, with all other components of measurement uncertainty (if there are such, e.g. Type A). There is no information on probability distributions, so some assumptions must be made on the method. However, in this particular case, even when assuming the 'worst-case scenario' of summing up all the values, it can be seen that the value of the combined standard uncertainty is dictated mostly by the two other components, as shown below. The 'summing' method is typically applied in commercial devices.

If a more realistic method of combining was used (explained later in this chapter), then, in this particular case, the contribution from the digital display resolution would become negligible, especially if the uncertainty was to be rounded to a single digit of precision (Taylor, 1997).

Hence, the two remaining values are 'reproducibility' and 'linearity'. Again, for a commercial device, it is sensible to assume that 100% of the values are expected to lie within the given limits, so either rectangular or triangular distribution can be employed.

The linearity (or rather non-linearity) factor has a greater value and because of the lack of knowledge of the shape of the non-linearity, it is probable that it can occur at any point of range, with its maximum value. Hence, it is prudent to presume rectangular distribution so that Equation 5.2 would be used. A trapezoidal distribution could also be an option (JCGM, 2008). The relevant values are summarised in Table 5.7.

The notation given by the manufacturer in Table 5.6 is somewhat confusing because the linearity value is given with the '±' sign, whereas the reproducibility is shown without it. A decision must be made whether the latter should be interpreted as ±0.01 g or ±0.005 g. If in doubt, it is recommended to contact the manufacturer in order to clarify any potential misunderstandings. However, because the specified resolution is 0.01 g, then in this case, it is probably acceptable to assume that the reproducibility is also ±0.01 g (because it is not possible to show 'half of a digit').

The values of the imported parameters may also be important in a reverse process, namely, the upper limit of uncertainty can be stated by some specification and then the appropriate instrument has to be used so that the limits are satisfied.

This is the case in such standards as IEC (2008), which, for example, gives the following specification: *A voltmeter responsive to r.m.s. values having an accuracy of ±0,2% or better shall be used'*. It is interesting to note that even in an *international standard*, the term 'accuracy' might be left undefined or not explained in the context of measurement *uncertainty*.

If the '±0.2% accuracy' is to be satisfied, then what should be the confidence interval: 68%, 95%, 99% or 100%? Because there is no other information available, a worst case should be assumed with 100% confidence interval with the implied rectangular probability distribution for a given instrument.

However, the value is expressed in relative units (%). Therefore, the manufacturer's technical specification would have to be converted to the relative units for a given voltage level expected for a given sample, flux density level and operating frequency (compare with Table 5.3). In such case, the limits must be calculated first before they can be compared to the specification of a given device.

5.4   MATHEMATICAL MODEL

Measurement is a process of quantifying an unknown physical value by comparing it with a known value. The known value can be obtained in a number of ways. A common way is by employing a measurement instrument that can provide such a comparison in a wide range of values.

Sometimes the measurement process is very simple, for example, the determination of a weight of a given object. With modern electronic instruments, it is usually sufficient to place the object on the scales and the result of the measurement is provided on a digital display. Therefore, there is a simple and direct relationship between the process of measuring and the measured value.

However, in many measurements, the relationship is more complex because some calculations are involved in the process of measurement, due to more than one variable that needs to be determined.

For instance, measuring the rectangular cross-sectional area of a sheet sample requires measuring two values of distance (e.g. thickness d and width l), which have to be multiplied in order to calculate the value of the area. Therefore, Equation 5.13 represents the *mathematical model* of such a measurement:

$$A = d \cdot l \quad (\text{m}^2) \tag{5.13}$$

The mathematical model is required here because it defines the mathematical relationship between all the necessary variables, and this in turn is necessary for calculating the exact way in which uncertainties of all of the measured (perhaps separately) variables contribute to the final measured value (Taylor, 1997; JCGM, 2008).

For example, in the fieldmetric technique (described in Chapter 2), the power loss can be calculated from Equation 2.13. There are a number of variables whose values must be measured or otherwise known before the calculation of the final result can take place: frequency, density of the material, two waveforms of B and two waveforms of H.

The purpose of the mathematical model is to help with the definition of *sensitivity coefficients* for each contribution to uncertainty (JCGM, 2008), as discussed in the following sections.

5.5 PROPAGATION OF UNCERTAINTY

The previous sections dealt only with the identification and calculation of single components of uncertainty. Almost always there are multiple components contributing to the overall uncertainty of the measurement, and they should be combined into a single value of *combined standard uncertainty*.

As mentioned above, according to the recommendation given in JCGM (2008), all components of uncertainty (Type A and Type B) can be treated equally, and therefore this provides the first step towards combining them into one overall value.

5.5.1 SENSITIVITY COEFFICIENTS

If the calculations are based on a mathematical model of measurement, then the *sensitivity coefficients* must also be defined mathematically.

For a very simple case of weighing a mass, the measurand is measured directly by the balance and therefore the uncertainty of the balance can represent directly the uncertainty of the measurement. We can say that in such a case, the sensitivity coefficient is equal to unity. This is not the case for more complex mathematical models.

For example, a hypothetical value V_m is to be found, and is defined by the following mathematical model:

$$V_m = h(v, \phi) = v \cdot \sin(\phi) \qquad \text{(unit of } v\text{)} \qquad (5.14)$$

where h represents the multi-variable function in the mathematical model.

But the only values that can be measured directly are v and ϕ, and V_m must be calculated from Equation 5.14. Therefore, the uncertainty of V_m will be impacted in a different way by the uncertainty of v and completely differently by the uncertainty of ϕ as dictated by the sensitivity coefficients. In such a case, the mathematical model is required because uncertainty estimation is not possible without it.

An assumption is made that the input variable uncertainty u_i represents a *small* deviation about the value of the measurand v. The sensitivity factors can be estimated by expanding the function of the mathematical model with a first-order Taylor series, defined for each input variable (JCGM, 2008), so that the generic equation for a *sensitivity coefficient* S of the ith input variable v_i in the measurand $V_m = h(v_1, v_2, ..., v_n)$ is

$$S_i = \frac{\partial(h(v_i))}{\partial v_i} \qquad \text{(units of } V_m/v_i\text{)} \qquad (5.15)$$

PRACTICAL COMMENT – EXAMPLE

Equation 5.15 is equivalent to the calculation of a partial derivative of the mathematical model with respect to the given input variable. Partial derivative is calculated by assuming that other components are constant around the operating point (hence the requirement of 'small' changes).

The value of the calculated derivative can be negative, but uncertainties are always non-negative. In the final calculation, the values are squared so the negative sign is eliminated. For simplicity, in the examples below, the values were chosen such that the negative sign does not occur.

Therefore, by means of example, for Equation 5.14, we get that

$$\sin(\phi) = \text{constant} \quad S_v = \frac{\partial h(v,\phi)}{\partial v} = \frac{\partial(v \cdot \sin(\phi))}{\partial v} = \sin(\phi)\frac{\partial v}{\partial v} = \sin(\phi) \quad \text{(unitless)}$$

(5.16)

$$v = \text{constant} \quad S_\phi = \frac{\partial h(v,\phi)}{\partial \phi} = \frac{\partial(v \cdot \sin(\phi))}{\partial \phi} = v \cdot \frac{\partial(\sin(\phi))}{\partial \phi} = v \cdot \cos(\phi) \quad \text{(units of } v\text{)}$$

(5.17)

An important conclusion can be drawn from Equations 5.16 and 5.17, namely, the sensitivity coefficients can be calculated only around the 'operating point', or the just-measured value of the measurand. The input uncertainties are scaled through the sensitivity coefficients to produce standard 'scaled' uncertainties $u_{s,S}$ as follows:

$$u_{s,S} = S \cdot u_s \quad \text{(units of measurand)}$$

(5.18)

If the measurement uncertainty was to be defined for a whole range of measurements, then the calculations would have to be defined for each measured point because they could differ significantly. This can be sometimes achieved by means of curve fitting, for instance, by the least-squares method (Taylor, 1997).

Let us assume that there were two cases, and in both, the measured values were $v = 10.351$ V with standard uncertainty $u_v = 0.023$ V. But in the first case, $\phi = 88°$ with standard uncertainty $u_\phi = 1°$, and in the second case, $\phi = 178°$ also with standard uncertainty $u_\phi = 1°$.

These assumed values with the corresponding calculations by means of Equations 5.16 through 5.18 are presented in Table 5.8.

As evident from Table 5.8, not only very similar measurements, carried out with *the same* input uncertainties (e.g. the same equipment), can produce significantly different combined uncertainties, but also the dominant contribution of the final uncertainty can shift from one variable to another (row 4 in Table 5.8). This is the

TABLE 5.8
Example of Sensitivity Coefficients for Two Cases, for Type B Components[a]

		Case 1	Case 2
1.	Measured input value	$v = 10.351$ V; $\phi = 88° \approx 1.536$ rad	$v = 10.351$ V; $\phi = 178° \approx 3.1067$ rad
2.	Standard uncertainty of input value	$u_v = 0.023$ V; $u_\phi = 1°\ (\approx 0.01745$ rad)	$u_v = 0.023$ V; $u_\phi = 1°\ (\approx 0.01745$ rad)
3.	Sensitivity coefficient (Equations 5.16 and 5.17)	$S_v = \sin(88°) \approx 0.9994$; $S_\phi = v \cdot \cos(88°) \approx 0.361$ V	$S_v = \sin(178°) \approx 0.0349$; $S_\phi = v \cdot \cos(178°) \approx 10.345$ V
4.	Scaled uncertainty (Equation 5.18)	$u_{s,scaled,v} = 0.9994 \cdot 0.023$ V $\approx \mathbf{0.023}$ V; $u_{s,scaled,\phi} = 0.361$ V $\cdot 0.01745$ ≈ 0.0064 V	$u_{s,scaled,v} = 0.0349 \cdot 0.023$ V $= 0.00080$ V; $u_{s,scaled,\phi} = 10.345$ V $\cdot 0.01745$ $= \mathbf{0.181}$ V
5.	Calculated V_m (Equation 5.14)	$V_m = 10.351$ V $\cdot 0.9994 = \mathbf{10.345}$ V	$V_m = 10.351$ V $\cdot 0.0349 = \mathbf{0.361}$ V
6.	Combined absolute standard uncertainty[b]	$u_{c,Vm} \approx \mathbf{0.024}$ V	$u_{c,Vm} \approx \mathbf{0.181}$ V
7.	Corresponding combined relative standard uncertainty[c]	$u_{c,Vm} \approx \underline{\mathbf{0.23\%}}$	$u_{c,Vm} \approx \underline{\mathbf{50\%}}$
8.	Calculated values	$\mathbf{10.345\ V \pm 0.024\ V}$	$\mathbf{0.361\ V \pm 0.181\ V}$
9.	Final reported value with absolute uncertainty	$\underline{\mathbf{10.34\ V \pm 0.03\ V}}$ standard uncertainty	$\underline{\mathbf{0.4\ V \pm 0.2\ V}}$ standard uncertainty
10.	Corresponding final reported value with percentage uncertainty[c]	$\mathbf{10.34\ V \pm 0.3\%}$ **standard uncertainty**	$\mathbf{0.4\ V \pm 50\%}$ **standard uncertainty**

a For clarity, the Type A component was neglected in calculations (infinitely many measurements averaged).

b Equations for combined uncertainty are explained later in the text (e.g. Equation 5.25), but were included here for better illustration.

c Relative uncertainty is not an additional uncertainty but the same as absolute uncertainty, just expressed in percent of the final value of the measurand V_m.

main reason why the standard measurement uncertainty should be assessed for each operating point separately.

This is especially true for the measurement of rotational power loss with the field-metric method. The power loss calculation depends on the average angle between the B and H vectors, and this angle can vary significantly between lower and higher excitation levels (Alinejad-Beromi, 1992; Sievert et al., 1995). At high excitation, the value of the B–H angle decreases, which can severely impact the measurement uncertainty at such a working point. Therefore, reproducibility will also be impacted, even though for lower excitations, it can be significantly better.

It is recommended by JCGM (2008) that the estimation of uncertainties by means of 'worst-case scenario' should be avoided because it may lead to significant over-estimation of uncertainties. Hence, estimation of uncertainty for the whole range of measured values should be avoided, and point-by-point calculations should be implemented instead (as evident from Table 5.8). These uncertainties can be later combined for a given range, if the estimated values and the employed methodology allow it.

5.5.2   PROPAGATION OF UNCERTAINTY – COMBINED UNCERTAINTY

The law of propagation of uncertainty is based on the consequences of the *central limit theorem*, which states that the probability distribution for a large number of independent variables will be approximately normal (bell-shaped, Figure 5.3a) even though the individual variables might not follow the normal distributions.

An example given in JCGM (2008) shows just three variables, each with the rectangular distribution, which itself is known to significantly differ from the normal distribution. But a combination of just three such variables leads to a close approximation of the normal distribution in terms of numerical values obtained from statistical calculations, as far as the value of the standard deviation is concerned (Cook, 2002).

For instance, for 95% confidence interval, the combined distribution is defined by $k_Q = 1.937$, and for the normal distribution, it is $k_Q = 1.960$, so the difference in the hypothetically estimated uncertainty would be only around 1%, whereas the uncertainty is normally rounded to at most two significant digits (JCGM, 2008).

Hence, in practice, if the combined uncertainty is not very strongly dominated by a single Type A evaluation or single rectangular Type B rectangular distribution, the combined uncertainty reasonably approximates normal distribution (JCGM, 2008), with coverage factors as given in Table 5.9.

The *law of propagation of uncertainty* utilises a sum of all *variances* of the input variables and their sensitivity coefficients (Taylor, 1997; JCGM, 2008).

Variance defines the level of scattering of data and in statistics it is a more basic variable than the standard deviation. The relationship between the two quantities is that the standard deviation σ_s is a square root of variance $_v\sigma_s^2$ (Equation 5.19).

$$\sigma_s = \sqrt{_v\sigma_s^2} \quad \text{(units of measurand)} \tag{5.19}$$

TABLE 5.9
Values of Coverage Factor k_Q for Normal Distribution

Level of Confidence (%)	Coverage Factor k_Q (Unitless)
68.27	1.000
90.00	1.645
95.00	1.960
95.45	2.000
99.00	2.576
99.73	3.000

Source: JCGM, Evaluation of measurement data, Guide to the expression of uncertainty in measurement, *Joint Committee for Guides in Metrology, JCGM 10: 2008*, 2008. Copyright 2008 Joint Committee for Guides in Metrology and Bureau International des Poids et Mesures. Reproduced with permission. Courtesy of Bureau International des Poids et Mesures.

PRACTICAL COMMENT

This is why standard deviation, and hence also uncertainty, always has a non-negative value because it is always calculated as a square root of a squared value, which is never negative for real numbers.

As mentioned above, the expansion in a first-order Taylor series can be carried out only around the 'operating point'. If the measurand V_m is described by a function $h(v_1, v_2, v_3, ..., v_n)$ with n of independent variables v, so that $V_m = h(v_1, v_2, v_3, ..., v_n)$, it can be written that (JCGM, 2008)

$$_v\sigma_s^2 = \sum_{i=1}^{n}\left[\left(\frac{\partial h}{\partial v_i}\right)^2 \cdot (u_{s,i})^2\right] + 2 \cdot \sum_{i=1}^{n-1}\left[\sum_{j=i+1}^{n}\left(\frac{\partial h}{\partial v_i} \cdot \frac{\partial h}{\partial v_j} \cdot u_{s,i} \cdot u_{s,j} \cdot \rho_{i,j}\right)\right]$$

(5.20)

(units of the measurand squared)

where $\rho_{i,j}$ is the *correlation coefficient* between v_i and v_j, which for two variables v_1 and v_2 can be calculated as (Taylor, 1997)

$$\rho_{1,2} = \frac{\sum_{i=1}^{n}\left[(v_{1,i} - \overline{v_1}) \cdot (v_{2,i} - \overline{v_2})\right]}{\sqrt{\sum_{i=1}^{n}(v_{1,i} - \overline{v_1})^2 \cdot \sum_{i=1}^{n}(v_{2,i} - \overline{v_2})^2}}$$

(unitless)   (5.21)

If the variables are independent (or in practice they are assumed to be independent), then the second term in Equation 5.20 is eliminated because the correlation coefficient $\rho_{i,j}$ is zero for all pairs of independent variables. In such cases, the law of propagation of uncertainties simplifies to

$$_v\sigma_s^2 = \sum_{i=1}^{n}\left[\left(\frac{\partial h}{\partial v_i}\right)^2 \cdot (u_{s,i})^2\right] \qquad \text{(units of the measurand squared)} \qquad (5.22)$$

As mentioned above, standard uncertainty is a square root of the variance – Equation 5.19. Therefore, for a case with independent variables, the combined uncertainty can be calculated as

$$u_c = \sigma_s = \sqrt{\sum_{i=1}^{n}\left[\left(\frac{\partial h}{\partial v_i}\right)^2 \cdot (u_{s,i})^2\right]} \qquad \text{(units of the measurand)} \qquad (5.23)$$

However, the partial derivatives in Equation 5.23 represent the sensitivity coefficients of Equation 5.15. Therefore, Equation 5.23 can be rewritten by using the relationship (5.18) so that the *combined standard uncertainty* u_c can be calculated from the scaled standard uncertainties:

$$u_c = \sqrt{\sum_{i=1}^{n}[(S_i)^2 \cdot (u_{s,i})^2]} \qquad \text{(units of the measurand)} \qquad (5.24)$$

$$u_c = \sqrt{u_{s,S,1}^2 + u_{s,S,2}^2 + u_{s,S,3}^2 + \cdots + u_{s,S,n}^2} \qquad \text{(units of the measurand)} \quad (5.25)$$

PRACTICAL COMMENT

Equations 5.24 and 5.25 are equivalent – the latter was given here just for better 'visual' illustration of the equation. Such combination of variables is called 'addition in quadrature' (Keightley, 2008) and was employed for the calculation of the values of combined uncertainties in Table 5.8.

The quadrature sum gives a result in which the contribution of the larger input values is favoured over the lower values. For example, if there are two input values of scaled uncertainties: 3.8 and 1, then the addition in quadrature produces a value of $\sqrt{(14.44 + 1)} = \sqrt{15.44} \approx 3.93$ (whereas normal addition would return 4.8). If such a value was rounded to a single significant digit (Taylor, 1997), then the result is 4 and the contribution of the lower value becomes negligible in practice. Of course, the calculations must be carried out anyway because it is not possible to know the final value beforehand.

If there is a large number of input components, then such differences might become non-negligible. For example, if there is one value of 3.8, and another five values each equal to 1, then the addition in quadrature gives $\sqrt{(14.44 + 1 + 1 + 1 + 1 + 1)} = \sqrt{19.44} = 4.41 \approx 5$, so the influence is significant enough to be taken into account (because the final uncertainties should be always rounded up).

This example also shows that in order to significantly improve the accuracy of the measurement, the highest contributions must be tackled first because the smaller ones will make very little practical difference.

5.5.3 Effective Degrees of Freedom

The concept of degrees of freedom was discussed with Student's distribution. In practical measurements, always fewer than 'infinitely many' observations are made. Therefore, according to Table 5.2, a correction has to be made in order to ensure that the acquired information represents standard uncertainty.

A similar concept applies to the value of standard combined uncertainty calculated from Equations 5.24 and 5.25. If some or all components in this equation were not derived from an 'infinite' amount of information, then the calculated value will not represent accurately the *standard* combined uncertainty. Therefore, a correction by an appropriate factor from Table 5.2 is required.

However, the difficulty lies in the fact that various components might have been acquired with various numbers of degrees of freedom so that a decision has to be made as to the applicable number to be used in Table 5.2.

For this reason, the concept of the *effective number of degrees of freedom* $_e d_f$ was introduced in the uncertainty analysis. It can be calculated from the Welch–Satterthwaite formula (JCGM, 2008)

$$_e d_f = \frac{u_c^4}{\sum_{i=1}^{n} \left[\frac{(S_i)^4 \cdot (u_{s,i})^4}{d_{f,i}} \right]} \quad \text{(unitless)} \quad (5.26)$$

where u_c is the combined uncertainty calculated from Equations 5.24 and 5.25, S_i is the sensitivity coefficient of each component calculated from Equation 5.15, $u_{s,i}$ is the standard uncertainty of each component and $d_{f,i}$ is the number of degrees of freedom for each component.

Therefore, the *expanded combined uncertainty* $_e u_c$ is calculated by bringing together all the relevant calculations shown above so that

$$_e u_c = u_c \cdot k_{Q,t} \quad \text{(unitless)} \quad (5.27)$$

where u_c is the combined uncertainty calculated from Equations 5.24 and 5.25 and $k_{Q,t}$ is Student's distribution correction factor assigned from Table 5.2 for the value of effective degrees of freedom calculated from Equation 5.26.

PRACTICAL COMMENT

If the calculated number of $_e d_f$ is not an integer, then it is advised to truncate it to the nearest *lower* whole number, rather than using ordinary mathematical rounding (JCGM, 2008). Such truncation towards lower values allows erring on the side of caution, especially for lower values.

The results of Equation 5.26 is such that the new calculated value will be always lower or equal to the sum of all degrees of freedom for all components (JCGM, 2008).

But if there are components assumed to have 'infinite' number of degrees of freedom and they contribute significantly to the combined uncertainty, then the calculated value can be very high, for example, even significantly greater than 100, so that values for 'infinity' can be taken from Table 5.2.

However, if the prevailing component of uncertainty has few degrees of freedom, then a substantial correction might be required as per Table 5.2. For this reason, Equation 5.26 should always be used because it cannot be known beforehand which components are contributing significantly.

5.5.4   Reporting Uncertainty

As mentioned above, the numerical result of a measurement should always be reported with the measured value, followed by the value of uncertainty, as well as the confidence interval. Also, any information immediately relevant to the uncertainty statement should also be supplied (e.g. type of distribution, critical measurement conditions, correlation matrix between variables).

The reported measured value should be rounded to the same precision as its uncertainty (JCGM, 2008). The values of uncertainties should be *rounded up* to at most two significant digits.

PRACTICAL COMMENT

In normal practice, uncertainties starting with '1' or '2' would be rounded up to two significant digits, and those starting with '3' and higher to just one significant digit.

For example, '10.47' would be rounded up to '11' and '0.003407' to '0.004', and so on. Rounding down is allowed for reasonable numbers, for example, '28.05' can be rounded down to '28' (Taylor, 1997; JCGM, 2008). Of course, during the calculation of each component of uncertainty, greater precision should be used in order to avoid any unnecessary rounding errors.

A typical simple example* (from JCGM, 2008) of reporting a measured value with *standard uncertainty* is given below (Copyright 2008 Joint Committee for Guides in Metrology and Bureau International des Poids et Mesures. Reproduced with permission. Courtesy of Bureau International des Poids et Mesures). Such a short statement contains quite a lot of information, namely, 1 – the value of measurement, 2 – the value of uncertainty, 3 – that the given uncertainty is *standard* and not *expanded* so

---

* These two examples from JCGM (2008) are modified slightly in order to conform to the symbols employed elsewhere in this book. Numerical values were also changed for better clarity, but otherwise the format is as quoted.

the confidence level is not 95% or some other value, 4 – that the measured value can be trusted to *at most* two significant digits of the uncertainty.

$m = (21.47 \pm 0.35)$ kg, where the number following the symbol $\pm$ is the numerical value of (the combined standard uncertainty) u_c and not the confidence interval.

PRACTICAL COMMENT

It should be noted that reporting the result as just '$m = (21.47 \pm 0.35)$ kg' is incomplete because it does not give *any* information about the level of confidence. In this case, the value '0.35' represents the *standard* uncertainty, for which by definition the level of confidence is only 68% (Figure 5.3a). This information should be conveyed as well.

For this reason, when reporting a value with *standard* uncertainty, JCGM (2008) recommends that the symbol '$\pm$' is avoided because it has traditionally been used to indicate an interval corresponding to a high level of confidence, and therefore it could be confused with *expanded*, rather than *standard*, uncertainty. If in doubt, the full statement should be used to avoid any confusion.

When reporting a value of *expanded uncertainty*, the example given in JCGM (2008) is as below. With such a statement, all the terms are either clear, or clearly defined so that confusion is avoided (Copyright 2008 Joint Committee for Guides in Metrology and Bureau International des Poids et Mesures. Reproduced with permission. Courtesy of Bureau International des Poids et Mesures):

$m = (21.47 \pm 0.79)$ kg, where the number following the symbol $\pm$ is the numerical value of an expanded uncertainty $_e u_c = k \cdot u_c$, with $_e u_c$ determined from a combined standard uncertainty $u_c = 0.35$ kg and a coverage factor $k = 2.26$ based on the t-distribution for $d_f = 9$ degrees of freedom, and defines an interval estimated to have a level of confidence of 95 percent.

PRACTICAL COMMENT

An important observation from such a statement is that it informs the reader directly about the value of the coverage factor ($k = 2.26$), which can be used to convert the value of expanded uncertainty to the *standard uncertainty*. This is required if such an uncertainty is to be used in further measurements as an imported uncertainty. This numerical value would be used for a *divisor*, as explained in the next section.

5.5.5 DIVISORS

It should be emphasised that Equations 5.24 and 5.25 for combined uncertainty are valid only if *all* the uncertainties of the input components represent *the same confidence level*. It is recommended (JCGM, 2008; UKAS, 2012) that the calculations are based on *standard uncertainties*.

If *expanded uncertainty* is to be reported, then the value can be easily calculated from the *combined standard uncertainty* (JCGM, 2008). All the input uncertainties must be evaluated, scaled or otherwise calculated to be representative of standard uncertainties.

PRACTICAL COMMENT

The need for finding out the value of *standard uncertainty* is the primary reason why the cumbersome conversions from the example given in Table 5.3 must be made so that values can be expressed as those shown in Tables 5.4 and 5.5. This is especially significant for the abovementioned imported uncertainties, which often represent the rectangular distribution with 100% confidence interval rather than a corresponding standard uncertainty.

In order to minimise the risk of making a mistake in calculations, the methodology presented, for example, in UKAS (2012) employs the concept of *divisors*. Producing a table with all the input components listed helps with pre-processing of all these components with calculations appropriate for each separate value.

A divisor is simply a number that scales the value of a given component of *expanded* uncertainty in order to convert it into *standard* uncertainty. If the input values are available as standard deviations, then the divisor is unity. If the distribution is rectangular or triangular, then it is $\sqrt{3}$ or $\sqrt{6}$, respectively (see Equations 5.2 and 5.3). If the imported value expresses an expanded uncertainty of normal distribution, then, depending on the level of confidence, the divisor is set according to Table 5.9. Other values will apply for differently shaped probability distributions.

5.5.6   Uncertainty Budget

The term *uncertainly budget* denotes a combination of all the uncertainty components that contribute in the given measurement. An example of a step-by-step approach for the calculation of the contributions was shown in Table 5.8.

In order to carry out the determination of the uncertainty of a given measurement, it is recommended, and indeed required in practice, to go through the following steps, as suggested in Cook (2002) (Copyright of National Association of Testing Authorities. Reproduced with permission. Courtesy of National Association of Testing Authorities):

1. *Make a model of the measurement.*
2. *List all the sources of uncertainties.*
3. *Calculate the standard uncertainties for each component using type A analysis for those with repeated measurements and type B for others.*
4. *Calculate sensitivity coefficients.*
5. *Calculate the combined uncertainty, and, if appropriate its effective degrees of freedom.*

6. *Calculate the expanded uncertainty. Use a nominal or a calculated coverage factor. Round the measured value and the uncertainty to obtain the reported values.*

The widely accepted approach for the estimation of an uncertainty budget for a given measurement is to use spreadsheet software, with calculations arranged as shown in Tables 5.10 and 5.11. With such an approach, the values will be recalculated automatically and can be repeated with ease, for example, for the next measurement.

The example calculated below is based on a hypothetical measurement of an RMS value of unknown current by means of measuring of voltage drop across a shunt resistor (see also Figure 3.41), whose calibrated nominal value is **0.4691 Ω**. All of the partial contributions to the uncertainty are treated as completely independent.

The input data must be first pre-processed in order to ensure that it is in the appropriate format, so that the final calculations can be carried out easily.

The voltage is measured with a digital voltmeter. The measurement was repeated five times as shown in Table 5.10, so the calculated Type A standard uncertainty was **0.0137 A**; by definition, the distribution is **normal**, so the divisor is **1** and also the sensitivity coefficient is **1** (unitless). This is because the sensitivity coefficient S_I for the current is

$$S_I = \frac{\partial I}{\partial I} = 1 \quad \text{(unitless)} \tag{5.28}$$

The voltmeter specification given by the manufacturer was '*0.25% ± 1 digit*' for the range of the measured voltage. Using the average value from Table 5.10 as the operating point, the applicable limits of the specification can be calculated as 2.315 V $\cdot$ 0.25%/100% + 0.001 V = **0.00679 V**.

In the absence of other information, for the specification of commercial instruments, it can be assumed that the distribution is most likely **rectangular**, so a divisor of $\sqrt{3}$ is applicable (Figure 5.3b and Equation 5.2).

The shunt resistor had a calibrated value of (0.4691 ± 0.0011) Ω with a confidence level of 95% at 21°C as specified by its calibration certificate. Therefore, the nominal

---

TABLE 5.10

Calculation of Type A Uncertainty

Reading	Measured V	Nominal R	I = V/R	Mean (Equation 5.4)	Standard Deviation σ_s (Equation 5.5)	Type A Uncertainty, u_A (Equation 5.7)
1	2.300 V	0.4691 Ω	4.904 A	I = **4.936 A**	0.02749 A	**0.0137 A**
2	2.309 V		4.923 A	V = **2.315 V**	0.001289 V	0.00645 V
3	2.333 V		4.974 A			
4	2.311 V		4.928 A			
5	2.323 V		4.953 A			

TABLE 5.11

Example of Uncertainty Budget Calculation, Including Type A and Type B Uncertainty Contributions

a	b	c	d	e	f	g	h
Type of Contribution	Absolute Uncertainty	Distribution	Divisor	Sensitivity Coefficient	Degrees of Freedom	Absolute Standard Uncertainty ($e \cdot b/d$)	Corresponding Relative Standard Uncertainty ($g/I) \cdot 100\%$
Voltmeter specification	0.00679 V	Rectangular	$\sqrt{3}$	2.132 ($1/\Omega$)	Infinity	0.00836 A	0.169%
Shunt due to calibration	0.0011 Ω	Normal	1.96	10.52 (V/Ω^2)	Infinity	0.00590 A	0.120%
Shunt due to temperature	0.000375 Ω	Triangular	$\sqrt{6}$	10.52 (V/Ω^2)	Infinity	0.00161 A	0.033%
Type A (repeatability)	0.0137 A	Normal	1	1 (unitless)	4	0.01370 A	0.278%
Combined standard uncertainty (Equations 5.24 and 5.25)						0.0172 A	0.349%
Value to be reported – standard uncertainty						**0.017 A**	**0.4%**
Effective degrees of freedom (Equation 5.33)					9		
Student's correction factor from Table 5.2, for 95% confidence level					2.26		
Value to be reported – expanded uncertainty, confidence level 95% (Equation 5.27)						**0.04 A**	**0.8%**

value of the resistor and expanded uncertainty are given directly as $R = $ **0.4691 Ω** and **0.0011 Ω**, respectively.

Because of the stated confidence level of 95%, it can be probably assumed that the distribution was **normal**. No degrees of freedom was stated, so in the absence of such information, the value of 'infinity' can be assumed and hence the applicable divisor would be **1.96**. The divisor would not be used as 2.00 because the confidence was stated as 95% and not as 95.45% (see also Table 5.2).

The manufacturer of the shunt resistor specified its temperature coefficient as '*400 ppm/°C*', which corresponds to (400 ppm/°C) · 0.4691 $\Omega/10^6$ ppm $= 0.000188$ $\Omega/°C$.

The measurement was carried out in a temperature-controlled laboratory, but it is known that the temperature can vary from 19°C to 23°C. This is $\pm 2°C$ from the calibrated temperature of the shunt resistor; therefore, it will cause additional uncertainty because the actual temperature of the shunt was not measured directly. Hence, the maximum expected deviation could be up to 2°C, which would correspond to a maximum change of resistance as 2°C · 0.000188 $\Omega/°C = $ **0.000375 Ω**.

For temperature-controlled environments, **triangular** distribution can be assumed (Figure 5.3c) and because of Equation 5.3, the divisor will be $\sqrt{6}$. (The triangular distribution can be assumed only under certain conditions. Sometimes, in practice, normal distribution might be more applicable in such a case, for example, if the temperature is actually measured (Carriage, 2017).)

The mathematical model for the measurement is

$$I = \frac{V}{R} \quad \text{(A)} \tag{5.29}$$

The sensitivity coefficient S_V for voltage contribution is calculated from Equation 5.15 as the derivative of the function with respect to voltage, treating the variable R as constant:

$$S_V = \frac{\partial I}{\partial V} = \frac{\partial}{\partial V}\left(\frac{V}{R}\right) = \frac{1}{R} \cdot \frac{\partial V}{\partial V} = \frac{1}{R} \quad (1/\Omega) \tag{5.30}$$

Thus, the sensitivity coefficient S_V around the 'operating point' of the nominal values (A2LA, 2014) is $S_V = 1/0.4691\ \Omega = $ **2.132 (1/Ω)**.

PRACTICAL COMMENT

It is *crucial* that the calculated contributions are expressed in correct SI units. For this reason, the sensitivity coefficients must be derived very carefully and it is always a good idea to double-check the calculations with proper units, especially for more complex functions. A very useful feature is the option of calculations with SI units in mathematical software such as commercial *Mathcad* (PTC, 2017) or freeware *Smath Studio* (Ivashov, 2017).

This is important because a popular spreadsheet software is designed to work only with numerical values, with the physical units completely ignored. Special provisions must be made individually by the user to make sure that the units are taken into account.

Even with the use of the spreadsheet software, it is easy to make a human error by using the units, for example, of 'mm' instead of 'm'. So the SI unit would be correct as such, but the values would be wrong by several orders of magnitudes. Therefore, it is a good habit to convert *all* the values into the base SI units, without any SI prefixes, especially when using spreadsheet software.

The sensitivity coefficient $S_{R,cal}$ for the resistance contribution due to resistor calibration is calculated as the derivative of the function with respect to resistance, treating the variable V as constant:

$$S_{R,cal} = \frac{\partial I}{\partial R} = \frac{\partial}{\partial R}\left(\frac{V}{R}\right) = V \cdot \frac{\partial}{\partial R}\left(\frac{1}{R}\right) = V\frac{1}{R^2} = \frac{V}{R^2} \quad (\text{V}/\Omega^2) \quad (5.31)$$

So the sensitivity coefficient $S_{R,cal}$ = 2.132 V/(0.4691 Ω)^2 = **2.132 (1/Ω)**.

The sensitivity coefficient $S_{R,temp}$ for the resistance contribution due to resistor temperature is the same as in Equation 5.31 because any change in resistance will contribute in the same mathematical way (compare Equations 5.32 and 5.31), so its numerical value will also be the same $S_{R,temp} = S_{R,cal} = $ **10.52 (V/Ω^2)**.

$$S_{R,temp} = \frac{\partial I}{\partial R} = \frac{V}{R^2} \quad (\text{V}/\Omega^2) \quad (5.32)$$

Therefore, all the values derived above can now be summarised in an uncertainty budget, as shown in Table 5.11.

The calculation of the effective degrees of freedom is carried out by using Equation 5.26. There are components with infinity in the denominator but '1/∞' tends to zero so they are eliminated. However, the information about their contribution is still carried in the value of the combined uncertainty, so that the calculated value of effective degrees of freedom is 9.8, which should be rounded *down* to **9**.

$$_e d_f = \frac{0.0172^4}{\dfrac{0.00835^4}{\infty} + \dfrac{0.00590^4}{\infty} + \dfrac{0.00161^4}{\infty} + \dfrac{0.01370^4}{4}} = \frac{0.0172^4}{\dfrac{0.01370^4}{4}} = 9.8 \quad (\text{unitless})$$

$$(5.33)$$

PRACTICAL COMMENT

Therefore, in this particular case, the use of Welch–Satterthwaite equation was actually required because the number of effective degrees of freedom was a considerably low value, and an appropriate correction had to be applied from Table 5.2.

It should be noted that the Student's distribution was taken into account for the whole final calculation and not just for the input value which was measured with 4 degrees of freedom.

The final value of the measurement should be reported with very minimum information as shown below – with the standard uncertainty value as

$$I = 4.936 \text{ A, standard uncertainty of } 0.017 \text{ A}$$

or with the expanded uncertainty value as

$$I = 4.94 \text{ A} \pm 0.04 \text{ A, where '0.04 A' means expanded uncertainty determined}$$
$$\text{with a coverage factor of 2.26, with 9 degrees of freedom and confidence level 95\%}$$

PRACTICAL COMMENT

In the second format, it is necessary to include the information about the number of degrees of freedom. This is because all the calculations are based on *standard* uncertainty. If just the confidence level of 95% was stated, then this would incorrectly imply that the distribution is normal. Therefore, a divisor or 1.96 would be assumed, instead of the correct value of 2.26.

Knowing just the value of the coverage factor, it is possible to look up the number of degrees of freedom from Table 5.2, and thus having precise information required for the Welch–Satterthwaite formula.

In the first format, the information is given directly that it is the *standard* deviation, so no further correction is necessary because by definition it will be that divisor = 1.

If the result was just reported as '$I = (4.94 \pm 0.04)$ A', it would be impossible to know if perhaps the values mean rectangular distribution with 100% confidence level. In such case, the assumed divisor would have to be $\sqrt{3} = 1.73$ and thus the estimated imported standard uncertainty would be 0.023 A, which differs from the correct value of 0.017 A by 35%.

So, the lack of information about the distribution would mean that a conservative approach would have to be taken by assuming the rectangular distribution, and this would lead to an overestimation of uncertainty. This is why rectangular distribution can be assumed in practice as a proxy for a 'worst-case' approach, if no other information is available for imported uncertainties.

It should be noted that the probability distribution of the combined uncertainty can be completely different from the normal bell curve (Cox et al., 2010). The estimated value of uncertainty can be similar, but the shape of the distribution might not be. This is why it is important to provide accurate information about the *standard* deviation (first example), or enough information so that the value can be extracted accurately (second example).

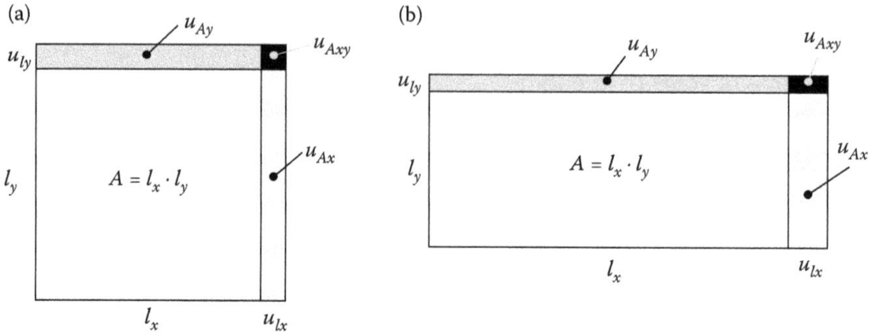

FIGURE 5.5   Illustration of first-order and higher-order uncertainties for (a) square area and (b) rectangular area. (Adapted from Cook R.R., *Assessment of Uncertainties of Measurement for Calibration & Testing Laboratories*, 2nd edition, National Association of Testing Authorities, 2002, ISBN 0-909307-46-6.)

5.5.7   HIGHER-ORDER TERMS

Higher-order terms of uncertainty are *always* present (Cook, 2002). It is just that for simpler cases that can be linearised around the 'operating point' their contribution is usually so small in practice that they can be neglected. This greatly simplifies calculations. The linearisation is a consequence of using only the first-order components from the Taylor expansion in Equation 5.20.

The concept of higher-order term is illustrated in Figure 5.5. A nominal area of the rectangle is dictated by two length values so that $A = l_x \cdot l_y$. The measurements of l_x and l_y are affected by their uncertainties u_{lx} and u_{ly}, respectively. Therefore, each of them contributes to the uncertainty of the area A, and the first-order contributions are u_{Ax} and u_{Ay}, respectively.

The higher-order term is the black area in Figure 5.5. As can be seen, its 'amplitude' (i.e. the area of the black rectangle) is proportional to the product of the two uncertainties, and therefore its area is much smaller in this case, and hence it can be 'safely' neglected (Cook, 2002).

PRACTICAL COMMENT

Let us assume that $l = l_x = l_y$ and $u_{lx} = u_{ly}$ and the uncertainties are not correlated (Figure 5.5a). Also, both $u_{lx} = u_{ly}$ are 1% of l. Therefore, according to Equations 5.24 and 5.25, the combined uncertainty for the area would be $\sqrt{(0.01^2 + 0.01^2)} = 0.0141$ or 1.41%. This value is not 2% because the input uncertainties are not correlated and the probability of both of them occurring with their full amplitude is appropriately lower.

It can be seen from Figure 5.5a that the area represented by the black square would be proportional to the product of both input uncertainties. The area in question would be at worst $0.01 \cdot 0.01 = 0.0001$ or 0.01%. Such value compared to 1.41% is comparable to the *third* significant digit and according to the JCGM (2008), it is normally rounded off anyway. But this example shows why it is recommended that the final digit should be rounded *up*.

The same applies if the input uncertainties have different amplitudes. If u_{lx} = 2% of l_x and u_{ly} = 0.5% of l_y and l_y = 0.5 $\cdot$ l_x, then the combined uncertainty would be $\sqrt{[(0.5 \cdot 0.02)^2 + (1 \cdot 0.005)^2]}$ = 0.0111 or 1.11%. The worst-case value for the black rectangle area would be 1 $\cdot$ 0.02 $\cdot$ 0.5 $\cdot$ 0.005/0.5 = 0 .0001 or 0.01%. Again, this is a change at the third significant digit, and hence can be neglected in practice (especially after rounding up).

The 'negligibility' is directly applicable for linear functions and indirectly to any non-linear function that can be linearised with first-order Taylor series by means of Equation 5.20.

As illustrated in Figure 5.6, such approach can produce sufficient approximation of the non-linear function so that, around the operating point of the nominal value, the problem can be treated as linear. However, such approach works *only* if both the following conditions are met:

1. The linear approximation is made around the nominal input value.
2. The uncertainty of the input value remains negligibly small as compared to the non-linearity of the function.

If these conditions are not met, then the non-linearity of the function is significant and higher-order contributions (Figure 5.5) must be taken into account (Cook, 2002; JCGM, 2008). The appropriate sensitivity coefficients (Equation 5.15) must be calculated for higher-order derivatives accordingly (Cook, 2002).

This can be explained by referring to the drawing in Figure 5.6. If the non-linear function around the operating point v_{nom} was approximated with a quadratic function or some other higher-order polynomial, then the differences between the curved approximation and the real function would be significantly reduced, as compared to a simple linear function. However, in most cases, this is not required.

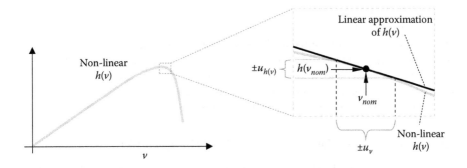

FIGURE 5.6 Linearisation of a non-linear function $h(v)$ around the 'operating point' of nominal value v_{nom}.

5.5.8 CORRELATION OF INPUT VARIABLES

So far, in all the above discussion, the implicit assumption was that the uncertainties of the input variables were independent of each other – they were *not correlated*. In many cases, such assumption can be made because the uncertainty of the measuring instrument (e.g. voltmeter) would be normally independent of the uncertainty of other variables (e.g. shunt resistor).

However, if there is a correlation between the input variables and their uncertainties, then this should be taken into account because it can result in a greater, but sometimes also smaller, uncertainty (Cook, 2002; JCGM, 2008). The calculations should then be carried out with all the terms in Equation 5.20, with the correlation coefficient calculated from Equation 5.21.

If the input quantities are correlated in a known way, then the underlying function can be rearranged and expressed in an alternative algebraic form so that the correlation is eliminated (Cox et al., 2010). However, this might not always be possible.

PRACTICAL COMMENT

In the example with Equation 5.13, there are two input variables. These two values could be measured by two *different* methods, for example, the thickness by means of a micrometer and the width by a digital vernier caliper. Therefore, two such uncertainties would not be correlated and they should be combined with quadratic summation.

However, if both values were measured by *the same* instrument, then the uncertainty should be added in an ordinary way rather than in quadrature. This is because if the instrument overestimated one dimension, it is much more probable that it could also overestimate the other dimension, and this can be explained by looking at Figure 5.5.

If the perturbation in the x direction is positive and in the y direction is zero (no correlation), then only one variable contributes to the uncertainty of the output variable. But if both directions are perturbed in the same way, both positive or both negative (positive correlation coefficient), then the resulting uncertainty will be higher.

It is also possible that the contributions could cancel each other out. If one perturbation was positive and the other negative (negative correlation coefficient), then the resulting uncertainty of the output variable would be decreased. In such case, adding the input uncertainties in quadrature would severely overestimate the uncertainty of the output variable.

5.6   OTHER METHODS OF UNCERTAINTY ESTIMATION

The mathematical model discussed in the previous section defines only the mathematical relationship between the contributing variables. In an ideal world, such mathematical functions could be used for quite direct estimation of the involved uncertainties.

However, in reality, analytic derivation of functions for sensitivity coefficients can be quite a laborious process. Sometimes, the function might be too complex to find an analytical solution for the derivatives.

There is also the difficulty of knowing the threshold at which the higher-order components become 'significant'. For linear functions, they can be safely ignored, but in more cases than not, the employed functions are non-linear and thus it cannot be arbitrarily stated that they can be ignored. To be absolutely certain, it would be necessary to estimate the magnitude of the higher-order terms in order to investigate if their influence is significant. But once they are calculated, they can be used in the calculations anyway.

If the input variables are correlated, then their correlation coefficient *must be known* because otherwise the values cannot be used properly in the propagation of uncertainties (Cox et al., 2010). Because of these difficulties, other ways of defining some or all of these values were devised.

5.6.1 Sequential Perturbation

If the mathematical model as well as all required input variables are known, then the process of uncertainty propagation can be approached with a different technique.

The differentiation of complex functions is difficult and prone to human errors. In technological areas where such functions are commonly used (e.g. fluid mechanics), the technique of *sequential perturbation* was introduced (Moffat, 1985; Manteufel, 2012; Recktenwald, 2017).

The concept of sequential perturbation can be explained by analysing Figure 5.6. The input variable is set to its nominal value v_{nom} so that the output value of the function can be calculated at such operating point, and it will be equal to $h(v_{nom})$.

The input variable can be perturbed by the amount of its standard uncertainty, for example, to a new value $v_{nom} + u_v$ and a new output value can be calculated from the function so that $h(v_{nom} + u_v)$. Therefore, the sensitivity $S_{h(v)}$ of the function $h(v)$ caused by the uncertainty u_v of the input variable v can be calculated by following the mathematical definition of a derivative:

$$S_{h(v)} = \frac{h(v_{nom} + u_v) - h(v_{nom})}{u_v} \quad \text{(units of } h(v)/v) \quad (5.34)$$

Therefore, there is no need for the analytical calculation of the function responsible for $S_{h(v)}$ because it can be computed numerically, and after all it is the value of the sensitivity coefficient that is important, not its mathematical form as such (Table 5.11). Such perturbation should be applied in the positive and negative directions and the results averaged.

This is illustrated in Figure 5.6, in which perturbation of v_{nom} by u_v corresponds to the change of $h(v)$ from $h(v_{nom})$ by a value of $u_{h(v)}$. The slope of the straight line approximation is the tangent of the function, which in effect is the derivative of such a function at that point and hence the sensitivity coefficient at that point.

If the main relationship is a function of several variables, then the calculation by Equation 5.34 is repeated *sequentially* for each of the variable (Recktenwald,

2017), so that all the components of uncertainty (column 'g' in Table 5.11) can be calculated one by one. They can be later combined in the usual way, by means of Equations 5.24 and 5.25. As always, it should be borne in mind that Equations 5.24 and 5.25 are applicable only for 'small' uncertainties, or for functions with sufficiently good linear approximation.

Therefore, this method allows complete avoidance of the analytical calculation of derivatives. It should also be possible to test the influence of correlated variables because they can be varied simultaneously as necessary.

5.6.2 EXPERIMENTAL PERTURBATION

The method of perturbation can also be applied experimentally. In a sufficiently controlled environment, it is possible to change just one input variable at a time, with other values being held constant, and measure the influence on the output variable. For instance, in the example from Table 5.11, it would be possible to keep the same specification of the voltmeter and the same nominal value of shunt, and to ensure that the temperature of the shunt is varied by a controlled amount. This would allow finding out the exact uncertainty of the shunt due to temperature change.

Indeed, such an approach is used even in accredited calibration laboratories (Drake, 2016). For some measurements, uncertainty is dictated not only by the measurement equipment. For example, for magnetic measurements, by means of the Epstein frame, single-sheet tester and toroidal sample method, the measurement uncertainty is related to the type of material under test (UKAS, 2016). NPL is an accredited laboratory in the United Kingdom and therefore all the sources of input uncertainties are meticulously controlled. Still, the best achieved uncertainty for the measurement of power loss under AC excitation is at the level of 0.65%, and for higher excitations and frequencies reaches 1.8% (UKAS, 2016).

The uncertainty analysis for these measurements was carried out by using the 'sequential perturbation' but carried out experimentally. The process is tedious and time consuming, but for a fixed apparatus, it has to be carried out just once for a given range and type of material under test (Drake, 2016).

PRACTICAL COMMENT

NPL is a renowned accredited laboratory and the so-called Calibration and Measurement Capability (CMC) with expanded uncertainty ($k = 2$) is at the level of 0.75% for a well-established measurement method (UKAS, 2016).

It can be therefore expected that for measurements of rotational power loss values, with a far-less-established method, the achievable expanded uncertainty will be *significantly* worse when the measurements are carried out by a non-accredited laboratory.

For example, calibration curves for H-coils carried out by the author exhibited dispersion of values at the level of 0.5% (Figure 3.46) (Zurek, 2005a). Such uncertainty combined for two H-coils (H_x and H_x) would therefore be at least $\sqrt{(0.005^2 + 0.005^2)} = 0.0071$ or 0.7%, which would consume almost the whole uncertainty budget if compared to the 0.75% value.

But of course there will be numerous other sources of uncertainties, as discussed later in this chapter. For example, as shown in Figure 3.18, *B*-coils accuracy would also be affected by the presence of drilled holes. In order to achieve influence at the level of 0.5% for a *B*-coil, its width would have to be 200 times greater than the hole diameter. For a hole of diameter 0.3 mm, this would mean a *B*-coil width of 60 mm, and for a hole of diameter 0.5 mm, it would be at least 100 mm (Zurek, in press). This immediately poses some minimum limits on the sample size.

There are two significant difficulties with employing the method of experimental perturbation. First, just one variable should be changed at a time, which means that all other variables should be held constant. This requires an extremely stable source of excitation because the procedure will be lengthy. Also, real power loss is dissipated in the sample so its temperature could change slightly, which would change its resistivity and thus impact on the power loss due to eddy currents.

PRACTICAL COMMENT

Similar problems arise when trying to control other input variables. For rotational magnetisation, it might be extremely difficult to both control a given variable and to introduce a perturbation. For example, how to ensure stable *uniformity* of magnetisation, and then how to introduce perturbation in it, so that *all other variables* are unaffected?

It is not possible to simply change the shape of the controlled *B* loci because this would by definition change both *B* and *H* waveforms. Probably the magnetising yoke would have to be constructed in such a way as to allow for perturbation in non-uniformity of magnetisation. For example, one way could be to use different electrical connections of windings (Figures 4.31 and 4.32) so that their configuration could be changed theoretically without affecting other components in the system. But such an approach would first require meticulous verification of the non-uniformity in each case so that a known perturbation was applied. This is why such a procedure is very time consuming, but it could be an excellent topic for a research project (Drake, 2016).

The other difficulty is the fact that it is necessary to detect the perturbation in the output variable, which itself is affected by noise and random dispersion due to Type A uncertainty (Section 5.3.1). However, even the average value does not give perfect information due to 'uncertainty of uncertainty' as shown in Table 5.1. Even after recording 50 readings, the uncertainty of the average value is still at the level of 10%. Therefore, a significant number of readings would have to be recorded to ensure that the uncertainty of the mean value is significantly reduced.

One method of extracting more 'stable' information from the noisy values is to employ appropriate curve fitting with some polynomial function with the least square method (Ahlers et al., 2000a). Then the 'mathematical model' can be defined by the approximating function, which will have a certain number of degrees of freedom related to the number of available data points. Hence, an appropriate uncertainty can be defined for such Type A evaluation and can be used for further propagation of uncertainties.

Such methods are tedious and time consuming but are used even in accredited laboratories so they are proven to be useful in such circumstances. Unfortunately, the exact details of such procedures are not published because of confidentiality issues (Drake, 2016).

5.6.3 Monte Carlo Method

Another way of introducing the perturbations is by using a probabilistic method of the Monte Carlo type. Such methods can be very powerful in the estimation of combined uncertainties, and in certain cases they even show better performance (Cox et al., 2010) than the classical GUM method (JCGM, 2008).

In the sequential perturbation approach (analytical, numerical or experimental), the changes are introduced to one input at a time. This helps in the identification of absolute standard uncertainty components (Table 5.11) and sensitivity coefficients, but the quadrature summation must still be used for the calculation of the combined uncertainty.

It would be incorrect to introduce perturbation equal to standard deviations to all the inputs at once because, for example, positive perturbations in all the inputs could just add together, so that the output would be disturbed proportionally to the ordinary sum of all the components and this would likely overestimate the actual standard deviation of the output variable.

However, it is possible to introduce the changes to all the inputs at once by simulating the assumed probability distribution for each input. Such a method can be thought of as *simultaneous perturbation* because all inputs are changed simultaneously, but in a random way as dictated by their individual probability distributions. All the input probabilities combine in a truly statistical way such that the resulting output perturbation corresponds to the actual behaviour of such a system, as confirmed by analytical statistical calculations. Even highly non-linear systems can be analysed, which would normally necessitate the use of difficult-to-calculate higher-order terms (Cox et al., 2010).

All the simulations are carried out with an appropriate software (for smaller sample populations, it is possible to use spreadsheet calculations). The simulations are designed around the mathematical model of a given measurement and hundreds or thousands of simulations can be easily repeated, thus giving very good modelling of statistical behaviour with large numbers.

The method works equally well for any type of input and output distributions: normal, rectangular, triangular, U-shaped, asymmetrical, multimodal, etc., and can give better results than the quadratic summation of uncertainties (Cox et al., 2010).

Simulations were carried by the author for the example discussed above, with the mathematical model described by Equation 5.29. The perturbations were applied as a randomly distributed 'noise' around the nominal operating point as shown in Table 5.10.

The graphs in Figure 5.7 show histogram plots of probability distributions for all four inputs of the previously shown example (Table 5.11). Their absolute uncertain-

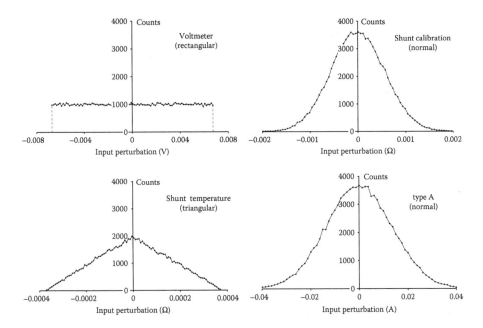

FIGURE 5.7 Probabilistic perturbations for each input in Monte Carlo simulation with 10,000 points and 100 intervals (see also Table 5.11).

ties were applied as required, as evident from the change of horizontal scale on each of the graphs. The data were simulated with 10,000 combinations (points) split into 100 intervals and sorted accordingly. Each point in the curve denotes the number of counts in a given interval. The curves are not smooth because of the finite number of samples used for simulation.

The dominating component was the Type A uncertainty; hence the shape of the histogram for the *output* variable very closely matches an ideal normal (Gaussian) distribution, as shown in Figure 5.8. This is especially visible with a higher number of points that smooth out the variations.

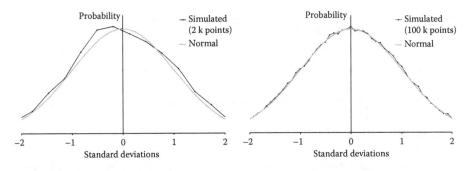

FIGURE 5.8 Simulated combined output probabilities, with 2000 points (2 k) and 100,000 points (100 k).

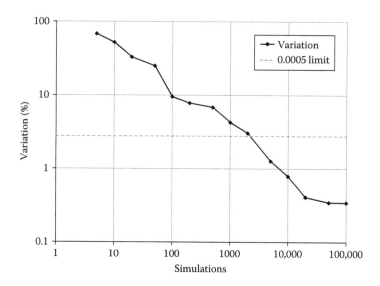

FIGURE 5.9   Required number of points for sufficient 'accuracy' of simulations (the limit 0.0005 corresponds to around 2.8% of the uncertainty value).

The data in Figure 5.8 (left) are a typical outcome for a smaller number of points. The curve is distorted and might be even somewhat asymmetrical.

The required minimum number of points was estimated by repeating the simulations with increased number of combinations (Figure 5.9). In this particular case, the limit was estimated at 0.0005 A, which is one-half of the second significant digit of the 'ideal' value of 0.017 A (Table 5.11). Because of the rounding to at most two significant digits, any variations in the simulated combined standard deviation lower than 0.0005 A will be insignificant.

As can be seen from Figure 5.9, a minimum of around 2000 points were required in order to reduce the variations to below the required level and the histogram at this threshold is shown in Figure 5.8 (left). A higher number of points gives lower variations and smoother histogram curves.

PRACTICAL COMMENT

It was mentioned above that a combination of just three rectangular input distributions produces a fairly close approximation of the normal distribution (Cook, 2002; JCGM, 2008; Cox et al., 2010).

The Monte Carlo technique can be used to demonstrate this effect. The four inputs in Table 5.11 are such that there are two normal, one rectangular and one triangular distribution. Their combination produces result as shown in Figure 5.8.

However, what happens if the simulations are changed to a hypothetical case in which all the inputs follow rectangular distributions with the corresponding standard deviations having the same values as before? It turns out that the

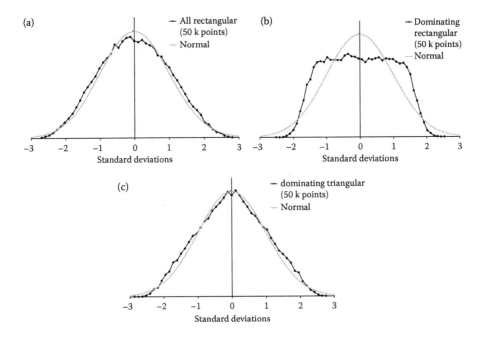

FIGURE 5.10 Simulated and normalised hypothetical combined output distributions with the input distributions: (a) all four inputs rectangular, (b) one dominating rectangular and (c) one dominating triangular (the normal distribution curve is shown for reference).

result is indeed quite similar to a normal distribution, as shown in Figure 5.10a. It should be emphasised that for simulation in Figure 5.10a, *all* four inputs were set to have *rectangular* distributions and yet the outcome is very close to normal.

The 'ideal' value of standard deviation for a normally distributed output was 0.01721, whereas for the data from Figure 5.10a, it was 0.01756. With normal rounding, the first value could be rounded down to 0.017 and the second to 0.018 (although it is 'safer' to always round up (JCGM, 2008)). The difference before rounding is therefore 0.00035, which is below the one-half digit assumed above. The difference after rounding is 0.001, which is around 6%. Therefore, we can see that the difference between the two calculated values of 0.01721 and 0.01756 was only around 2% whereas the difference from rounding was 6%. Moreover, it should be borne in mind that the 'uncertainty of uncertainty' is actually quite high as shown in Table 5.1, which is why such values should not be expressed with too many significant digits.

The data in Figure 5.10b was simulated with one hypothetical input uncertainty to be rectangular and dominating in value over the other inputs. The rectangular-like shape is clearly dominating. In such a case, the ordinary quadrature summation might return incorrect results (JCGM, 2008), whereas with the Monte Carlo approach, the standard deviation will be correctly calculated as long as a sufficient number of random combinations are simulated (Cox et al., 2010).

The data in Figure 5.10c were simulated with one input uncertainty to be triangular and dominating in value over the other inputs. The triangular-like shape is still clearly recognisable. However, in this case, the difference with respect to the normal distribution is much smaller. In any case, the Monte Carlo simulation would correctly calculate the resulting standard deviation.

5.7 THEORETICAL SENSITIVITY COEFFICIENTS FOR ROTATIONAL POWER LOSS

With the fieldmetric technique, the value of rotational power loss can be calculated from the measured quantities with the following equation:

$$P_{rot} = \frac{f}{D} \cdot \int\limits_0^T \left(\frac{dB_x(t)}{dt} \cdot H_x(t) + \frac{dB_{y(t)}}{dt} \cdot H_y(t) \right) dt \qquad \text{(W/kg)} \qquad (5.35)$$

Under ideally controlled circular magnetising conditions, the B_x and B_y waveforms are purely sinusoidal and cosinusoidal. Therefore, higher harmonics in H_x and H_y are eliminated from the $(dB/dt \cdot H)$ product.

PRACTICAL COMMENT

The elimination of higher harmonics from the $(dB/dt) \cdot H$ product occurs in a similar way as in the Fourier transformation. During decomposition into Fourier series, the arbitrary signal is multiplied by sine and cosine waveforms, one harmonic at a time. Only the component of the input signal that contains a given harmonic produces a non-zero average proportional to the amplitude of the given harmonic. All higher and lower harmonics produce zero average and thus are 'eliminated' for that given harmonic. The same mechanism applies here.

The waveforms are assumed to be perfectly *controlled* so that the influence of the control is neglected. This is because we want to analyse the hypothetical influence just of the *measured* quantities. Therefore, the amplitudes of sine/cosine signals can be assumed to be affected only by calibration constants used for the sensors (and other similar factors). Hence, the uncertainty of the effective area of a B-coil will influence only the amplitude of its signal. The same applies for an H-coil, whose signal will be scaled with the calibration constant.

It is also assumed that there is no correlation between the input variables because in this example, we wish to analyse only the *detection* of the various quantities rather than the actual physical behaviour of the material (Drake, 1982).

Under such conditions, it can be shown that Equation 5.35 is equivalent to the following (see section 8.3 in Chapter 8 for full derivation):

$$P_{rot} = \frac{\pi \cdot f}{D} \cdot [B_x \cdot H_x \cdot \sin(\phi_x) + B_y \cdot H_y \cdot \sin(\phi_y)] \qquad \text{(W/kg)} \qquad (5.36)$$

where ϕ_x, ϕ_y is the angle by which the B waveform lags the corresponding H waveform.

We can distinguish eight specific variables that can influence the P_{rot} value: $D, f,$ $B_x, B_y, H_x, H_y, \phi_x$ and ϕ_y. Uncertainties in the measurement of these input variables will translate into the combined uncertainty of the P_{rot} measurement according to the principles described above.

Therefore, we can use Equation 5.36 as the mathematical model of the measurement and derive analytically the corresponding sensitivity coefficients as shown in Table 5.12.

All uncertainties should be evaluated individually and if necessary with a multilevel approach. This means that if a measurement of length is required for the calculation of the effective area of a B-coil, then first the uncertainty of the length-measuring instrument must be known, before the uncertainty of the length measurement can be estimated, which can be next used for the calculation of the uncertainty of the effective area. Only then can the combined uncertainty of a B-coil be carried

---

TABLE 5.12
Sensitivity Coefficients for Equation 5.36

Component	Sensitivity Coefficient[a]	
D	$\dfrac{\partial P_{rot}}{\partial D} = \pi \cdot f \cdot [B_x \cdot H_x \cdot \sin(\phi_x) + B_y \cdot H_y \cdot \sin(\phi_y)] \cdot \dfrac{1}{D^2}$ (W·m³/kg²)	(5.37)
f	$\dfrac{\partial P_{rot}}{\partial f} = \dfrac{\pi}{D} \cdot [B_x \cdot H_x \cdot \sin(\phi_x) + B_y \cdot H_y \cdot \sin(\phi_y)]$ (W·s/kg)	(5.38)
B_x	$\dfrac{\partial P_{rot}}{\partial B_x} = \dfrac{\pi \cdot f}{D} \cdot H_x \cdot \sin(\phi_x)$ (W/(kg·T))	(5.39)
B_y	$\dfrac{\partial P_{rot}}{\partial B_y} = \dfrac{\pi \cdot f}{D} \cdot H_y \cdot \sin(\phi_y)$ (W/(kg·T))	(5.40)
H_x	$\dfrac{\partial P_{rot}}{\partial H_x} = \dfrac{\pi \cdot f}{D} \cdot B_x \cdot \sin(\phi_x)$ (W·m/(kg·A))	(5.41)
H_y	$\dfrac{\partial P_{rot}}{\partial H_y} = \dfrac{\pi \cdot f}{D} \cdot B_y \cdot \sin(\phi_y)$ (W·m/(kg·A))	(5.42)
ϕ_x	$\dfrac{\partial P_{rot}}{\partial \phi_x} = \dfrac{\pi \cdot f}{D} \cdot B_x \cdot H_x \cdot \cos(\phi_x)$ (W/kg)	(5.43)
ϕ_y	$\dfrac{\partial P_{rot}}{\partial \phi_y} = \dfrac{\pi \cdot f}{D} \cdot B_y \cdot H_y \cdot \cos(\phi_y)$ (W/kg)	(5.44)

[a] Full derivations of the equations are given in Section 8.3 of Chapter 8.

out, by taking into account its uncertainties for effective area, air flux, voltage detection and so on.

Therefore, multiple levels of the uncertainty estimation will be required before the final uncertainty budget can be defined. This is discussed in more detail in the following sections.

5.7.1   IDEALISED EXAMPLE OF ROTATIONAL POWER LOSS UNCERTAINTY

The sensitivity coefficient for **density D** in Table 5.12 is given just for completeness. In reality, this value is not 'measured' in the same sense during magnetic measurement as the other variables. It is a value specified by the manufacturer of the material and the same numerical value would be used by any laboratory performing the magnetic measurements. Therefore, it can be assumed that from the measurement viewpoint, the uncertainty for D is actually **zero** (unless there is some information to the contrary).

With magnetic measurements, the **frequency f** does not have to be measured but can be taken from the specification of the signal generator. Let us assume that the data acquisition device PCI-6052E was used (NI, 2005) (this device was used by the author for carrying out the research presented in Zurek (2005a)).

The base clock accuracy is given as $\pm 0.01\%$, which at 50 Hz corresponds to **0.005 Hz**. Therefore, absolute uncertainty 0.005 Hz can be assumed with a **rectangular** distribution. (But of course the frequency can be measured directly and then also the specification of the frequency meter will need to be taken into account.)

For simplicity and clarity, let us assume that **flux density B** measurement is affected only by the number of turns and the cross-sectional area of the B-coil and voltmeter accuracy. Both B-coils were characterised in the same way, so the uncertainties will be very similar, although not identical.

In this example, the nominal point for the calculations is 1 T. The number of turns is known precisely (1 turn in this case) so the uncertainty is **zero**. The thickness of the sample was measured with a digital micrometer with 1 µm resolution. Therefore, the uncertainty of the digital display was 0.29 µm (see Equation 5.12). The thickness was measured at many* different locations on the sample and the mean value was 0.4781 mm with Type A standard uncertainty of 1.1 µm.

However, according to the specification of the manufacturer of electrical steel, there is a coating of '*approximately 3*µm' thickness (Cogent, 2002a). Therefore, a value of 6 µm (coating on both sides of the sample) should be subtracted from the mean and the corrected nominal thickness is thus **0.4721 mm**.

The word 'approximately' in the thickness specification implies that there is also some uncertainty in the manufacturer data, which unfortunately is not defined. Using the practical insight (or double-checking with the manufacturer), a variation of not more than 25% can be expected in such industrial processes, which translates into an uncertainty of 0.75 µm (rectangular), which in turn translates into 0.43 µm of a standard uncertainty (divisor $\sqrt{3}$).

---

* For simplicity and brevity we are neglecting here the correction resulting Student distribution and Welch-Satterthwaite formula (Equation 5.26). Under normal conditions these would have to be applied for *each* level of calculation.

Hence, the standard uncertainty of the thickness should be evaluated by using all three[*] above-listed components (Type A, digital display and industrial variation) so that $\sqrt{(0.29^2 + 1.1^2 + 0.43^2)} = $ **1.3 μm (normal)**.

The width of the coil (distance between the edges of holes) was measured with a digital vernier caliper with a resolution of 0.01 mm; hence the uncertainty of a digital display was 2.9 μm.

For the B_x coil the mean value of the readings was **19.901 mm** with a standard deviation of 13 μm. Hence, by quadrature summing[†] of the contribution of the digital display, the standard uncertainty was $\sqrt{(2.9^2 + 13^2)} = $ **14 μm (normal)**.

Therefore, the effective area of B-coil for the x direction can be estimated as $0.4721 \cdot 19.901 = $ **9.3953 mm²**. The estimation of the standard uncertainty for the area requires using sensitivity coefficients so that $\sqrt{((19.901 \cdot 0.0013)^2 + (0.4721 \cdot 0.014)^2)} = 0.027$ mm² (around 0.3% in relative terms). This uncertainty translated through its sensitivity coefficient would correspond to **2.87 mT (normal)**.

The voltmeter accuracy is specified for the lowest range (NI, 2005), for a '1 year calibration' (similar concept as shown in Table 5.3) is 0.041% of the reading, with 9.7 μV offset, 1.9 μV averaged noise and quantisation, 0.0006%/°C temperature drift.[‡] The DC offset can be neglected because it was numerically removed from the signal (only AC component was measured). The measurements were performed at a temperature difference not greater than 20°C from the original calibration, so the expected uncertainty would be 0.0006%/°C · 20°C = 0.012% (rectangular).

At nominal values of 50 Hz and 1 T, the voltage induced in the B-coil would be $V_{RMS} = \sqrt{2} \cdot \pi \cdot f \cdot B_p \cdot N \cdot A = 2.087$ mV. So the uncertainty due to calibration is 0.041% · 2.087 mV = 0.856 μV, due to noise and quantisation is 1.9 μV and due to temperature is 0.250 μV (all rectangular). By using divisors of $\sqrt{3}$, the combined standard uncertainty of voltage measurement is **1.21 μV (normal)**. This corresponds to uncertainty in B as 1.21 μV/$(\sqrt{2} \cdot \pi \cdot f \cdot N \cdot A) = $ **0.58 mT** (or 0.06%).

There is an additional effect that occurs due to the presence of the holes drilled in the sample (see Figure 3.18). The detected B level can be influenced almost directly proportionally to the ratio between the hole size and the width of the coil. The simulations and experiments show that this always leads to overestimation of values so it could be treated as an offset and thus in theory subtracted from the uncertainty budget. But the effect is non-linear and its influence on rotational power loss is unknown. For this reason, it will be added here as an uncertainty rather than a systematic offset.

The holes were drilled as 0.3 mm with 19.9 mm effective spacing. Therefore, the ratio is around 0.3 mm/19.9 mm = 1.5% and we shall treat it as having rectangular distribution. The resulting standard uncertainty due to this effect is thus **8.7 mT**. This value of uncertainty comprises some uncertainty in it because the effect is unknown (Figure 3.18c). But deeper analysis will not be carried out here.

Other effects on B-coil are neglected in this example. Therefore, the combined standard uncertainty for the B_x coil would be $\sqrt{(2.9^2 + 0.58^2 + 8.7^2)} = $ **9.2 mT (normal)**.

---

[*] Neglecting effective degrees of freedom.

[†] Again, neglecting here the effective degrees of freedom.

[‡] These values were taken directly from the manufacturer specification, similarly as they were previously in Table 5.3.

The calculations will not be repeated for the B_y coil, but due to rounding in the intermediate steps, the same result of **9.2 mT (normal)** was achieved for the initial values of 19.987 mm and 15 μm (normal). It can be seen that the final value (9.2 mT) is very strongly influenced by the hole-presence effect (8.7 mT).

For simplicity, let us assume that the measured sample was an isotropic material, so that $H_x = H_y = 113.4$ A/m and also $\phi_x = \phi_y = 45.1°$. The nominal value of rotational power loss was 3.32 W/kg (Zurek, 2017c).

The H-coils were calibrated in a known source of field and their calibration curves were available (e.g. as shown in Figure 3.46). Similar sub-level uncertainty evaluation would have to be followed also for the H-coils, but only final values derived from the calibration curves will be used here.

As can be seen from Figure 3.46b, an uncertainty of 0.5% (rectangular, divisor $\sqrt{3}$) can be assumed, so the standard uncertainty for the contribution due to the calibration of the sensor constant for both H_x and H_y would be **0.29%**.

The actual calibration constants $N \cdot A$ were 0.003319387 m² and 0.00236372 m², for H_x and H_y coils, respectively. Therefore, the voltage induced for sinusoidal H waveforms with $H_x = H_y = 113.4$ A/m would be 105.08 and 74.826 μV, respectively.

It was stated above that the voltmeter uncertainty due to calibration was 0.041%, which translates into 0.0249 and 0.0177 μV of standard uncertainty, respectively, for x and y.

There is 1.9 μV due to noise and quantisation which applies to both channels, but scaled to a standard uncertainty of 1.097 μV. There is further 0.012% due to temperature, which for x and y means 0.00727 and 0.00518 μV, respectively.

Therefore, the combined uncertainty for H_x voltage is $\sqrt{(0.0249^2 + 1.097^2 + 0.00727^2)} = 1.10$ μV and for H_y voltage is $\sqrt{(0.0177^2 + 1.097^2 + 0.00518^2)} = 1.10$ μV, and it is practically the same for both channels.[*] Propagating these values through their sensitivity coefficients results in **1.2 A/m** for both channels.

The remaining uncertainties are those of the angles. The datasheet of the voltmeter does not give any direct information about phase errors. The specified small-signal bandwidth limit (−3 dB point) is for 480 kHz and it is not possible to convert this information to a phase uncertainty at 50 Hz. If it is assumed that the signal chain behaves like a first-order filter, then at 50 Hz, it can be expected to have a phase delay of 6 angular minutes or 0.1°, which individually would be a non-negligible number.

However, correlation between the signals must be taken into account. All the signals are measured by the same voltmeter hardware so that the phase of all waveforms will be affected by the same amount. As a result, there will be no influence on the angle *between* B and H because the angular offset will compensate itself out.[†]

However, this data acquisition device (PCI-6052E) does not have simultaneous sampling, so it uses a multiplexer. This definitely introduces a non-negligible phase delay between the subsequent input channels (Zurek et al., 2005d). At 50 Hz, the

---

[*] It can be seen that in both cases, it is the internal noise and quantisation error of the data acquisition device that dominate the uncertainty.

[†] This is why it was stated before, and in his research, the author decided not to use any additional amplification, for which differences between channels could result in unknown phase errors.

number of points per cycle would be 1000. Therefore, the equivalent 'angular distance' between two subsequent points would be 360°/1000 points $= 0.36°$.

This is the maximum phase error that can be expected between any two channels due to multiplexing. If the order in which the channels are sampled is known, then this phase error can be largely compensated with the post-triggering technique (Figure 4.68) (Zurek et al., 2005d). However, the compensation is not perfect, and an assumption can be made that in practice a maximum of 10% of the phase value will not be compensated; thus the uncertainty in the phase will be $0.36° \cdot 10\%/100\% = 0.036°$ (rectangular).*

The Type A uncertainty was obtained for repeating the measurements 5 times (hence equivalent to **4 degrees of freedom**[†]), on different samples. The resulting standard uncertainty of Type A was **0.020 W/kg**, which at the nominal value of 3.32 W/kg corresponded to around 0.6% (Zurek, 2005a).

Therefore, we identified all the components for building the uncertainty budget, as summarised in Table 5.13.

It should be borne in mind that this calculation was presented only for a very simplified example, with many components of uncertainty *not taken into account*. Therefore, the values from Table 5.13 are likely to be *underestimated* because of the neglected unknown influence of the magnetising apparatus.

Nonetheless, it can be seen in Figure 2.18 that the differences between the curves measured by two different laboratories using the fieldmetric method for the same sample are indeed rather small around 1 T excitation level. The reported values for that particular sample were 30.23 mJ/kg for Wolfson Centre for Magnetics and 29.32 mJ/kg for INRiM (Ragusa et al., 2008), which is a difference of 0.91 mJ/kg and corresponds to 3.1% (as referred to the lower value). This is comparable to the expanded uncertainty of 3% estimated in Table 5.13 for a confidence level of 99%.

However, it should be noted that in the same paper, the data at 1.7 T are 64.27 mJ/kg and 55.69 mJ/kg, respectively, which constitutes a difference of 8.58 mJ/kg or 15.4%, which is significantly greater. This is because the uncertainty of the method is related to the excitation level, and estimation like in Table 5.13 must be carried out accordingly for each range.

PRACTICAL COMMENT

This is recognised also in the NPL approach of experimental sequential perturbation, for which different uncertainties are given for Epstein frame for lower (0.65%) and higher (0.75%) excitation levels (UKAS, 2016).

The detailed calculations will not be repeated here, but uncertainty was re-estimated for an excitation level of 1.9 T – the results are summarised in Table 5.14.

---

* The units of degrees rather than radians are used here for better illustration of the involved amplitudes. Of course, the calculations must be carried out accordingly – for example, most spreadsheet softwares such as Excel or Open Office require the angle to be input in radians, *not* degrees.

† For the final value the effective degrees of freedom will be used, in order to illustrate how the calculations are performed.

TABLE 5.13

Example of Uncertainty Budget Calculation for an Idealised and Simplified Case of Rotational Power Loss, Under Circular Rotational $B_R = 1.0$ T, $H_x = H_y = 113.4$ A/m, $\phi_x = \phi_y = 45.1°$, $P_{rot,nom} = 3.32$ W/kg

a	b	c	d	e	f	g	h
Variable	Absolute Uncertainty	Distribution	Divisor	Sensitivity Coefficient (Table 5.12)	Degrees of Freedom	Absolute Standard Uncertainty (e · b/d)	Corresponding Relative Standard Uncertainty ($g/P_{rot,nom} \cdot 100\%$)
D	0 kg/m³	Unknown	1	0.00043 (W · m³/kg²)	∞	0 W/kg	0%
f	0.005 Hz	Rectangular	√3	0.0664 (W · s/kg)	∞	0.000192 W/kg	0.0058%
B_x	0.0087 T	Normal	1	1.649 (W/(kg · T))	∞	0.0143 W/kg	0.43%
B_y	0.0087 T	Normal	1	1.649 (W/(kg · T))	∞	0.0143 W/kg	0.43%
H_x	1.2 A/m	Normal	1	0.0145 (W · m/(kg · A))	∞	0.0175 W/kg	0.53%
H_y	1.2 A/m	Normal	1	0.0145 (W · m/(kg · A))	∞	0.0175 W/kg	0.53%
ϕ_x	0.036°	Rectangular	√3	0.000363 (W/kg)	∞	0.0000075 W/kg	0.0002%
ϕ_y	0.036°	Rectangular	√3	0.000363 (W/kg)	∞	0.0000075 W/kg	0.0002%
Type A	0.020 W/kg	Normal	1	1 (unitless)	4	0.020 W/kg	0.60%
Combined standard uncertainty (Equations 5.24 and 5.25)						0.038 W/kg	1.14%
Standard uncertainty						**0.04 W/kg**	1.2%
Effective degrees of freedom (Equation 5.26)					50		
Student's correction factor (e.g. from Table 5.2)						2.01 for 95%	
						2.68 for 99%	
Expanded uncertainty, confidence level 95% (Equation 5.27)						**0.08 W/kg**	2.3%
Expanded uncertainty, confidence level 99.9% (Equation 5.27)						**0.10 W/kg**	3%

TABLE 5.14

Example of Uncertainty Budget for an Idealised Case of Rotational Power Loss, at Higher Excitation of Circular Rotational

$B_R = 1.7$ T, $H_x = 3814$ A/m, $H_y = 5516$ A/m, $\phi_x = \phi_y = 0.359°$, $P_{rot,nom} = 6.813$ W/kg

a	b	c	d	e	f	g	h
Variable	Absolute Uncertainty	Distribution	Divisor	Sensitivity Coefficient	Degrees of Freedom	Absolute Standard Uncertainty (e · b/d)	Corresponding Relative Standard Uncertainty ($g/P_{rot,nom} \cdot 100\%$)
D	0 kg/m³	Unknown	1	0.000759 (W · m³/kg²)	∞	0 W/kg	0%
f	0.005 Hz	Rectangular	√3	0.116 (W · s/kg)	∞	0.000393 W/kg	0.0058%
B_x	0.010 T	Normal	1	0.41 (W/(kg · T))	∞	0.0049 W/kg	0.072%
B_y	0.010 T	Normal	1	0.60 (W/(kg · T))	∞	0.0071 W/kg	0.104%
H_x	2.0 A/m	Normal	1	0.000185 (W · m/(kg · A))	∞	0.00044 W/kg	0.0064%
H_y	1.7 A/m	Normal	1	0.000185 (W · m/(kg · A))	∞	0.00037 W/kg	0.0055%
ϕ_x	0.036°	Rectangular	√3	0.000363 (W/kg)	∞	0.0000075 W/kg	0.00011%
ϕ_y	0.036°	Rectangular	√3	0.000363 (W/kg)	∞	0.0000075 W/kg	0.00011%
Type A	0.028 W/kg	Normal	1	1 (unitless)	4	0.028 W/kg	0.41%
Combined standard uncertainty (Equations 5.24 and 5.25)						0.029 W/kg	0.43%
Standard uncertainty						**0.029 W/kg**	**0.5%**
Effective degrees of freedom (Equation 5.26)					4		
Student's correction factor from Table 5.2						2.78 for 95%	
						4.60 for 99%	
Expanded uncertainty, confidence level 95% (Equation 5.27)						**0.08 W/kg**	**1.2%**
Expanded uncertainty, confidence level 99,9% (Equation 5.27)						**0.14 W/kg**	**2.0%**

TABLE 5.15

Example of Uncertainty for Conventional Grain-Oriented Electrical Steel

B_R	1.0 T	1.7 T
$P_{rot,nom}$	2.274 W/kg	3.064 W/kg
Absolute Type A standard uncertainty	0.056 W/kg	0.095 W/kg
Corresponding relative Type A standard uncertainty	2.5%	3.1%
Degrees of freedom	4	
Student's correction factor	2.78 for 95%	
	4.60 for 99%	
Expanded relative uncertainty, 95% confidence level	7%	9%
Expanded relative uncertainty, 99% confidence level	12%	15%

As can be seen, the absolute uncertainty increased somewhat for the standard and expanded uncertainty for 99% confidence level.

It should be noted that all contributions in the uncertainty budget either reduced or remained insignificant, as compared to the value from Table 5.13. The interesting observation is the value of effective degrees of freedom reduced from 50 to just 4, which indicates that the non-repeatability became a far more important factor.

The final value of 2.0% falls significantly short of the difference of 15.4% between two laboratories (Ragusa et al., 2008). Therefore, there must be further uncertainty contributions at play, which were not taken into account in this example. Such severe underestimation of uncertainty renders this measurement *incorrect* and shows how important it is to compare the results between different laboratories.

The uncertainty becomes significantly worse for the measurement of highly anisotropic samples. The calculations will not be repeated there, but for example, for conventional grain-oriented electrical steel, the Type A uncertainty values are as shown in Table 5.15.

Even if the uncertainty related to the B-coil doubles (because of a thinner sheet of the sample), the final result will still be dominated by the Type A uncertainty, and in any case cannot be *smaller* than the largest contribution.

Thus, we can use these values to roughly estimate what should be the expected expanded uncertainty. For this particular measurement, the values are quite large, and as shown in Table 5.15, the expanded uncertainty for 95% confidence level is up to 9%, but for 99% confidence level even up to 15%.

This very strongly indicates that all contributing factors were not taken into account, or the measurement was not carried out under sufficiently controlled conditions because repeatability is quite poor and dominates the uncertainty budget. It should be stressed that most laboratories showed similar scattering of the data during the round-robin testing (Sievert et al., 1995).

PRACTICAL COMMENT

The value of Type A uncertainty can be reduced by performing more measurements so that for the same standard deviation the value reduces proportionally to $\sqrt{n}$ (Equation 5.6). However, with such magnetic mea-surements, this is difficult because it requires repeating the measurement on *various* samples that might not be physically available. For this reason, such measurements are typically endowed with a low value of degrees of freedom.

Values similar to those estimated in Table 5.15 for 95% confidence level were estimated by other researchers for their systems, for example, a value of 8.7% is given in Pluta (2001).

5.8 DIFFICULTY WITH MATHEMATICAL MODEL

The mathematical model defined in Equation 5.36 was over-simplistic. Many assump-tions were made during derivation, which cannot be met during actual measurement.

More importantly though, there was an explicit assumption for the derivation of the sensitivity coefficients (Table 5.12) that all the input variables were completely independent. This is obviously not the case in real life because the B and H vectors and the phase lag between them are strongly correlated through permeability and power loss. Also, the H waveforms are highly distorted because of the anisotropy of the sample and the phase angle between the B and H vectors varies rapidly (some examples are shown in Chapter 7) even if B waveforms are very well controlled – but in reality the waveshape control cannot be infinitely precise.

The covariance between these input variables can be calculated from the input data, but it will be very strongly linked with the anisotropy and crystallographic texture of the sample under test (see also Figure 1.22). Therefore, it is difficult to gen-eralise the data over various materials or even various grades of similar materials.

The method of experimental sequential perturbation (Section 5.6.2) cannot be easily applied because it is not possible to separate the variation of one variable from the next one. However, it can be argued that some useful information about the cor-relation can be obtained by perturbation of one input and registration of the change in the correlated inputs. For example, if the B waveform is well controlled, then perturbation of the B amplitude by some small amount will result in changes in the amplitude of the H waveform as well as the associated phase shift.

But there are other phenomena that are even more difficult to be taken into account. Some of them have not been explained sufficiently yet in the literature, so the way they should be accounted for mathematically is simply unknown at this stage.

An example of such effect can be noticed in Figure 2.31 (there are also more results in Chapter 7). The effect occurs even when measuring the materials with low anisotropy, at lower level of excitation, where the repeatability is quite good, namely, the values of rotational power loss measured under rotation in the clockwise direc-tion are different than the value measured for anticlockwise rotation.

Such big differences are pertinent to the fieldmetric technique because with the thermometric method, they are not observed (Ragusa, 2013), at least not to such a large extent (Sievert et al., 1995).

Recent work by the author shows that the off-axis sensitivity of H-coils (see also Figure 3.50) can be responsible for such effects (Zurek, 2017d).

However, in order to include such effects in the mathematical model of Equation 5.35, the laboratory would need to know quite precisely the amplitude and/or phase variations of the appropriate sensor signals. This has not been researched sufficiently well and there is no detailed information in the literature, or at best the information is very limited and given in a very general form. Therefore, the problem remains unknown and thus it is not possible to include it in the mathematical model. More work is required in this matter.

5.9   DIFFICULTY WITH MEASUREMENT DEFINITION

From the description of the difficulties presented above, it can be concluded that in its current state, the rotational power loss suffers from the *measurement problem definition* (Taylor, 1996; JCGM, 2008). To quote (JCGM, 2008):

> *The objective of a measurement is to determine the value of the measurand, that is, the value of the particular quantity to be measured. A measurement therefore begins with an* **appropriate specification of the measurand, the method of measurement, and the measurement procedure.**

The measurement method should be defined sufficiently well so that the potential additional inaccuracies are controlled or at least taken into account. This may simply mean good definition of the measurement conditions so that the measurement procedure could be *reproduced* at a different time, with different instrumentation or in a different laboratory (JCGM, 2008).

Therefore, such measurement should be described not only by the measurement result, its uncertainty and the confidence level, but also by all these conditions and quantities that are important for *reproducibility* of the measurement: ambient temperature, type of magnetising yoke, orientation of the sample, specification of the instruments and so on.

Without such complete information, the measurement cannot be reproduced with similar uncertainty and therefore the measured result offers far less usefulness.

Uncertainty of the measurement method is relevant to the rotational power loss measurement in multiple ways. One of the most important problems is the controlled shape of magnetisation, for example, circular loci of flux density. If the magnetising conditions are not well controlled, they can have a direct impact on the measured value, even though all the constituent variables in Equation 5.35 were measured with satisfactory accuracy as such (Sievert et al., 1995).

The same applies to the uniformity of magnetisation. As discussed in Chapters 3 and 4, the sample should be magnetised as uniformly as possible. Often, a value of 1% uniformity is used as an arbitrary target, without justification of how such a value could propagate through the mathematical model and impact the actual measurement uncertainty.

This is why the *measurement problem definition* is also very important and sufficiently good reproducibility cannot be achieved without it.

For instance, the method and procedure of measurement are quite well specified for standardised measurements such as Epstein frame (IEC, 2008; ASTM, 2014) and SST (IEC, 2010). For Epstein frame, the sample should have a certain minimum weight, the construction of the magnetising apparatus guarantees repeatable uniformity of magnetisation, the waveshape of dB/dt should be well controlled, the secondary winding has to be compensated for air flux and so on. All these conditions specify the method as well as the procedure in which the measurement should be performed.

So far, there is no agreement on the acceptable shape and size of the magnetising yoke for rotational measurements. Therefore, at least the *measurement procedure* is not adequately defined, so significantly larger discrepancies can be expected than it is the case for the standardised methods.

Nevertheless, many sources of uncertainties, for example, those directly related to the sensors, can be defined with quite good precision and thus they can be included into the final uncertainty calculation.

However, the whole problem appears much more complex because it should also take into account the *entire* measurement procedure and other factors that are likely to affect it. In the case of rotational power loss, the problem is so complex that it might not be possible to formulate the mathematical model so that it contains all the necessary components.

There are several components that can affect the measurement procedure for rotational power loss. It is difficult to generalise because many of them would be related to the specific type of magnetising apparatus used in the measurement. However, a few of the most important ones will be discussed below, by the way of example.

5.9.1 FEEDBACK AND WAVESHAPE DISTORTION

As discussed above, the B-coil sensors introduce some uncertainty in the measurement of B and this can be easily taken into account from the viewpoint of measured values (see Table 5.12).

However, the measured B waveforms are also used as an input variable to the digital feedback for controlling the excitation. Therefore, if the B values were measured incorrectly, this will have an immediate impact on the value of the nominal operating point. In theory, this is not different from the standardised alternating methods such as Epstein frame and SST. But in practice, the magnetisation is applied as rotating through all directions and it is much more difficult to attain the same level of waveshape control as it is the case for alternating conditions.

For alternating magnetisation, the form factor FF can be driven to very low values and less than 0.2% can be achieved routinely, and THD can also be driven below 1%. With such good control, the dB/dt waveform appears visually as 'pure sine' (see also Figures 4.84 and 4.87) (Zurek, 2017a).

It should be borne in mind that under alternating magnetisation, the distortion in the dB/dt waveform comes predominantly from saturation effects (above the 'knee' of the B–H magnetisation curve). Therefore, the shape of the distorted waveform follows a similar pattern for any magnetically soft material, namely, the induced voltage becomes more 'peaky' if the waveshape control (feedback) is not employed. The anisotropy of the sample is almost irrelevant as far as the shape is concerned because it mainly affects the amplitude.

This is not the case for rotational magnetisation, for which the distortions are highly influenced by the anisotropy of the sample (see also Figure 1.22). As a result, for highly anisotropic materials, it is much more difficult to keep equally precise waveshape control. There can be visible distortions in the *B* waveform – this is evident, for example, in Figure 2.27a. And such distortions in B waveforms mean even greater distortions in *dB/dt*.

In general, for rotational magnetisation, *FF* is *not* as good a criterion as it is for alternating magnetisation (Sievert et al., 1995) because distortions do not appear only due to saturation of the sample. In the worst case, a significant amplitude of higher harmonic can be present (up to 20%) and the *FF* value would not detect it, as evident from Figure 4.76. Therefore, the THD should be used as a more reliable method, or at the very least, the maximum permissible *FF* would have to be specified as a much lower value, for example, less than 0.1% (Figure 4.76).

Both the *FF* and THD values translate the full spectrum of harmonics into a single number representing the distortion of the waveform, so it is possible that different combinations of harmonics can result in the same numerical value. Therefore, a universally applicable mathematical model linking an *FF* or THD value to the actual distortion at a given harmonic cannot be derived.

PRACTICAL COMMENT

If the mathematical model is known then it could be possible to derive the sensitivity coefficient for a given harmonic propagating through the uncertainty budget. However, under general conditions, it is not known which harmonics are present and to what extent if only the *FF* or THD values are available. Therefore, it is also impossible to estimate the uncertainty contribution just from the *FF* or THD values. Some attempts were made to do so but the conclusion was that unrealistically high values would be achieved in such a way (Sievert et al., 1995). As evident from Figure 4.76, a given harmonic could have up to 20% amplitude and *FF* could be still below 1%.

Additional uncertainty can be caused by the mains hum (Figure 4.52) and mains frequency mismatch (Figures 4.53 and 4.54). These effects can present themselves in just the signal acquisition (bogus signal), in just power loss (affected magnetising current) as well as in both simultaneously.

The same applies to DC offset in the magnetising current (Drake, 1982; Krah et al., 2002). If such DC offset is not eliminated completely (e.g. by a separating transformer or in-series capacitor), then there will be some uncertainty associated with the fact that the resulting magnetisation will be asymmetrical to some extent.

Larger sample size (discussed in the next section) will also be helpful because the produced signals will be larger and thus the relative errors should be reduced.

However, estimation of uncertainty due to such effects is very difficult, especially if they should be propagated through the whole mathematical model.

Another problem is the so-called space harmonics that are normally created by the irregularity of the magnetic circuit. For example, a stator-like yoke with 24 slots will tend to introduce a multiple of the 24th harmonic in the rotating field (Gorican et al., 2000a; Anuszczyk et al., 2009). These harmonics will be suppressed to a large degree by the active waveshape control (feedback), but some component will remain and the harmonic value will be very much yoke-dependent. It is not possible to quantify this effect through purely analytical calculations.

Detailed uncertainty analysis shows that a significant contribution to the uncertainty is made by the estimated distortion from sine (Ahlers et al., 2000a).

This is an *important* conclusion because even for the *very well-controlled* alternating magnetisation, the effect of the distortion is still *significant*. Therefore, for rotational magnetisation of anisotropic materials, the effect can be expected to be much worse, due to increased difficulty with the waveshape control.

Interestingly, intercomparison of alternating power loss measurements with Epstein frame and SST between three standard European laboratories (Germany, Italy and the United Kingdom) reveals that the reproducibility of the methods is between 0.7% and 1.1% for 95% confidence level (depending on the method and amplitude of excitation) (Sievert et al., 2000). Similar values were obtained in a much wider study with 10 laboratories: Italy, Germany ($\times$3), the United Kingdom, China ($\times$3), France, Sweden and the United States (Appino et al., 2015). Therefore, for a far less defined method of rotational magnetisation, much larger uncertainties should be expected.

5.9.2 Sample Size

Repeatability of measurement in a given rotational system is problematic due to the fact that grain-oriented electrical steel has relatively large grains. These are comparable in size to a typical area of 'uniform' magnetisation in rotational apparatus (typically around 20–50 mm) and hence there are large variations even between subsequent samples cut from the same raw sheet of material, which can cause variability of up to 20% of the measured 'uniform' B (Zurek, in press). Examples of such variations for grain-oriented electrical steel are shown in Chapter 7.

It is suspected that this could be the reason why there are such large variations in repeated measurements (Table 5.15) even though other sources of uncertainties appear to be much smaller. Indeed, it is known that the magnetic sample must be of sufficient volume so that 'magnetic averaging' takes place over larger effective volume of the sample and the localised effects of grains are averaged out.

For standardised measurements, the sample should be *'at least 240 g'* (IEC, 2008) or *'at least 20 strips'* (ASTM, 2014). This is equivalent to an active sheet area of around 1500 cm^2 and this is the main reason why the same minimum size (30 cm $\cdot$ 50 cm = 1500 cm^2) is also required for the standardised SST (IEC, 2010).

The typical 'active' area in rotational measurements is only around 2 cm $\cdot$ 2 cm = 4 cm^2, which is around 400 times smaller, and even a larger size of 6 cm $\cdot$ 6 cm = 36 cm^2 is still at least 40 times smaller than the permissible *minimum* for the other methods.

Such small active area cannot give the same spatial averaging as the large samples of Epstein and SST. Therefore, almost by definition, it is not possible to attain sufficient *reproducibility* because slightly different distribution of the largest grains can result in different local magnetic properties.

At best, materials with relatively small grains can be measured under such conditions, whereas the grain-oriented steel would require significantly greater area of uniform magnetisation.

If the area is reduced by an order of magnitude, then this would still call for 150 cm^2 or at least a square of $12 \text{ cm} \cdot 12 \text{ cm}$ of uniformly magnetised area of the sheet. For a circular sample, which gives better uniformity, this would mean a minimum diameter of 17 cm (due to the diagonal dimension). However, because only around 75% of the sample diameter can be thought of as being 'uniformly' magnetised (Gorican et al., 2002; Wanjiku et al., 2015a), the real sample size would need to have a diameter of 23 cm.

It will be difficult in practice to magnetise such a large sample with controlled magnetising conditions, especially if the material is grain-oriented electrical steel. The rotational yoke is an open magnetic circuit that requires a lot of apparent power to drive the sample up to the required excitation level (Zurek, 2005a).

However, even standardised loss measurements are *typically* carried out only up to 1.8 T (UKAS, 2016). Also, by using a compact round yoke with an appropriate winding and configuration (Figure 4.93), the power drive requirements should be optimised. An additional improvement could be obtained from using the 'vertical shield' arrangement, which in simulations appears to give significantly lower requirements for the magnetising current (Figure 4.28b). A combination of all these optimisations should allow achieving sufficiently high excitation. However, it might be possible that some compromise would have to be made between the largest possible sample size and the power required for its magnetisation.

Larger samples also give an additional benefit of reduced demagnetising factor because the diameter of the sample will increase but its thickness would remain the same. Therefore, the effects of H normal to the sample should be reduced and thus the errors associated with it (Zurek, 2017c).

PRACTICAL COMMENT

The problem of different properties for different samples is recognised in international standards for magnetic measurements, which aim at the reproducibility of some average properties by means of a significant volume of the tested sample (IEC, 2008; ASTM, 2009).

The thickness of electrical steel laminations is typically specified to have a tolerance of around 10% (JFE, 2016). Therefore, the thickness could vary for samples cut from different places of a large sheet. The cross-sectional area of the magnetised sample depends directly on its thickness, so such uncertainty would immediately impact the measured result – if it was assumed to be equal for all samples. This problem is greatly reduced by deriving the effective cross-sectional area from the *weight* of the sample, rather than its thickness.

However, even if the thickness is measured precisely, the crystallisation process could be affected by the local thickness of the sample, so that grains with different sizes are produced. It is well known that the size of grains impacts the specific loss of the material; hence the magnetic properties could differ (Hilzinger et al., 2013).

With the current state of the knowledge, the influence of such effects on the final measurement uncertainty is impossible to estimate; hence they cannot be mathematically taken into account. Therefore, the only practical way forward is to increase the active volume of the sample under test, so that averaging from a larger area is achieved and such effects are minimised to a 'negligible' level. Only then can round-robin testing between many laboratories show if the effect can indeed be assumed to be much smaller than other sources of uncertainties (Sievert et al., 1995).

An example of a slightly larger sample size is given in Mori et al. (2015). A comparison between circular (99 mm) and square sample (150 mm) was presented for non-oriented electrical steel. The agreement between the circular and square sample was better than 2% for excitation between 1.5 and 1.7 T even though the sensing area was only 20 mm · 20 mm. Higher-controlled circular B could not be attained for the square sample. The hole diameter for B-coil wires was not specified, although the wires were 0.1 mm.

However, even with such relatively small sensors, the uniformity of magnetisation was definitely improved simply because of the sample size. Even just increasing the distance between the magnetising poles in the same setup is beneficial because local gradients are reduced (Krah et al., 2002).

Larger sample size would allow employing the larger sensors that encompass larger effective area. For the example given above, the B-coil width could be therefore increased to around 190 mm. Hence, drilling 0.3 mm holes would constitute uncertainty of B measurement due to this effect of around 0.15%, or less for higher excitation (Zurek, in press). This is a fairly low value, but still definitely not negligible. Some researchers chose to use more than one search coil so that multiple pairs of holes were drilled (Brix et al., 1982; Alinejad-Beromi, 1992; Enokizono, 1992; Kedous-Lebouc et al., 1992). Such an approach allows better averaging of B over the area of interest. But it should be borne in mind that multiple holes would create multiple distortions, which could worsen the effect to even a larger degree.

The additional benefit of larger coils would be smaller uncertainty of orthogonality between B_x and B_y sensors, for at least three reasons. First, the holes should be drilled with the help of a positioning table with specific precision (or a similar device). Assuming a typical positioning precision of 0.01 mm, the relative angular error would be significantly smaller. This is because for a 190 mm coil, it would be atan(0.01/190) = 0.17° and, for instance, for a 60 mm coil drilled with the same precision, atan(0.01/60) = 0.55°.

Second, a similar angular error improvement would apply to the actual positioning of the wire in the holes, as illustrated in Figure 3.15. Third, the improved positioning will apply to the sample itself (for similar reasons as shown in Figure 4.16b).

Larger B-coils will mean large induced signals, which is beneficial because the signal-to-noise ratio will be appropriately improved.

Of course, larger H-coils should also be used and therefore better signal-to-noise conditions could also be obtained. However, larger H-coils could also be problematic for at least two reasons.

Even a fairly small 20-mm H-coil can comprise hundreds or even thousands of turns (Brix et al., 1982). More turns would mean greater self-inductance and self-capacitance of the coil, which could lead to unwanted phase shift problems. The surface area of such H-coil would be significantly greater and hence the parasitic capacitance between the sample and such sensor could also contribute to additional phase shifts.

It was also shown that a wire-wound H-coil is not completely immune to off-axis components of H penetrating it (Zurek, 2017c). However, such an H-coil could be large enough so that this unwanted sensitivity could be significantly reduced by using different techniques than wire winding – for example, by using printed circuit board technology. The main concept of such sensor is shown in Figure 3.50 and in Chapter 6.

PRACTICAL COMMENT

It was mentioned above that it should be possible to have a sample whose uniformly magnetised area is around an on order of magnitude smaller than of the standardised methods, or 150 cm² versus 1500 cm², respectively.

In many random processes (such as averaging of noise), the improvement is proportional to the square root of the value, rather than directly to the value. This means that a reduction of size by a factor of 10 should theoretically correspond to an increase of 'noise' by √10, so around 3 times.

The standardised methods have a reproducibility of power loss measurement at the level of around 1%. Let us assume here hypothetically that the repeatability due to sample variations is the main contributor to the overall uncertainty. Hence, if other factors could be neglected, then the reproducibility of around 3% could be obtained for the rotational power loss measurements. This would be quite encouraging if such values were achievable in reality.

It should be noted that if the uniform area is 2 cm · 2 cm = 4 cm², then this is 375 times smaller and therefore, by using the rough estimation of the square root approach, the expected repeatability would be √375 or around 20 times worse. Hence, scattering of 20% could be expected when referred to the 1% for the standardised methods.

This is of course a far over simplistic estimation, but it does show that on the one hand, the sample size must be increased, but on the other hand, maybe the samples do not have to be as large as for the other standardised methods.

5.9.3 CLOCKWISE/ANTICLOCKWISE DIFFERENCES

It is well known that the fieldmetric method produces different results of power loss for clockwise CW and anticlockwise ACW rotation (Sievert et al., 1995; Gorican et al., 2002; Zurek et al., 2006a; Maeda et al., 2008). The non-physical shapes of the CW and ACW loss characteristics (see Figure 3.49, with negative losses) signify a substantial uncertainty if just each individual curve was to be taken into account.

Non-orthogonality of sensors directly contributes to the CW–ACW differences (Sievert, 1990; Mori et al., 2005; Zurek et al., 2006a). The mechanism is illustrated in Figure 5.11a. Ideal orthogonal H sensors should detect only the components along the x and y axes. However, real H sensors will not be perfectly orthogonal and will be positioned at some apparent axes x' and y' deviating, respectively, by angles θ_x and θ_y from the reference x and y axes. Therefore, the actual components detected by the sensors will not be H_x and H_y, but some other values of H'_x and H'_y, so the amplitude and angle of the detected H vector will be somewhat incorrect. The error arises because the assumption is made that these components lie at the x and y axes, but they do not, hence $\sqrt{(H_x^2 + H_y^2)} \neq \sqrt{(H_x'^2 + H_y'^2)}$.

If the angles θ_x and θ_y were known precisely, then the correct values could be calculated by using the following equations (which can be derived by following the definitions of Figure 5.11a):

$$H_x = \frac{H'_x \cdot \cos(\theta_y) - H'_y \cdot \sin(\theta_x)}{\cos(\theta_x) \cdot \cos(\theta_y) + \sin(\theta_x) \cdot \sin(\theta_y)} \quad \text{(A/m)} \quad \quad (5.45)$$

$$H_y = \frac{H'_y \cdot \cos(\theta_x) + H'_x \cdot \sin(\theta_y)}{\cos(\theta_x) \cdot \cos(\theta_y) + \sin(\theta_x) \cdot \sin(\theta_y)} \quad \text{(A/m)} \quad \quad (5.46)$$

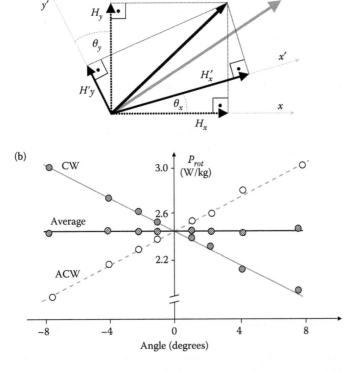

FIGURE 5.11 Non-orthogonal positioning of B and H sensors (a) (Adapted from Zurek S. et al., *IEE Proceedings, Science, Measurement and Technology*, 153 (4), 2006a, 147.) and their contribution to the CW-ACW differences (b) Cancellation of CW–ACW differences by averaging. (Adapted from Sievert J., *IEEE Transactions on Magnetics*, 26 (5), 1990, 2553.)

The same applies for B sensors, which can also be non-orthogonal and positioned at some different effective angles (see Figures 3.14 and 3.15). Calculations similar to Equations 5.45 and 5.46 could be used for the correction of the relative angular position between the B and H sensors.

As can be seen, Equations 5.45 and 5.46 have significant complexity and an attempt to calculate the appropriate partial derivatives to estimate the propagation of uncertainties through Equation 5.35 would lead to rather convoluted arithmetic.

Additionally, it can be seen immediately from Equations 5.45 and 5.46 that there is a cross-correlation between the x and y components, which would require the calculation of the covariance between the two variables, which at the same time would also be correlated to both B components. This would further increase the level of difficulty for such calculations.

However, in reality, the non-orthogonality of B or H sensors cannot be known precisely, so Equations 5.45 and 5.46 cannot be used for the elimination of such errors, at least not with sufficiently good precision. However, mathematical analysis shows that some of these types of errors produce CW–ACW differences that have exactly the same value but opposite sign, so the averaging from CW and ACW values cancels them out (Zurek et al., 2006a). Therefore, there is no need to calculate either the sensitivity coefficients or the covariance because the uncertainty components *do* have opposing signs and thus a simple average leads to self-cancellation of such differences. This is quite fortunate.

The same applies to several other sources of these CW–ACW differences, so that calculating an average of the CW and ACW reduces them to a great extent (Sievert et al., 1995; Zurek, 2017d).

However, there appears the question whether the average of CW–ACW cancels the differences *exactly*. Is there any remaining uncertainty that contributes to the uncertainty of the average?

There is at least one component that would not average out, namely, there will always be some uncertainty of the orthogonality of B-coils. Therefore, the information used by the digital feedback will be imprecise and also the controlled B_x and B_y waveforms will have some uncertainty as to their phase relationship, which should be exactly 90° but because of the uncertainty it will not be.

Therefore, the controlled excitation will not be *exactly* circular, but it will comprise a certain amount of uni-axial magnetisation so it will be elliptical. The effect on the measurement might be fairly small as compared to other sources (Figure 5.12), but it *will* introduce some *unknown* uncertainty. And it is recognised from other studies that measurements under elliptical rotation (rather than purely circular) *does* influence the measured losses (Figure 2.36).

It was already shown that uni-directional magnetisation in the rotational yoke also produces asymmetrical effects that do not cancel each other out (see Figure 2.37) (Zurek et al., 2009a). It is most likely that the errors in the angular positioning of the sensors and/or their off-axis sensitivity contribute to such effects (Zurek, 2017c).

In any case, imperfect orthogonality of the sensors cannot be taken into account completely. This will lead to asymmetrical control of the magnetising conditions, which will produce physical differences in the measured loss, rather than just artefacts of signal processing. Such differences are shown in Figure 7.9.

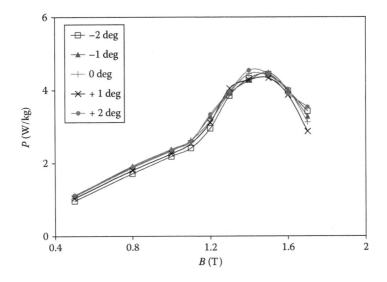

FIGURE 5.12 Rotational power loss versus change of phase angles between B_x and B_y signals (nominal value is $90° + 0°$). (Adapted from Zurek S., *Two-Dimensional Magnetisation Problems in Electrical Steels*, PhD thesis, Wolfson Centre for Magnetics Technology, Cardiff University, Cardiff, the United Kingdom, 2005a.)

5.9.4 Cross-Sectional Area of the Sample

The determination of the cross-sectional area in the standardised method such as Epstein frame and SST is based on sample weight because the whole active area of the sample can be assumed to be magnetised uniformly. Such an approach guarantees less uncertainty in the definition of the area because, for example, a 0.1 g change in 240 g of sample mass means a relative resolution 0.04%. The measurement of mass is quick, convenient and accurate.

This is *not* the case for any system in which there is non-negligible air gap and when the local detection of B is employed. Ordinary digital vernier callipers have a resolution of 0.01 mm, which with the coil width of 20 mm would have a similar relative resolution of 0.05%. But such measurement is much more difficult to be performed in practice. This is especially true if the holes are drilled with a small diameter, for example, below 0.5 mm.

As an alternative, the hole positioning would have to be relied on the drilling positioning, which usually has similar precision. Another approach could be to use an appropriately calibrated traveling microscope. Either way, the measurement of the effective coil width is much more complicated.

An even greater problem is with the measurement of the thickness. A typical micrometer will have a resolution of around 0.001 mm, and with a sample thickness of 0.27 mm, the relative resolution is therefore around 0.37%, which is around an order of magnitude worse than the length or mass measurement. Additionally, usually there will be some coating on the sample so the micrometer will measure the full thickness, including the coating. It is not possible to remove the coating in a truly non-invasive way, so the manufacturer information must be used instead. This can represent quite a significant uncertainty because a typical coating thickness could be around 3 μm (0.003 mm) (Cogent, 2002a,b), but there will be a coating layer on each side of the sheet. The exact thickness will depend on the industrial processes used, so even from the same manufacturer, it could vary from 0.5 to 10 μm, depending on the grade of the steel (ArcelorMittal, 2011; Voestalpine, 2011).

All these values must be taken into account accordingly by the measurement of *dimensions* rather than with the help of the *mass* of the sample.

5.9.5   CORRELATION

The derivation of appropriate equations that take into account the correlation between various variables is difficult due to the mathematical complexity of the involved equations.

It is not possible to give a general form for such an equation because it will depend on the type of the magnetising system, types of sensors, number of measuring instruments and so on. Additionally, the exact procedures are subjected to a certain amount of confidentiality. For this reason, the details of such procedures and the exact calculation methods are rarely published (Drake, 2016), even if the final uncertainty figure should be made available for the accredited laboratories (UKAS, 2016).

As a result, there are limited examples published in the literature for alternating magnetic measurements (Ahlers et al., 2000a, b), but there are no examples so far for rotational magnetisation (at the time of writing). Therefore, no such equations can be shown in this book for rotational magnetisation.

However, an example for alternating magnetisation in Epstein frame is given, for instance, in Ahlers et al. (2000a) as

$$P_s = \frac{\dfrac{N_1}{N_2} \cdot \dfrac{U_{1,fund} \cdot R_{meas} \cdot U_2 \cdot \cos(\phi)}{m} \cdot \dfrac{4 \cdot l}{l_m}}{\beta \cdot \left(\dfrac{f}{f_{pred}}\right)^2 \cdot \left(\dfrac{F}{1.1107}\right)^2 + (1-\beta) \cdot \dfrac{f}{f_{pred}}} \cdot \left[\dfrac{f \cdot N_2 \cdot m \cdot \hat{J}_{pred}}{\bar{U}_2 \cdot l \cdot \rho_m \cdot \left(1 - \dfrac{\mu_0 \cdot \hat{H} \cdot A_t}{\hat{J} \cdot A} \cdot \gamma\right)}\right]^{\alpha} - \Delta T \cdot \alpha_T$$

(W/kg)

(5.47)

where $\Delta T = T - T_{pred}$ (T) the symbols are as defined in Ahlers et al. (2000a): P_s is the mathematical model for specific power loss (W/kg), N_1 is the number of primary turns (unitless), N_2 is the number of secondary turns (unitless), $U_{1,fund}$ is the RMS value of the fundamental harmonic of the voltage across the primary shunt resistor (V), R_{meas} is the value of shunt resistor (Ω), U_2 is the RMS value of secondary voltage (V), ϕ is the angle between fundamental components of the voltage across the primary shunt resistor and secondary voltage (rad), m is the mass of the sample (kg), l is 0.94 m, l_m is the length of the strips (m), β is the factor for the eddy current portion of the loss (unitless), f is the nominal frequency (Hz), f_{pred} is the actual frequency (Hz), F is the form factor of secondary voltage (unitless), $\hat{J}_{pred}$ is the corrected peak polarisation (T), $\bar{U}_2$ is the average of rectified secondary voltage (V), ρ_m is the resistivity of the sample ($\Omega \cdot$ m), μ_0 is the permeability of free space (H/m), $\hat{H}$ is the peak value of magnetic field strength (A/m), A_t is the cross-sectional area of the mutual inductor (m²), A is the cross-sectional area of the Epstein frame (m²), γ is the factor for deviation of flux compensation (unitless), α is the relative slope of $P_s = f(\hat{J})$ curve at the operating point (unitless), T is the temperature (K), T_{pred} is the corrected temperature (K) and α_T is the temperature loss coefficient (W/(kg $\cdot$ K)).

Equation 5.47 represents the semi-empirical model of the measurement. It takes into account uncertainties arising from air flux compensation, experimental perturbation procedure, imperfect waveshape control, etc., but only for uni-directional excitation.

5.9.6 MECHANICAL FORCES

As discussed in Chapter 2, mechanical stresses acting on the sample can significantly change its magnetic properties such as power loss or permeability.

This problem is recognised in the international standards, which define the mechanical arrangements for the sample and the yoke. The force acting on the double-lapped corners of Epstein frame sample should be around 1 N per each corner (IEC, 2008). In the SST method, the heavy upper yoke should have its own suspension so that the force on the sample should be 100–200 N.

And even with such precise recommendations, there is a much larger discrepancy of apparent power loss measured especially by SST (Sievert et al., 2000; Appino et al., 2015).

In the rotational measurements, the sample will have to be put in a sample holder. Certain amount of clamping force would need to be employed because otherwise the sample could vibrate during magnetisation, which can adversely affect the stability and hence repeatability of the measurements.

The clamping force cannot be too great because it will change the magnetic properties of the material. But on the other hand, it must also be applied in a controlled and uniform way. This will be especially important for thin samples. In recent years, not only the grain-oriented but also the non-oriented steel is available in ever-thinner grades of around 0.2 mm and less (ArcelorMittal, 2011; AK Steel, 2013). Amorphous ribbons are even thinner (<0.05 mm).

Such thin samples would have to be clamped delicately but uniformly over a large area. In general, the effect of such mechanical clamping will be unknown. It is possible to apply the method of experimental perturbation, but this has to be done deliberately and over a number of specimens before any conclusions about the order of magnitude of such influences can be drawn with sufficient confidence.

5.9.7 SUMMARY

As mentioned above, not all Type B components of uncertainty can be taken into account. This is the case not only because it is difficult to measure, calculate or even estimate the order of magnitude of a given effect but there will also be numerous sources of uncertainties that are not recognised yet in the measurement procedure. Some of these can be identified and *quantified* by detailed research (Zurek, 2017c,d, in press), but some of them will remain as 'unknown unknowns'.

For this reason, it is best to design the magnetising setup in the optimum way so that most of the obvious sources of uncertainties are controlled or at least minimised. By careful design, some of the sources of uncertainty can be eliminated altogether. For example, if signal pre-amplification is not used, then all the uncertainty components related to those pre-amplifiers will drop out from the uncertainty budget, which is advantageous.

The known uncertainty components will have to be meticulously taken into account. Most of them will need to undergo a multi-level uncertainty assessment, so that sub-level uncertainty budgets will have to be produced. Even if some components will be found to be negligibly small, at the very least, they should be estimated to be able to judge whether their contribution can or cannot be neglected.

As an example, a simplified (!) list of such multi-level assessment is as detailed below.

The highest level represents the final combined uncertainty for the measurement of rotational power loss P_{rot}. The lower levels in the list represent deeper levels of uncertainty analysis that have to be included in the whole process.

I. Measurement of signals
    A. Flux density B
        1. Voltage measurement
            a. Voltmeter accuracy (including calibration)
                i.   Input range
                ii. Phase shifts
                iii. Quantisation effects
                iv. Internal noise
                v.   Influence of temperature
                vi. Loading by internal impedance
                vii. Temperature effects
            b. Additional signal amplification
                i.   Gain linearity versus bandwidth
                ii. Phase shifts
                iii. Internal noise
                iv. DC offset
                v.   Temperature effects
                vi. Power supply noise rejection
        2. Digital signal integration
            a. Rounding errors
                i.   Integration method (e.g. trapezoidal)
                ii. Numerical precision
            b. DC offset elimination
            c. Initial conditions
        3. B-coil active cross-sectional area

      a. Sample thickness
      b. Coating thickness
      c. *B*-coil width
  4. Air flux
      a. Measurement of *B* instead of *J*
      b. *B*-coil area versus sample area
      c. Wire twisting
  5. Effect of holes
  6. *B*-coil positioning
      a. Angular alignment
      b. Non-orthogonal positioning
      c. Non-central positioning
  7. Two sensors (B_x and B_y)
      a. Similar for both (but should be assessed separately)
B. Magnetic field strength *H*
  1. Voltage measurement
      a. Similar as for *B*-coils (but must be assessed separately)
  2. Signal integration
      a. Similar as for *B*-coils (but must be assessed separately)
  3. H-coil calibration
      a. Magnetic field source (e.g. solenoid)
      b. Magnetising current source
      c. Voltmeter uncertainty
      d. Loading by voltmeter impedance
      e. Positioning during calibration
          i. Angular positioning
         ii. Linear positioning
      f. Repeatability of measurement (Type A)
  4. Other *H*-coil parameters
      a. Positioning
          i. Distance from sample surface (extrapolation of *H*)
         ii. Angular alignment
        iii. Non-orthogonal sensors
        iv. Non-central placement
         v. Precision of sensor holder
        vi. Durability of sensor holder
      b. Parasitic effects
          i. Wire twisting
         ii. Temperature effects
        iii. Cross-talk between H_x and H_y sensors
      c. Sensitivity to off-axis *H*
  5. Two sensors (H_x and H_y)
      a. Similar for both *H*-coils (but should be assessed separately)
C. Frequency *f*
  1. Stability of signal generator (internal clock)
  2. Non-integer points per cycle

    D. Material density D
        1. Uncertainty of manufacturer data
    E. Temperature T
        1. Changing resistivity of the sample
    F. Type A uncertainty
        1. Repeatability of measurements for the same sample
        2. Repeatability of measurements between various samples
        3. Repeatability effects due to different user
II. Method and procedure
    A. Positioning of the sample
        1. Precision of sample shape
        2. Spatial positioning
        3. Angular positioning
        4. Clamping force on sample
    B. Demagnetisation of the sample
    C. Quality of sample cutting and annealing (residual mechanical stress)
    D. Quality of magnetisation
        1. DC offset in magnetising current
        2. Volume averaging over grains
        3. Uniformity of magnetisation
        4. Quality of waveshape control (FF and THD)
        5. Amplitude control (B_{peak} stability)
        6. Space harmonics
        7. Mains hum and other noise
III. Uncertainty calculations
    A. Correlation between B and H (and other variables)
    B. Higher order terms of non-linear functions
    C. Contribution due to B-H product of higher harmonics (non-ideal wave-shape control)

    The author is thankful for all the state-of-the-art knowledge distributed over the many technical papers published by scientists and researchers worldwide. Even the papers showing large measurement errors (some of them published by the author) were extremely useful because they showed which practical solutions work better and which are not so good. Not all the papers on the topic could be referenced in this book – which is a pity in the author's opinion.

    However, the collective knowledge paints a picture of the optimum hardware configuration that can be used for a measurement system capable of measuring the rotational power loss with the smallest uncertainty possible.

    Such an optimum configuration is discussed in detail, and its summary is given at the end of Chapter 4 (Figure 4.93 and Table 4.2). All these recommendations are perfectly possible in practice – they just need to be combined and applied in an experimental apparatus. An international intercomparison between a few laborato-ries would be an excellent continuation of the previous study (Sievert et al., 1995), and it would allow assessing the progress in our abilities to measure the rotational power loss in a better-defined and more precise way.

The 'best possible' uncertainty for rotational measurements could be still much greater as compared to alternating magnetisation. However, even if the uncertainties are an order of magnitude worse than the standardised methods, then an agreement better than 10% should still be achievable, which should begin to be useful for industrial applications.

Hopefull, with the help of this book, y this will be the case in the near future.

6 Examples of Measurement Equipment

This chapter presents photographs and images of some experimental apparatus used for rotational measurements (as well as 1D, 2D and 3D properties) by several laboratories in various countries. They show only a very small selection of all the equipment used by the many laboratories worldwide. They are not presented in any particular order. The author is extremely grateful to all the colleagues from universities, companies and research institutions worldwide for making these images available.

FIGURE 6.1 Magnetic measurement system capable of performing the measurements with Epstein frame, custom single-sheet and single-strip testers (visible on the left of the image in wooden boxes), as well as a rotational yoke for a square sample (on the right.) The rotational yoke could accommodate 100 mm square samples. The *H*-coils were connected to pre-amplifiers with selectable gain. The data acquisition and generation had 12 bit resolution. The system was used by the author during his MSc studies at Czestochowa University of Technology, Poland. (Adapted from Zurek S., *Rotational Magnetisation in Flat Soft Magnetic Materials* (in Polish: *Przemagnesowanie obrotowe w materialach magnetycznie miekkich plaskich*), MSc thesis, Czestochowa University of Technology, Poland, 2000.)

FIGURE 6.2   Round rotational yoke for a circular sample used by the author during his rotational studies. The internal diameter of the yoke was 84 mm. The windings were made as a simple two-phase (without the 'sinusoidal' distribution). The lamination stack was recovered from an old induction motor stator with the height of the stack reduced to around 40 mm. The laminations were welded in several places outside. The yoke was kindly supplied by Slawomir Zurek and used by the author in the laboratory of Wolfson Centre for Magnetics, Cardiff University, Cardiff, the United Kingdom. (Zurek, 2005a).

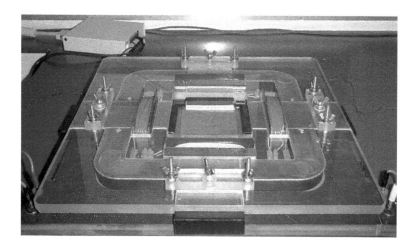

FIGURE 6.3   Horizontal rotational yoke for 100 mm square samples. The 'knee' parts were movable and the pole pieces replaceable with narrower ones, so the same construction was also suitable for 80 mm square samples. The base was made from 20-mm-thick Tufnol, and the top from 15-mm-thick Plexiglas. All the magnetic parts of the yoke were made from 0.27-mm grain-oriented electrical steel. The two-phase magnetising coils had 80 turns of 1.5 mm wire. The white box in the top left corner of the image is the connection to a 16 bit data acquisition device NI DAQ PCI-6052E. (Zurek, 2005a). The yoke was supplied jointly by KBR Magneto and Czestochowa University of Technology, Poland. (Reproduced from Tumanski S., *Handbook of Magnetic Measurements*, CRC Press, 2011, © Taylor & Francis, ISBN 9781439829516. With permission.)

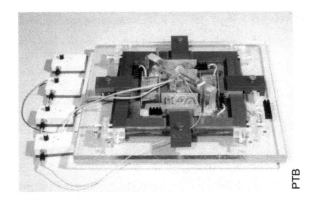

FIGURE 6.4 Horizontal yoke with a square sample (Sievert, 1992a). (Photograph courtesy of Johannes Sievert, National Metrology Institute (PTB), Germany.)

FIGURE 6.5 Vertical yoke suitable for measurements on sheets larger than the size of the yoke (see also Figure 4.20b) (Sievert et al., 1992b). (Photograph courtesy of Johannes Sievert, National Metrology Institute (PTB), Germany.)

FIGURE 6.6   Stator-like core of a three-phase induction motor used as a magnetising yoke to generate the required 2D field in the plane of a lamination placed in a vacuum chamber (visible inside the stator). The setup was used for thermometric measurements (see also Figure 2.16) (Ragusa C. et al., 2008.) (Courtesy of Carlo Appino and Carlo Ragusa, Electromagnetics Division, National Institute of Metrological Research (INRiM), Italy.)

FIGURE 6.7 Experimental round yokes used for rotational measurement over a wide frequency range (de la Barriere et al., 2015) and an alternative design for lower frequencies. (Photographs courtesy of Carlo Appino and Carlo Ragusa, Electromagnetics Division, National Institute of Metrological Research (INRiM), Italy). (a) High-frequency rotational yoke, (b) top view of the high-frequency yoke with the circular sample inside, (c) details of the magnetising winding of the high-frequency yoke, (d) normal frequency rotational yoke, (e) top view of the normal frequency yoke without the sample, and (f) vacuum chamber for thermometric measurements.

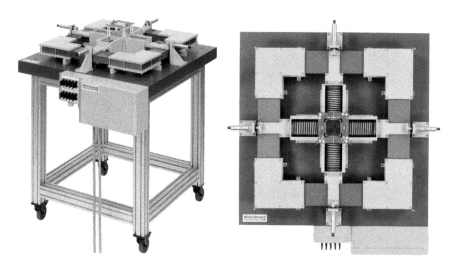

FIGURE 6.8   General and top view of a commercially available square rotational yoke, showing its structure (Borg Bartolo et al., 2015). (Photographs courtesy of Piotr Klimczyk, Brockhaus Measurements, Germany.)

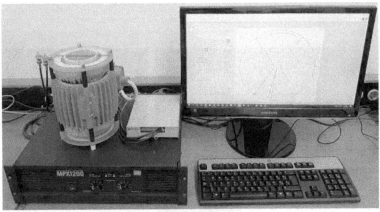

FIGURE 6.9   Yoke made of an electric motor with two-phase 'sinusoidal' windings (Figure 4.32) and the magnetising system with a power amplifier, data acquisition device and the user interface on the computer screen (Gorican et al., 2000a). (Photographs courtesy of Victor Gorican, Maribor University, Slovenia.)

FIGURE 6.10 Vertical yoke with a cross-shaped sample, assembled and disassembled showing the cross-shaped sample with *B*-coils (Fonteyn et al., 2009). (Photographs courtesy of Paavo Rasilo, Aalto University, Finland.)

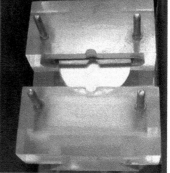

FIGURE 6.11 Anisometer for torquemetric measurements on small disc samples (20 mm diameter). The large electromagnet can drive the sample close to saturation. The sample is held in a sample holder with a *B*-coil placed with its sensing axis perpendicular to the direction of magnetisation. (Photographs released under CC-BY-3.0 license, courtesy of Wojciech Pluta, Czestochowa University of Technology, Poland.)

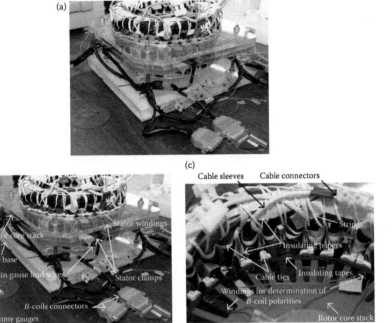

FIGURE 6.12 Round yoke for rotational and 2D measurements of magnetostriction: (a) overview of the yoke, (b) photograph with names of the components and (c) close-up view of the inside of the yoke and magnetising windings (Somkun et al., 2010b; Moses et al., 2015). (Photographs courtesy of Sakda Somkun, formerly of Wolfson Centre for Magnetics, Cardiff University, Cardiff, the United Kingdom.)

FIGURE 6.13 Three-phase yoke with a hexagonal sample (Hasenzagl et al., 1996). (Photograph courtesy of Georgi Shilyashki, Vienna University of Technology, Austria.)

(a)

(b)

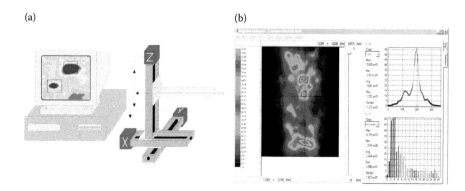

FIGURE 6.14   *Magnetovision* system capable of scanning the amplitude of *H* with an AMR sensor directly above the surface of the sample: drawing showing the concept of the device (a) and screenshot of the output image during scanning (b). (From Slawomir Tumanski, Warsaw University of Technology, Poland. Copyright © 2007 *Przeglad Elektrotechniczny (Electrical Review)*. Reproduced with permission from Tumanski S., *Przeglad Elektrotechniczny (Electrical Review)*, R. 83 (NR 1/2007), 2007b, 108.)

(a)

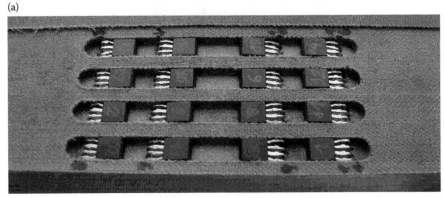

(b)

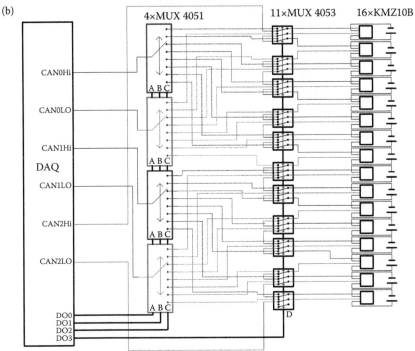

FIGURE 6.15 An array of magnetoresistive sensors (a) and the diagram of electric connections of the sensors (b) used in an experimental single-strip tester. (From Slawomir Tumanski, Warsaw University of Technology, Poland. Copyright © 2007 *Przeglad Elektrotechniczny (Electrical Review)*. Reproduced with permission from Tumanski S. et al., *Przeglad Elektrotechniczny (Electrical Review)*, R. 83 (NR 4/2007), 2007c, 46.)

FIGURE 6.16 Electromagnetic transducer with rotational excitation for the evaluation of fatigue and stress in steel samples: (a) close-up view of the miniature two-channel magnetising yoke whose longest dimension is just 19 mm, (b) 3D drawing of the yoke and (c) underside view of the measurement head with the yokes and the electronic chip visible in the middle (Chady et al., 2009; Psuj et al., 2014; Psuj et al., 2015). (Images courtesy of Grzegorz Psuj and Tomasz Chady, West Pomeranian University of Technology, Poland.)

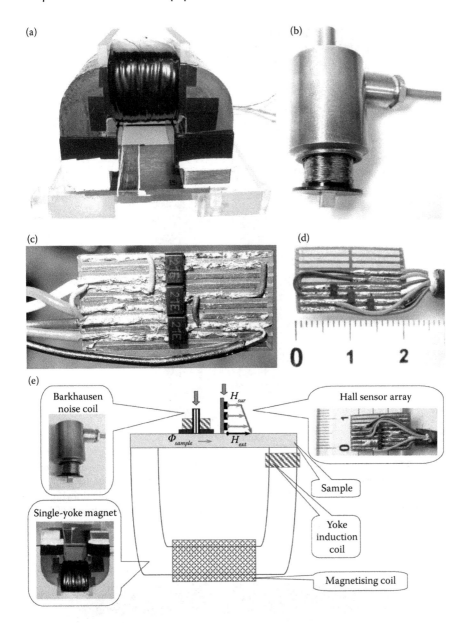

FIGURE 6.17 System for the measurement of Barkhausen noise and extrapolation of H towards the sample surface: (a) single C-core magnetising yoke with the sample under test, (b) Barkhausen noise sensor, (c) array of three Hall effect sensors for the extrapolation of H, (d) different design of array of Hall effect sensors and (e) diagram of all the components (Stupakov et al., 2009a; Stupakov et al., 2009b; Stupakov, 2013a, Stupakov A. et al., 2016). (Photographs and images courtesy of Alexandr Stupakov, Institute of Physics of the Czech Academy of Sciences, Czech Republic.)

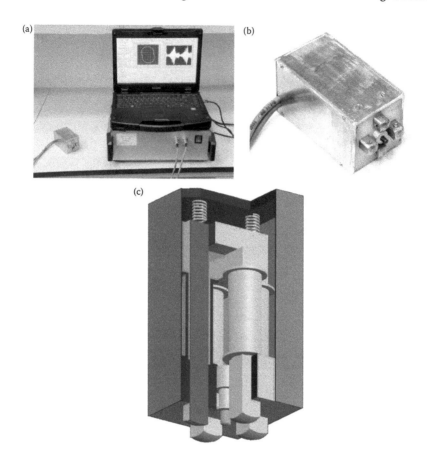

FIGURE 6.18   Portable system for the characterisation of stress anisotropy in the samples by means of magnetisation applied at arbitrary direction: (a) the whole measurement setup, (b) close-up of the prototype of the measurement head and (c) cut-away 3D drawing of the measurement head (Piotrowski et al., 2015). (Photographs and images courtesy of Leszek Piotrowski, Gdansk University of Technology, Poland.)

FIGURE 6.19   Magnetising setup for magnetic measurements under stress applied perpendicularly to the surface of the sheet. Compressive stress up to 20 MPa could be applied simultaneously with controlled magnetisation up to 1.7 T at 50 Hz (Yamamoto et al., 2014). (Photograph courtesy of Ken-ichi Yamamoto, University of Ryukyus, Japan.)

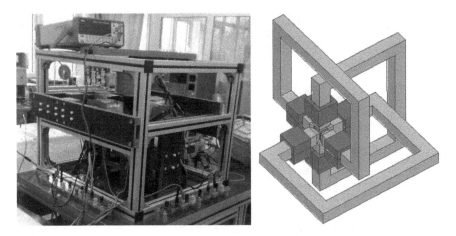

FIGURE 6.20 Three-dimensional magnetisation system for cubic samples (Li et al., 2014b, 2016). (Photograph and drawing courtesy of Yongjian Li, Province-Ministry Joint Key Laboratory of EFEAR, Hebei University of Technology, China.)

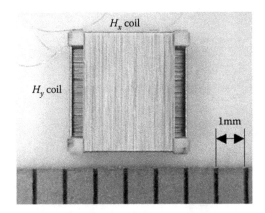

FIGURE 6.21 Double *H*-coils (for the measurement of H_x and H_y components) wound on a 4-mm-wide former. The thickness of the coil was around 0.6 mm. Each coil had 350 turns of 0.014 mm enamelled wire. (Photograph courtesy of Shigeru Aihara, Nishi Nippon Electric Wire & Cable Co. Ltd, Japan, and Takahashi Todaka, Oita University, Japan. Copyright © 2009 *Przeglad Elektrotechniczny (Electrical Review).* Reproduced with permission from Aihara S. et al., *Przeglad Elektrotechniczny (Electrical Review),* R. 87 (9b/2011), 2011, 73.)

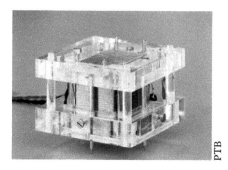

FIGURE 6.22   Side and bottom view of a sensor holder comprising two pairs of needle probes for the measurement of B, a flat double H-coil and two Rogowski–Chattock potentiometers for the measurement of H (Xu et al., 1997). (Photographs courtesy of Johannes Sievert, National Metrology Institute (PTB), Germany.)

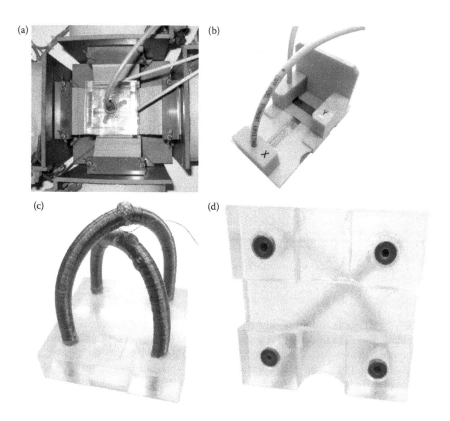

FIGURE 6.23 B and H sensors in commercially available rotational measurement system: (a) top view on a sample-and-sensor holder with 'pocket' B-coils and diagonal RCPs, (b) 'pocket' B-coils disassembled for insertion of a sample, (c) side view of a pair of orthogonal but diagonal RCPs and (d) bottom view of diagonal RCPs. (Courtesy of Piotr Klimczyk, Brockhaus Measurements, Germany.)

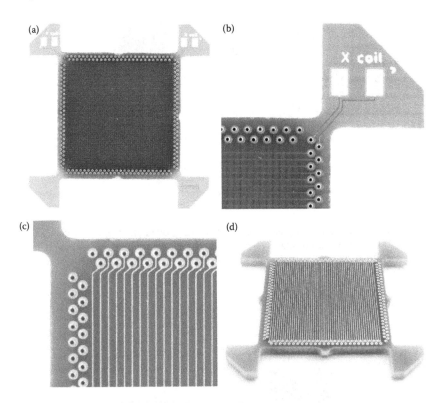

FIGURE 6.24 'Ideal' double *H*-coil made with PCB technology (see also Figure 3.50), 1.6 mm thick, 20 × 20 mm area: (a) top view, (b) connection pads for soldering wires, (c) parallel tracks and (d) bottom view. The PCB layout was designed by the author and the prototypes were kindly provided by Piotr Klimczyk of Brockhaus Measurements, Germany. (Photographs by Joanna Kaczmarzyk, CC-BY-3.0, *Encyclopedia Magnetica*.)

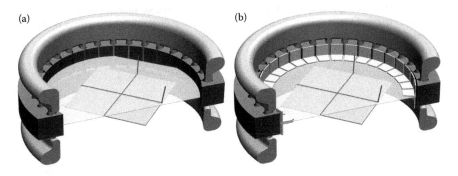

FIGURE 6.25 Cut-away view of 'optimum' round 2D yoke: (a) large sample diameter (>150 mm) with trimmed straight edges for sample positioning, large sensing area (*B*-coils >100 mm), PCB-based 'ideal' *H*-coils with grooves for *B*-coils; (b) the same but with 'vertical' shielding. (Images courtesy of Andrew Gilham, Megger Instruments Ltd, the United Kingdom.)

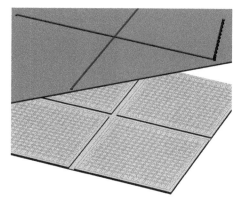

FIGURE 6.26   The same as in Figure 6.25 but with exploded and magnified view showing the grooves in the PCB-based 'ideal' *H*-coil for accommodating the *B*-coil wires. (Courtesy of Andrew Gilham, Megger Instruments Ltd, the United Kingdom.)

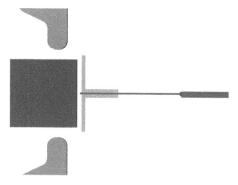

FIGURE 6.27   Cross-sectional view of the 'vertical' shielding (left-hand side only shown). The light-grey angled parts have vertical parts parallel to the surface of the yoke, and horizontal parts parallel to the surface of the sample (see also Figure 6.25b). (Courtesy of Andrew Gilham, Megger Instruments Ltd, the United Kingdom.)

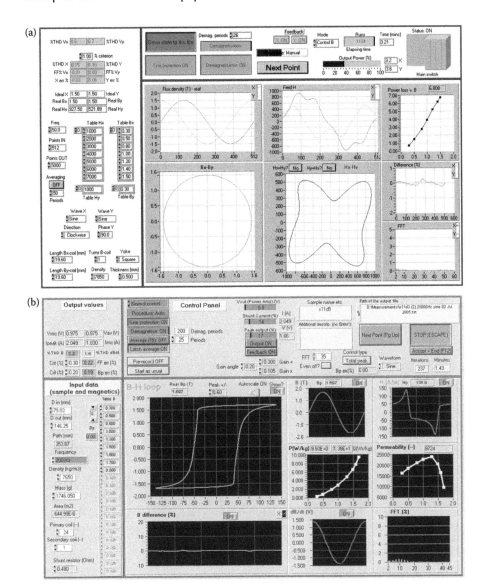

FIGURE 6.28 User interface developed in LabVIEW 8.5 software – (a) Screenshot of a rotational measurement on a non-oriented electrical steel sample under controlled circular *B*, with a number of variables controlled, including the B_{peak} error for both waveforms (B_x and B_y below 0.1%) and form factor error (below 0.1%); the measurements were performed with the square yoke from Figure 6.3, but also the round yoke from Figure 6.2 could be used interchangeably. (Zurek, 2005a.) (b) Similar interface but for uni-directional AC measurements on toroidal samples. The software was designed, programmed and used by the author during his PhD studies at Wolfson Centre for Magnetics, Cardiff University, Cardiff, the United Kingdom.

7 Examples of Measured Data

This chapter presents some experimental results included here mainly in order to illustrate typical capabilities of the methods and equipment described in the previous chapters. Most of these results were acquired by the author during his studies on uni-direction and rotational magnetisation. The results are not presented in any specific order.

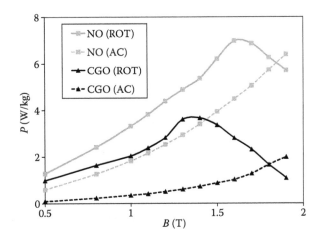

FIGURE 7.1 Comparison of alternating (AC) and rotational (ROT) losses for grain-oriented (CGO) and non-oriented (NO) electrical steels at 50 Hz. (Data re-plotted from Zurek S., Some interesting observations of loss and magnetism in soft magnetic materials under non-standard conditions, Magnetics, Energy Efficiency and the Environment, *Presented at UK Magnetics Society Seminar*, Cardiff University, Cardiff, the United Kingdom 2007c.)

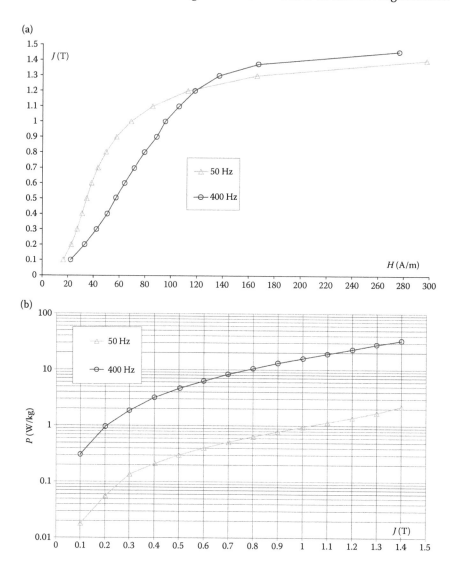

FIGURE 7.2   *J–H* curves (a) and $P = f(J)$ curves (b) measured under sinusoidal *J* at 50 Hz in non-standard single-sheet tester with magnetic path length 100 mm and sample width 100 mm. The sample was cut from non-oriented electrical steel sheet EP330-35A (EP12). The characteristics show that the material can exhibit higher permeability at higher frequency, even though the loss is much higher at 400 Hz than at 50 Hz. (Adapted from Zurek S., *Rotational Magnetisation in Flat Soft Magnetic Materials* (in Polish: *Przemagnesowanie obrotowe w materialach magnetycznie miekkich plaskich*), MSc thesis, Czestochowa University of Technology, Poland, 2000.)

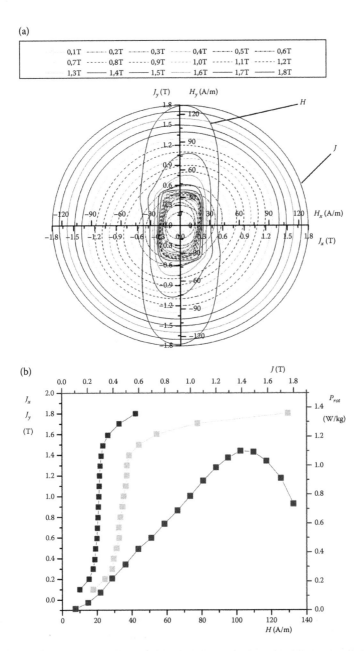

FIGURE 7.3 Loci of J and H (a) and the corresponding $J = f(H)$ and $P = f(J)$ characteristics (b). The data were measured at 50 Hz under controlled rotational circular J (ACW rotation) with the measurement system shown in Figure 6.1, on a square sample of Vacodur 50 (49% Co–Fe alloy). Rolling direction of the sample was aligned with the x axis. (Adapted from Zurek S., *Rotational Magnetisation in Flat Soft Magnetic Materials* (in Polish: *Przemagnesowanie obrotowe w materialach magnetycznie miekkich plaskich*), MSc thesis, Czestochowa University of Technology, Poland, 2000.)

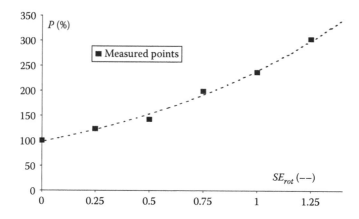

FIGURE 7.4 Normalised rotational loss P versus aspect ratio of elliptical B ($SE_{rot} = B_y/B_x$) measured at 50 Hz for Megaperm (40% Ni–Fe alloy). (Adapted from Zurek S., *Rotational Magnetisation in Flat Soft Magnetic Materials,* (in Polish: *Przemagnesowanie obrotowe w materialach magnetycznie miekkich plaskich*), MSc thesis, Czestochowa University of Technology, Poland, 2000.)

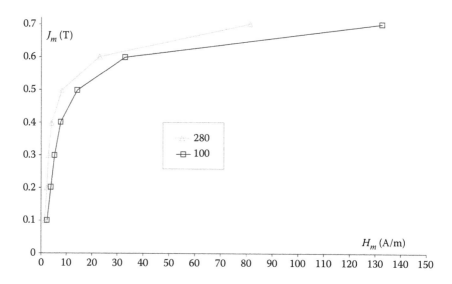

FIGURE 7.5 Comparison of J–H curves measured under alternating controlled sinusoidal J in a small prototype single-sheet/strip testers with magnetic path length of 100 mm (sample width 100 mm) and 280 mm (sample width 30 mm). Both measurements were carried out at 50 Hz, for 0.22-mm-thick Mumetall (77% Ni–Fe alloy), cut at 45° with respect to the rolling direction. The data show that yoke imperfection influences the error of H measurement due to the effective value of the magnetic path length and the presence of a small air gap between the sample and the yoke, especially for such high-permeability material. (Adapted from Zurek S., *Rotational Magnetisation in Flat Soft Magnetic Materials,* (in Polish: *Przemagnesowanie obrotowe w materialach magnetycznie miekkich plaskich*), MSc thesis, Czestochowa University of Technology, Poland, 2000.)

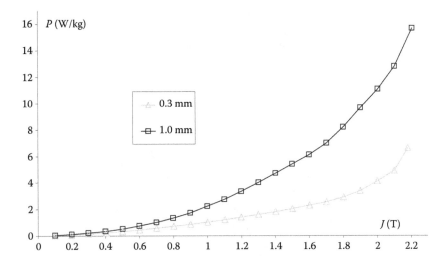

FIGURE 7.6 Power loss under alternating magnetisation increases with the sample thickness due to eddy currents. The data were measured at 50 Hz on samples of Vacoflux 50 (49% Co–Fe alloy) in a non-standard single-sheet tester with magnetic path length 100 mm and sample width 100 mm. (Adapted from Zurek S., *Rotational Magnetisation in Flat Soft Magnetic Materials* (in Polish: *Przemagnesowanie obrotowe w materialach magnetycznie miekkich plaskich*), MSc thesis, Czestochowa University of Technology, Poland, 2000.)

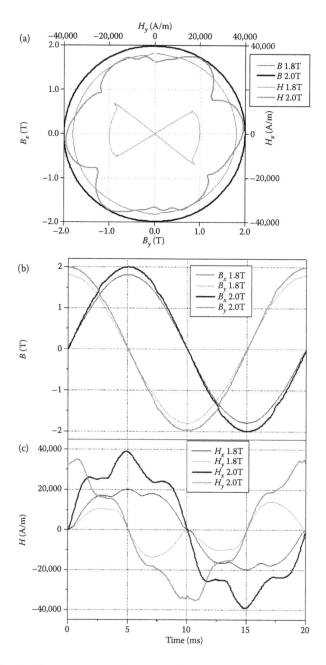

FIGURE 7.7   Results measured at 50 Hz for grain-oriented electrical steel M4 in a round rotational yoke (see also Figure 6.2): (a) – controlled circular B and resulting H loci with the rolling direction of the sample aligned with the vertical axis, (b and c) – the corresponding controlled sinusoidal B_x and B_y waveforms producing circular rotational B and resulting H_x and H_y waveforms. (Copyright © 2005 IEEE. Reprinted with permission from Zurek S. et al., *IEEE Transactions on Magnetics*, 41 (11), 2005b, 4242.)

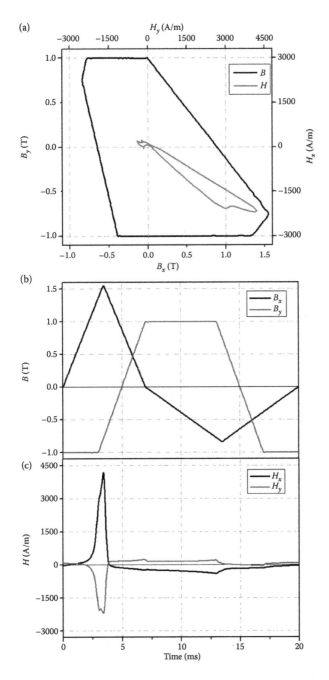

FIGURE 7.8 Results measured under arbitrary rotational conditions in a round yoke at 50 Hz for a non-oriented electrical steel sample: (a) rotational B and H loci, (b) the B_x wave-form controlled as asymmetrical triangular and B_y as symmetrical trapezoidal and (c) the corresponding H_x and H_y waveforms. (Copyright © 2005 IEEE. Reprinted with permission from Zurek S. et al., *IEEE Transactions on Magnetics*, 41 (11), 2005b, 4242.)

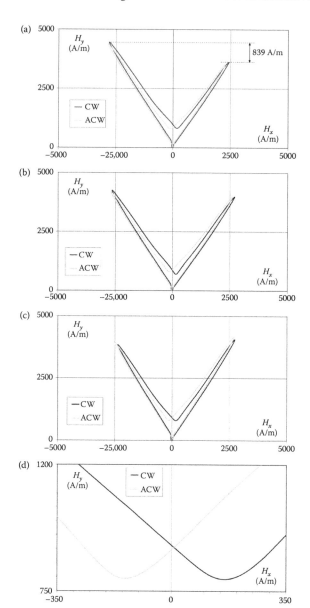

FIGURE 7.9   Asymmetry of rotational H loci due to angular misalignment of the B sensors causing skewing of the circularity of the B control. The data were measured for grain-oriented electrical steel at 50 Hz in a round yoke. Only the top half of each graph is shown for clarity, with the positions of the sensors as $B_x = 0°$ and (a) $B_y = 91°$, (b) $B_y = 90°$, (c) $B_y = 89°$ and (d) magnified view of the asymmetry as the curves cross the vertical axis (ACW minimum always on the left, CW minimum always on the right). (Adapted from Zurek S., *Two-Dimensional Magnetisation Problems in Electrical Steels*, PhD thesis, Wolfson Centre for Magnetics Technology, Cardiff University, Cardiff, the United Kingdom, 2005a.)

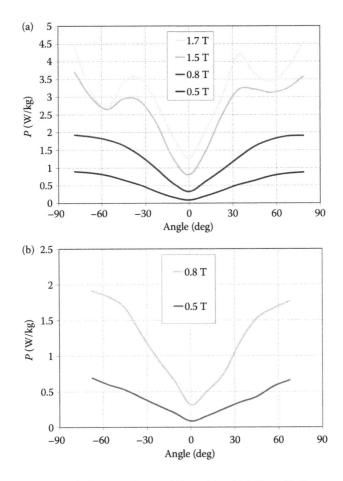

FIGURE 7.10 Loss measured under alternating sinusoidal B at 50 Hz at various inclination angles to the rolling direction for a circular sample (Figure 6.2): (a) Conventional grain-oriented electrical steel. (Reproduced from Zurek S. et al., *IEE Proceedings, Science, Measurement and Technology*, 153 (4), 2006b, 152. With permission of the Institution of Engineering & Technology) (b) High-permeability grain-oriented electrical steel. (Copyright © 2009 *Przeglad Elektrotechniczny (Electrical Review)*. Reproduced with permission from Zurek S. et al., *Przeglad Elektrotechniczny (Electrical Review)*, R. 85 (1/2009), 2009a, 16.) The asymmetry of the curves is most likely caused by angular misalignment of the $B–H$ sensors (Zurek, 2005a; Zurek et al., 2009a) and/or off-axis sensitivity of H-coils (Zurek, 2017c).

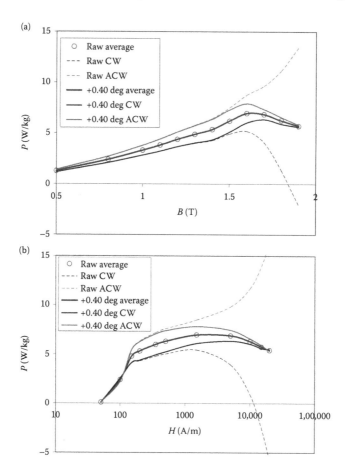

FIGURE 7.11   Rotational power loss measured for non-oriented electrical steel sheet at 50 Hz under: (a) controlled circular *B* and (b) controlled circular *H* (*raw* – as measured, +0.40° – recalculated for such angle between *B* and *H* sensors). The recalculation was carried out by employing Equations 5.45 and 5.46, by assuming that the *B* sensors were positioned exactly, and both *H* sensors were rotated away by the angle given in the figure. (Adapted from Zurek S., *Two-Dimensional Magnetisation Problems in Electrical Steels*, PhD thesis, Wolfson Centre for Magnetics Technology, Cardiff University, Cardiff, the United Kingdom, 2005a.)

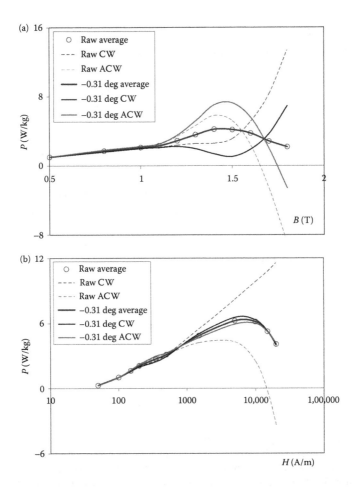

FIGURE 7.12 Rotational power loss measured for grain-oriented electrical steel sheet at 50 Hz under: (a) controlled circular B and (b) controlled circular H (*raw* – as measured, $-0.31°$ – recalculated for such angle between B and H sensors). (Adapted from Zurek S., *Two-Dimensional Magnetisation Problems in Electrical Steels*, PhD thesis, Wolfson Centre for Magnetics Technology, Cardiff University, Cardiff, the United Kingdom, 2005a.)

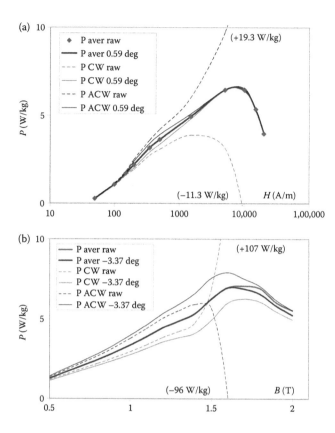

FIGURE 7.13 Very high positive and negative loss values can be caused by larger angular misalignment of the B and H sensors. The values in brackets correspond to those at the maximum excitation of 30 kA/m for controlled circular H for conventional grain-oriented electrical steel (a) and 2 T at controlled circular B of non-oriented electrical steel (b). (Adapted from Zurek S., *Two-Dimensional Magnetisation Problems in Electrical Steels*, PhD thesis, Wolfson Centre for Magnetics Technology, Cardiff University, Cardiff, the United Kingdom, 2005a.)

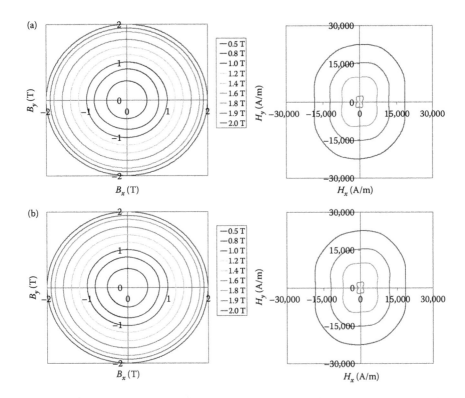

FIGURE 7.14 Rotational *B* and *H* loci for non-oriented electrical steel sheet measured under circular *B* at 50 Hz: (a) clockwise and (b) anticlockwise. (Adapted from Zurek S., *Two-Dimensional Magnetisation Problems in Electrical Steels*, PhD thesis, Wolfson Centre for Magnetics Technology, Cardiff University, Cardiff, the United Kingdom, 2005a.)

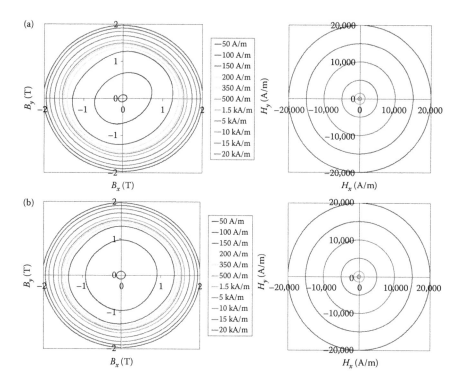

FIGURE 7.15   Rotational B and H loci for non-oriented electrical steel sheet measured under circular H at 50 Hz: (a) clockwise and (b) anticlockwise. (Adapted from Zurek S., *Two-Dimensional Magnetisation Problems in Electrical Steels*, PhD thesis, Wolfson Centre for Magnetics Technology, Cardiff University, Cardiff, the United Kingdom, 2005a.)

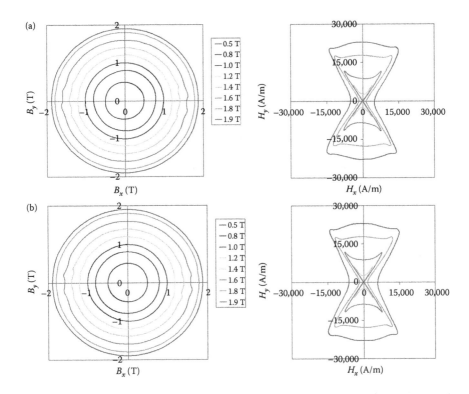

FIGURE 7.16 Rotational B and H loci for conventional grain-oriented electrical steel sheet measured under circular B at 50 Hz: (a) clockwise and (b) anticlockwise. (Adapted from Zurek S., *Two-Dimensional Magnetisation Problems in Electrical Steels*, PhD thesis, Wolfson Centre for Magnetics Technology, Cardiff University, Cardiff, the United Kingdom, 2005a.)

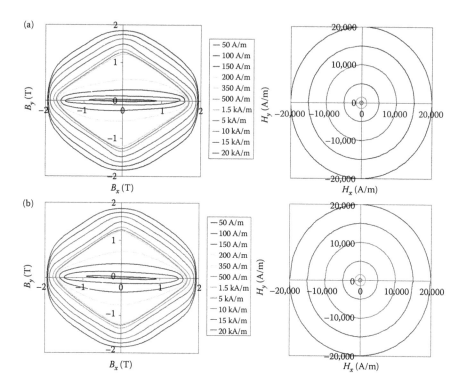

FIGURE 7.17   Rotational *B* and *H* loci for conventional grain-oriented electrical steel sheet measured under circular *H* at 50 Hz: (a) clockwise and (b) anticlockwise. (Adapted from Zurek S., *Two-Dimensional Magnetisation Problems in Electrical Steels*, PhD thesis, Wolfson Centre for Magnetics Technology, Cardiff University, Cardiff, the United Kingdom, 2005a.)

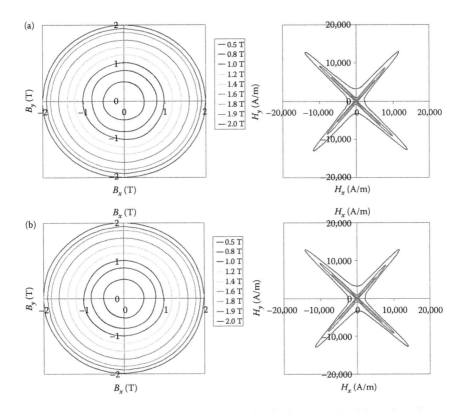

FIGURE 7.18 Rotational B and H loci for double grain-oriented electrical steel sheet measured under circular B at 50 Hz: (a) clockwise and (b) anticlockwise. (Adapted from Zurek S., *Two-Dimensional Magnetisation Problems in Electrical Steels*, PhD thesis, Wolfson Centre for Magnetics Technology, Cardiff University, Cardiff, the United Kingdom, 2005a.)

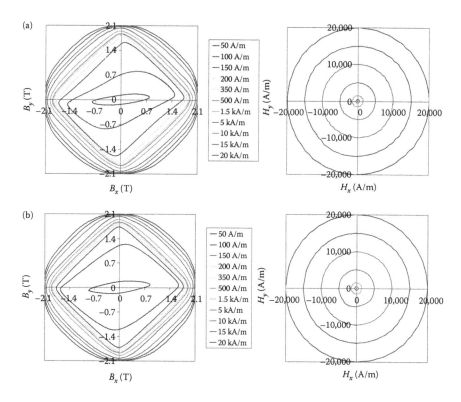

FIGURE 7.19   Rotational *B* and *H* loci for double grain-oriented electrical steel sheet measured under circular *H* at 50 Hz: (a) clockwise and (b) anticlockwise. (Adapted from Zurek S., *Two-Dimensional Magnetisation Problems in Electrical Steels*, PhD thesis, Wolfson Centre for Magnetics Technology, Cardiff University, Cardiff, the United Kingdom, 2005a.)

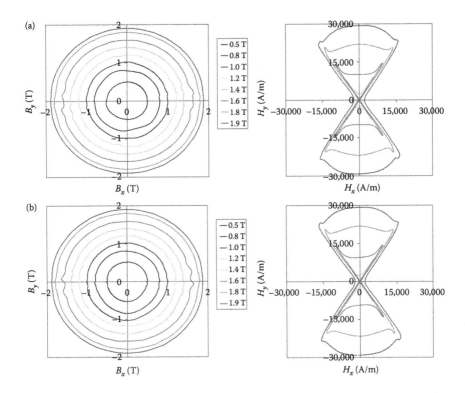

FIGURE 7.20 Rotational *B* and *H* loci for high-permeability grain-oriented electrical steel sheet measured under circular *B* at 50 Hz: (a) clockwise and (b) anticlockwise. (Adapted from Zurek S., *Two-Dimensional Magnetisation Problems in Electrical Steels*, PhD thesis, Wolfson Centre for Magnetics Technology, Cardiff University, Cardiff, the United Kingdom, 2005a.)

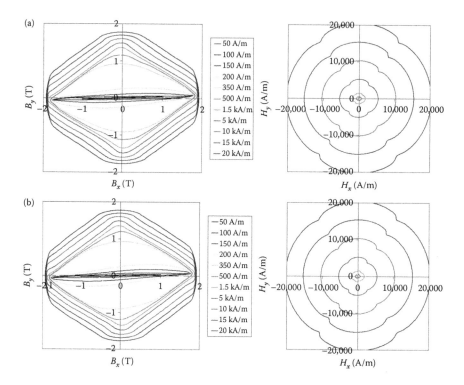

FIGURE 7.21   Rotational B and H loci for high-permeability grain-oriented electrical steel sheet measured under semi-controlled H (only the magnetising currents were set to sinusoidal shapes) at 50 Hz: (a) clockwise and (b) anticlockwise. (Adapted from Zurek S., *Two-Dimensional Magnetisation Problems in Electrical Steels*, PhD thesis, Wolfson Centre for Magnetics Technology, Cardiff University, Cardiff, the United Kingdom, 2005a.)

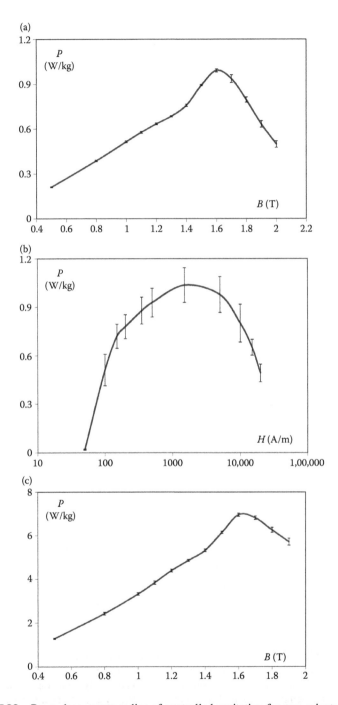

FIGURE 7.22 Power loss versus radius of controlled excitation for non-oriented electrical steel sheet measured under: (a) circular B at 10 Hz, (b) circular H at 10 Hz, (c) circular B at 50 Hz. (*Continued*)

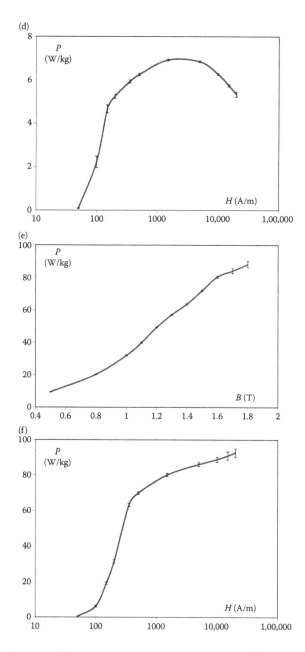

FIGURE 7.22 (Continued)   Power loss versus radius of controlled excitation for non-oriented electrical steel sheet measured under: (d) circular *H* at 50 Hz, (e) circular *B* at 250 Hz and (f) circular *H* at 250 Hz. Each measurement was repeated by removing the sample from the yoke and reassembling again. The curves show the average from five measurements; the vertical bars show the corresponding standard deviation. (Adapted from Zurek S., *Two-Dimensional Magnetisation Problems in Electrical Steels*, PhD thesis, Wolfson Centre for Magnetics Technology, Cardiff University, Cardiff, the United Kingdom, 2005a.)

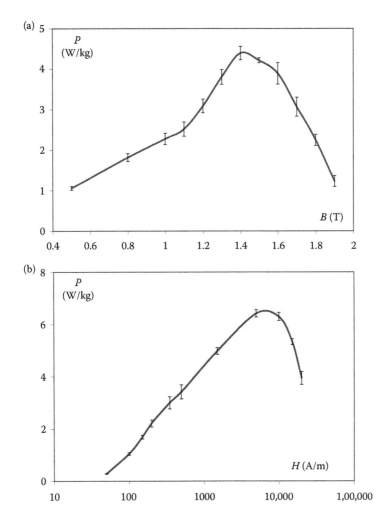

FIGURE 7.23 Power loss versus radius of controlled circular B (a) and circular H (b) for conventional grain-oriented electrical steel sheet measured at 50 Hz. (Adapted from Zurek S., *Two-Dimensional Magnetisation Problems in Electrical Steels*, PhD thesis, Wolfson Centre for Magnetics Technology, Cardiff University, Cardiff, the United Kingdom, 2005a.)

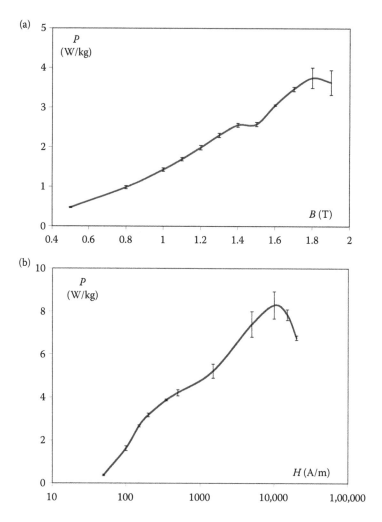

FIGURE 7.24 Power loss versus radius of controlled circular *B* (a) and circular *H* (b) for double grain-oriented electrical steel sheet measured at 50 Hz. (Adapted from Zurek S., *Two-Dimensional Magnetisation Problems in Electrical Steels*, PhD thesis, Wolfson Centre for Magnetics Technology, Cardiff University, Cardiff, the United Kingdom, 2005a.)

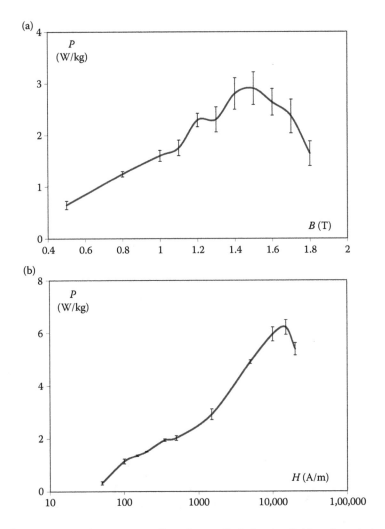

FIGURE 7.25 Power loss versus radius of controlled circular B (a) and semi-circular H (b) for high-permeability grain-oriented electrical steel sheet measured at 50 Hz. (Adapted from Zurek S., *Two-Dimensional Magnetisation Problems in Electrical Steels*, PhD thesis, Wolfson Centre for Magnetics Technology, Cardiff University, Cardiff, the United Kingdom, 2005a.)

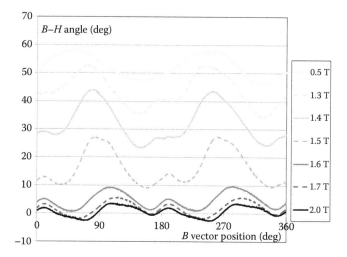

FIGURE 7.26 *B*–*H* angle trajectories for non-oriented electrical steel at 50 Hz, circular *B* (positive values mean that *H* leads *B*), the angles are calculated point-by-point by using Equation 2.15 for the data from Figure 7.14a (clockwise).

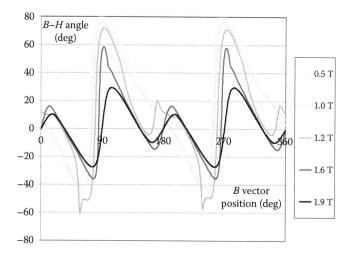

FIGURE 7.27 *B*–*H* angle trajectories for conventional grain-oriented electrical steel at 50 Hz, circular *B* (positive values mean that *B* lags *H*), calculated with Equation 2.15 for the data from Figure 7.16a (clockwise).

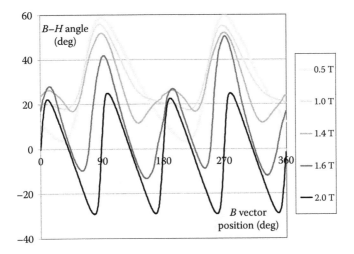

FIGURE 7.28 *B–H* angle trajectories for double-oriented electrical steel at 50 Hz, circular *B* (positive values mean that *B* lags *H*), calculated with Equation 2.15 for the data from Figure 7.18a (clockwise).

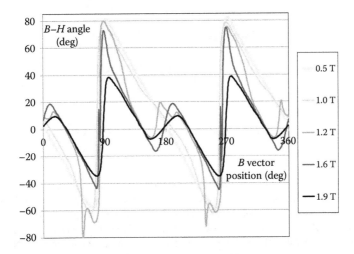

FIGURE 7.29 *B–H* angle trajectories for high-permeability grain-oriented electrical steel at 50 Hz, circular *B* (positive values mean that *B* lags *H*), calculated with Equation 2.15 for the data from Figure 7.20a (clockwise).

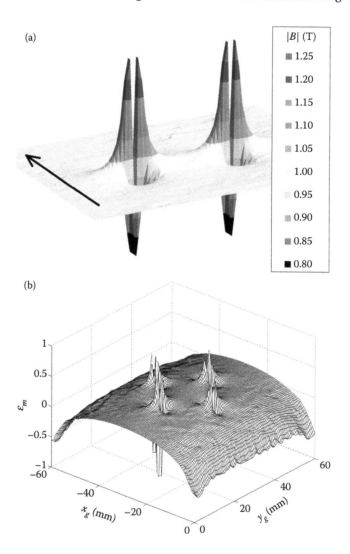

FIGURE 7.30   Typical distortion in *B* distribution around holes drilled in the sample for making *B*-coils, calculated with 3D FEM: (a) For uni-axial magnetisation in single-strip tester. (Copyright © 2017 *Przeglad Elektrotechniczny (Electrical Review)*. Reproduced with permission from Zurek S., *Przeglad Elektrotechniczny (Electrical Review)*, in press.) (b) For square sample in 2D tester. (From James Borg Bartolo, formerly of University of Nottingham, the United Kingdom. Reprinted from Borg Bartolo J. et al., *International Journal of Applied Electromagnetics and Mechanics*, 48 (2–3), 2015, 225. Copyright 2015, with permission from IOS Press.)

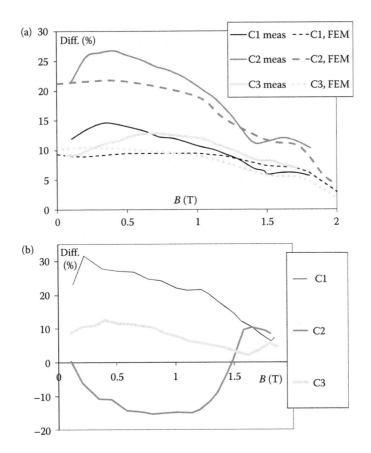

FIGURE 7.31 *B*-coil measurement errors caused by the presence of the holes drilled for passing the wires: (a) comparison of FEM simulations and measurement of *B*-coil errors for non-oriented electrical steel and (b) example of measured results of *B*-coil errors for grain-oriented electrical steel, in which at lower excitation large local grains can have greater impact than the discontinuity caused by the holes. (Copyright © 2017 *Przeglad Elektrotechniczny (Electrical Review)*. Reprinted with permission from Zurek S., *Przeglad Elektrotechniczny (Electrical Review)*, in press.)

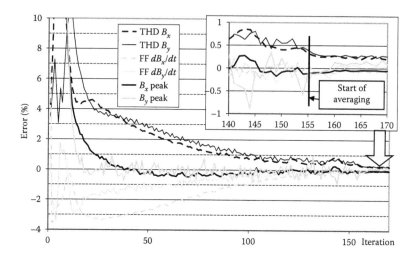

FIGURE 7.32   Typical digital feedback convergence for a rotational measurement at 50 Hz for conventional grain-oriented electrical steel, with circular flux density started at zero and controlled at 1.5 T; the inset shows the magnification of the final part of the convergence at the last 15 iterations. (Adapted from Zurek S. et al., *IEEE Transactions on Magnetics*, 41 (11), 2005b, 4242.)

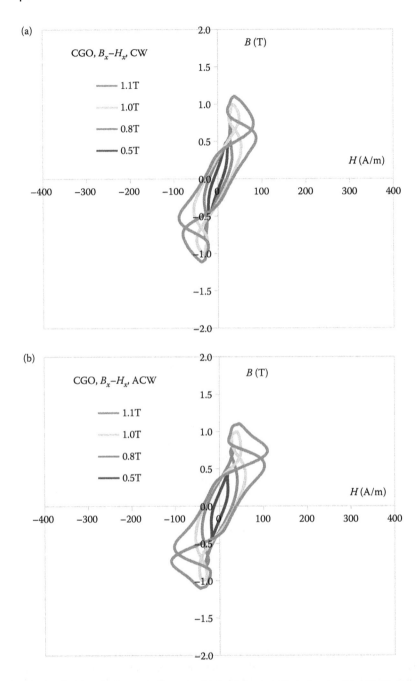

FIGURE 7.33 Typical B_x–H_x loops measured under controlled circular B at 50 Hz for conventional grain-oriented electrical steel, corresponding to the data shown in Figure 7.16: (a) CW lower range, (b) CW medium range (Copyright © 2017 IEEE. Reprinted with permission from Zurek S., *IEEE Transactions on Magnetics*, 50 (4), 2014a, 1.). *(Continued)*

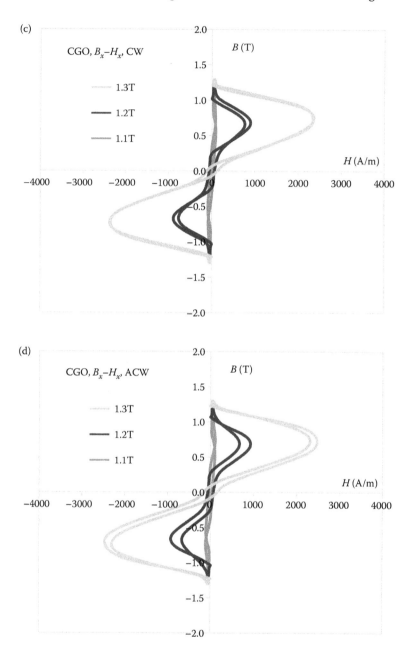

FIGURE 7.33 (Continued)   Typical B_x–H_x loops measured under controlled circular B at 50 Hz for conventional grain-oriented electrical steel, corresponding to the data shown in Figure 7.16: (c) CW high range, (d) ACW lower range.   *(Continued)*

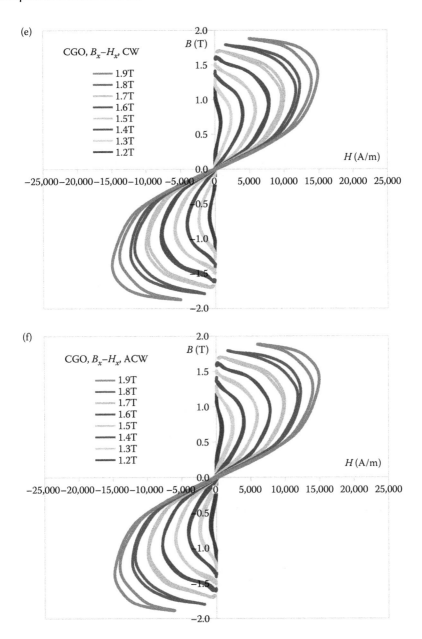

FIGURE 7.33 (Continued) Typical B_x–H_x loops measured under controlled circular B at 50 Hz for conventional grain-oriented electrical steel, corresponding to the data shown in Figure 7.16: (e) ACW medium range, (f) ACW high range.

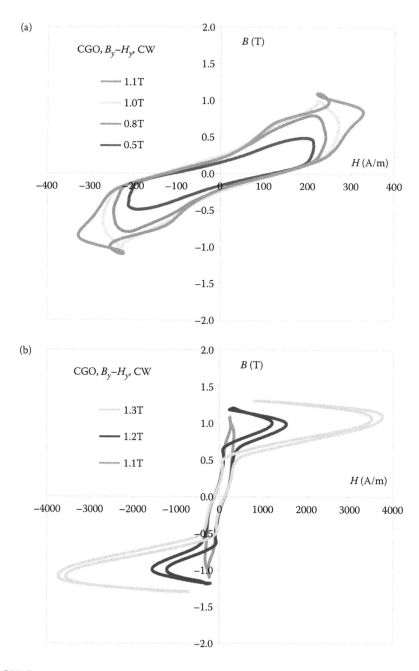

FIGURE 7.34 Typical B_y–H_y loops measured under controlled circular B at 50 Hz for conventional grain-oriented electrical steel, corresponding to the data shown in Figure 7.16: (a) CW lower range, (b) CW medium range, (Copyright © 2017 IEEE. Reprinted with permission from Zurek S., *IEEE Transactions on Magnetics*, 50 (4), 2014a, 1.)   (*Continued*)

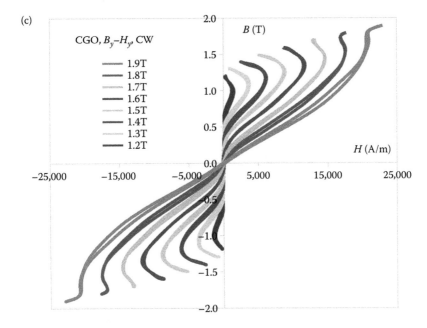

(c)

CGO, B_y–H_y, CW

— 1.9T
— 1.8T
— 1.7T
— 1.6T
— 1.5T
— 1.4T
— 1.3T
— 1.2T

FIGURE 7.34 (Continued)   Typical B_y–H_y loops measured under controlled circular B at 50 Hz for conventional grain-oriented electrical steel, corresponding to the data shown in Figure 7.16: (c) CW high range.

8 Additional Materials

8.1 NUMERICAL INTEGRATION OF WAVEFORMS

As discussed in Chapter 3, in some cases, the signals must be integrated, for example, when calculating the B waveform from the V waveform, or H from a signal measured from an H-coil or RCP.

The general concept of numerical integration is shown in Figure 3.7c. However, there are several details that need to be taken into account when applying the numerical integration. In this section, the integration of a waveform is described, which is equivalent to an analytical *indefinite integral* because the input as well as the output variable is a waveform (function, rather than a single value).

An example of a sinusoidal waveform represented digitally is shown in Figure 8.1. For clarity and simplicity, only 8 points per cycle were used in this particular case, and the amplitude and time were normalised to a value $V = 1$ V and $T = 1$ s ($f = 1$ Hz), respectively. In this case, it was also assumed that the indexing of points i starts from 0 rather than 1, but of course the numbering can start from any number.

It should be borne in mind that any non-real DC offsets should be removed from the signals in order to avoid the effect of ramping in the numerical integration, as illustrated in Figure 3.8.

As illustrated in Figure 3.7 and defined in Equations 3.7 and 3.8, it is possible to calculate the partial area under the curve for each pair of neighbouring points. The value of this partial area is assigned to the end of the interval over which the partial area is calculated. Therefore, in order to calculate the partial area associated with the 0th point ($i = 0$), it would be required to know the value of the point before it ($i = -1$).

In the digital waveform as shown in Figure 8.1, there is no information about the previous point; therefore, the initial conditions cannot be set beforehand, and thus it can be assumed that this partial area is equal to zero, and the 'initial conditions' will be calculated *after* the integration was carried out.

The partial areas are calculated for each pair of points as shown in Table 8.1, in column 4. Then a 'raw' integral is computed, by summing up all the previous areas for each point, as per Equation 3.10. The way the summation by Equation 3.10 works is shown in Table 8.1 for one cell (bold font and arrows in column 3.).

The instantaneous value of the integral is a sum of all the previous values of the partial areas. In spreadsheet software, this can be easily calculated as the sum of the previous value of the integral (previous instantaneous value of the integral, arrow above) and the next value of the partial area (arrow on the left). This is repeated for each cell in column 5, accordingly.

The 'raw' integral is calculated without any knowledge about the initial conditions. However, as resulting from Faraday's law of induction (Equations 3.1 and 3.2), for exactly one cycle of periodic excitation, the induced voltage cannot have any DC offset. Hence, the apparent DC offset in the 'raw' integral is an artefact of mathematical operation (see Equation 3.5) rather than a real offset in the signal.

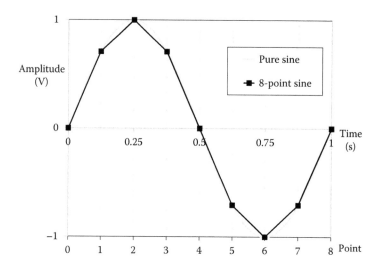

FIGURE 8.1 Example of a sinusoidal waveform represented numerically with 8 points.

Therefore, this DC offset can be calculated for the raw integral, as an average of all points, but with the last point excluded (column 6). The *interval* between points 7 and 8 in Figure 8.1 belongs to the same cycle, so its partial area is also used for integration. However, the *value* of the last point belongs to the next cycle (because as can be seen from Figure 8.1, for a periodic waveform, the point 8 is synonymous with the point 0). Therefore, the value of the last point must be omitted in the calculation of the DC offset because otherwise this value would be included twice in the calculation of the offset and thus the average would be incorrect (as illustrated in column 6, with the bottom value).

This has important implications because if the data acquisition device is set to acquire exactly one cycle of periodic excitation, then it will usually record just the points from 0 to 7, rather than from 0 to 8. However, for correct numerical integration, the value for the point 8 *must be* provided, either by acquiring $n + 1$ points in the first place, or at the very least, its value can be 'simulated' by copying the value of the $i = 0$ point to the $i = 8$ point (because the waveform should be continuous and periodic, so these two points should be equal).

And conversely – if the data acquisition is such that it provides $n + 1$ points (there are 9 points in Table 8.1), then it must be remembered to calculate the DC offset as in column 6, by omitting the last point to avoid errors.

In practice, this problem does not create as big errors as shown in column 6, because there are usually many more points per cycle, so that the contribution of the last point being included or excluded is small and thus more difficult to see any influence on the calculations. However, for fewer points per cycle, this value might become definitely non-negligible (as evident from the large difference between the two values in column 6) and thus attention should be paid so that correct mathematical calculations are always carried out.

An integral of $h'(t) = \sin(\omega \cdot t)$ is $h(t) = -\cos(\omega \cdot t)/\omega$, and as can be seen in column 7 of Figure 8.2, the waveform representing the final corrected integral is

TABLE 8.1

Example of Numerical Integration of $V(t)$ to $B(t)$, for $f = 1$ Hz, $T = 1/f = 1$ s, Points $n = 8$, Time Step $\Delta t = T/n = 0.125$ s, Number of Turns of B-Coil $N = 1$, Cross-Sectional Area of B-Coil $A = 1$ m^2

1.	2.	3.	4.	5.	6.	7.	8.	9.
		Sine	Partial Area, Equation 3.7,	'Raw' Integral = Sum of	Offset (Mean of All Points Apart from	Integral = Raw Integral−Offset	Integral Scaled	Flux Density
	Time, t	Voltage	$PA_i = \Delta t \cdot (V_{i-1} + V_i)/2$	Areas, Equation 3.10	the Last Point)		by $2 \cdot \pi$	B = Integral/$(N \cdot A)$
Point, i	(s)	(V)	(V · s)	(V · s)	(V · s)	(V · s) ≡ (T · m^2)	(a.u.)	(T)
0	0.000	0.000	0ᵃ	0.000	Average (from 0 to 7) = 0.151	−0.151	−0.948	−0.151
1	0.125	0.707	0.044	0.044		−0.107	−0.670	−0.107
2	0.250	1.000 ↘	0.107	0.151	Averageᶜ (from 0 to 8) = 0.134	0.000	0.000	0.000
3	0.375	0.707 →	0.107ᵇ	0.258		0.107	0.670	0.107
4	0.500	0.000	0.044	0.302		0.151	0.948	0.151
5	0.625	−0.707	−0.044	0.258 ↓		0.107	0.670	0.107
6	0.750	−1.000	−0.107 →	0.151ᵇ		0.000	0.000	0.000
7	0.875	−0.707	−0.107	0.044		−0.107	−0.670	−0.107
8	1.000	0.000	−0.044	0.000		−0.151	−0.948	−0.151

ᵃ Assumed zero if no initial conditions are known.

ᵇ The arrows show which two input values are used to calculate the value in the target cell.

ᶜ An incorrect DC offset calculation is shown as well for better illustration.

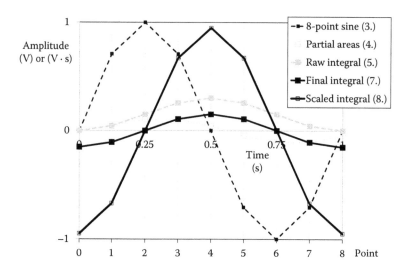

FIGURE 8.2   Plot of values from Table 8.1 (the numbers in the legend correspond to the columns in Table 8.1).

indeed negative cosine, whose amplitude is scaled down by a factor of $2 \cdot \pi$ (because $\omega = 2 \cdot \pi \cdot f$, but as evident in Figure 8.1, the value of f was set to 1 Hz due to T being 1 s). The phase of the integrated waveform is shifted by *precisely* $-90°$ (there is no phase error, even though the waveform was based on just 8 points).

Therefore, if the final integral was re-scaled by the factor of $2 \cdot \pi$ (column 8), then in this case, the resulting waveform should have an amplitude of unity. However, as can be seen in column 8 and in Figure 8.2, the peak of the integrated waveform deviates from unity by some amount, in this case 0.948 instead of 1.000, so the difference is quite large, amounting to around 5%.

This difference arises because the integral was calculated by using the trapezoidal method for the calculation of the numerical integral (see also Figure 3.7). Approximation of a continuous function that is convex (i.e. 'bulges' away from the horizontal axis) means that *each* partial area of a trapezoid will be *underestimated*, as compared to the area under the smooth curve (Figure 8.3). Therefore, the total sum of all such areas is also smaller for the trapezoidal approximation than it is for the continuous waveform. This underestimation has a direct impact on the amplitude of the integrated waveform, which for such function will *always* be smaller by some amount (e.g. if the input waveform was a form of a sinusoidal signal).

For this reason, the Simpson method can give better accuracy because it approximates the convex shape in a much closer way, so that the difference in the partial areas is much smaller (Figure 8.3). However, the calculations involved with the interpolation of a polynomial function over three (or more) points are more complex (and will not be described here).

The error resulting from the trapezoidal approximation can be quantified for a pure sine. The numerical integral was repeated for a sinusoidal waveform represented from 4 to 1000 points and the peak value of the integral was compared with

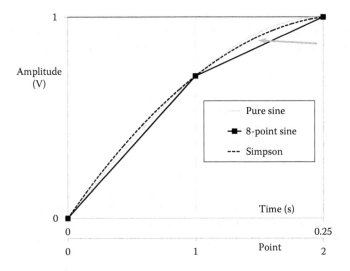

FIGURE 8.3 Explanation of the integration error due to approximation by the trapezoidal method (only two initial points are shown for clarity) with the Simpson method shown for comparison; the arrow shows the difference in the areas of the pure sine and the trapezoidal approximation between points 1 and 2.

the ideal value for a pure sine. The resulting difference was expressed in percent and the results are shown in Figure 8.4.

As can be expected, the relationship is proportional to the inverse square of the number of points per cycle used in the calculations. The amplitude error reduces quite quickly, and for 100 points per cycle, it is only around 0.03% but it is present and it will affect the calculated B waveform, by *always* underestimating it.

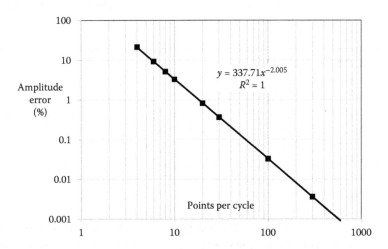

FIGURE 8.4 Percentage error of amplitude of numerically integrated pure sine versus number of points per cycle used in calculations; the error percentage values follow an approximated function $337.71 \cdot n^{-2.005}$ (%), where n denotes a number of points per cycle.

Therefore, the digital feedback will compensate the excitation in order to bring this value up to the required ideal level as dictated by calculations, and thus the actual B in the sample would be *always higher* than the values indicated by the measurement based on digital integration.

At 1000 points per cycle, the error is very small, around 0.00033%, which is a value normally insignificant as compared to other sources of errors, and even 0.03% for 100 points per cycle is a rather small value. Therefore, in practice, the Simpson method would not be required because sufficient accuracy can be obtained just from the trapezoidal approximation. For example, 200 points/cycle should guarantee that this error would be below 0.01% (Figure 8.4).

However, when using slower data acquisition devices or higher frequencies, so that the number of points per cycle is reduced, this error might be non-negligible. At 20 points per cycle, the error is around 0.8% (Figure 8.4). Therefore, such values should be taken into account in the overall uncertainty budget.

It should be noted that the unit of the values for the integral in column 7 of Table 8.1 is (V · s), which is synonymous with (T · m²). Therefore, the waveform should be divided by the number of turns and cross-sectional area of the B-coil in order to calculate the corresponding instantaneous values of flux density B (column 9), as defined in Equations 3.5 and 3.11.

It should be very strongly emphasised here that there is a minus in Equations 3.1 and 3.2. However, this minus applies to the electromotive force *EMF* induced *inside* the coil (due to Lenz's law). The *induced EMF* produces voltage across the voltmeter connected to such coil, and this voltage has a *positive* sign (Lubelski, 1982). Therefore, the calculation of B based on the *measured voltage V* should be carried out *without* this additional minus, as per Equation 3.2. For this reason, the B values in column 9 have the same polarity as the calculated integral in column 7 (rather than being negated).

To summarise, if the input voltage V was sinusoidal (column 3) so that $V(t) = V \cdot \sin(\omega \cdot t)$, then the numerically integrated flux density B will be negative cosine so that $B(t) = -B \cdot \cos(\omega \cdot t)$ as shown in column 9 of Table 8.1.

PRACTICAL COMMENT

This negative sign in Equations 3.1 and lack of it in Equation 3.2 is a common source of confusion for electrical engineers. It is interesting to note that it even found its way to the international IEC standard for the toroidal sample method (IEC, 2003), where the integral for the calculation of B shows the negative sign, even though it is based on a *measured* voltage. It is easy to show that this is incorrect in practice. Let us use the example of a toroidal sample method (see also Figures 2.19 and 8.6), so that the magnetic circuit has no air gap that could cloud the analysis.

The polarity of the voltage *measured* across the primary and secondary windings is the same, as long as the windings are wound the same way around the core. This is evident from the operation of an autotransformer, which can step up the input voltage V_{in} to some higher voltage V_{out} (Figure 8.5a). The polarity in the bottom part of the winding ('A') has to be the same as in the top part ('B') because, otherwise, the output voltage would not be greater than the

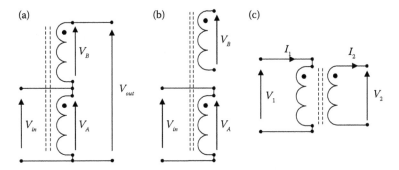

FIGURE 8.5 Illustration of Equation 3.2: (a) autotransformer with a two-part winding configured such that $V_{in} = V_A$ and $V_{out} = V_A + V_B$, (b) the two parts of the winding are separated but they keep the same voltage polarity and (c) definition of currents and voltages in the primary and secondary winding; the dashed double lines denote electromagnetic coupling through a magnetic core.

input voltage. As illustrated in Figure 8.5a, by definition, it must be that $V_A = V_{in}$ (assuming ideal windings, no losses, etc.).

The top part of the winding must have the same polarity of voltage even if it is not connected to the bottom part (Figure 8.5b). Hence, we can see that if both coils are wound around the core in the same way, then the polarity of voltage in the secondary winding is the same as that of the primary winding. Assuming that there is no leakage, the same flux density would penetrate both windings.

Let us assume that the magnetising current waveform $I(t)$ in the primary winding is

$$I(t) = I_p \cdot \sin(\omega \cdot t + \theta) \qquad (A) \qquad (8.1)$$

where I_p is the peak current (A), $\omega = 2 \cdot \pi \cdot f$ is the angular frequency (rad/s) and θ is the angle by which the $H(t)$ waveform leads the $B(t)$ waveform.

The $H(t)$ waveform is directly proportional to the $I(t)$ waveform, so that both have by definition the same phase, but with the amplitude scaled by some constant (see Equations 1.3 and 3.19); hence

$$H(t) = H_p \cdot \sin(\omega \cdot t + \theta) \qquad (A/m) \qquad (8.2)$$

For normal, lossy magnetic material, the $B(t)$ waveform will have its phase delayed with respect to $H(t)$. We defined above that $H(t)$ will lead $B(t)$ by the angle θ, so we can write that

$$B(t) = B_p \cdot \sin(\omega \cdot t) \qquad (T) \qquad (8.3)$$

For an ordinary lossy material, the angle θ will be between 0° and 90°. If it was less than 0°, then the material would generate energy instead of consuming it. If it was greater than 90°, then the load would be capacitive (which is not an ordinary case for windings below the resonant point caused by the parasitic self-capacitance).

(a) (b)

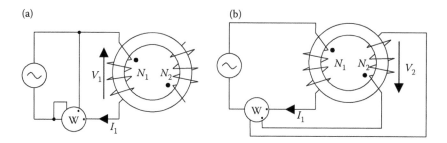

FIGURE 8.6   Illustration showing the use of a wattmetric method in which the measured power must always be positive: (a) when measuring the primary voltage V_1 and (b) when measuring the secondary voltage V_2; the dot at the respective terminal of the wattmeter denotes polarity of connection. The arrow of the secondary voltage points down because this is dictated by the direction the coil is wound (compare also to Figure 8.5 observing the position of the dots).

For sinusoidal waveforms, the value of power measured by the wattmeter in Figure 8.6a can be analytically calculated as

$$P = I_{1,RMS} \cdot V_{1,RMS} \cdot \cos(\phi_{I,V}) \qquad (\text{W}) \qquad (8.4)$$

where $I_{1,RMS}$ is the RMS value of the primary current (A), $V_{1,RMS}$ is the RMS value of the primary voltage (V) and $\phi_{I,V}$ is the angle between the current and voltage.

Both RMS values in Equation 8.4 are always positive (apart from the trivial case when they are zero). Therefore, for the value of power to be positive, the $\cos(\phi_{I,V})$ must also be positive, and this occurs only if $-90° < \phi_{I,V} < 90°$. It is evident from Figure 8.5 that it does not matter whether the primary or secondary voltage is measured because they are equal so that $V = V_1 = V_2$, with the same phase and amplitude (assuming the same numbers of turns, lossless windings and ideal wattmeter). For better illustration of waveforms (shown Figure 8.7), let us set that $H(t)$ leads $B(t)$ by 45°.

The B waveform is defined above as $B(t) = B_p \cdot \sin(\omega \cdot t)$. The derivative of sine is cosine, so by employing Equation 3.2 it means that the voltage across the winding should be $V(t) = V_p \cdot \cos(\omega \cdot t)$; hence the waveforms are as plotted in Figure 8.7a. For clarity, the instantaneous power $P(t)$ is also plotted, which is a point-by-point multiplication of the instantaneous values of $I(t)$, and the instantaneous values of $V(t)$. As can be seen from Figure 8.7a, the average of the instantaneous power curve $P(t)$ would be positive.

The value of $\theta = 45°$ was used between the waveforms of B and H in Figure 8.7a, which meant that the current *lagged* the voltage by 45°; hence $\phi_{I,V} = -45°$. If all waveforms are scaled so that their peak is unity, then the loss calculated from Equation 8.4 would be $P = 0.707\ \text{A} \cdot 0.707\ \text{V} \cdot \cos(-\pi/4) = +0.354\ \text{W}$.

However, if Equation 3.1 was used for defining the relationship between the *measured* voltage and the flux density, then it would have to be (incorrectly) assumed that $V(t) = -V_p \cdot \cos(\omega \cdot t)$. The primary current is the same as before so the phase of $H(t)$ is still $+45°$ and phase of $B(t)$ still 0° (sine). But then according

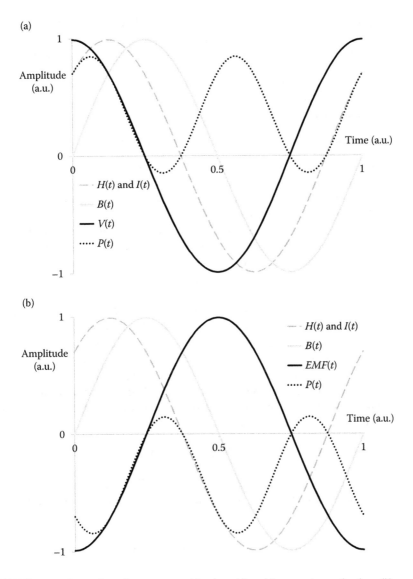

FIGURE 8.7   Illustration of a correct positive loss (a) and incorrect negative loss (b).

to Equation 3.1 the phase of voltage would have to be −cosine (−90°) so phase between voltage and current $\phi_{I,V}$ would have to be +135°, which would produce *negative* loss. This case is plotted in Figure 8.7. As can be seen, the average value of the power waveform would be negative in most instants of time, indicating negative loss of value $P = 0.707 \text{ A} \cdot 0.707 \text{ V} \cdot \cos(3 \cdot \pi/4) = -0.354 \text{ W}$.

This negative value is of course physically correct, but it must be written as the *internally induced EMF(t)* rather than the *measured V(t)*. This is because the internally induced *EMF* is in fact a source for the energy transferred through the electromagnetic coupling, so its power is negative because it supplies the energy. But the *measured* voltage must be taken with a *positive* sign for calculations of power loss dissipated in the material.

For this reason, the equation given in IEC (2003) is incorrect and Equation 3.2 is used for the numerical integral example given in Table 8.1.

Of course, the same rules apply when the integral is calculated for H sensors such as H-coil or RCP.

8.2   NUMERICAL INTEGRATION OF POWER LOSS

The power loss in the fieldmetric method can be calculated from Equation 2.13, which is also repeated as Equation 8.7. However, for simplicity and clarity, let us consider a uni-directional case, as defined by Equation 1.32, which is applicable to the toroidal sample method (Figures 2.19 and 8.6).

The numerical integral used for this calculation is equivalent to an analytical *definite integral* because a single value is produced as the output (rather than a waveform). The calculations are quite similar as for the indefinite numerical integral shown in Table 8.1, namely, there is also an input waveform for which the partial areas are calculated.

First, the $dB(t)/dt$ waveform needs to be derived. Luckily, owing to Equation 3.2, there is a direct relationship between the measured $V(t)$ and $dB(t)/dt$. Therefore, as far as the power loss calculation is concerned, there is no need to first integrate $V(t)$ into $B(t)$, only to later again differentiate $B(t)$ into $dB(t)/dt$. The voltage waveform can be used directly in the power loss calculation; it just needs to be scaled by the constant factor as dictated by Equation 3.2. In fact, this direct relationship is exploited in all the wattmetric methods because the primary current and secondary voltage are connected to the wattmeter, and the measured result can be scaled just by a single constant of proportionality.

Therefore, the $dB(t)/dt$ can be simply calculated as

$$dB(t)/dt = V(t)/(N \cdot A) \qquad (T/s) \equiv (V/m^2) \qquad (8.5)$$

Then the input waveform for the under-the-integral function $h(t)$ can be calculated. The newly produced $dB(t)/dt$ can be multiplied on a point-by-point basis by the $H(t)$ values, so that

$$h(t) = H(t) \cdot dB(t)/dt \qquad (A \cdot V/(m \cdot m^2)) \equiv (W/m^3) \qquad (8.6)$$

PRACTICAL COMMENT

If the information about $H(t)$ comes from an H-coil or RCP (or a similar sensor), then the output signal must be first integrated, as described in the previous section, because it is not possible to use $dH(t)/dt$ directly in the same way as $dB(t)/dt$ in Equation 8.5.

An example of such a calculation is shown in Table 8.2. The $dB(t)/dt$ waveform in column 4 is derived from column 3 (but in this particular case the factor of proportionality is unity, because $A = 1$ m^2 and also $N = 1$). The magnetic field strength $H(t)$

TABLE 8.2

Example of Numerical Integration of Power Loss, for $f = 1$ Hz, $T = 1/f = 1$ s, Points $n = 8$, Time Step $\Delta t = T/n = 0.125$ s, Number of Primary Turns $N_1 = 1$, Number of Turns of B-Coil $N = 1$, Cross-Sectional Area of B-Coil $A = 1$ m^2, $l = 1$ m; $H(t)$ Is Assumed to Lead $B(t)$ by $\pi/8$ (22.5°) Which Is Equivalent to $H(t)$ Lagging $dB(t)/dt$ by $3 \cdot \pi/8$ (67.5°)

1.	2.	3.	4.	5.	6.	7.	8.
		Sine Voltage	$dB(t)/dt =$	Magnetic Field Strength	Function under Integral	Partial Areas	Integral (Sum of All
Point, i	Time, t (s)	$V(t) = V \cdot \sin(\omega \cdot t)$ (V)	$V(t)/(N \cdot A)$ (V/m^2)	$H(t) = H \cdot \sin(\omega \cdot t - 3 \cdot \pi/8)$ (A/m)	$h(t) = H(t) \cdot dB(t)/dt$ (A · V/m^3)	$PA_i = \Delta t \cdot (h_{i-1} + h_i)/2$ (s · A · V/m^3) $\equiv$ (J/m^3)	Values from Column 7) (s · A · V/m^3) $\equiv$ (J/m^3)
0	0.000	0.000	0.000	-0.924	0.000	0.000	
1	0.125	0.707	0.707	-0.383	-0.271 ↘	-0.017	
2	0.250	1.000	1.000	0.383	0.383 →	0.007[a]	
3	0.375	0.707	0.707	0.924	0.653	0.065	
4	0.500	0.000	0.000	0.924	0.000	0.041	0.191
5	0.625	-0.707	-0.707	0.383	-0.271	-0.017	
6	0.750	-1.000	-1.000 →	-0.383 →	0.383[a]	0.007	
7	0.875	-0.707	-0.707	-0.924	0.653	0.065	
8	1.000	0.000	0.000	-0.924	0.000	0.041	

[a] The arrows show which two input values are used to calculate the value in the target cell.

in column 5 has a phase angle assumed such that it would be leading the corresponding $B(t)$ waveform by 22.5° (arbitrarily set value, for better illustration).

The under-the-integral point-by-point function $h(t) = H(t) \cdot dB(t)/dt$ is calculated in column 6 as a point-by-point product of the values in the two previous columns. The partial areas are calculated in column 7, and the final integral in column 8 With the definite integral, there is no need to remove any DC components, and the final result of the integrated power loss is simply the sum of all values in column 7, namely, 0.191 J/m³. (Of course, any unwanted DC offsets should be removed from *both* input waveforms *before* the integration.)

It should be noted that the final value has the unit of (J/m³). If the specific power loss is to be expressed, for example, in the unit of (W/kg), then it is simply required to multiply the result by the frequency f and divide by the sample density D, as defined, for example, in Equation 8.7.

We can see that the corresponding voltage from column 3 of Table 8.2 has the value of $V_{RMS} = 0.707$ V. The corresponding primary current would be $I_{RMS} = 0.707$ A. The phase shift between the voltage and the current is $\phi_{I,V} = 90° - 22.5° = 67.5°$. Hence, using Equation 8.4, we can calculate that $P = 0.707$ A $\cdot$ 0.707 V $\cdot$ cos(67.5°) $= 0.191$ J/m³, which is the same value as calculated in Table 8.2 in column 8.

The interesting behaviour of such a definite numerical integral is that the power loss calculated from the $h(t) = H(t) \cdot dB(t)/dt$ function does *not* suffer from the same approximation errors as it was the case for the indefinite integral, even if very few points per cycle are used for the waveforms. This is because if the input waveforms do not contain even harmonics, then the partial areas are underestimated by exactly the same amount for the 'convex' or positive parts of the function under the integral, as they overestimated for the 'concave' or negative parts, which is evident from Figure 8.8.

However, it should be stressed that the *exact* value is obtained only if neither $dB(t)/dt$ nor $H(t)$ contains *any even harmonics*. All odd harmonics will be calculated *precisely*.

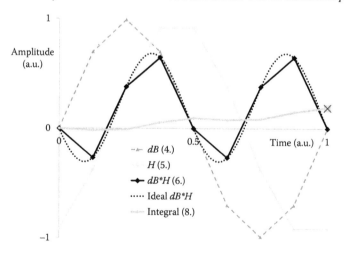

FIGURE 8.8   Definite numerical integral for the calculation of power loss (data from Table 8.2 are plotted); the large square symbol at the end of the thick grey curve of 'integral' denotes the final value of the power loss integral (0.191).

It can be expected that with significant even harmonics, the calculation of their contribution to power loss will suffer from similar inaccuracies as for the indefinite integral (see Figure 8.2). For few points per cycle, this contribution should be taken into account in the uncertainty budget (described in detail in Chapter 5).

8.3 DERIVATION OF EQUATION FOR ROTATIONAL LOSS UNDER IDEAL CONDITIONS

An example of a simplified case of uncertainty calculation for an idealised rotational condition was presented in Chapter 5, based on Equation 5.36. This equation was derived as follows (see also Table 8.3):

$$P_{rot} = \frac{f}{D} \cdot \int_0^T \left(\frac{dB_x(t)}{dt} \cdot H_x(t) + \frac{dB_{y(t)}}{dt} \cdot H_y(t) \right) dt \quad \text{(W/kg)} \tag{8.7}$$

$$P_{rot} = P_x + P_y \quad \text{(W/kg)} \tag{8.8}$$

$$P_x = \frac{f}{D} \cdot \int_0^T \left(\frac{dB_x(t)}{dt} \cdot H_x(t) \right) dt \quad \text{(W/kg)} \tag{8.9}$$

$$P_y = \frac{f}{D} \cdot \int_0^T \left(\frac{dB_y(t)}{dt} \cdot H_y(t) \right) dt \quad \text{(W/kg)} \tag{8.10}$$

TABLE 8.3
Equations for CW and ACW Direction of Rotation

For Clockwise (CW) (Assuming Starting Point of $B_x(0) = 0$ and $B_y(0) = +B_y$)		For Anticlockwise (ACW) (Assuming Starting Point of $B_x(0) = 0$ and $B_y(0) = -B_y$)		Units
$B_x(t) = B_x \cdot \sin(\omega \cdot t)$	(8.11)	$B_x(t) = B_x \cdot \sin(\omega \cdot t)$	(8.12)	(T)
$B_y(t) = B_y \cdot \cos(\omega \cdot t)$	(8.13)	$B_y(t) = -B_y \cdot \cos(\omega \cdot t)$	(8.14)	(T)
$H_x(t) = H_x \cdot \sin(\omega \cdot t + \phi_x)$	(8.15)	$H_x(t) = H_x \cdot \sin(\omega \cdot t + \phi_x)$	(8.16)	(A/m)
$H_y(t) = H_y \cdot \cos(\omega \cdot t + \phi_y)$	(8.17)	$H_y(t) = -H_y \cdot \cos(\omega \cdot t + \phi_y)$	(8.18)	(A/m)
$\dfrac{dB_x(t)}{dt} = B_x \cdot \cos(\omega \cdot t) \cdot \omega$	(8.19)	$\dfrac{dB_x(t)}{dt} = B_x \cdot \cos(\omega \cdot t) \cdot \omega$	(8.20)	(T/s)
$\dfrac{dB_y(t)}{dt} = -B_y \cdot \sin(\omega \cdot t) \cdot \omega$	(8.21)	$\dfrac{dB_y(t)}{dt} = B_y \cdot \sin(\omega \cdot t) \cdot \omega$	(8.22)	(T/s)

$$P_x = \frac{f}{D} \cdot \int_0^T \left(\frac{dB_x(t)}{dt} \cdot H_x(t) \right) dt \quad (8.23) \qquad \text{(W/kg)}$$

$$P_y = \frac{f}{D} \cdot \int_0^T \left(\frac{dB_x(t)}{dt} \cdot H_x(t) \right) dt \quad (8.24) \qquad \text{(W/kg)}$$

For clockwise (CW):

$$P_{x,CW} = \frac{f}{D} \cdot \int_0^T (B_x \cdot \cos(\omega \cdot t) \cdot \omega \cdot H_x \cdot \sin(\omega \cdot t + \phi_x)) dt \quad \text{(W/kg)} \quad (8.25)$$

$$P_{y,CW} = \frac{f}{D} \cdot \int_0^T (-B_y \cdot \sin(\omega \cdot t) \cdot \omega \cdot H_y \cdot \cos(\omega \cdot t + \phi_y)) dt \quad \text{(W/kg)} \quad (8.26)$$

$$P_{x,CW} = \frac{f}{D} \cdot B_x \cdot \omega \cdot H_x \cdot \int_0^T (\cos(\omega \cdot t) \cdot \sin(\omega \cdot t + \phi_x)) dt \quad \text{(W/kg)} \quad (8.27)$$

$$P_{y,CW} = -\frac{f}{D} \cdot B_y \cdot \omega \cdot H_y \cdot \int_0^T (\sin(\omega \cdot t) \cdot \cos(\omega \cdot t + \phi_y)) dt \quad \text{(W/kg)} \quad (8.28)$$

$$P_{x,CW} = \frac{f}{D} \cdot B_x \cdot \omega \cdot H_x$$
$$\cdot \frac{1}{2} \cdot \int_0^T (\sin(\omega \cdot t + \phi_x + \omega \cdot t) + \sin(\omega \cdot t + \phi_x - \omega \cdot t)) dt \quad \text{(W/kg)} \quad (8.29)$$

$$P_{y,CW} = -\frac{f}{D} \cdot B_y \cdot \omega \cdot H_y$$
$$\cdot \frac{1}{2} \cdot \int_0^T (\sin(\omega \cdot t + \omega \cdot t + \phi_y) + \sin(\omega \cdot t - \omega \cdot t - \phi_y)) dt \quad \text{(W/kg)} \quad (8.30)$$

$$P_{x,CW} = \frac{f \cdot B_x \cdot \omega \cdot H_x}{2 \cdot D} \cdot \int_0^T (\sin(2 \cdot \omega \cdot t + \phi_x) + \sin(\phi_x)) dt \quad \text{(W/kg)} \quad (8.31)$$

$$P_{y,CW} = -\frac{f \cdot B_y \cdot \omega \cdot H_y}{2 \cdot D} \cdot \int_0^T (\sin(2 \cdot \omega \cdot t + \phi_y) + \sin(-\phi_y)) dt \quad \text{(W/kg)} \quad (8.32)$$

$$P_{x,CW} = \frac{f \cdot B_x \cdot \omega \cdot H_x}{2 \cdot D} \cdot \left[\int_0^T (\sin(2 \cdot \omega \cdot t + \phi_x)) dt + \int_0^T (\sin(\phi_x)) dt \right] \quad \text{(W/kg)} \quad (8.33)$$

$$P_{y,CW} = -\frac{f \cdot B_y \cdot \omega \cdot H_y}{2 \cdot D} \cdot \left[\int_0^T (\sin(2 \cdot \omega \cdot t + \phi_y)) dt - \int_0^T (\sin(\phi_y)) dt \right] \quad \text{(W/kg)} \quad (8.34)$$

$$P_{x,CW} = \frac{f \cdot B_x \cdot \omega \cdot H_x}{2 \cdot D} \cdot \left[\frac{\cos(\phi_x) - \cos(\phi_x + 2 \cdot \omega \cdot T)}{2 \cdot \omega} + \sin(\phi_x) \cdot T \right] \quad \text{(W/kg)} \quad (8.35)$$

$$P_{y,CW} = -\frac{f \cdot B_y \cdot \omega \cdot H_y}{2 \cdot D} \cdot \left[\frac{\cos(\phi_y) - \cos(\phi_y + 2 \cdot \omega \cdot T)}{2 \cdot \omega} - \sin(\phi_y) \cdot T \right] \quad \text{(W/kg)} \quad (8.36)$$

$$P_{x,CW} = \frac{f \cdot B_x \cdot \omega \cdot H_x}{2 \cdot D} \cdot \left[\frac{\cos(\phi_x) - \cos(\phi_x + 4 \cdot \pi)}{2 \cdot \omega} + \sin(\phi_x) \cdot T \right] \quad \text{(W/kg)} \quad (8.37)$$

$$P_{y,CW} = -\frac{f \cdot B_y \cdot \omega \cdot H_y}{2 \cdot D} \cdot \left[\frac{\cos(\phi_y) - \cos(\phi_y + 4 \cdot \pi)}{2 \cdot \omega} - \sin(\phi_y) \cdot T \right] \quad \text{(W/kg)} \quad (8.38)$$

$$P_{x,CW} = \frac{f \cdot B_x \cdot \omega \cdot H_x}{2 \cdot D} \cdot \left[\frac{\cos(\phi_x) - \cos(\phi_x)}{2 \cdot \omega} + \sin(\phi_x) \cdot T \right] \quad \text{(W/kg)} \quad (8.39)$$

$$P_{y,CW} = -\frac{f \cdot B_y \cdot \omega \cdot H_y}{2 \cdot D} \cdot \left[\frac{\cos(\phi_y) - \cos(\phi_y)}{2 \cdot \omega} - \sin(\phi_y) \cdot T \right] \quad \text{(W/kg)} \quad (8.40)$$

$$P_{x,CW} = \frac{f \cdot B_x \cdot \omega \cdot H_x}{2 \cdot D} \cdot \sin(\phi_x) \cdot T \quad \text{(W/kg)} \quad (8.41)$$

$$P_{y,CW} = \frac{f \cdot B_y \cdot \omega \cdot H_y}{2 \cdot D} \cdot \sin(\phi_y) \cdot T \quad \text{(W/kg)} \quad (8.42)$$

$$P_{x,CW} = \frac{B_x \cdot \omega \cdot H_x}{2 \cdot D} \cdot \sin(\phi_x) \quad \text{(W/kg)} \quad (8.43)$$

$$P_{y,CW} = \frac{B_y \cdot \omega \cdot H_y}{2 \cdot D} \cdot \sin(\phi_y) \quad \text{(W/kg)} \quad (8.44)$$

$$P_{x,CW} = \frac{\pi \cdot f}{D} \cdot B_x \cdot H_x \cdot \sin(\phi_x) \quad \text{(W/kg)} \quad (8.45)$$

$$P_{y,CW} = \frac{\pi \cdot f}{D} \cdot B_y \cdot H_y \cdot \sin(\phi_y) \quad \text{(W/kg)} \quad (8.46)$$

$$P_{rot,CW} = \frac{\pi \cdot f}{D} \cdot [B_x \cdot H_x \cdot \sin(\phi_x) + B_y \cdot H_y \cdot \sin(\phi_y)] \quad \text{(W/kg)} \quad (8.47)$$

And for anticlockwise (ACW):

$$P_{x,ACW} = \frac{f}{D} \cdot \int_0^T (B_x \cdot \cos(\omega \cdot t) \cdot \omega \cdot H_x \cdot \sin(\omega \cdot t + \phi_x)) dt = P_{x,CW} \quad \text{(W/kg)} \quad (8.48)$$

$$P_{y,ACW} = \frac{f}{D} \cdot \int_0^T (-B_y \cdot \sin(\omega \cdot t) \cdot \omega \cdot H_y \cdot \cos(\omega \cdot t + \phi_y)) dt = P_{y,CW} \quad (\text{W/kg}) \quad (8.49)$$

$$P_{x,ACW} = \frac{\pi \cdot f}{D} \cdot B_x \cdot H_x \cdot \sin(\phi_x) \quad (\text{W/kg}) \quad (8.50)$$

$$P_{y,ACW} = \frac{\pi \cdot f}{D} \cdot B_y \cdot H_y \cdot \sin(\phi_y) \quad (\text{W/kg}) \quad (8.51)$$

$$P_{rot,ACW} = \frac{\pi \cdot f}{D} \cdot [B_x \cdot H_x \cdot \sin(\phi_x) + B_y \cdot H_y \cdot \sin(\phi_y)] \quad (\text{W/kg}) \quad (8.52)$$

$$P_{rot,ACW} = P_{rot,CW} \quad (\text{W/kg}) \quad (8.53)$$

8.4 SENSITIVITY COEFFICIENTS

The sensitivity coefficients were used in Table 5.12 in Chapter 5. They were derived by using Equation 8.47 or equivalent Equation 8.52 as the mathematical model. for D:

$$\frac{\partial P_{rot}}{\partial D} = \frac{\partial}{\partial D} \left\{ \frac{1}{D} \cdot \pi \cdot f \cdot [B_x \cdot H_x \cdot \sin(\phi_x) + B_y \cdot H_y \cdot \sin(\phi_y)] \right\} \quad (\text{W} \cdot \text{m}^3/\text{kg}^2) \quad (8.54)$$

$$\frac{\partial P_{rot}}{\partial D} = \pi \cdot f \cdot [B_x \cdot H_x \cdot \sin(\phi_x) + B_y \cdot H_y \cdot \sin(\phi_y)] \cdot \frac{\partial}{\partial D}\left(\frac{1}{D}\right) \quad (\text{W} \cdot \text{m}^3/\text{kg}^2) \quad (8.55)$$

$$\frac{\partial P_{rot}}{\partial D} = \pi \cdot f \cdot [B_x \cdot H_x \cdot \sin(\phi_x) + B_y \cdot H_y \cdot \sin(\phi_y)] \cdot \frac{1}{D^2} \quad (\text{W} \cdot \text{m}^3/\text{kg}^2) \quad (8.56)$$

for f:

$$\frac{\partial P_{rot}}{\partial f} = \frac{\partial}{\partial f} \left\{ f \cdot \frac{\pi}{D} \cdot [B_x \cdot H_x \cdot \sin(\phi_x) + B_y \cdot H_y \cdot \sin(\phi_y)] \right\} \quad (\text{W} \cdot \text{s/kg}) \quad (8.57)$$

$$\frac{\partial P_{rot}}{\partial f} = \frac{\pi}{D} \cdot [B_x \cdot H_x \cdot \sin(\phi_x) + B_y \cdot H_y \cdot \sin(\phi_y)] \frac{\partial}{\partial f}(f) \quad (\text{W} \cdot \text{s/kg}) \quad (8.58)$$

$$\frac{\partial P_{rot}}{\partial f} = \frac{\pi}{D} \cdot [B_x \cdot H_x \cdot \sin(\phi_x) + B_y \cdot H_y \cdot \sin(\phi_y)] \quad (\text{W} \cdot \text{s/kg}) \quad (8.59)$$

for B_x:

$$\frac{\partial P_{rot}}{\partial B_x} = \frac{\partial}{\partial B_x} \left\{ f \cdot \frac{\pi}{D} \cdot B_x \cdot H_x \cdot \sin(\phi_x) + f \cdot \frac{\pi}{D} \cdot B_y \cdot H_y \cdot \sin(\phi_y) \right\} \quad (\text{W/(kg} \cdot \text{T)}) \quad (8.60)$$

$$\frac{\partial P_{rot}}{\partial B_x} = \frac{\partial}{\partial B_x}\left\{\frac{\pi \cdot f}{D} \cdot B_x \cdot H_x \cdot \sin(\phi_x)\right\} + \frac{\partial}{\partial B_x}\left\{f \cdot \frac{\pi}{D} \cdot B_y \cdot H_y \cdot \sin(\phi_y)\right\} \quad (\text{W/(kg·T)})$$

(8.61)

$$\frac{\partial P_{rot}}{\partial B_x} = \frac{\pi \cdot f}{D} \cdot H_x \cdot \sin(\phi_x) \cdot \frac{\partial}{\partial B_x}(B_x) \quad (\text{W/(kg·T)}) \qquad (8.62)$$

$$\frac{\partial P_{rot}}{\partial B_x} = \frac{\pi \cdot f}{D} \cdot H_x \cdot \sin(\phi_x) \quad (\text{W/(kg·T)}) \qquad (8.63)$$

for B_y:

$$\frac{\partial P_{rot}}{\partial B_y} = \frac{\partial}{\partial B_y}\left\{f \cdot \frac{\pi}{D} \cdot B_x \cdot H_x \cdot \sin(\phi_x) + f \cdot \frac{\pi}{D} \cdot B_y \cdot H_y \cdot \sin(\phi_y)\right\} \quad (\text{W/(kg·T)}) \quad (8.64)$$

$$\frac{\partial P_{rot}}{\partial B_y} = \frac{\partial}{\partial B_y}\left\{\frac{\pi \cdot f}{D} \cdot B_x \cdot H_x \cdot \sin(\phi_x)\right\} + \frac{\partial}{\partial B_y}\left\{f \cdot \frac{\pi}{D} \cdot B_y \cdot H_y \cdot \sin(\phi_y)\right\} \quad (\text{W/(kg·T)})$$

(8.65)

$$\frac{\partial P_{rot}}{\partial B_y} = \frac{\pi \cdot f}{D} \cdot H_y \cdot \sin(\phi_y) \cdot \frac{\partial}{\partial B_y}(B_y) \quad (\text{W/(kg·T)}) \qquad (8.66)$$

$$\frac{\partial P_{rot}}{\partial B_y} = \frac{\pi \cdot f}{D} \cdot H_y \cdot \sin(\phi_y) \quad (\text{W/(kg·T)}) \qquad (8.67)$$

for H_x:

$$\frac{\partial P_{rot}}{\partial H_x} = \frac{\partial}{\partial H_x}\left\{f \cdot \frac{\pi}{D} \cdot B_x \cdot H_x \cdot \sin(\phi_x) + f \cdot \frac{\pi}{D} B_y \cdot H_y \cdot \sin(\phi_y)\right\} \quad (\text{W·m/(kg·A)}) \quad (8.68)$$

$$\frac{\partial P_{rot}}{\partial H_x} = \frac{\partial}{\partial H_x}\left\{\frac{\pi \cdot f}{D} B_x \cdot H_x \cdot \sin(\phi_x)\right\} + \frac{\partial}{\partial H_x}\left\{f \cdot \frac{\pi}{D} B_y \cdot H_y \cdot \sin(\phi_y)\right\} \quad (\text{W·m/(kg·A)})$$

(8.69)

$$\frac{\partial P_{rot}}{\partial H_x} = \frac{\pi \cdot f}{D} \cdot B_x \cdot \sin(\phi_x) \cdot \frac{\partial}{\partial H_x}(H_x) \quad (\text{W·m/(kg·A)}) \qquad (8.70)$$

$$\frac{\partial P_{rot}}{\partial H_x} = \frac{\pi \cdot f}{D} \cdot B_x \cdot \sin(\phi_x) \quad (\text{W·m/(kg·A)}) \qquad (8.71)$$

for H_y:

$$\frac{\partial P_{rot}}{\partial H_y} = \frac{\partial}{\partial H_y}\left\{f \cdot \frac{\pi}{D} \cdot B_x \cdot H_x \cdot \sin(\phi_x) + f \cdot \frac{\pi}{D} \cdot B_y \cdot H_y \cdot \sin(\phi_y)\right\} \quad (\text{W·m/(kg·A)})$$

(8.72)

$$\frac{\partial P_{rot}}{\partial H_y} = \frac{\partial}{\partial H_y}\left\{\frac{\pi \cdot f}{D} \cdot B_x \cdot H_x \cdot \sin(\phi_x)\right\} + \frac{\partial}{\partial H_y}\left\{f \cdot \frac{\pi}{D} \cdot B_y \cdot H_y \cdot \sin(\phi_y)\right\} \quad (\text{W·m/(kg·A)})$$

$$(8.73)$$

$$\frac{\partial P_{rot}}{\partial H_y} = \frac{\pi \cdot f}{D} \cdot B_y \cdot \sin(\phi_y) \cdot \frac{\partial}{\partial H_y}(H_y) \qquad (\text{W·m/(kg·A)}) \qquad (8.74)$$

$$\frac{\partial P_{rot}}{\partial H_y} = \frac{\pi \cdot f}{D} \cdot B_y \cdot \sin(\phi_y) \qquad (\text{W·m/(kg·A)}) \qquad (8.75)$$

for ϕ_x:

$$\frac{\partial P_{rot}}{\partial \phi_x} = \frac{\partial}{\partial \phi_x}\left\{f \cdot \frac{\pi}{D} \cdot B_x \cdot H_x \cdot \sin(\phi_x) + f \cdot \frac{\pi}{D} \cdot B_y \cdot H_y \cdot \sin(\phi_y)\right\} \qquad (\text{W/kg}) \quad (8.76)$$

$$\frac{\partial P_{rot}}{\partial \phi_x} = \frac{\partial}{\partial \phi_x}\left\{\frac{\pi \cdot f}{D} \cdot B_x \cdot H_x \cdot \sin(\phi_x)\right\} + \frac{\partial}{\partial \phi_x}\left\{f \cdot \frac{\pi}{D} \cdot B_y \cdot H_y \cdot \sin(\phi_y)\right\} \qquad (\text{W/kg}) \quad (8.77)$$

$$\frac{\partial P_{rot}}{\partial \phi_x} = \frac{\pi \cdot f}{D} \cdot B_x \cdot H_x \cdot \frac{\partial}{\partial \phi_x}(\sin(\phi_x)) \qquad (\text{W/kg}) \qquad (8.78)$$

$$\frac{\partial P_{rot}}{\partial \phi_x} = \frac{\pi \cdot f}{D} \cdot B_x \cdot H_x \cdot \cos(\phi_x) \qquad (\text{W/kg}) \qquad (8.79)$$

for ϕ_y:

$$\frac{\partial P_{rot}}{\partial \phi_y} = \frac{\partial}{\partial \phi_y}\left\{f \cdot \frac{\pi}{D} \cdot B_x \cdot H_x \cdot \sin(\phi_x) + f \cdot \frac{\pi}{D} \cdot B_y \cdot H_y \cdot \sin(\phi_y)\right\} \qquad (\text{W/kg}) \quad (8.80)$$

$$\frac{\partial P_{rot}}{\partial \phi_y} = \frac{\partial}{\partial \phi_y}\left\{\frac{\pi \cdot f}{D} \cdot B_x \cdot H_x \cdot \sin(\phi_x)\right\} + \frac{\partial}{\partial \phi_y}\left\{f \cdot \frac{\pi}{D} \cdot B_y \cdot H_y \cdot \sin(\phi_y)\right\} \qquad (\text{W/kg}) \quad (8.81)$$

$$\frac{\partial P_{rot}}{\partial \phi_y} = \frac{\pi \cdot f}{D} \cdot B_y \cdot H_y \cdot \frac{\partial}{\partial \phi_y}(\sin(\phi_y)) \qquad (\text{W/kg}) \qquad (8.82)$$

$$\frac{\partial P_{rot}}{\partial \phi_y} = \frac{\pi \cdot f}{D} \cdot B_y \cdot H_y \cdot \cos(\phi_y) \qquad (\text{W/kg}) \qquad (8.83)$$

References

A2LA, *G104 – Guide for Estimation of Measurement Uncertainty in Testing*, American Association for Laboratory Accreditation, 2014.

Ahlers H. et al., Uncertainties of magnetic properties measurements of electrical steel sheet, *Proceedings of 6th 1&2DM Workshop*, Bad Gastein, Austria, 2000a, paper I-4, p. 26.

Ahlers H. et al., The uncertainties of magnetic properties measurements of electrical sheet steel, *Journal of Magnetism and Magnetic Materials*, 215–216, 2000b, 711.

Aihara S. et al., Characteristic evaluation of 4mm-square-sized double H-coil, *Przeglad Elektrotechniczny (Electrical Review)*, R. 87 (9b/2011), 2011, 73.

AK Steel, *Grain Oriented Electrical Steels, Product Data Bulletin*, 2013.

Alatawneh N. et al., Rotational core loss and permeability measurements in machine laminations with reference to permeability asymmetry, *IEEE Transactions on Magnetics*, 48 (4), 2012a, 1445.

Alatawneh N. et al., Test specimen shape considerations for the measurement of rotational core losses, *IEEE Transactions on Energy Conversion*, 27 (1), 2012b, 151.

Alatawneh N. et al., Design of a novel test fixture to measure rotational core losses in machine laminations, *IEEE Transactions on Industry Applications*, 48 (5), 2012c, 1467.

Alinejad-Beromi Y., *Rotational Power Loss Measurement System under Controlled Magnetization*, PhD thesis, University of Wales, Cardiff, UK, 1992.

Alnigrad K., *Digital Waveform Control under Rotational Magnetisation*, PhD thesis, Wolfson Centre for Magnetics Technology, Cardiff University, Cardiff, UK, 1999.

Allegro, *A1302, Continuous-Time Ratiometric Linear Hall Effect Sensor ICs*, Data sheet, 2013a.

Allegro, *A3144, Sensitive Hall-Effect Switches for High-Temperature Operation*, Data sheet, 2013b.

Amp-Line, *Model AL-1000-CR-H/A AC Constant Current Power Amplifiers*, Data sheet, 2016.

Analogue Associates, *Analogue/Crown MT600, MT1200, MT2400 Power Amplifiers, Reference Manual*, 2001.

Analog Devices, *Ultraprecision Operational Amplifier OP177*, Data sheet, Rev. G, 2012.

Anderson P., A universal DC characterisation system for hard and soft magnetic materials, *Journal of Magnetism and Magnetic Materials*, 320 (20), 2008, e589.

Anderson P., Constant dB/dt DC characterisation through digital control of magnetic field, *Przeglad Elektrotechniczny (Electrical Review)*, R. 87 (9b), 2011, 77.

Ando B. et al., Adding noise to improve measurement, *IEEE Instrumentation & Measurement Magazine*, 56, 2001, 24.

Anuszczyk J., *Method and Apparatus for Measurement of Power Loss in Ferromagnetics Under Rotational Magnetisation* (in Polish: *Sposob i urzadzenie do pomiaru strat mocy w ferromagnetykach przemagnesowanych obrotowo*), Patent application, PRL, Poland, P127–489, 1984.

Anuszczyk J., *Analysis of Localisation of Flux Density and Power Loss in Rotational Magnetisation in Magnetic Circuits of Electric Machines* (in Polish: *Analiza rozkladu indukcji i strat mocy przy przemagnesowaniu obrotowym w obwodach magnetycznych maszyn elektrycznych*), Zeszyty Naukowe Politechniki Lodzkiej, Nr 629, Rozprawy Naukowe z. 158, Lodz, 1991.

Anuszczyk J.W. et al., *Soft Ferromagnets in Rotational Fields* (in Polish: *Ferromagnetyki miekkie w polach obrotowch*), Wydawnictwa Naukowo-Techniczne, Warsaw, Poland, 2009, ISBN 978-83-204-3484-2.

Aouli R. et al., Behaviour and characterization of ferromagnetic sheets under rotating flux density conditions in electrical machines, *Proceedings of 6th 1&2DM Workshop*, Bad Gastein, Austria, 2000, p. 302.

Appino C. et al., The energy loss components under alternating, elliptical and circular flux in non-oriented alloys, *Proceedings of 5th 1&2DM Workshop*, Grenoble, France, 1997, p. 55.

Appino C. et al., One-dimensional/two-dimensional loss measurements up to high inductions, *Journal of Applied Physics*, 105, 2009, 07E718.

Appino C. et al., International comparison on SST and Epstein frame measurements in grain-oriented Fe-Si sheet steel, *International Journal of Applied Electromagnetics and Mechanics*, 48, 2015, 123.

ArcelorMittal, *Guide to Electrical Steels and Magnetic Circuit Cores*, 2011.

ASTM, *Standard Test Methods for Alternating-Current Magnetic Properties of Materials at Power Frequencies Using Sheet-Type Test Specimens*, A804/A804M-04, 2009.

ASTM, *Standard Test Method for Alternating-Current Magnetic Properties of Materials at Power Frequencies Using Wattmeter-Ammeter-Voltmeter Method and 25-cm Epstein Test Frame*, A343/A343M-14, 2014.

ASTM Standard A772, *Standard Test Method for AC Magnetic Permeability of Materials Using Sinusoidal Current*, 2011.

Astrom K.J. et al., *Feedback Systems: An Introduction for Scientists and Engineers*, Princeton University Press, Princeton, NJ, 2010, ISBN 9781400828739.

ATI, *Grain-Oriented Electrical Steel*, Technical data sheet, Allegheny Technologies Incorporated, 2014.

Augutis V. et al., Determination of metal surface hardened layer depth using magnetic Barkhausen noise, *Materials Science*, 12 (1), 2006, 84.

Aumala O., Fundamentals and trends of digital measurement, *Measurement*, 26, 1999, 45.

Auty F. et al., *Beginner's Guide to Measurement in Electronic and Electrical Engineering, Good Practice Guide No 132*, National Physical Laboratory, The Institution of Engineering and Technology, Teddington, UK, 2014.

Avrunin J., *Not so High on High-Resolution*, 2013, accessed online 6 June 2013, http://tek.com/blog/not-so-high-high-resolution

Baily F.G., The hysteresis of iron and steel in a rotating magnetic field, *Philosophical Transactions of Royal Society of London, Series A*, 187, 1896, 715.

Bajorek R. et al., *Magnetic Measurement System MAG–TD100*, Operation manual, R&J Measurement, Wroclaw, Poland, 1999.

Bajorek J. et al., Non-destructive method of magnetic permeability control of non-magnetic machine elements, *Journal of Magnetism and Magnetic Materials*, 215-215, 2000a, 726.

Bajorek R. et al., A compact single sheet tester for investigation of both alternating and rotational magnetisation, *Proceedings of 6th 1&2 Workshop*, Bad Gastein, Austria, 2000b, p. 80.

Baker B., *Use an RC Filter to Deglitch A DAC*, 2011, accessed online 7 February 2016, http://edn.com/electronics-blogs/bakers-best/4368256/use-an-rc-filter-to-deglitch-a-dac

Bakshi U.A. et al., *Control systems*, Technical Publications, Pune, India, 2010, ISBN 9788184317770.

Baldwin J.A. Jr., A controlled-flux hysteresis loop tracer, *Review of Scientific Instruments*, 41, 1970, 468.

Ballou R. et al., Helimagnetism in the cubic Laves phase YMn2, *Journal Magnetism and Magnetic Materials*, 70 (1–3), 1987, 129.

Baran J.W. et al., *Electrical Engineer's Guide, Vol. 1,* (in Polish: *Poradnik Inzyniera Elektryka, Tom 1*), Wydawnictwa Naukowo-Techniczne, Warsaw, Poland, 1996, ISBN 8320419662.

Baranowski S. et al., Comparison of digital feedback methods of the control of flux -density shape, *Przeglad Elektrotechniczny (Electrical Review)*, R. 85 (Nr 1/2009), 2009, 93.

Barbisio E. et al., Accurate measurement of magnetic power losses and hysteresis loops under generic induction waveforms with minor loops, *Proceedings of Soft Magnetic Materials Conference*, Dusseldorf, Germany, 2003, p. 257.

Basak et al., Influence of stress on rotational loss in silicon iron, *Proceedings of the IEE*, 152 (2), 1978, 165.

Basak A., Rowe D.M., Anayi F.J., Thin film senses magnetic flux and loss in rotary electric motors, *IEEE Transactions on Magnetics*, 33 (5), 1997, 3382.

Beattie R., *The Hysteresis of Nickel and Cobalt in a Rotating Magnetic Field*, Philosophical Magazine, 1 (6), 1901, p. 642.

Beckley P., *Electrical Steels, a Handbook for Producers and Users*, European Electrical Steels, Newport, United kingdom, 2000a, ISBN 0-9540039-0-X.

Beckley P., Industrial magnetic measurements, *Journal of Magnetism and Magnetic Materials*, 215-216, 2000b, 664.

Beckley P. et al., Single sided non-enwrapping power loss testers, *Proceedings of 6th 1&2DM*, Bad Gastein, Austria, 2000c, p. 19.

Benedetti G., Askoll, Italy, *Private Communication*, 2012.

Bertotti G., General properties of power losses in soft ferromagnetic materials, *IEEE Transactions on Magnetics*, 24 (1), 1988, 621.

Bertotti G. et al., Loss measurements on amorphous alloys under sinusoidal and distorted induction waveform using a digital feedback technique, *Journal of Applied Physics*, 73 (10), 1993, 5375.

Birkelbach G. et al., Very low frequency magnetic hysteresis measurements with well-defined dependence of the flux density, *IEEE Transactions on Magnetics*, 22 (5), 1986, 505.

Blundell M.G. et al., A new method of measuring power loss of magnetic materials under sinusoidal flux conditions, *Journal of Magnetism and Magnetic Materials*, 19, 1980, 243.

Boon C.R. et al., Rotational hysteresis loss in single-crystal silicon-iron, *Proceedings of the IEE*, 111 (3), 1964, 605.

Boon C.R. et al., Alternating and rotational power loss at 50 c/s in 3% silicon-iron sheets, *Proceedings of the IEE*, 112 (11), 1965, 2147.

Borg Bartolo J. et al., An investigation into the geometric parameters affecting field uniformity in four pole magnetisers, *International Journal of Applied Electromagnetics and Mechanics*, 48 (2–3), 2015, 225.

Bottauscio O. et al., Space and time distribution of magnetic field in 2D magnetizers, *Przeglad Elektrotechniczny (Electrical Review)*, R. 81 (Nr 5/2005), 2005, 8.

Bozorth R.M., *Ferromagnetism*, IEEE Press, New Jersey, 2003, ISBN 0-7803-1032-2.

Brailsford F., Rotational hysteresis loss in electrical sheet steels, *Journal IEE*, 83, 1938, 566.

Brailsford F., Investigation of the eddy-current anomaly in electrical sheet steels, *Journal IEE*, 95 (II), 1948, 38.

Braisford F., *Magnetic Materials*, Methuen & Co. Ltd., London, UK, 1954.

Brix W. et al., Method for the measurement of rotational power loss and related properties in electrical steel sheets, *IEEE Transactions on Magnetics*, 18 (6), 1982, 1469.

Brix W. et al., Improved method for the investigation of the rotational magnetization process in electrical steel sheets, *IEEE Transactions on Magnetics*, 20 (5), 1984, 1708.

Brockhaus Measurements, *Private Communication*, 2014.

Brockhaus Measurements, *Power Amplifier PA 100 Series*, Data sheet, 2016.

Brockhaus Measurements, *Domain Viewer DV90*, 2017, accessed online 17 February 2017, http://brockhaus.com

Burr-Brown, *PGA 204, PGA205, Programmable Gain Instrumentation Amplifier*, Data sheet, 1993.

Buschow K.H.J. et al., *Physics of Magnetism and Magnetic Materials*, Plenum Publishers, New York, 2003, ISBN 0306474212.

Carriage T., Deputy Head of Calibration Laboratory at Megger Instruments Ltd, UK, *Private Communication*, 2017.

Chady T. et al., Electromagnetic transducer with rotational excitation field for evaluation of fatigue and stress loaded steel samples, *IEEE Transactions on Magnetics*, 45 (10), 2009, 3897.

Chattock A.P., On magnetic potentiometer, *Philosophical Magazine, Ser. 5*, 24 (146), 1887, 23.

Chen L. et al., Measurement of rotational magnetic properties of nanocrystalline alloys by a modified B-H sensor, *AIP Advances*, 7 (5), 2017, 056614, DOI: 10.1063/1.4973595.

Chikazumi S., *Physics of Ferromagnetism*, 2nd edition, Oxford Science Publications, Oxford, 2002, ISBN 0-19-851776-9.

Chukwuchekwa N. et al., Barkhausen noise in grain-oriented 3% Si-Fe at 50 Hz, *Journal of Electrical Engineering*, 61 (7/s), 2010, 69.

Chukwuchekwa N., *Investigation of Magnetic Properties and Barkhausen Noise of Electrical Steel*, PhD thesis, Wolfson Centre for Magnetics, Cardiff University, Cardiff, UK, 2011.

Cochran P., *Polyphase Induction Motors, Analysis, Design and Application*, Marcel Dekker, New York, 1989, ISBN 0-8247-8043-4.

Cogent, *Electrical Steel Grain Oriented Unisil*, Unisil-H, Product Catalogue, 2002a.

Cogent, *Electrical Steel Non Oriented Fully Processed*, Product Catalogue, 2002b.

Colliss D. et al., Evidence for superparamagnetism within a molecular sieve: Magnetic and Mossbauer effect studies, *IEEE Transactions on Magnetics*, 4 (3), 1968, 470.

Cook R.R., *Assessment of Uncertainties of Measurement for Calibration & Testing Laboratories*, 2nd edition, National Association of Testing Authorities, Australia, 2002, ISBN 0-909307-46-6.

Cooper B.A. et al., Night vision and thermal imaging equipment, *Proceedings of National Avian – Wind Power Planning Meeting III*, San Diego, California, May 1998, p. 164.

Cox J.F., *Fundamentals of Linear Electronics: Integrated and Discrete*, Delmar / Thomson Learning, 2002, ISBN 0-7668-3018-7.

Cox M.G. et al., *Software Support for Metrology, Best Practice Guide No. 6*, NPL Report MS 6, NPL, 2010.

Crevecoeur G. et al., Analysis of the local material degradation near cutting edges of electrical steel sheets, *IEEE Transactions on Magnetics*, 44 (11), 2008, 3137.

Crowell L.B., *Quantum Fluctuations of Spacetime, World Scientific Series in Contemporary Chemical Physics, Vol. 25*, World Scientific Publishing, Singapore, 2005, ISBN 981-256-515-9.

Crown, *Macro-Tech MA-3600VZ, Power Amplifier, Service Manual*, 2000.

Crown, *Macro-Tech MA-3600VZ, Power Amplifier, Operation Manual*, 2007.

Crown, 2016, accessed online 1 February 2016, http://crownaudio.com

Czeija E. et al., Apparatus for measuring the change of magnetic flux or the change of flux in ferromagnetic material from the induced voltage (in German: *Vorrichtung zum Messen des Wechselinduktionsflusses oder der Flussanderung in ferromagnetischen Materialien aus der Induktionsspannung*), Osterreichisches Patentamt Patentschrift Nr. 180990, 1955.

Dami M.A. et al., New approach to characterisation of electrical alloys under rotating magnetic flux density, *IEE Proceedings, Science, Measurements and Technology*, 143 (6), 1996, 399.

de la Barriere O. et al., A novel magnetizer for 2D broadband characterization of steel sheets in soft magnetic composites, *International Journal of Applied Electromagnetics and Mechanics*, 48 (2–3), 2015, 239.

De Wulf M. et al., Real-time controlled arbitrary excitation for identification of electromagnetic properties of non-oriented steel, *Journal de Physique IV, France*, 8, 1998, Pr2-705.

Derebasi N. et al., Computerised DC bridge method of thermistor measurement of localised power loss in magnetic materials, *IEEE Transactions on Magnetics*, 28 (5), 1992, 2467.

Dhar A. et al., Magnetizing frequency dependence of magneto-acoustic emission in pipeline steel, *IEEE Transactions on Magnetics*, 28 (2), 1992, 1003.

Dieterly D., DC permeability testing of Epstein samples with double-lap joints, *Proceedings of 51st Annual Meeting of ASTM*, Detroit, MI, 1948, p. 39.

Drake A.E., Precise total power loss measurements on electrical sheet steels, *Journal of Magnetism and Magnetic Materials*, 26, 1982, 181.

Drake T., Magnetic Section at National Physical Laboratory, United Kingdom, *Private Communication*, 2016.

Dytran, *Dual Element Accelerometer, Model Number 7705A1, Performance Specifications*, 2013.

Dytran, *Introduction to Charge Mode Accelerometers*, 2014, accessed online 3 December 2014, http://dytran.com/assets/pdf/introduction%20to%20charge%20mode%20accelerometers.pdf

Ebrahimi H. et al., Coupled magneto-mechanical analysis considering permeability variation by stress due to both magnetostriction and electromagnetism, *IEEE Transactions on Magnetics*, 49 (5), 2013, 1621.

Elektrisola, *Enamelled Copper Wire acc to IEC – Europe*, 2016, accessed online 25 February 2016, http://elektrisola.com/enamelled-wire/enamelled-wire-types/iec/europe.html

Elmore W.C. et al., *Physics of Waves*, Dover Publications, New York, 1985, ISBN 0-486-64926-1.

Emura M. et al., Angular dependence of magnetic properties of 2% silicon electrical steel, *Journal of Magnetism and Magnetic Materials*, 226–230, 2001, 1524.

Engdahl S.G. et al., Measurements and modelling of 2-D magnetization and magnetoelasticity in silicon iron, *Proceedings of 5th 2DM Workshop*, Grenoble, France, 1997, p. 1.

Enokizono M. et al., Rotational power loss of silicon steel sheet, *IEEE Transactions on Magnetics*, 26 (5), 1990a, 2562.

Enokizono M. et al., Measurement of dynamic magnetostriction under rotating magnetic field, *IEEE Transactions on Magnetics*, 26 (5), 1990b, 2067.

Enokizono M. et al., Optimum yoke construction for rotational power loss measurements apparatus, *Anales de Fisica, Series B*, 86, 1990c, 320.

Enokizono M., Studies on two-dimensional magnetic measurement and properties of electrical steel sheets at Oita University, *Proceedings of 1st 1&2DM Workshop*, Braunschweig, Germany, 1992, p. 82.

Enokizono M. et al., The problem of simplified two-dimensional magnetic measurement apparatus, *Proceedings of 5th 2DM Workshop*, Grenoble, France, 1997, p. 87.

Enokizono M. et al., Chaotic behavior of Barkhausen noise induced in silicon steel sheets, *IEEE Transactions on Magnetics*, 35 (5), 1999, 3421.

Enokizono M. et al., Two-dimensional magnetic property under DC-biased field, *Proceedings of 6th 1&2DM Workshop*, Bad Gastein, Austria, 2000, p. 215.

Enokizono M. et al., A measurement system for two-dimensional DC-biased properties of magnetic materials, *Journal Magnetism and Magnetic Materials*, 254–255, 2003, 39.

Ewing J.A., *Magnetic Induction in Iron and Other Metals*, 3rd edition, Van Nostrand Company, London, 1900.

Fiorillo F. et al., Extended induction range analysis of rotational losses in soft magnetic materials, *IEEE Transactions on Magnetics*, 24 (2), 1988, 1960.

Fiorillo F. et al., Power losses under sinusoidal, trapezoidal and distorted induction waveform, *IEEE Transactions on Magnetics*, 26 (5), 1990, 2559.

Fiorillo F. et al., The measurement of rotational losses at I.E.N.: Use of the thermometric method, *Proceedings of 1st 1&2DM Workshop*, Braunschweig, Germany, 1992a, p. 162.

Fiorillo F., A phenomenological approach to rotational power losses in soft magnetic laminations, *Proceedings of 1st 1&2DM Workshop*, Braunschweig, Germany, 1992b, p. 11.

Fiorillo F., *Measurement and Characterization of Magnetic Materials*, Elsevier Series in Electromagnetism, Elsevier Academic Press, Amsterdam, 2004, ISBN 0122572513.

Fiorillo F. et al., Soft magnetic materials, in: Webster J. (ed.), *Wiley Encyclopedia of Electrical and Electronics Engineering*, John Wiley & Sons, New York, 2016, DOI: 10.1002/047134608X.W4504.pub2.

Flanagan W.M., *Handbook of Transformer Design & Applications*, 2nd edition, McGraw Hill, 1993, ISBN 0070212910.

Flanders P.J., A Hall sensing magnetometer for measuring magnetization, anisotropy, rotational loss and time effects, *IEEE Transactions on Magnetics*, 21 (5), 1985, 1584.

FLIR, *Thermal Imaging Camera X8000sc*, Data sheet, 2014.

FLIR, *Advanced Thermal Imaging Cameras for R&D and Science Applications*, 2017, accessed online 17 February 2017, http://flir.com

Fonteyn K. et al., Measurement of magnetic properties under alternating field with a vertical rotational single sheet tester, *Przeglad Elektrotechniczny (Electrical Review)*, R. 85 (Nr 1/2009), 2009, 52.

Fonteyn K.A., *Energy-Based Magneto-Mechanical Model for Electrical Steel Sheet*, PhD thesis, Aalto University, Espoo, Finland, 2010.

Fornasini P., *The Uncertainty in Physical Measurements, an Introduction to Data Analysis in the Physics Laboratory*, Springer Science & Business Media, New York, 2008, ISBN 978-0-387-78649-0.

Fulmek P.L. et al., Energetic model of ferromagnetic hysteresis: Magnetization of grain oriented steel sheets in asymmetric directions, *Proceedings for 5th 1&2DM Workshop*, Grenoble, France, 1997, p. 17.

Garcia J.A. et al., A quasi-static magnetic hysteresis loop measurement system with drift correction, *IEEE Transactions on Magnetics*, 42 (1), 2006, 15.

Garshelis I.J. et al., Recovery of magnetostriction values from the stress dependence of Young's modulus, *IEEE Transactions on Magnetics*, 22 (5), 1986, 436.

Gaworska-Koniarek D. et al., Magnetic field strength sensor, *Symposium of Magnetic Measurements and Modeling*, Czestochowa-Siewierz, Poland, 2016.

Geirinhas Ramos H. et al., Studies of two-dimensional magnetic phenomena in electrical steel sheets at LME/IST, *Proceedings of 4th 2DM Workshop*, Cardiff, UK, 1995, paper 8.

Goldman A., *Handbook of Modern Ferromagnetic Materials*, Kluwer Academic Publishers, Boston/Dordrecht/London, 1999, ISBN 0-412-14661-4.

Gorican V. et al., 2-D measurements of magnetic properties using a round RSST, *Proceedings of 6th 1&2DM Workshop*, Bad Gastein, Austria, 2000a, p. 66.

Gorican V. et al., The measurement of power losses at high magnetic field densities or at small cross-section of test specimen using the averaging, *Journal of Magnetism and Magnetic Materials*, 215–216, 2000b, 693.

Gorican V. et al., The influence of terrestrial magnetism on power loss measurements in case of annealed amorphous magnetic material. *Proceedings of 6th 1&2DM Workshop*, Bad Gastein, Austria, 2000c, p. 204.

Gorican V. et al., Performance of round rotational single sheet tester (RRSST) at higher flux densities in the case of GO materials, *Proceedings of 7th 1&2DM Workshop*, Ludenscheid, Germany, 2002, p. 143.

Gorican V. et al., Unreliable determination of vector B in 2-D SST, *Journal of Magnetism and Magnetic Materials*, 254–255, 2003, 130.

Goss N.P., *Electrical Sheet and Method and Apparatus for Its Manufacture and Test*, 1934, US Patent 1,965,559.

Goss N.P., New development in electrical strip steels characterized by fine grain structure approaching the properties of a single crystal, *Transactions of American Society for Metals*, 23, 1935, 511.

Gozdur R., Determination of quasi-static hysteresis loop of electrical steel (in Polish: *Wyznaczanie quasi-statycznej petli histerezy blach elektrotechnicznych*), *Przeglad Elektrotechniczny (Electrical Review)*, R. 80 (Nr 2/2004), 2004, 147.

Gozdur R. et al., Power loss measurements of nanocrystalline and amorphous magnetic cores, *Zeszyty Problemowe, Maszyny Elektryczne* (Nr 100/2013 cz. II), 2013, 175.

Grimwood G.C. et al., Rotational hysteresis in polycrystalline alloys, *IEEE Transactions on Magnetics*, 14 (5), 1978, 359.

Grossinger R. et al., Accurate measurement of the magnetostriction of soft magnetic materials, *Proceedings of 6th 1&2DM Workshop*, Bad Gastein, Austria, 2000, p. 35.

Grote N. et al., Measurement of the magnetic properties of soft magnetic material using digital real-time current control, in: Kose V. and Sievert J. (eds.), *Non-linear Electromagnetic Systems*, IOS Press, Amsterdam, 1998, p. 56.

Gubbins D. et al., *Encyclopedia of Geomagnetism and Paleomagnetism*, Springer, Dordrecht, 2007, ISBN 978-1-4020-3992-8.

Gubbiotti G. et al., Magnetostatic interaction in arrays of nanometric permalloy wires: A magneto-optic Kerr effect and a Brillouin light scattering study, *Physical Review B*, 72 (224413), 2005, 224413-1.

Gumaidh A.M. et al., Measurement and analysis of rotational power loss in soft magnetic materials, *Proceedings of 1st 2DM Workshop*, Braunschweig, Germany, 1993a, p. 173.

Gumaidh A.M. et al., Characterisation of magnetic materials under two-dimensional excitation, *IEEE Intermag Conference Digest*, 1993b, paper CD-06.

Guo Y. et al., Measurement and modeling of rotational core losses of soft magnetic materials used in electrical machines: A review, *IEEE Transactions on Magnetics*, 44 (2), 2008, 279.

Guo Y. et al., Investigation of magnetic properties of magneto-rheological fluids under rotating magnetic excitation, *Proceedings of 14th 1&2DM Workshop*, Tianjin, China, 2016, p. 135.

Hadoud S.A., *Three Dimensional Rotational Power Loss Measurement Using Computerised Data Acquisition System*, PhD Thesis, Cardiff University, Cardiff, UK, 1998.

Hadoud S.A. et al., Characterisation of electrical steels under three-dimensional excitation, *Proceedings of 5th 1&2DM Workshop*, Grenoble, France, 1997, p. 127.

Hall J., *Evaluation of Residual Stress in Electrical Steel*, PhD thesis, School of Engineering, Cardiff University, Cardiff, UK, 2001.

Hall J., Assessment of the effect of lift-off on a magnetic flux injection technique for detection of residual curvature in electrical steel, *Przeglad Elektrotechniczny (Electrical Review)*, R. 87 (No. 9b/2011), 2011a, 79.

Hall B., Don't forget the degrees of freedom: Evaluating uncertainty from small numbers of repeated measurements, *Presented at ANAMET Meeting*, October, 2011b.

Hall M. et al., Obtaining the d.c. properties of soft magnetic materials using an open circuit measurement technique, *Przeglad Elektrotechniczny (Electrical Review)*, R. 85 (Nr 1/2009), 2009, 28.

Hanks J., *Five Critical Accuracy Technologies for Data Acquisition, National Instruments*, 2016, accessed online 20 February 2016, http://dataweek.co.za/news.aspx?pklnewsid=13677

Harrison S.A. et al., Rotational hysteresis losses in isotropic media, *IEEE Transactions on Magnetics*, 35 (5), 1999, 3962.

Hartmann K., *Relationships between Barkhausen Noise, Power Loss and Magnetostriction in Grain-oriented Silicon Iron*, PhD thesis, Wolfson Centre for Magnetics, Cardiff University, Cardiff, UK, 2003.

Hasan S. et al., Effect of DC bias on magnetization current waveforms of single phase power transformer, *Proceedings of Power Engineering and Optimization Conference (PEDCO) Melaka*, Malaysia, 2012, p. 496.

Hasenzagl A. et al., A 3-phase excited test system for simultaneous studies of field vectors, losses, magnetostriction and domains for multi-directional magnetisation, *Proceedings of 4th 2D Workshop*, Cardiff, UK, 1995, paper 7.

Hasenzagl A. et al., Novel 3-phase excited single sheet tester for rotational magnetization, *Journal of Magnetism and Magnetic Materials*, 160 (1), 1996, 180.

Hashi S. et al. Study on the deformation of 3% Si-Fe single crystal with magnetic field being deviated from [001], *IEEE Transactions on Magnetics*, 32 (5), 1996, 4848.

Hempel K.H., Final discussion, *Proceedings of 1st 1&2DM Workshop*, Braunschweig, Germany, 1992, p. 207.

Hempel K.A. et al., A phenomenological description of the anisotropic and non linear properties of electrical sheet under general two-dimensional magnetic excitation by means of reluctance tensor, *Proceedings of 5th 2DM Workshop*, Grenoble, France, 1997, p. 93.

Henney K. (ed.), *Radio Engineering Handbook*, McGraw-Hill Book Company, 1950.

Hilzinger R. et al., *Magnetic Materials, Fundamentals, Products, Properties, Applications*, Publicis Publishing, Erlangen, 2013, ISBN 9783895783524.

Hoganas, *Powder News*, Special issue PowderMet 2009, Las Vegas, USA, 2009.

Hollitscher H., Core losses in magnetic materials at very high flux densities when the flux is not sinusoidal, *IEEE Transactions on Magnetics*, 5 (3), 1969, 642.

Honeywell, *1, 2 and 3 Axis Magnetic Sensors HMC1051/HMC1052/HMC1053*, Data sheet, 2006.

Honeywell, *1- and 2-Axis Magnetic Sensors HMC1001/1002/1021/1022*, Data sheet, 2008.

Horowitz P. et al., *The Art of Electronics*, 3rd edition, Cambridge University Press, 2015, ISBN 978-0521809269.

Hubert A. et al., *Magnetic Domains: The Analysis of Magnetic Microstructures*, Springer, Berlin Heidelberg, New York, 1998, ISBN 9783540641087.

Hubert O. et al., A new experimental set-up for the characterisation of magneto-mechanical behaviour of materials submitted to biaxial stresses: Application to FeCo alloys, *Przeglad Elektrotechniczny (Electrical Review)*, R. 81 (Nr 5/2005), 2005, 19.

Hurley W.G. et al., *Transformers and Inductors for Power Electronics: Theory, Design and Applications*, John Wiley & Sons, West Sussex, Uinted Kingdom, 2013, ISBN 9781118544679.

Ichijo N. et al., A new 2-D magnetic measurement method with vertical yokes, *Proceedings of 7th 1&2DM Workshop*, Bad Gastein, Austria, 2000, p. 197.

IEC 60404-1:2000, *Magnetic Materials, Part 1: Classification*, 2000.

IEC 60404-2:1998+A1:2008, *Magnetic Materials, Part 2: Methods of Measurement of the Magnetic Properties of Electrical Steel Strip and Sheet by Means of an Epstein Frame*, 2008.

IEC 60404-3:2010, *Magnetic Materials, Part 3: Methods of Measurement of the Magnetic Properties of Electrical Steel Strip and Sheet by Means of a Single Sheet Tester*, 2010.

IEC 60404-6:2003, *Magnetic Materials, Part 6: Methods of Measurement of the Magnetic Properties of Magnetically Soft Metallic and Powder Materials at Frequencies in the Range 20 Hz to 200 kHz by the Use of Ring Specimens*, 2003.

Imamura M. et al. Magnetization process and magnetostriction of a four percent Si-Fe single crystal close to (110)[001], *Transactions on Magnetics*, 17 (5), 1981, 2497.

Ionita V. et al., Experimental validation of electromagnetic field computation in magnetic materials, *IEEE Transactions on Magnetics*, 44 (6), 2008, 882.

Ishihara Y., Measuring method of magnetic characteristics in any direction for silicon steel, *Proceedings of 1st 1&2DM Workshop*, Braunschweig, Germany, 1992, p. 204.

Ishihara Y. et al., Magnetic properties of electrical steel sheets measured by RSST and SST, *Proceedings of 3rd 2DM Workshop*, Turin, Italy, 1993.

Ishihara Y. et al., Comparison of two averaging methods for improving the measurement accuracy of power loss, *Journal of Magnetism and Magnetic Materials*, 215–216, 2000, 696.

Ishikawa S. et al., AC magnetic properties of electrical steel sheet under two-dimensional DC-biased magnetization, *IEEE Transactions on Magnetics*, 48 (4), 2012, 1413.

Ivanyi A. et al., 2D/3D models for a three-phase fed single sheet tester, *IEEE Transactions on Magnetics*, 34 (5), 1998, 3004.

Ivashov A., *Smath Studio, Units*, 2017, accessed online 27 January 2017, http://smath.info/wiki/units.ashx

Jacob J.M., *Advanced AC Circuits and Electronics: Principles & Applications*, Thomson / Delmar Learning, 2004, ISBN 0-7668-2330-X.

Jander A. et al., Magnetoresistive sensors for nondestructive evaluation, *Presented at 10th International Symposium on Nondestructive Evaluation for Health Monitoring and Diagnostics*, 2005.

JCGM, Evaluation of measurement data, Guide to the expression of uncertainty in measurement, *Joint Committee for Guides in Metrology, JCGM 10: 2008*, 2008.

Jerri A.J., The Gibbs phenomenon in Fourier analysis, splines and wavelet approximations, Springer Science + Business Media, 1998, ISBN 978-1-4419-4800-7.

Jesenik M. et al., Enlargement of the rotational field homogeneity area in a two-phase round rotational single sheet tester, *Informacije MIDEM*, 32 (2), 2002, 100.

Jesenik M. et al., Field homogeneity in a two-phase rotational single sheet tester with square sample, *IEEE Transactions on Magnetics*, 39 (3), 2003a, 1495.

Jesenik M. et al., Calculation of the rotational magnetic fields in the sample of the rotational single sheet testers, *Przeglad Elektrotechniczny (Electrical Review)*, R. LXXIX (12/2003), 2003b, 920.

Jezierski E., *Transformers* (in Polish: *Transformatory*), Wydawnictwa Naukowo-Techniczne, 1975.

JFE, *Electrical Steel Sheets, JFE G-Core, JFE N-Core, JFE Steel Corporation*, 2016, accessed online 10 May 2016, http://jfe-steel.co.jp/en/products/electrical/catalog/f1e-001.pdf

Jiles D., *Introduction to Magnetism and Magnetic Materials*, Taylor & Francis, New York, NY, 1998, ISBN 0412798603.

Johnk C.T.A., *Engineering Electromagnetic Fields & Waves*, Willey International Edition, New York, NY, 1975, ISBN 9780471442905.

Kai Y. et al., Influence of stress on vector magnetic property under alternating magnetic flux conditions, *IEEE Transactions on Magnetics*, 47 (10), 2011a, 4344.

Kai Y., Measurement of vector magnetic property under stress along arbitrary direction in non-oriented electrical steel sheet, *Przeglad Elektrotechniczny (Electrical Review)*, R. 87 (No. 9b/2011), 2011b, 101.

Kai Y. et al., Influence of shear stress on vector magnetic properties of non-oriented electrical steel sheet, *Proceedings of 12th 1&2DM*, Vienna, Austria, 2012, p. 65.

Kai Y. et al., Measurement of two-dimensional magnetostriction of a non-oriented electrical steel sheet under shear stress, *International Journal of Applied Electromagnetics and Mechanics*, 48 (2–3), 2015, 233.

Kanada T. et al., Distributions on localised iron loss of three-phase amorphous transformer model core by using two-dimensional magnetic sensor, *IEEE Transactions on Magnetics*, 32 (5), 1996, 4797.

Kanada T. et al., Magnetic properties of soft magnetic materials under tensile and compressive stress, *Przeglad Elektrotechniczny (Electrical Review)*, R. 87 (No. 9b/2011), 2011, 93.

Kaplan A., Magnetic core losses resulting from a rotating flux, *Journal of Applied Physics*, 32 (3), 1961, 320.

Kedous-Lebouc A. et al., On the magnetization process in electrical steel in unidirectional and rotational field, *Proceedings of 1st 2DM Workshop, PTB*, Braunschweig, Germany, 1992, p. 36.

Kedous-Lebouc A. et al., Measurements of 2D magnetic properties at LEG: Comparison between a large and a small RSST, *Proceedings of 4th 1&2DM Workshop*, Cardiff, UK, 1995, paper 3.

Keightley J., *A Brief Workshop on Measurement Uncertainty*, Radionuclide Calibrator Users' Forum, National Physical Laboratory, Teddington, UK, 2008.

Khachan J., Computer-assisted magnetic core loss measurements on single strips of metallic glass, *Review of Scientific Measurements*, 63 (5), 1992, 3222.

Kepco, *Operator's Manual, BOP (M) (D) 100W, 200W, 400W Bipolar Power Supply*, 2015.

Kepco, *KLN Series*, 2016, accessed online 3 February 2016, http://kepcopower.com/kln.htm

KERN, *PFB, Operating Manual, Precision Balance*, Technical specification, 2013.

Keysight Technologies, *3410A and 34411A Multimeters*, Data sheet, 2015.

Khanlou A. et al., A non-enwrapping, single yoke on-line magnetic testing system for use in the steel industry, *Proceedings of Intermag*, 1992, p. 161.

Kidd M.L., Watch out for those thermoelectric voltages! Metrology 101, *Cal Lab: The International Journal of Metrology*, 19 (2), 2012, 18.

Kim K.U. et al., *High Frequency Measuring System for Magnetic Properties of Materials*, US Patent 6100685, 2000.

Kimura Y. et al., Investigation of a measurement method of 2-D magnetic properties by means of 1-D single sheet tester, *1&2DM Workshop*, Tianjin, China, 2016.

Kistler, *8630C & 8636C PiezoBEAM Accelerometers*, Data sheet, 2012.

Kitz E. et al., Sheet sensor design for flux density distribution analysis in machine cores, *Presented at Soft Magnetic Materials Conference*, Dusseldorf, Germany, 2003.

Klimczyk P., *Novel Techniques for Characterisation and Control of Magnetostriction in G.O.S.S.*, PhD thesis, Wolfson Centre for Magnetics, Cardiff University, Cardiff, UK, 2012.

Klimczyk P. et al., Comparison of uniaxial and rotational magnetostriction of non-oriented and grain-oriented electrical steel, *Przeglad Elektrotechniczny (Electrical Review)*, R. 87 (No. 9b/2011), 2011, 33.

Kollar M. et al., Stressed amorphous ribbons under in-plane magnetizing field, *Proceedings of 6th 1&2DM Workshop*, Bad Gastein, Austria, 2000, p. 248.

Konadu S.N., *Non-Destructive Testing and Surface Evaluation of Electrical Steels*, PhD thesis, Wolfson Centre for Magnetics, Cardiff University, Cardiff, UK, 2006.

Kondo T. et al., The iron loss characteristics of the magnetic material under the distorted flux waveforms, *Proceedings of Soft Magnetic Materials Conference*, Dusseldorf, Germany, 2003, p. 293.

Konishi S. et al., Domains and domain-wall motion in grain-oriented 50-percent Ni-Fe tapes, *IEEE Transactions on Magnetics*, 6 (1), 1970, 105.

Kornetzki M. et al., On the theory of hysteresis loss in rotating magnetic field (in German: *Zur Theorie der Hystereseverluste im magnetischen Drehfeld*), *Zeitschrift fur Physik*, Bd., 142, 1955, 70.

Kottler V. et al., Imaging of magnetic domains in thin Co/Pt and CoNi/Pt multilayers by near field magneto-optical circular dichroism, *IEEE Transactions on Magnetics*, 34 (4), 1998, 2012.

Krah J. et al., Experimental verification of alternating audio frequency 2D magnetization set-up for soft magnetic sheets and foils, *Proceedings of 7th 1&2DM Workshop*, Ludenscheid, Germany, 2002, p. 103.

Krell C. et al., Rotational single sheet testing on sample of arbitrary size and shape, *Proceedings of 6th 1&2DM Workshop*, Bad Gastein, Austria, 2000, p. 96.

Krell C. et al., Magnetostriction of different types of Fe-based soft magnetic alloys under rotational magnetization, *Przeglad Elektrotechniczny (Electrical Review)*, R. 81 (Nr 5/2005), 2005, 47.

Krismanic G. et al., A hand-held sensor for analyses of local distributions of magnetic fields and losses, *Journal of Magnetism and Magnetic Materials*, 215–216 (2), 2000, 720.

Kryder M.H. et al., Kerr effect imaging of dynamic processes, *IEEE Transactions on Magnetics*, 26 (6), 1990, 2995.

Kubach R.W., *A Hysteresisgraph for Plotting Magnetization Curves*, Dayton University, National Technical Information Service, 1966.

Kuczmann M., Design of 2D RRSST system by FEM with T, Φ-Φ potential formulation, *Pollack Periodica, An International Journal for Engineering and Information Sciences*, 3 (1), 2008, 67.

Kuepferling M. et al., Rate dependence of the magnetocaloric effect in La-Fe-Si compounds, *Joint European Magnetic Symposia*, 40, 2013, 06010.

Kulite, *Kulite Sensing Technology*, 2014, accessed online 23 January 2014, http://kulite.com/docs/transducer_handbook/section2.pdf

LakeShore, *Fluxmeter model 480, User's Manual*, 2004.

LakeShore, *Model 460 3-Channel Gaussmeter, User's Manual*, 2009.

Lancarotte M.S. et al., Improving the magnetizing device design of the single sheet tester of two-dimensional properties, *Journal of Magnetism and Magnetic Materials*, 269 (3), 2004, 346.

Langford-Smith F. (ed.), *Radiotron Designer's Handbook*, Amalgamated Wireless Valve Company, Wireless Press, 1953.

Leicht J. et al., Hysteresis loss component under non-sinusoidal flux waveforms, *Journal of Magnetism and Magnetic Materials*, 320 (20), 2008, e608.

Leite J.V. et al., Analysis of a rotational single sheet tester using 3D finite element model taking into account hysteresis effect, *COMPEL International Journal of Computations and Mathematics in Electrical and Electronic Engineering*, 26 (4), 2007, 1037.

LEM, *CTSR 0.3; 0.6-P & -TP, Non-Contact and Accurate Milliamps Measurement Single Supplied*, 2014a, accessed online 11 May 2014, http://lem.com

LEM, *ITN 12-TP, Made for High Accuracy on Printed Circuits*, 2014b, accessed online 11 May 2014, http://lem.com

Leonowicz M. et al., *Modern Magnets, Technologies, Coercivity Mechanisms, Applications*, (in Polish: *Wspolczesne magnesy, technologie, mechnizmy koercji, zastosowania*), Wydawnictwa Naukowo-Techniczne, Warszawa, 2005, ISBN 83-204-3049-6.

Li Y. et al., Measurement of soft magnetic composite material using an improved 3-D tester with flexible excitation coils and novel sensing coils, *IEEE Transactions on Magnetics*, 46 (6), 2010, 1971.

Li Y. et al., Study on rotational hysteresis and core loss under three-dimensional magnetization, *IEEE Transactions on Magnetics*, 47 (10), 2011, 3520.

Li Y. et al., A novel 3-D magnetic properties tester with C-type cores, *Proceedings of 13th 1&2DM Workshop*, Turin, Italy, 2014a, p. 62.

Li Y. et al., Design and analysis of a novel 3-D magnetization structure for laminated silicon steel, *IEEE Transactions on Magnetics*, 50 (2), 2014b, 7009504.

Li Y. et al., Rotational core loss of silicon steel laminations based on three-dimensional magnetic properties measurement, *IEEE Transactions on Applied Superconductivity*, 26 (4), 2016, 8201205.

Lin G.C. et al., Direct measurement of the magnetocaloric effect in $La_{0.67}Ca_{0.33}MnO_3$, *Journal of Magnetism and Magnetic Materials*, 300 (2), 2006, 392.

Loisos G., *Novel Flux Density Measurement Methods of Examining the Influence of Cutting on Magnetic Properties of Electrical Steels*, PhD thesis, Wolfson Centre for Magnetics, Cardiff University, Cardiff, UK, 2002.

Loisos G. et al., Critical evaluation and limitations of localised flux density measurements in electrical steels, *IEEE Transactions on Magnetics*, 37 (4), 2001, 2755.

Loizos G., Schneider Electric, *Private Communication*, 2012.

Lord H.W., Dynamic hysteresis loop measuring equipment, *Electrical Engineering*, 71 (6), 1952, 518.

Lubelski K., *Theoretical Electrical Engineering, Part II, Edition III* (in Polish: *Elektrotechnika teoretyczna, cz. II, wydanie III*), Skrypt Uczelniany, Politechnika Czestochowska, Czestochowa, Poland, 1982.

Ludke J. et al., Hybrid control used to obtain sinusoidal curve form for the power loss measurement on magnetic materials, *Materials Science Forum*, 373–376, 2001, 469.

Lukacs E., *Probability and Mathematical Statistics: An Introduction*, Academic Press, New York and London, 2014, ISBN 9781483269207.

Lundgren A., *On Measurement and Modelling of 2D Magnetization and Magnetostriction of SiFe Sheets*, PhD thesis, Royal Institute of Technology, Electric Power Engineering, Stockholm, Sweden, 1999.

Lyke R.F., Modern magnetic testing equipment for electrical steels, *IEEE Transactions on Magnetics*, 18 (6), 1982, 1466.

Maeda Y. et al., Study of the counterclockwise/clockwise (CCW/CW) rotation problem with the measurement of 2-dimensional magnetic properties, *Przeglad Elektrotechniczny (Electrical Review)*, R. 83 (Nr 04/2007), 2007, 18.

Maeda Y. et al., Rotational power loss of magnetic steel sheet in a circular rotational magnetic field in CCW/CW directions, *Journal of Magnetism and Magnetic Materials*, 320 (20), 2008, e567.

Magni, A., Stroboscopic investigation of domain wall motion at high frequencies, *Proceedings of 13th 1&2DM Workshop*, Turin, Italy, 2014, p. 108.

Majocha A. et al., Bridge method for measurements of power loss in magnetically soft materials (in Polish: *Mostkowa metoda pomiaru stratnosci materialow magnetycznie miekkich*), *Przeglad Elektrotechniczny (Electrical Review)*, R. 86 (Nr 4/2010), 2010, 79.

Makaveev D. et al., Measurement system for 2D magnetic properties of electrical steel sheets: Design and performance, *Proceedings of 6th 1&2DM Workshop*, Bad Gastein, Austria, 2000a, p. 48.

Makaveev D. et al., Waveform control algorithm for rotational single sheet testers using system identification techniques, *Journal of Applied Physics*, 87 (9), 2000b, 5983.

Makaveev D. et al., Controlled circular magnetization of electrical steel in rotational single sheet tester, *IEEE Transactions on Magnetics*, 37 (4), 2001, 2740.

Malkinski L. et al., Simultaneous measurement of magnetic anisotropy and Barkhausen effect in Fe-Si steel discs rotating in a constant magnetic field, *Proceedings of 1&2DM Workshop*, Torino, 1993, p. 43.

Mancini R., *Op Amps for Everyone, Design Reference*, Texas Instruments, 2002.

Manteufel R.D., Sequential perturbation uncertainty perturbation in thermal-fluid applications, *Proceedings of ASME International Mechanical Engineering Congress and Exposition*, 2012, paper No. IMECE2012-88316, p. 281.

Marketos P., *Construction of Transformer Cores Using Consolidated Stacks of Electrical Steel*, PhD thesis, Wolfson Centre for Magnetics, Cardiff University, Cardiff, UK, 2002.

Marketos P. et al., Novel transformer core design using consolidated stacks of electrical steel, *IEEE Transactions on Magnetics*, 42 (10), 2006, 2821.

Marketos P. et al., A method for defining the mean path length of the Epstein frame, *IEEE Transactions on Magnetics*, 43 (6), 2007, 2755.

Mason H., *Basic Introduction to the Use of Magnetoresistive Sensors, Application Note 37*, Zetex Semiconductors, 2003.

Masui H., Influence of stress condition on initiation of magnetostriction in grain oriented silicon steel, *IEEE Transactions on Magnetics*, 31 (2), 1995, 930.

Mathcad, *Using Vectors Instead of Range Variables, PTC University Learning Exchange*, 2014, accessed online 5 March 2014, http://learningexchange.ptc.com

Matheisel Z., *Cold-Rolled Electrical Steels* (in Polish: *Blachy elektrotechniczne walcowane na zimno*), Wydawnictwa Naukowo-Techniczne, Warszawa, Poland, 1975.

MathWorks, *Cumulative Trapezoidal Numerical Integration, MatLab, Documentation Center*, 2013, accessed online 6 June 2013, http:// mathworks.co.uk

MathWorks, *Vectorisation, MatLab, Documentation Center*, 2014, accessed online 5 May 2014, http://mathworks.co.uk

MathWorks, *Data Acquisition Toolbox, Connect to Data Acquisition Cards, Devices, and Modules*, 2016, accessed online 25 February 2016, http://uk.mathworks.com/products/daq/index.html?s_tid=gn_loc_drop

Matsubara K. et al., Effects of impedance of primary winding and mutual inductor on distortion of flux waveform in single sheet tester, *IEEE Transactions on Magnetics*, 31 (6), 1995a, 3382.

Matsubara K. et al., Acceleration technique of waveform control for single sheet tester, *IEEE Transactions on Magnetics*, 31 (6), 1995b, 3400.

Matsubara K. et al., Effect of staircase output voltage waveform of a D/A converter on iron losses measured using an H coil, *Journal of Magnetism and Magnetic Materials*, 160, 1996, 185.

Matsuo H. et al., Measurement system of DC-biased magnetic properties, *Proceedings of Japanese-Bulgarian-Macedonian Joint Seminar on Applied Electromagnetics*, Ohrid, Macedonia, 2000, p. 65.

Matsuo T. et al., Measurements of transient vector hysteretic property of steel sheet, *Przeglad Elektrotechniczny (Electrical Review)*, R. 85 (Nr 1/2009), 2009, 60.

Matthews M., Head of Calibration Laboratory at Megger Instruments Ltd, UK, *Private Communication*, 2017.

Maxim Integrated, *Deglitching Techniques for High-Voltage R-2R DACs*, Application Note 583, 2001.

Maxim Integrated, *Implementing Cold-Junction Compensation in Thermocouple Applications*, Application Note 4026, 2007.

Maxim Integrated, *How to Select the Best Audio Amplifier for Your Design*, 2016, accessed online 2 February 2016, http://maximintegrated.com/en/app-notes/index.mvp/id/5590

Mazurek R., *Effects of Burrs on a Three Phase Transformer Core Including Local Loss, Total Loss and Flux Distribution*, PhD thesis, Cardiff University, Cardiff, UK, 2012.

Mazzetti P. et al., Electronic hysteresisgraph holds dB/dt constant, *Review of Scientific Instruments*, 37 (5), 1966, 548.

McQuade F.A. et al., Domain observations under two dimensional magnetisation, *Proceedings of 4th 2DM Workshop*, Cardiff, UK, 1995, paper. 6.

Measurement Specialties, *ELHS Load Cell*, Data sheet, 2012.

Meeker D.C., *Finite Element Method Magnetics, FEMM*, 2015, accessed online 8 June 2015, http:/femm.info

Mehnen L. et al., 2D magnetisation control by means of evolutionary algorithms, *Proceedings of 6th 1&2DM Workshop*, 2000, Bad Gastein, Austria, p. 122.

Melexis, *Triaxis Position Sensor IC MLX90360*, Data sheet, 2013.

Metglas (R), *2605 SA1 iron based alloy, material safety data sheet*, 2011a, accessed online 11 October 2011, http://metglas.com

Metglas, *Amorphous Alloys for Transformer Cores, Materials Mag!c, Hitachi Metals*, 2011b.

Micro-Measurements, *General Purpose Strain Gages – Linear Pattern*, Data sheet, 2010.

Micro-Measurements, *Strain Gage Selection: Criteria, Procedures, Recommendations*, Tech Note TN-505-4, 2014.

Microsemi, *Park, Inverse Park and Clark, Inverse Clark Transformations MSS Software Implementation, User Guide*, 2013.

Mimura M. et al., Examination of precise measurement of DC magnetic properties of Permalloy under low flux density more than a few mT, *IEEE Transactions on Magnetics*, 48 (11), 2012, 3614.

Minov B. et al., Development of magnetic after-effect setup and application in the study of relaxation processes in Fe-C, Fe-Cu and Fe-Cr alloys, *Przeglad Elektrotechniczny (Electrical Review)*, R. 87 (Nr 9b), 2011, 120.

Mistras, *Micro30 Sensor, Product Data Sheet*, Physical Acoustics Corporation, 2011.

Miura N., *Physics of Semiconductors in High Magnetic Fields*, Oxford University Press, Oxford, New York 2008, ISBN 9780198517566.

Miyagi D. et al., Study on measurement method of 2 dimensional magnetic properties of electrical steel using diagonal exciting coil, *Przeglad Elektrotechniczny (Electrical Review)*, R. 85 (Nr 1/2009), 2009, 47.

Miyagi D. et al., Measurement of magnetic properties of nonoriented electrical steel sheet at liquid nitrogen temperature using single sheet tester, *IEEE Transactions on Magnetics*, 46 (2), 2010, 314.

Miyamoto M. et al., Measurements of vector hysteretic property of silicon steel sheets at liquid nitrogen temperature, *Przeglad Elektrotechniczny (Electrical Review)*, R. 87 (No. 9b/2011), 2011, 111.

Moffat R.J., Using uncertainty analysis in the planning of an experiment, *Journal of Fluids Engineering*, 107 (2), 1985, 173.

Moghadam A.T. et al., Comparison of flux distribution in three-phase transformer cores assembled from amorphous material and grain-oriented silicon iron, *IEEE Transactions on Magnetics*, 25 (5), 1989, 3964.

Moghadam A.J. et al., Localised power loss measurement using remote sensors, *IEEE Transactions on Magnetics*, 29 (6), 1993, 2998.

Morales R. et al., Magnetization reversal processes in amorphous and polycrystalline Co-Si patterned nanowires, *IEEE Transactions on Magnetics*, 38 (5), 2002, 2565.

Mori K. et al., *2-D Magnetic Rotational Loss of Electrical Steel at High Magnetic Flux Density, IEEE Transactions on Magnetics*, 41 (10), 2005, 3310.

Mori T. et al., Comparison of rotational single sheet testers for evaluating 2-D magnetic properties of electrical steel sheet, *International Journal of Applied Electromagnetics and Mechanics*, 48, 2015, 219.

Morishita M. et al., Examination of magnetic properties of several magnetic materials at high temperature, *Przeglad Elektrotechniczny (Electrical Review)*, R. 87 (N. 9b), 2011, 106.

Moses A.J. et al., Power loss and flux density distributions in the T-joint of a three phase transformer core, *IEEE Transactions on Magnetics*, 8 (4), 1972, 785.

Moses A.J. et al., Measurement of rotating flux in silicon iron laminations, *IEEE Transactions on Magnetics*, 9 (4), 1973, 651.

Moses A.J., Importance of rotational losses in rotating machines and transformers, *Journal of Materials Engineering and Performance*, 1 (2), 1992, 235.

Moses A.J. et al., Aspects of the cut-edge effect stress on the power loss and flux density distribution in electrical steel sheets, *Journal of Magnetism and Magnetic Materials*, 215–216, 2000, 690.

Moses A.J. et al., Characterising electrical steels under complex magnetising conditions, *Journal of Magnetism and Magnetic Materials*, 254–255, 2003, 54.

Moses A.J. et al., A novel instrument for real-time dynamic domain observation in bulk and micromagnetic materials, *IEEE Transactions on Magnetics*, 41 (10), 2005, 3736.

Moses A.J. et al., Real time dynamic domain observation in bulk materials, *Journal of Magnetism and Magnetic Materials*, 304 (2), 2006a, 150.

Moses A.J. et al., AC Barkhausen noise in electrical steels: Influence of sensing technique on interpretation of measurements, *Journal of Electrical Engineering*, 57 (8/S), 2006b, 3.

Moses A.J. et al., Effect of DC voltage on AC magnetisation of transformer core steel, *Presented at 13th ISEM*, East Lansing, MI, September 2007.

Moses A.J. et al., Losses due to the transverse component of flux density in grain oriented electrical steels, *Journal of Optoelectronics and Advanced Materials*, 10 (5), 2008, 1110.

Moses A.J., Energy efficient electrical steels: Magnetic performance prediction and optimization, *Scripta Materialia*, 67, 2012, 560.

Moses A.J. et al., Modeling 2-D magnetostriction in nonoriented electrical steels using a simple magnetic domain model, *IEEE Transactions on Magnetics*, 51 (5), 2015, 6000407.

Mthombeni T.L. et al., New Epstein frame for lamination core loss measurements under high frequencies and high flux densities, *IEEE Transactions on Energy Conversion*, 22 (3), 2007, 614.

Murata, *NTC Thermistors, Cat. No. R44E-15*, Data sheet, 2012.

Nakano M. et al., Improvements of single sheet testers for measurement of 2-D magnetic properties up to high flux density, *IEEE Transactions on Magnetics*, 35 (5), 1999, 3965.

Nakata T. et al., Improvement of measuring accuracy of magnetic field strength in single sheet testers by using two H coils, *IEEE Transactions on Magnetics*, 23 (5), 1987, 2596.

Nakata T. et al., Improvements of measuring equipments for rotational power loss, *Proceedings of 1st 1&2DM Workshop*, Braunschweig, Germany, 1992, p. 191.

Nakata T. et al., Measurement of magnetic characteristics along arbitrary directions of grain-oriented silicon steel up to high flux densities, *IEEE Transactions on Magnetics*, 29 (6), 1993, 3544.

Narita K. et al., Rotational hysteresis loss in silicon-iron single crystal with (001) surfaces, *IEEE Transactions on Magnetics*, 10 (2), 1974, 165.

National Grid, *Frequency Response Services*, 2016, accessed online 10 February 2016, http://www2.nationalgrid.com/uk/services/balancing-services/frequency-response

Nencib N. et al., 2D analysis of rotational loss tester, *IEEE Transactions on Magnetics*, 31 (6), 1995, 3388.

Nencib N. et al., Performance evaluation of a large rotational single sheet tester, *Journal of Magnetism and Magnetic Materials*, 160 (1996), 1996, 174.

Neonark, *SMOKE (Surface Magneto-Optical Kerr Effect) Measuring Equipment, BH-S620*, 2013.

Ng D.H.L. et al., Effect of biaxial stress on magnetoacoustic emission from nickel, *IEEE Transactions on Magnetics*, 28 (5), 1992, 2214.

Ng L.H.D. et al., The dependence of magnetoacoustic emission on magnetic induction and specimen thickness, *IEEE Transactions on Magnetics*, 30 (6), 1994, 4857.

NI, *National Instruments, Strain Gauge Measurement – A Tutorial, Application Note 078*, 1998.

NI, *National Instruments, DAQ NI 6115/6120 User Manual, Multifunction I/O Devices for PCI/PXI/CompactPCI Bus Computers*, 2002.

NI, *National Instruments, NI 6052E Family Specification*, 2005.

NI, *National Instruments, SCXI-1503 User Manual*, 2007a.

NI, *National Instruments, NI USB-9237, User Guide and Specifications*, 2007b.

NI, *National Instruments, NI 6711/6713/DAQCard-6715 Specifications*, 2007c.

NI, *National Instruments, NI 4071 Specifications, 7½-Digit FlexDMM™ and 1.8 MS/s Isolated Digitizer*, 2008a.

NI, *National Instruments, NI 6115/6120 Specifications*, 2008b.

NI, *National Instruments, NI USB-621x User Manual*, 2009.

NI, *National Instruments, Integral x(t)*, LabVIEW 2011 Help, Full Development System, 2011.

NI, *National Instruments, Top Five Considerations for Taking the Stress Out of Strain Measurements*, 2012, accessed online 13 March 2012, http://ni.com/white-paper/11488/en

NI, *National Instruments, NI USB-6210*, Data sheet, 2013a.

NI, *National Instruments, NI USB-6120*, Data sheet, 2013b.

NI, *National Instruments, Analog Waveform VIs and Functions, LabVIEW 2013 Help*, 2014a, accessed online 5 March 2014, http://ni.com

NI, *National Instruments, NI 9211, ±80 mV Thermocouple Input, 14 S/s, 4 Ch Module*, Data sheet, 2014b.

NI, *National Instruments, NI 9239, Specifications*, 2014c.

NI, Technical Support of National Instruments, *Private Communication*, 2014d.

NI, National Instruments, *Reducing Glitches on the Analog Output of MIO DAQ Devices*, 2016a, accessed online 6 February 2016, http://digital.ni.com/public.nsf/allkb/9b13e25 b0b0ec197862562b4007d2e33

NI, *National Instruments, Butterworth Filter VI*, 2016b, accessed online 26 March 2016, http://zone.ni.com/reference/en-xx/help/371361h-01/lvanls/butterworth_filter

Nicholson J.W., *The Chemistry of Polymers*, 3rd edition, Royal Society of Chemistry, Cambridge, 2006, ISBN 0-85404-684-4.

NIST, *ITS-90 Type K Thermocouple*, 2014, accessed online 21 November 2014, http://srdata. nist.gov/its90/main/its90_main_page.html

NPL, *Guide to the Measurement of Pressure and Vacuum*, National Physics Laboratory, London 1998, ISBN 0-904457-29-X.

Oka H. et al., Flux control and flux distribution in a ferrite orthogonal core, *IEEE Transactions on Magnetics*, 25 (5), 1989, 3839.

Oka M. et al., Evaluation of four types of rotational magnetic flux sensor using signal-to-noise ratio, *Proceedings of 3rd International Workshop on ENDE*, IOS Press, 1998a, ISBN 9051993757, p. 196.

Oka M. et al., Comparison of the various types of rotational magnetic flux sensors with search coils using the signal-to-noise ratio, *Journal of Technical Physics*, 39 (3–4), 1998b, 593.

Okazaki Y. et al., 2-dimensional magnetic measurement for rectangular single strip of non oriented electrical steel, *Proceedings of 6th 1&2DM Workshop*, Bad Gastein, Austria, 2000, p. 76.

Oledzki, J.S., Validation of the needle method for magnetic measurements, *Proceedings of 7th 1&2DM Workshop*, Braunschweig, Germany, 2003, p. 203.

Omega, *Strain Gage Installation, Technical Data*, 2014, accessed online 23 January 2014, http://omega.com/techref/pdf/strain_gage_technical_data.pdf

Optris, *LaserSight, Operators Manual, Optris Infrared Thermometers*, 2006.

Owzareck M., Measurement method for normal magnetization curve of soft magnetic composites with high magnetization currents, *International Journal of Applied Electromagnetics and Mechanics*, 48, 2015, 181.

Parakka A.P. et al., Effect of surface mechanical changes on magnetic Barkhausen emissions, *IEEE Transactions on Magnetics*, 33 (5), 1997, 4026.

Parviainen A. et al., 2D measurement of core loss using a 3-phase excitation apparatus, *Proceedings of 6th 2DM Workshop*, Bad Gastein, Austria, 2000, p. 117.

Passadis K. et al., Neural-network-based single-sided non-enwrapping power loss tester, *Journal of Magnetism and Magnetic Materials*, 254–255, 2003, 385.

Patel H.V. et al., A new adaptive automated feedback system for Barkhausen signal measurement, *Sensors and Actuators A*, 129, 2006, 112.

Patel H.V. et al., The dependence of AC Barkhausen noise measurement on data acquisition parameters, *Przeglad Elektrotechniczny (Electrical Review)*, R. 83 (Nr 4/2007), 2007, 99.

Patel H.V., *Relationship Between AC Barkhausen Noise and Losses in Electrical Steel*, PhD thesis, Wolfson Centre for Magnetics, Cardiff University, Cardiff, UK, 2009.

Peak Group, *Test Probes*, 2014, accessed online 6 June 2014, http://www.thepeakgroup.com

Pelgrom M.J.M., *Analog-to-Digital Conversion*, 2nd edition, Springer Science & Business Media, 2012, ISBN 9781461413714.

Permiakov V. et al., 2D magnetic measurements under 1D stress, *Przeglad Elektrotechniczny (Electrical Review)*, R. 81 (Nr 5/2005), 2005, 68.

Perov N.S. et al., A vibrating sample anisometer, *Proceedings of 6th 1&2 DM Workshop*, Bad Gastein, Austria, 2000, p. 104.

Petit M. et al., A suitable method for characterization of ferrofluids or low permeability materials and their anisotropic behaviour, *Proceedings of 13th 1&2D Workshop*, Torino, Italy, 2014, p. 106.

Pfutzner H., Rotational magnetization and rotational losses of grain-oriented silicon steel sheets – Fundamental aspects and theory, *IEEE Transactions on Magnetics*, 30 (5), 1994, 1208.

Pfutzner H. et al., The needle method for induction tests: Sources of error, *IEEE Transactions on Magnetics*, 40 (3), 2004, 1610.

Pfutzner H. et al., A study on possible sources of errors of loss measurement under rotational magnetisation, *Przeglad Elektrotechniczny (Electrical Review)*, R. 83 (Nr 4/2007), 2007, 9.

Pfutzner H. et al., Rotational magnetisation in transformer cores – A review, *IEEE Transactions on Magnetics*, 47 (11), 2011, 4523.

Pfutzner H. et al., Magnetic dummy sensors – A novel concept for interior flux distribution tests, *Proceedings of 13th 1&2DM Workshop*, Turin, Italy, 2014, p. 59.

Phillips R. et al., Domain configuration under rotational flux and applied stress conditions in silicon iron, *IEEE Transactions on Magnetics*, 10 (2), 1974, 168.

Phophongviwat T., *Investigation of the Influence of Magnetostriction and Magnetic Forces on Transformer Core Noise and Vibration*, PhD thesis, Wolfson Centre for Magnetics, Cardiff University, Cardiff, UK, 2013.

Phway P.P.T. et al., Frequency dependence of magnetostriction for magnetic actuators, *Journal of Electrical Engineering*, 55 (10/S), 2004, 7.

Phway P.P.T., *Magnetostrictively Induced Mechanical Resonance of Electrical Steel Strips*, PhD thesis, Wolfson Centre for Magnetics, Cardiff University, Cardiff, UK, 2007.

Phway P.P.T. et al., Magnetostriction trend of non-oriented 6.5% Si-Fe, *Journal of Magnetism and Magnetic Materials*, 320 (20), 2008, e611.

Piotrowski L. et al., Stress anisotropy characterisation with the help of Barkhausen effect detector with adjustable magnetic field direction, *International Journal of Applied Electromagnetics and Mechanics*, 48 (2–3), 2015, 163.

Platil A. et al., Two-channel ADC system for hysteresisgraph, *Journal of Electrical Engineering*, 50 (8/S), 1999, 56.

Platil A. et al., Factors limiting the accuracy of sampling hysteresisgraph, *Presented at Baltic Electronics Conference*, Tallinn, Estonia, 2000.

Platil A. et al., Sampling measurements with digital hysteresisgraph, *Journal of Magnetism and Magnetic Materials*, 254–255, 2003, 108.

Pluta W. *Influence of Magnetic Anisotropy of Electrical Steels on Power Loss under Rotational Magnetisation* (in Polish: *Wpływ anizotropii magnetycznej blach elektrotechnicznych na straty mocy przy przemagnesowaniu obrotowym*), PhD thesis, Lodz University of Technology, Lodz, Poland, 2001.

Pluta W., Czestochowa University of Technology, Poland, *Private Communication*, 2013.

Polytec, *Single Point Vibrometers*, 2014, accessed online 7 January 2014, http://polytec.com/int/products/vibration-sensors/single-point-vibrometers

Prabhu N.S. et al., Residual stress analysis in surface mechanical attrition treated (SMAT) iron and steel component materials by magnetic Barkhausen emission technique, *IEEE Transactions on Magnetics*, 48 (12), 2012, 4713.

Pro-Power, *Copper Enamelled Wire*, Data sheet, 2015, accessed online 15 December 2015, http://farnell.com/datasheets/1942907.pdf

Psuj G., Fusion of multiple parameters of signals obtained by vector magnetic flux observation for evaluation of stress loaded steel samples, *International Journal of Applied Electromagnetics and Mechanics*, 49 (1), 2015, 17.

Psuj G. et al., Stress evaluation in non-oriented electrical steel samples by observation of vector magnetic flux under static and rotating field conditions, *International Journal of Applied Electromagnetics and Mechanics*, 44 (3–4), 2014, 339.

PTC Mathcad, *Converting Your Engineering Units with PTC Mathcad*, 2017, accessed online, 27 January 2017, http://ptc.com/engineering-math-software/mathcad/worksheets/engineering-unit-converter

PUI Audio, *Electret Condenser Microphone Basics*, 2008, accessed online 15 August 2008, http://digikey.com

Pysz W., *Galvanometer, Wikimedia Commons*, 2017, accessed online 17 February 2017, http://commons.wikimedia.org/wiki/file:galvanometer_late_20cent.jpg

Qu Q-C., Precise magnetic properties measurement on electrical sheet steels under deep saturation, *IEEE Transactions on Magnetics*, 20 (5), 1984, 1717.

Qiu Z.Q. et al., Surface magneto-optic Kerr effect (SMOKE), *Journal of Magnetism and Magnetic Materials*, 200, 1999, 664.

Radley G.S., Domain observation and wall velocity measurement under rotating field in grain-oriented silicon-iron, *Proceedings of 1st 1&2DM Workshop*, Braunschweig, Germany, 1992, p. 25.

Radley B. et al., Apparatus for experimental simulation of magnetic flux and power loss distribution in a turbogenerator stator core, *IEEE Transactions on Magnetics*, 17 (3), 1981, 1311.

Ragusa C., Politecnico di Torino, Italy, *Private Communication*, 2013.

Ragusa C. et al., A three-phase single sheet tester with digital control of flux loci based on the contraction mapping principle, *Journal of Magnetism and Magnetic Materials*, 204, 2006, e568.

Ragusa C. et al., An intercomparison of rotational loss measurements in non-oriented Fe–Si alloys, *Journal of Magnetism and Magnetic Materials*, 320 (20), 2008, e623.

Ranjan R. et al., Magnetic properties of decarburized steels: An investigation of the effects of grain size and carbon content, *IEEE Transactions on Magnetics*, 23 (3), 1987, 1869.

Rasilo P. et al., Iron losses, magnetoelasticity and magnetostriction in ferromagnetic steel laminations, *IEEE Transactions on Magnetics*, 49 (5), 2013, 2041.

Recktenwald G., *Uncertainty Estimation and Calculation*, 2017, accessed online 29 January 2017, http://web.iitd.ac.in/~pmvs/courses/mel705/uncertainty2.pdf

Reisinger E., Measurement of iron losses due to alternating and rotating magnetization, *Proc. Electric Energy Conference*, Adelaide, Australia, 1987, p. 388.

Ripka P. et al., *Master Book on Sensors, Part B*, Czech Technical University, Skoda Auto, Prague, 2003.

Rogowski W., *Measuring Magnetic Properties of Materials and Apparatus Therefor*, US Patent 1,204,489, 1916.

Rogowski W. et al., Measurement of magnetic tension (in German: Die Messung der magnetischen Spannung), *Archiv für Elektrotechnik*, 1912, 1 (4), 141.

Sakata S. et al., The magnetic characteristics of non-oriented magnetic steel sheet under the control of distorted elliptic rotational flux, *Presented at 16th Soft Magnetic Materials Conference*, Dusseldorf, Germany, 2003.

Salz W., A two-dimensional measuring equipment for electrical steel, *IEEE Transactions on Magnetics*, 30 (3), 1994, 1253.

Salz W. et al., Which field sensors are suitable for a rotating flux apparatus? *Proceedings of 1st 1&2DM Workshop*, Braunschweig, Germany, 1992, p. 117.

Sankaran P. et al., Use of a voltage follower to ensure sinusoidal flux in a core, *IEEE Transactions on Magnetics*, 19 (4), 1983, 1572.

Sasaki Y., An approach estimating the number of domain walls and eddy current losses in grain-oriented 3% Si-Fe tape wound cores, *IEEE Transactions on Magnetics*, 16 (4), 1980, 569.

Sasaki T. et al., Measurement of rotational power losses in silicon-iron sheets using wattmeter method, *IEEE Transactions on Magnetics*, 21 (5), 1985, 1918.

Sato T. et al., Measurement of iron loss in wound iron core by using visualising iron loss distribution system, *Proceedings of 13th 1&2DM Workshop*, Turin, Italy, 2014, p. 47.

Scholes R., Application of operational amplifiers to magnetic measurements, *IEEE Transactions on Magnetics*, 6 (2), 1970, 289.

Sentron A.G., *Integrated 3-Axis Hall Element*, Data sheet, 2000.

Shilyashki G. et al., Automatic 3-dimensional flux analyses of a 3-phase model transformer core, *Proceedings of 13th 1&2DM Workshop*, Turin, Italy, 2014, p. 102.

Shilyashki G. et al., Pin sensor for interior induction measurements in transformer cores, *IEEE Transactions on Magnetics*, 51 (1), 2015a, 4001904.

Shilyashki G. et al., A tangential induction sensor for 3D-analyses of peripheral flux distributions in transformer cores, *IEEE Transactions on Magnetics*, 51 (6), 2015b, 4003306.

Shimamura M. et al., Approach to 2-dimensional high frequency magnetic characteristic measurement with high speed & accuracy "Vector (2D) hysteresis analyzer system", *Proceedings of 6th 1&2DM Workshop*, Bad Gastein, Austria, 2000, p. 138.

Shimoji H. et al., Core loss distribution measurement of electrical steel sheets using a thermographic camera, *Przeglad Elektrotechniczny (Electrical Review)*, R. 87 (No. 9b/2011), 2011, 65.

SI, *The International System of Units (SI)*, 8th edition, Organisation Intergouvernementale de la Convention du Metre, 2006.

Siebert S., Industrial test apparatuses for magnetic loss measurements – Standardized and customized systems, *Proceedings of 6th 1&2DM Workshop*, Bad Gastein, Austria, 2000, p. 9.

Sievert J.D., Determination of AC magnetic power loss of electrical steel sheet: Present status and trends, *IEEE Transactions on Magnetics*, 20 (5), 1984, 1702.

Sievert J., Recent advances in the one- and two-dimensional magnetic measurement technique for electrical sheet steel, *IEEE Transactions on Magnetics*, 26 (5), 1990, 2553.

Sievert J., Studies on the measurement of two-dimensional magnetic phenomena in electrical sheet steel at PTB, *Proceedings of 1st 1&2DM Workshop*, Braunschweig, Germany, 1992a, p. 102.

Sievert J., Two-dimensional magnetic measurements – History and achievements of the workshop, *Przeglad Elektrotechniczny (Electrical Review)*, R. 87 (No. 9b/2011), 2011, 2.

Sievert J. et al., The measurement of rotational power loss in electrical sheet steel using a vertical yoke system, *Journal of Magnetism and Magnetic Materials*, 112 (1–3), 1992b, 91.

Sievert J. et al., *Intercomparison of Measurements of Magnetic Losses in Electrical Sheet Steel Under Rotation Flux Conditions*, Commission of the European Communities, Report EUR 16255 EN, EC Brussels, Luxembourg, 1995.

Sievert J. et al., European intercomparison of measurements of rotational power loss in electrical sheet steel, *Journal of Magnetism and Magnetic Materials*, 160 (1996), 1996, 115.

Sievert J. et al., Magnetic measurements on electrical steels using Epstein and SST methods – Euromet comparison between standard laboratories, *Proceedings of 6th 1&2DM Workshop*, Bad Gastein, Austria, 2000, paper II-5, p. 194.

SIGMA-NOT, *Przeglad Elektrotechniczny (Electrical Review)*, 2011, accessed online 23 August 2011, http://pe.org.pl

Singh Y., *Electro Magnetic Field Theory*, Pearson Education India, 2011, ISBN 9788131760611.

Soinski M., Application of the anisometric method for determining polar curves of induction, apparent core loss, and core loss in cold-rolled electrical sheets of Goss texture, *IEEE Transactions on Magnetics*, 20 (1), 1984a, 172.

Soinski M., The use of electrical sheets in construction of cryoresistive transformers, *Cryogenics*, 24 (3), 1984b, 133.

Soinski M., Anisotropy of electrical resistivities in electrical sheets, *Journal of Magnetism and Magnetic Materials*, 53 (1–2), 1985, 54.

Soinski M., The anisotropy of coercive force in cold-rolled Goss-texture electrical sheets, *IEEE Transactions of Magnetics*, 23 (6), 1987, 3878.

Soinski M., Anisotropy of DC magnetostriction in cold-rolled electrical sheets of Goss texture, *IEEE Transactions on Magnetics*, 25 (4), 1989, 3166.

Soinski M., Demagnetization effect of rectangular and ring-shaped samples made of electrical sheets placed in a stationary magnetic field, *IEEE Transactions on Instrumentation and Measurement*, 39 (5), 1990, 704.

Soinski M., *Magnetic Materials in Technology* (in Polish: *Materialy magnetyczne w technice*), Biblioteka Centralnego Osrodka Szkolenia i Wydawnictw SEP, Warszawa, 2001, ISBN 83-915103-5-2.

Soinski M., Nanocrystalline block cores for high-frequency chokes, *IEEE Transactions on Magnetics*, 50 (11), 2014, 2801904.

Soinski M. et al., Anisotropy in iron-based soft magnetic materials, in: K. Buschow (ed.), *Handbook of Magnetic Materials*, Chapter 4, North-Holland, 1995, p. 347.

Soken: Soken Tester, 2014, accessed online 12 January 2014, http://otdl.com/ir02.html

Somkun S., *Magnetostriction and Magnetic Anisotropy in Non-oriented Electrical Steels and Stator Core Laminations*, PhD thesis, Wolfson Centre for Magnetics, Cardiff University, Cardiff, UK, 2010.

Somkun S., Naresuan University, Phitsanulok, Thailand, *Private Communication*, 2014.

Somkun S. et al., Comparisons of AC magnetostriction of non-oriented electrical steels measured in Epstein and disc samples, *Journal of Electrical Engineering*, 61 (7/s), 2010, 89.

Son D. et al., Core loss measurements including higher harmonics of magnetic induction in electrical steel, *Journal of Magnetism and Magnetic Materials*, 160, 1996, 65.

Spornic S.A. et al., Numerical waveform control for rotational single sheet testers, *Journal de Physique IV, France*, 8, 1998, Pr2-741.

Stupakov A. et al., A system for controllable magnetic measurements of hysteresis and Barkhausen noise, *IEEE Transactions on Magnetics*, 65 (5), 2016, 1087.

Stauffer L.H., *Methods of and Devises for Determining the Magnetic Properties of Specimens of Magnetic Material*, 1958, Patent US 2828467.

Steentjes S., *Efficiently Modeling Soft Magnetic Materials for Transformers, Actuators and Rotating Electrical Machines*, PhD thesis, RWTH Aachen University, Aachen, Germany, 2017.

Stefan Mayer Instruments, *Magnetic Field Sensor FLC 100*, Data sheet, 2013.

Stranges N. et al., Measurement of rotational iron losses in electrical sheet, *IEEE Transactions on Magnetics*, 36 (5), 2000, 3457.

Strattan R.D. et al., Iron losses in elliptically rotating fields, *Journal of Applied Physics*, 33 (3), 1962, 1285.

Struers, *RotoPol-31, RotoPol-35, Instruction Manual*, 2008.

Stupakov A., Controllable magnetic hysteresis measurement of electrical steels in a single-yoke open configuration, *IEEE Transactions on Magnetics*, 48 (12), 2012, 4718.

Stupakov O. et al., Measurement of Barkhausen noise and its correlation with magnetic permeability, *Journal of Magnetism and Magnetic Materials*, 320 (3–4), 2008, 204.

Stupakov A. et al., Evaluation of ductile cast iron microstructure by magnetic hysteresis and Barkhausen noise methods, *Studies in Applied Electromagnetics and Mechanics*, 32, 2009a, 232.

Stupakov A. et al., Governing conditions of repeatable Barkhausen noise response, *Journal of Magnetism and Magnetic Materials*, 321 (2009), 2009b, 2956.

Stupakov O., Local non-contact evaluation of the ac magnetic hysteresis parameters of electrical steels by the Barkhausen noise technique, *Journal of Nondestructive Evaluation*, 32 (405–412), 2013, 405.

Stupakov O. et al., Three-parameter feedback control of amorphous ribbon magnetization, *Journal of Electrical Engineering*, 64 (3), 2013, 166.

Sugimoto S. et al., A new measurement device for two-dimensional vector-magnetic property in high magnetic flux density ranges, *Przeglad Elektrotechniczny (Electrical Review)*, R. 81 (Nr 5/2005), 2005, 27.

Suss S. et al., *Method for Testing a Transformer and Corresponding Test Device*, 2006, Patent US 6, 987, 390.

Suzuki Y. et al., A new measurement system of the surface magneto-optic Kerr effect (SMOKE), *IEEE Translation Journal on Magnetics in Japan*, 5 (4), 1990, 300.

Takada S. et al., Influence of magnetization pause on the magnetic properties of an electrical iron sheet, *Journal of Magnetism and Magnetic Materials*, 160, 1996, 33.

Tamaki T. et al., Hole effect on magnetic field uniformity in single sheet tester, *Przeglad Elektrotechniczny (Electrical Review)*, R. 85 (Nr 1/2009), 2009, 71.

Tan K.S. et al., Rotational loss in thin gage soft magnetic materials, *IEEE Transactions on Magnetics*, 21 (5), 1985, 1921.

Tang Q. et al., Measurement of magnetic properties of electrical steels at high flux densities using an improved single sheet tester, *Presented at 18th International Symposium on High Voltage Engineering*, Seoul, Korea, 2013.

Taylor J.R., *An Introduction to Error Analysis, The Study of Uncertainties in Physical Measurements*, 2nd edition, University Science Books, Sausalito, California, 1997, ISBN 0-935702-75-X.

Telcon, *PCB Mounting Hall Effect Current Transformer Type HTP25NP*, Data sheet, 2014a.

Telcon, *Open Loop Hall Effect Current Transformer Type HOY*, Data sheet, 2014b.

Tiunov V.F., On the ratio of magnetic losses in Fe-3% Si single crystals in rotating and linear-polarized magnetic fields, *The Physics of Metals and Metallography*, 113 (12), 2012, 1146.

ThyssenKrupp, *Further Processing Hints*, 2015, accessed online 26 March 2015, http://tkes.com

Tokyo Sokki Kenkyujo, *Foil Strain Gauge, Series "F"*, Data sheet, 2014, accessed online 26 May 2014, http://www.tml.jp/e/product/strain_gauge/catalog_pdf/fseries.pdf

Tompkins R.E. et al., New magnetic core loss comparator, *Journal of Applied Physics*, 29 (3), 1958, 502.

Trietley H.L., *Transducers in Mechanical and Electronic Design*, New York and Basel, Marcel Dekker, 1986, ISBN 0-8247-7598-8.

Tumanski S., Non-destructive testing of the stress effects in electrical steel by magnetovision method, *Non-Linear Electromagnetic Systems, ISEM 99*, IOS Press, 2000a, p. 273.

Tumanski S., Investigations of 2D parameters of electrical steels, *Proceedings of 6th 1&2DM Workshop*, Bad Gastein, 2000b, p. 185.

Tumanski S. *Principles of Electrical Measurement*, CRC Press, New York, London, 2006, ISBN 9780750310383.

Tumanski S., Induction coil sensors – A review, *Measurement Science and Technology*, 18 (3), 2007a, R31.

Tumanski S., Scanning of magnetic field as a method of investigations of the structure of magnetic materials, *Przeglad Elektrotechniczny (Electrical Review)*, R. 83 (Nr 1/2007), 2007b, 108.

Tumanski S., *Handbook of Magnetic Measurements*, CRC Press, 2011, ISBN 9781439829516.

Tumanski S. et al., Single strip tester of magnetic materials with array of magnetoresistive sensors, *Przeglad Elektrotechniczny (Electrical Review)*, R. 83 (Nr 4/2007), 2007, 46.

Tyco, *Tyco Electronics, Thick Film Power Resistors, Type BDS250/400 Series*, Data sheet, 2006.

UKAS, *M3003, The Expression of Uncertainty and Confidence in Measurement*, 3rd edition, United Kingdom Accreditation Service, Middlesex, 2012.

UKAS, *Schedule of Accreditation, 0478, NPL Management Ltd, Issue No. 072*, 2016, accessed online 30 December 2016, http://ukas.org/calibration/schedules/actual/0478calibration%20multiple.pdf

Usak E., The measurement of magnetisation curves with defined flux density waveform at low frequencies, *Journal of Electrical Engineering*, 53 (10/S), 2002, 77.

Vacumschmelze GmbH & Co. KG, 2011, accessed online 14 October 2011, http://vacuum-schmelze.com

Vishay, *2381 640 10.../NTCLE101E3...SB0, NTC Thermistors, Radial Leaded Special Accuracy*, Data sheet, 2009.

Voestalpine, *Isovac: Electrical Steel Strip, Technical Terms of Delivery*, 2011.

Von Musil R., Rotational magnetic excitation of the armature lamination in electrical machine, *Proceedings of 1st 1&2DM Workshop*, Braunschweig, Germany, 1992, p. 48.

Wachter A. et al., *Compendium of Theoretical Physics*, Springer, 2006, ISBN 978-0387-25799-0.

Wang S.C. et al., PC-based apparatus for characterising high frequency magnetic cores, *IEE Proceedings of Science, Measurement and Technology*, 146 (6), 1999, 304.

Wang G.F., *Magnetic and Calorimetric Study of the Magnetocaloric Effect in Intermetallics Exhibiting First-Order Magnetostructural Transitions*, Coleccion de Estudios de Fisica Vol. 99, Universidad de Zaragoza, 2012, ISBN 9788415538295.

Wanjiku J. et al., Design considerations of 2-D magnetizers for high flux density measurements, *IEEE Energy Conversion Congress and Exposition ECCE*, 2014, 4204, DOI: 10.1109/ECCE.2014.6953973.

Wanjiku J. et al., Investigating the sources of non-uniformity in 2-D core loss measurements, *International Journal of Applied Electromagnetics and Mechanics*, 48 (2,3), 2015a, 255.

Wanjiku J. et al., Shielding of the z-component of the magnetic field in a 2-D magnetizer with a deep yoke, *Presented at Electrical Machines & Drives Conference*, Coeur d'Alene, ID, 2015b.

Wan Mahadi W.N.L., *Design and Development of a Novel Two-Dimensional Magnetic Measurement System for Electrical Steels*, PhD thesis, *Wolfson Centre for Magnetics Technology*, Cardiff University, Cardiff, UK, 1996.

Werner E., *Magnetising Device for Measuring Magnetic Properties of Metal Sheets Magnetised with Alternating Current* (in German: *Einrichtung zur Messung magnetischer Eigenschaften von Blechen bei Wechselstrommagnetisierung*), Osterreichisches Patentamt Patentschrift Nr. 191015, 1957.

Westphalen A. et al., Vector and Bragg magneto-optical Kerr effect for the analysis of nanostructured magnetic arrays, *Review of Scientific Instruments*, 78 (121301), 2007, 121301-1.

White M.A., *Physical Properties of Materials*, 2nd edition, CRC Press, 2011, ISBN 9781439866511.

White S. et al., A multichannel magnetic flux controller for periodic magnetizing conditions, *IEEE Transactions on Instrumentation and Measurements*, 61 (7), 2012, 1896.

Wikipedia, *Soft Magnetic Materials Conference*, 2011, accessed online 8 October 2011, http://en.wikipedia.org/wiki/soft_magnetic_materials_conference

Wikipedia, *Crowbar (circuit)*, 2016, accessed online 28 February 2016, http://en.wikipedia.org/wiki/crowbar_(circuit)

Williams P., *Wolfson Centre for Magnetics*, Cardiff University, Cardiff, UK, *Private Communication*, 2006.

Wilson J.W. et al., Magneto-acoustic emission and magnetic Barkhausen emission for case depth measurement in En36 gear steel, *IEEE Transactions on Magnetics*, 45 (1), 2009, 177.

Wood R. et al., Divergence of flux in a grain-oriented electrical steel sheet locally magnetised by a single-yoke system, *Przeglad Elektrotechniczny (Electrical Review)*, R. 85 (Nr 1/2009), 2009, 31.

Xu J., Recent experiments on rotational power loss measurement at PTB, *Proceedings of 4th 2DM Workshop*, Cardiff, UK, 1995, p. 9.

Xu J. et al., On the reproducibility, standardization aspects and error sources of the fieldmetric method for the determination of 2D magnetic properties of electrical sheet steel, *Proceedings of 5th 1&2DM Workshop*, Grenoble, France, 1997, p. 43.

Xu X.T. et al., A comparison of magnetic domain images using a modified bitter pattern technique and the Kerr method on grain-oriented electrical steel, *IEEE Transactions on Magnetics*, 47 (10), 2011, 3531.

Yamamoto K.-I. et al., Magnetic properties of non-oriented electrical steel under compressive stress normal to their surface, *Przeglad Elektrotechniczny (Electrical Review)*, R. 87 (No. 9b/2011), 2011, 97.

Yamamoto K. et al., Effects of compressive stress normal to the surface of non-oriented electrical steel sheets, *International Journal of Applied Electromagnetics and Mechanics*, 44, 2014, 271.

Yamazaki K. et al., Loss analysis of permanent-magnet motor considering carrier harmonics of PWM inverter using combination of 2-D and 3-D finite-element method, *IEEE Transactions on Magnetics*, 41 (5), 2005, 1980.

Yanase S. et al., 2D magnetic rotational loss of electrical steel at high magnetic flux density, *Przeglad Elektrotechniczny (Electrical Review)*, R. 83 (Nr 04/2007), 2007, 31.

Yanase S. et al., AC magnetic properties of electrical steel sheet under DC-biased magnetization, *Przeglad Elektrotechniczny (Electrical Review)*, R. 87 (No. 9b), 2011, 52.

Yokogawa, *CT2241 Current Transformers*, Data sheet, 2011.

Yoshida T. et al., Development of measuring equipment of DC-biased magnetic properties using of open type single sheet tester, *Proceedings of Intermag*, San Diego, CA, 2006, p. 355.

Young P., *Spin Glasses and Random Fields*, World Scientific Publishing, Singapore, 1998, ISBN 9810232403, p. 100.

Young F.J. et al., Iron losses due to elliptically polarised magnetic fields, *Journal of Applied Physics*, 31 (5), 1960, 194.

Young F.J. et al., Method for measuring iron losses in elliptically polarized magnetic fields, *Journal of Applied Physics*, 37 (3), 1966, 1210.

Zaiontz C., *Real Statistics Using Excel*, 2016, accessed online 3 May 2016, http://real-statistics.com/other-key-distributions/gamma-function

Zbroszczyk J. et al., Angular distribution of rotational hysteresis losses in Fe-3.25%Si single crystals with orientations (001) and (011), *IEEE Transactions on Magnetics* 17 (3), 1981, 1275.

Zemanek I., Exciting signal generator for SSTs, *Journal of Magnetism and Magnetic Materials*, 254–255, 2003, 73.

Zeng J. et al., 2D vector magnetic hysteresis property measurement of magneto-rheological fluid material, *Proceedings of 12th 1&2D Workshop*, Vienna, Austria, 2012, p. 49.

Zhang Y. et al., Precise AC magnetic measurement under sinusoidal magnetic flux by using digital feedback of harmonic compensation, *Journal of Magnetism and Magnetic Materials*, 312, 2007, 443.

Zhang Y. et al., Measurement and computation of rotational core loss considering higher harmonic components of magnetic flux density waveform, Presented at COMPUMAG, Montreal, Canada, 2015, paper PC5.

Zhang C. et al., Comparison of two types 3D magnetic tester considering the normal magnetic flux of the silicon steel, *Proceedings of 14th 1&2DM Workshop*, Tianjin, China, 2016, p. 79.

Zhao Z. et al., Measurement and calculation of iron loss inside silicon steel lamination under DC biasing, *Proceedings of 14th Biennial IEEE Conference on Electromagnetic Field Computation*, Chicago, IL, 2010, p. 1.

Zhong J.J. et al., Improved measurement of magnetic properties with 3D magnetic fluxes, *Journal of Magnetism and Magnetic Materials*, 290–291 (2), 2005, 1567.

Zhu J.G., *Numerical Modelling of Magnetic Materials for Computer Aided Design of Electromagnetic Devices*, PhD thesis, University of Technology, Sydney, Australia, 1994.

Zhu J.G. et al., Two dimensional measurement of magnetic field and core loss using a square specimen tester, *IEEE Transactions on Magnetics*, 29 (6), 1993, 2995.

Zhu J.G. et al., Rotational core losses with circular H and/or B, *Presented at ISEM*, Braunschweig, Germany, 1997a.

Zhu J.G. et al., Measurement and modelling of losses under two dimensional excitation in rotating electrical machines, *Proceedings of 5th International Workshop on Two-Dimensional Magnetization Problems*, Grenoble, France, 1997b, p. 63.

Zhu B. et al., Multifunctional magnetic Barkhausen emission measurement system, *IEEE Transactions on Magnetics*, 37 (3), 2001, 1095.

Zhu J.G. et al., Measurement of magnetic properties under 3-D magnetic excitations, *IEEE Transactions on Magnetics*, 39 (5), 2003, 3429.

Zhu J.G. et al., 3D Measurement and modelling of magnetic properties of soft magnetic composite, *Przeglad Elektrotechniczny (Electrical Review)*, R. 85 (Nr 1/2009), 2009, 11.

Zirka S.E. et al., Measurement and modeling of B-H loops and losses of high silicon nonoriented steels, *IEEE Transactions on Magnetics*, 42 (10), 2006, 3177.

Zirka S. et al., Dynamic magnetization models for soft ferromagnetic materials with coarse and fine domain structures, *Journal of Magnetism and Magnetic Materials*, 394, 2015, 229.

Zurek S., *Rotational Magnetisation in Flat Soft Magnetic Materials* (in Polish: *Przemagnesowanie obrotowe w materialach magnetycznie miekkich plaskich*), MSc thesis, Czestochowa University of Technology, Poland, 2000.

Zurek S., *Two-Dimensional Magnetisation Problems in Electrical Steels*, PhD thesis, *Wolfson Centre for Magnetics Technology*, Cardiff University, Cardiff, UK, 2005.

Zurek S., Some interesting observations of loss and magnetism in soft magnetic materials under non-standard conditions, *Magnetics, Energy Efficiency and the Environment*, *Presented at UK Magnetics Society Seminar, Cardiff University*, Cardiff, 2007.

Zurek S., Qualitative analysis of Px and Py components of rotational power loss, *IEEE Transactions on Magnetics*, 50 (4), 2014, 1.

Zurek S., Qualitative 3D FEM study of B and H distribution in circular isotropic samples for two-dimensional loss measurements, *Przeglad Elektrotechniczny (Electrical Review)*, R. 91 (Nr 12/2015), 2015a, 43.

Zurek S., Effect of shunt resistor value on the measurement of magnetic properties, *International Journal of Applied Electromagnetics and Mechanics*, 48, 2015b, 187.

Zurek S., Demagnetisation of grain-oriented electrical steels (GOES), *Transformers Magazine*, 2 (4), 2015c, 32.

Zurek S., Rotational magnetisation: A phenomenon in three-phase three-limb transformer cores, *Transformers Magazine*, 2 (2), 2015d, 44.

Zurek S., Practical implementation of universal digital feedback for characterisation of soft magnetic materials under controlled AC waveforms, *Przeglad Elektrotechniczny (Electrical Review)*, R. 93, Nr 7/2017, 2017a, p. 16.

Zurek S., Example of vanishing anisotropy at high rotational magnetisation of grain-oriented electrical steel, *Przeglad Elektrotechniczny (Electrical Review)*, R. 93, Nr 7/2017, 2017b, p. 13.

Zurek S., Systematic errors of local B-coils, *Przeglad Elektrotechniczny (Electrical Review)*, in press.

Zurek S., Sensitivity to off-axis vector components of typical wire wound flat H-coil configurations, *IEEE Sensors Journal*, 17 (13), 2017c, 4021

Zurek S., Effect of off-axis H-coil sensitivity on clockwise-anticlockwise differences of rotational power loss, submitted to IET Science, Measurement & Technology, 2017d.

Zurek S., *H-coil with Improved Off-axis Immunity, IEEE Sensors Journal*, 2017e.

Zurek S., Theoretical concept and FEM simulations of improved shielding for round horizontal yokes for rotational power loss measurement, *Journal of Electrical Engineering*, 68 (No. 4) 2017f, 267, DOI: 10.1515/jee-2017–0038.

Zurek S. et al., Digital feedback controlled RSST system, *Proceedings of Soft Magnetic Materials Conference*, Dusseldorf, Germany, 2003, p. 365.

Zurek S. et al., Control of arbitrary waveforms in magnetic measurements by means of adaptive iterative digital feedback algorithm, *Przeglad Elektrotechniczny (Electrical Review)*, 02/2004, 2004, 122.

Zurek S. et al., Use of novel adaptive digital feedback for magnetic measurements under controlled magnetizing conditions, *IEEE Transactions on Magnetics*, 41 (11), 2005a, 4242.

Zurek S. et al., Adaptive iterative digital feedback algorithm for measurements of magnetic properties under controlled magnetising conditions over a wide frequency range, *Przeglad Elektrotechniczny (Electrical Review)*, (Nr 12/2005), 2005b, 5.

Zurek S. et al., Correction of triggering and interchannel delay for alternating and two-dimensional measurements of magnetic properties, *Przeglad Elektrotechniczny (Electrical Review)*, R. 81 (No. 5), 2005c, 78.

Zurek S. et al., Errors in the power loss measured in clockwise and anticlockwise rotational magnetisation. Part 1: Mathematical study, *IEE Proceedings, Science, Measurement and Technology*, 153 (4), 2006a, 147.

Zurek S. et al., Errors in the power loss measured in clockwise and anticlockwise rotational magnetisation. Part 2: Physical phenomena, *IEE Proceedings, Science, Measurement and Technology*, 153 (4), 2006b, 152.

Zurek S. et al., Rotational power losses and vector loci under controlled high flux density and magnetic field in electrical steel sheets, *IEEE Transactions on Magnetics*, 42 (10), 2006c, 2815.

Zurek S. et al., A novel capacitive flux density sensor, *Sensors and Actuators A*, 129, 2006d, 121.

Zurek S. et al., Influence of digital resolution of measuring equipment on the accuracy of power loss measured in Epstein frame, *Przeglad Elektrotechniczny (Electrical Review)*, R. 83 (Nr 4/2007), 2007a, 50.

Zurek S. et al., Practical applications and influence of PWM magnetisation parameters on iron losses in electrical steel, *Presented at Soft Magnetic Materials Conference*, Cardiff, UK, 2007b.

Zurek S. et al., Analysis of twisting of search coil leads as a method reducing the influence of stray fields on accuracy of magnetic measurements, *Sensors and Actuators A*, 142, 2008a, 569.

Zurek S. et al., Anomalous B-H behaviour of electrical steels at very low flux density, *Journal of Magnetism and Magnetic Materials*, 320 (20), 2008b, 2521.

Zurek S. et al., Measurements at very low flux density and power frequencies, *Journal of Electrical Engineering*, 59 (7/s), 2008c, 7.

Zurek S. et al., Asymmetry of magnetic properties of grain-oriented electrical steel, *Przeglad Elektrotechniczny (Electrical Review)*, R. 85 (1/2009), 2009a, 16.

Zurek S., Static and dynamic rotational losses in non-oriented electrical steel, *Przeglad Elektrotechniczny (Electrical Review)*, R. 85 (Nr 1/2009), 2009b, 89.

Zurek S. et al., Correlation between surface magnetic field and Barkhausen noise in grain-oriented electrical steel, *Przeglad Elektrotechniczny (Electrical Review)*, R. 85 (Nr 1/2009), 2009c, 111.

Index

Milton Keynes UK
Ingram Content Group UK Ltd.
UKHW021930071024
449327UK00022B/1754